£12

Royal Flying Corps
Military Wing

Lt W. C. Cambray, MC, the observer of F.E.2d A6516 of No.20 Squadron, demonstrates his perilous firing position when the rear Lewis had to be used in combat. The gun is on a telescopic pillar on an Anderson arch behind the observer's seat; the front gun is on a Clark mounting, officially known as an F.E.2B and D No.4 Mark IV mounting. The pilot, Capt F. D. Stevens, can be seen behind his fixed Lewis gun, which is on a Dixon-Spain mounting, and the camera is carried externally on the port side. The undercarriage has undergone the Trafford Jones modification.

The Aeroplanes
of the
Royal Flying Corps
Military Wing

J M Bruce

PUTNAM

BY THE SAME AUTHOR

British Aeroplanes 1914–1918

© J M Bruce 1982 and 1992

Second revised edition 1992

Published in Great Britain 1992 by
Putnam Aeronautical Books, an imprint of
Conway Maritime Press Ltd,
101 Fleet Street,
London EC4Y 1DE

British Library Cataloguing-in Publication Data
Bruce, J M (John McIntosh) *1923–*
Aeroplanes of the Royal Flying Corps (military wing).
I. Title
623.7460941

ISBN 0 85177 854 2

All rights reserved. Unauthorised duplication contravenes applicable laws.

Printed and bound in Great Britain by
WBC Print, Bridgend, Mid-Glamorgan

CONTENTS

Acknowledgements	x
Introduction	xi
The Numbering of Military Wing Aircraft	xviii
The Official System of Aircraft Nomenclature, 1914	xxii
Abbreviations used	xxiii
Measurements used	xxvi

The Aeroplanes of the Military Trials, 1912	**1**
The Aerial Wheel	4
Avro Type G	6
The Blériots	8
Borel monoplane	10
The Breguets	11
The Bristol biplanes	13
Bristol-Coanda monoplanes	14
Cody IV monoplane	15
Cody V biplane	16
Coventry Ordnance Works biplanes	16
The Deperdussins	18
D.F.W. Mars monoplane	21
Maurice Farman biplane	22
Flanders biplane	23
Handley Page Type F	25
The Hanriot monoplanes	26
Harper monoplane	28
Lohner biplane	29
Martin-Handasyde monoplane	30
Mersey monoplane	31
Piggott biplane	32
Vickers No.6 monoplane	33

The Aeroplanes of the Royal Flying Corps (Military Wing)

Aeronautical Syndicate Valkyries	35
Airco	
D.H.1	38
D.H.2	41
D.H.3	46
D.H.4	49
D.H.5	61
D.H.6	66
D.H.9	72
D.H.9A	78
D.H.10 Amiens	83
A.R.1	88
Armstrong Whitworth	
F.K.2	90
F.K.3	93
F.K.7 and F.K.8	100
F.K.9 and F.K.10	106
Avro	
Types E and Es	110
504	112
521	120
Beatty-Wright biplanes	123
Blériot	
XI	128
XII	133
XXI	135
Parasol monoplane	138
The Breguet biplanes	142
Bristol	
Boxkite	148
Prier monoplane	151
Coanda monoplane	153
T.B.8	156
G.B.75	159
Scouts B, C and D	161
S.2A	168
F.2A	169
F.2B	174
M.1A, 1B and 1C	180
Caproni Ca.1	185

Caudron
- 45 hp biplanes ... 189
- G.3 ... 191
- G.4 ... 194
- R.11 ... 196

Cody V biplane ... 198

Curtiss
- JN-3 ... 201
- C ... 204
- JN-4A ... 209
- JN-4(Can) ... 211

The Deperdussin monoplanes ... 215

Dunne D.8 ... 221

Henry Farman
- Type militaire, 1910 ... 223
- biplane ... 226
- *Wake-up, England* biplane ... 227
- F.20 ... 229
- F.27 ... 234

Maurice Farman
- Série 7 (Longhorn) ... 238
- Série 11 (Shorthorn) ... 241
- S.11 seaplane ... 247

F.B.A. Flying-boat ... 247

Flanders F.4 ... 249

The Grahame-White transaction ... 252

Grahame-White
- School Biplane ... 252
- Popular Biplane Type VII ... 254
- Popular Passenger Biplane, Type VIIc ... 255
- Type VIII ... 256
- Pusher Biplane ... 258
- Type XV ... 259

Handley Page O/400 ... 261

Howard Wright biplane ... 269

Martin-Handasyde monoplane ... 271

Martinsyde
- S.1 ... 272
- G.100 and G.102 ... 276
- F.3 and F.4 ... 280

Morane-Saulnier
- Types G and H ... 285

Type L	289
Type LA	292
Types N, I and V	296
Type BB	303
Type P	306
Type AC	312
Nieuport monoplanes	315
11 and 21	319
12 and 20	320
16	326
17 and 23	328
24 and 24bis	334
27	337
Paulhan biplane	338
Royal Aircraft Factory	
B.E.1	341
B.E.2, 2a and 2b	344
B.E.2c, 2d, 2e, 2f and 2g	354
B.E.3 and B.E.4	370
B.E.7	375
B.E.8 and B.E.8a	377
B.E.9	381
B.E.12, 12a and 12b	385
F.E.2a	397
F.E.2b, 2c and 2h	403
F.E.2d	419
F.E.4	426
F.E.8	431
F.E.9	436
R.E.1	440
R.E.5	445
R.E.7	451
R.E.8	458
S.E.2	464
S.E.4a	468
S.E.5 and S.E.5a	471
Short	
School Biplane	482
Tractor Biplane	484
S.62	486
827	487
Bomber	489

Sopwith
- 80 hp Biplane 491
- Tabloid 493
- LCT, the 1½ Strutter 499
- Sparrow 509
- Pup 512
- Triplane 519
- F.1 Camel 522
- 5F.1 Dolphin 535
- 7F.1 Snipe 544
- T.F.2 Salamander 550

Spad
- 7 553
- 12 559
- 13 561

Vickers
- Boxkite 564
- The early Gun-carrying Biplanes 566
- F.B.5 569
- F.B.7 574
- E.S.1 576
- F.B.9 582
- F.B.12 585
- F.B.14 590
- F.B.19 597
- F.B.27 Vimy 602

Voisin LA and LA.S 606
Wright biplanes 611
Appendix: Aircraft Manufacturers 617
Indexes
- Aircraft 621
- Aero-engines 627
- Armament 631
- People 635

Acknowledgements

For a substantial part of the textual contents of this book I am indebted to those men of the Royal Flying Corps and Britain's wartime aircraft industry who wrote the documents that survive in the official files now housed in the Public Record Office at Kew. To the staff of the P.R.O. I am more immediately indebted for the ready efficiency with which they made available the small proportion of their vast documentary holdings that time and opportunity permitted me to study.

My reading and interpretation of these historic papers was greatly aided by the generous help I have received, during a lifelong fascination with the formative years of aviation history, from many friends who have given or lent photographs or documents, written personal reminiscences, or conducted specialised research. Foremost among these is Stuart Leslie, whose wholehearted support and unstinting collaboration I have always had; indeed without it I should have accomplished little. Others who have always given more than merely generous help include Chaz Bowyer, Ed Ferko, Ken Molson, Eric Morgan, Jean Noël, Harald Penrose, OBE, FRAeS, Bruce Robertson, and many members of Cross & Cockade, the organisation devoted to the study and recording of the aviation of the 1914–18 war. It has been my privilege to be Vice-President of Cross & Cockade (Great Britain) since 1971.

To all who are named in the following pages as providers of photographs used as illustrations I gratefully record my thanks, and I particularly acknowledge my indebtedness to the Imperial War Museum; the Royal Aircraft Establishment (especially Hugh Colver and Bob Lawrie of the Public Relations Office); Ron Moulton of *Aeromodeller*; Stephen Piercey of *Flight International*; the Director of Public Relations (Royal Air Force) in the Ministry of Defence; Peter Liddle of Sunderland Polytechnic (whose 1914–18 Personal Experience Archives disgorged many invaluable photographs); Dr John Tanner, CBE, FRAeS, FRHistS, FLA, Director of the Royal Air Force Museum; and the infinitely patient staff of the Aviation Records Department of that great institution.

For permission to reproduce some passages from *The War in the Air* I am indebted to the Oxford University Press, for that from *The Clouds Remember* to Messrs Gale & Polden, and for the quotation from *Sagittarius Rising* to its author, Cecil Lewis.

The tedious task of typing much of the manuscript was swiftly and efficiently undertaken by Mrs Sally Franklin, and by my wife, who has unfailingly provided at all stages ever-present support and encouragement, combined with patience that has been truly superhuman.

One name remains. Wing Commander N. H. F. (John) Unwin, MBE, was himself of the RFC, went on to serve in the Royal Air Force, and worked for many years in the Royal Aircraft Establishment at Farnborough. While there he took an active and practical interest in the RAE's history, creating virtually single-handed the RAE Museum and ensuring the preservation of much that would have been destroyed or lost. All aviation historians are indebted to him, I more than most, and it is my greatest regret that John did not live to see the completion of this book, for he contributed to it more, perhaps, than he ever knew. I hope that he and his gallant contemporaries would have approved of it.

J. M. B.

Aircraft of the Military Wing on Farnborough Common in 1913. Visible are three Henry Farman F.20s, one of which is 274; behind the tree the B.E.4 No.204; a Gnome-powered Breguet, probably 211; three B.E.2as, of which 272 can be specifically identified; and on the right the Maurice Farman Longhorn 270. (*Ministry of Defence H1952*)

Introduction

It might fairly be claimed that the earliest British pioneers in the flying of aeroplanes for military purposes were Capt Bertram Dickson, Capt John Duncan Bertie Fulton, and Lt Launcelot Dwarris Louis Gibbs, all officers of the Royal Field Artillery. Lt Gibbs, indeed, was said to have been the seventh man to fly; certainly he had the earliest connection with any aircraft sponsored by the War Office, for he had participated in Lt J. W. Dunne's experiments at Blair Atholl in 1908. Capt Fulton was stationed at Bulford camp in 1909 and was inspired by Louis Blériot's cross-Channel flight to start building a monoplane of his own. However, impatient to fly, he bought from Claude Grahame-White a Blériot monoplane with a 28 hp Anzani engine and began to teach himself to fly on Salisbury Plain. Capt Dickson, after seeing the flying at Rheims in August 1909, acquired a Henry Farman biplane and learned to fly at Châlons.

Dickson joined the British & Colonial Aeroplane Co in the following year, and in September 1910 he took a Bristol Boxkite to the army's autumn manoeuvres. As companions he had Robert Loraine on another Boxkite and Lt Gibbs on his own clipped-wing Farman. Several of Dickson's flights clearly demonstrated the value of aircraft as reconnaissance vehicles: the cavalry were not amused.

In that year of 1910 the War Office put up a shed at Larkhill on Salisbury Plain for the use of the Hon Charles Rolls, who had undertaken to instruct army officers in flying. Some time after Rolls's untimely death at Bournemouth on 12 July, 1910, the shed was assigned to Capt Fulton, who did much flying at Larkhill, as did G. B. Cockburn. Of governmental support they received none, for the administration of the day needed no urging to accept the advice of the Committee of Imperial Defence

xi

A Henry Farman F.20 at Larkhill in 1913. (*RAF Museum PC 71/14/13*)

'. . . that the experiments with aeroplanes . . . should be discontinued, but that advantage should be taken of private enterprise in this branch of aeronautics.'*

Against this background of complacent parsimony it seems astonishing that on 28 February, 1911, an Army Order was issued, creating the Air Battalion of the

The War in the Air, by Sir Walter Raleigh, Vol I, page 141

This photograph is said to depict B.E.2a No.229 after a successful forced landing at Horsham, Sussex, in 1914. No.229 was first allotted to the Military Wing on 16 September, 1913, and was briefly with No.4 Squadron in October. By January 1914 it was with No.2 Squadron at Montrose and remained with that unit at least until 31 July, 1914. It went to France with the Aircraft Park on 14 August, 1914, but subsequently crashed at Dover on 27 August and was written off. (*RAF Museum P3655*)

xii

Royal Engineers. The official existence of this new unit began on 1 April, 1911, whereupon it stood to inherit the seven aircraft purchased (but not, at that date, all delivered) for a total of £6,303 during the year that ended on 31 March, 1911, plus a Blériot XII, the Hon Charles Rolls's Wright and Geoffrey de Havilland's second biplane, later named F.E.1. The Air Battalion was commanded by Brevet Major Sir Alexander Bannerman with Capt Philip William Lilian Broke-Smith, RE, as adjutant. This modest unit was divided into two Companies, of which No.2 Company flew such aeroplanes as were available at Larkhill. The Company's commander was Capt J. D. B. Fulton, Royal Field Artillery, and with him were Capt C. J. Burke, Royal Irish Regiment, Lt B. H. Barrington-Kennett, Grenadier Guards, Lt R. A. Cammell, Royal Engineers, and Lt D. G. Conner, Royal Artillery. They were soon joined by Capt E. B. Loraine, Grenadier Guards, Lt G. B. Hynes, Royal Garrison Artillery, and Lt H. R. P. Reynolds, Royal Engineers; others soon followed.

Among those who followed was Capt Henry Robert Moore Brooke-Popham, Oxfordshire & Buckinghamshire Light Infantry (later Air Chief Marshal Sir Robert Brooke-Popham, GCVO, KCB, CMG, DSO, AFC), who was one of those who

Avro 504As on the sunlit aerodrome at Netheravon in 1915. (*RAF Museum*)

foresaw clearly that aerial combat would occur in future conflicts. In a lecture delivered in 1910 he declared that he saw no reason why aviators should not shoot at one another while flying. Yet of incomparably greater historic significance was the profound but unrecognised influence he was to exert on the RFC in France during the 1914–18 war. Just how greatly the RFC benefited from Brooke-Popham's incisive and intuitive grasp of its needs, his instinctive, intelligent and seemingly instantaneous reactions in dealing with the multitudinous problems and difficulties that were his inescapable portion as Deputy Adjutant and Quartermaster General, above all from his total devotion to these duties, has never been assessed or acknowledged. Late in April 1912 Capt Brooke-Popham took over from Capt Fulton the command of No.2 (Aeroplane) Company. He lost little time in trying out his ideas on arming aircraft for the combats he had foreseen: he had a rifle fired from an aeroplane in flight, and is reported to have tried to install a machine-gun in one of his unit's aircraft.

In July 1911 the appearance of the German gunboat *Panther* at Agadir alerted at least some of Europe's statesmen to Kaiser Wilhelm's intentions, and in November Britain's Prime Minister, the Rt Hon Herbert Asquith,

'. . . . requested the standing sub-committee of the Committee of Imperial Defence, under the chairmanship of Lord Haldane, to consider the future

In an F.E.8 of No.41 Squadron, Lt Stephen Hay patrols the air above the Ypres Salient, 2 January, 1917. (*RAF Museum*)

development of aerial navigation for naval and military purposes, and the measures which might be taken to secure to this country an efficient aerial service.'*

Among the material considered by the sub-committee was a memorandum submitted by Capt Bertram Dickson that included this trenchant passage:

> 'In the case of a European war between two countries, both sides would be equipped with large corps of aeroplanes, each trying to obtain information of the other, and to hide its own movements. The efforts which each would exert in order to hinder or prevent the enemy from obtaining information . . . . would lead to the inevitable result of a war in the air, for the supremacy of the air, by armed aeroplanes against each other. This fight for the supremacy of the air in future wars will be of the first and greatest importance, and when it has been won the land and sea forces of the loser will be at such a disadvantage that the war will certainly have to terminate at a much smaller loss in men and money to both sides.'

The sub-committee reported with commendable alacrity and unanimity, recommending the formation of an aerial service, to be designated 'The Flying Corps' and to consist of a Naval Wing, a Military Wing and a Central Flying School that was to train pilots for both Wings. By Royal Warrant the Royal Flying Corps was constituted on 13 April, 1912; exactly one month later, on 13 May, the Air Battalion and its reserve were assimilated into the Military Wing of the new Corps.

The formation of squadrons followed, No.2 (Aeroplane) Company of the Air Battalion becoming No.3 Squadron, RFC, while a nucleus of aeroplane pilots at Farnborough became No.2 Squadron, RFC, and No.1 (Airship) Company became No.1 Squadron. Official determination to obtain the best aircraft for the aeroplane squadrons underlay the requirements of the Military Aeroplane Competition that was held at Larkhill in August 1912. These produced a field of

*The War in the Air, Vol I, page 198

entrant aircraft that were remarkable for their variety of size and configuration if not for any real military potential; while the various tests to which they were subjected led to the astonishing selection of the Cody biplane as the winner. The War Office had more sense than to order the Cody in quantity, wisely preferring the excellent B.E.2 designed by Geoffrey de Havilland and flown *hors concours* at Larkhill during the Competition.

A few weeks after the ending of the Military Trials the Military Wing lost four of its officers in two crashes during the Army's autumn manoeuvres. Both aircraft were monoplanes that had participated in the Military Trials (*see* pages 155 and 217); the accidents occurred on 6 and 10 September, 1912; on Saturday 14 September the War Office issued an instruction suspending the use of monoplanes by pilots of the Military Wing. This action constituted the so-called monoplane ban, and led to much adverse criticism of the War Office. The Admiralty took no such action in relation to the Naval Wing, and it consequently became commonplace for self-appointed experts to condemn the War Office for its hasty short-sightedness and to praise the Admiralty for its calm good sense. No one, it seems, found it possible to think of the War Office's action as one of humane prudence and of the Admiralty's as foolhardy bravado, but later generations more accustomed to stringent observation of air-safety considerations may take a more objective view.

What was unreasonable was the subsequent abandonment by the War Office of several monoplane types, apparently without structural testing or analysis. This inevitably increased the stridency of those critics who could find nothing good to say about anything done by the War Office or created by the Royal Aircraft Factory. Foremost among these was the late C. G. Grey, at that time editor of *The Aeroplane*, whose hatred of the Factory and all its works and employees was given tediously frequent and heavily acidulated expression in his journal. Much of what he wrote was little less than venomous character assassination, delivered secure in the knowledge that none of those whom he pilloried could reply in any

An unidentified F.E.2b goes about its warlike occasions. (*RAF Museum*)

way. Grey forfeited all right to respect or credibility as a commentator when he published in *The Aeroplane* of 31 January, 1917, what purported to be an obituary on Maj Frank Goodden, who had been killed on 28 January in the crash of the second prototype S.E.5. It was in fact a despicable denigration of Goodden, and was in such monstrously offensive bad taste that the British aviation world stood aghast.

In similar vein but more bombastic were the accusations and philippics of Noel Pemberton Billing, MP, who seemed to regard his election to Parliament in 1916 as an appointment to be keeper of the nation's aeronautical conscience. His barrages of questions and his participation in debates generally wasted much Parliamentary time, and his allegations were a major reason for the setting-up of the Bailhache enquiry into the Royal Flying Corps in 1916, which diverted much official effort into totally unproductive channels at a critical time in the war. Examination of what Billing brought as evidence before the Bailhache tribunal revealed the hollow nature of his exaggerated allegations, and the report of the enquiry summarily dismissed his charges as 'an abuse of language and entirely unjustifiable'.

Sopwith Camels of No.45 Squadron at Istrana in December 1917, shortly after the squadron's transfer to Italy from the Western Front. (*Air Cdre R. J. Brownell, CBE, MC, MM*)

Little more than two years after its formation the Military Wing of the RFC had to go to war. June of 1914 saw its few squadrons brought together, in what was known as a Concentration Camp, at Netheravon. There they practised what were believed to be the likely tasks to be undertaken in war: at that time, as Sir Walter Raleigh recorded,

> 'The single use in war for which the machines of the Military Wing of the Royal Flying Corps were designed and the men trained was (let it be repeated) reconnaissance.'*

On 31 July, 1914, Squadrons No.2, 3 and 4 had between them twenty-two B.E.2as, five Blériot XI monoplanes and six Henry Farman F.20s; No.5 Squadron had no serviceable aircraft but had hopes of eight Henry Farmans (two of which were to come from No.3 Squadron) and four B.E.8s. Additionally, three B.E.2as, three Blériots and four Maurice Farmans were hoped for, and two wireless-equipped B.E.2as were to be added to No.4 Squadron. What the four squadrons took to France on 13–15 August, 1914, were: No.2 Sqn, twelve B.E.2as; No.3 Sqn, six Blériot XIs, one Blériot Parasol, four Henry Farman F.20s; No.4 Sqn, eleven B.E.2as; No.5 Sqn, four Henry Farman F.20s and three Avro 504s.

Those were the aircraft that arrived; others had been lost on the way.

*The War in the Air, Vol I, page 260

The squadrons were followed to France by the Aircraft Park, which was the progenitor of the great organisation of Aircraft Depots and repair and supply depots that grew up as the war progressed, supporting the RFC's multiplying squadrons on the Western Front. The Park's reserve of spare aircraft in August 1914 comprised nine B.E.2as, the B.E.2c prototype, three B.E.8s, three Henry Farmans and four Sopwith Tabloids.

None of the men of the RFC who went to France in that fateful August could know what lay in the future: doubtless many of them shared the widely held belief that the war would be over by Christmas. They had to be followed by hundreds of thousands more; their handful of puny aircraft were the precursors of tens of thousands that flew and fought in France, Italy, Mesopotamia, Macedonia, Palestine, Africa and India, as well as in defence of the homeland. Men and aircraft had to undertake tasks undreamt of at the Concentration Camp of 1914. Herculean feats of production, design, experiment, development and maintenance were performed in these islands and in the fighting services and territories of Britain's allies: their sum sufficed to gain eventual victory, but at catastrophic cost. For the war that was fought on land, principally in Europe, the relevant aeronautical part of that massive effort of the British Empire was inspired by, intended for, developed by, or operated by the Royal Flying Corps. In considering what was attempted and achieved before and during the war of 1914–18 it has to be remembered that the whole combined existence of the Air Battalion and the Royal Flying Corps (Military Wing) spanned no more than seven years.

In the following pages a conscientious attempt has been made to record the histories of the types of aeroplanes acquired or operated by the Air Battalion, R.E., and the Military Wing of the Royal Flying Corps. These histories are presented primarily in an RFC context, but many of the aircraft saw service equally significant and varied in the hands of RNAS pilots; others were not tested in battle until after the RFC and RNAS had been brought together on 1 April, 1918, to form the Royal Air Force. It will be seen in a surprising number of cases that these histories are not quite what they have hitherto been believed or reported to be, though no claim is made that they are now complete or exhaustive.

Camel B6238, aircraft 'C' of No.45 Squadron, photographed at Istrana early in 1918. Normally flown by Capt R. J. Brownell, MC, MM, this Camel crashed at Biancade on 22 February, 1918. (*Air Cdre R. J. Brownell, CBE, MC, MM*)

But what is more important by far is that the whole story of these seven brief, eventful and fateful years can be seen to reflect the honest endeavours of capable and resourceful men in the Royal Flying Corps, the War Office, and the infant aircraft industry; all striving to create, develop, exploit, and employ, vast new fields of technology against the inexorable time scale of war. The marvel is not that there were some set-backs and failures in that brave, doomed septennium but that the sum of these men's efforts was so successful. And if this can be seen to be so the discerning reader will not fail to recognise the baseness of the campaigns of accusation so vindictively waged against them by irresponsible critics; campaigns that in other countries less reluctant to recognise unwarrantable abuse of the freedom of speech might have been regarded as treasonable.

Of the men of the RFC who flew, in peace and war, in the aeroplanes created and serviced by other men (and not a few women), nothing does them greater honour than Sir Walter Raleigh's simple yet comprehensive affirmation:

'They made courage and devotion the rule, not the exception.'

(*The War in the Air*, Vol I, page 7)

The Numbering of Military Wing Aircraft

Soon after the conclusion of the Military Trials in 1912 a new system of numbering the aeroplanes of the Military Wing was introduced. This superseded the Air Battalion's twin series of numbers in which tractor aircraft were numbered from B1 onwards, pushers from F1. As the numbers 1–200 were allotted to the Naval Wing for its aircraft, the Military Wing's series started at 201.

On 14 October, 1912, Mervyn O'Gorman, the Superintendent of the Royal Aircraft Factory, suggested that as new aircraft for the Military Wing were initially delivered to the Royal Aircraft Factory they should be numbered there. Apparently this was agreed verbally by Maj F. H. Sykes, O.C. the Military Wing. On 3 February, 1913, it was recorded that the allocations of serial numbers then stood thus:

1–200	Naval Wing aeroplanes
201–400	Military Wing aeroplanes
401–600	Central Flying School aeroplanes
601–800	Royal Aircraft Factory aeroplanes

The Military Wing's allocation was sub-divided between the four aeroplane squadrons, Nos.2, 3, 4 and 5, so that 201–250 were allotted for No.2 Squadron's aircraft, 251–300 for No.3's, 301–350 for No.4's, and 351–400 for No.5's. Thus, considered overall, numbering was not consecutive: for instance, 255 was taken up by a No.3 Squadron aircraft long before No.2 Squadron reached 245.

As the RFC expanded, the system had to adapt to conform. On 4 October, 1913, Sykes wrote:

'I should be glad to have a group allotted for No.6 Squadron, so that I can number some of the B.E. machines to be delivered for that squadron.'

At this the Assistant Director of Military Aeronautics (ADMA) demurred, opining that 'allotment of groups to squadrons seemed unnecessary and might be inconvenient', but Sykes returned to the matter on 15 October, 1913, when he wrote to the Director-General of Military Aeronautics:

xviii

'. . . machines for the Military Wing have hitherto been numbered according to the squadron to which they are allotted, and not consecutively as received. The object . . . was to facilitate the identification of the Squadron to which a machine belonged in the case of a forced landing or similar occurrence. Such a method of identification appears advisable but, in order to continue it, the allotment of a fresh group of numbers for No.6 Squadron is necessary.'

This request was not approved by Maj W. Sefton Brancker, only recently appointed as G.S.O.2 in charge of M.A.2, who ruled on 21 October, 1913, that service aircraft were to be numbered 201—399, with the block 501—600 to be reserved for training aircraft. Sykes accepted this on 8 November, 1913, and on 6 March, 1914, he wrote to ADMA:

'Aeroplane No.608, type R.E.1 (No.2), having been handed over to the Military Wing as a service machine has now been allotted the new number 362.'

This brought Brancker back to the matter, and on 9 March he wrote:

'. . . no numbers are considered necessary for the Royal Aircraft Factory, and . . . the following will now be the allocation
    1–200   Naval Wing aeroplanes
    201–400   Military Wing aeroplanes
    401–600   Central Flying School aeroplanes
    601–upwards Military Wing aeroplanes.'

More specifically, M.A.2 wrote to the O.C., RFC (Military Wing) next day to say that in view of the allocation of numbers from 601 onwards the R.E.1 should retain its original identity as 608.

In a separate communication of 10 March, M.A.2 intimated that thenceforth the allotment of numbers to aeroplanes would be made by the Chief Inspector of Aircraft, which appointment was at that time held by Maj J. D. B. Fulton. The last number allotted by the Military Wing was 338, the aircraft being a Maurice Farman Shorthorn.

On 5 May, 1914, Brancker wrote again to the O.C., RFC (Military Wing) to advise

'. . . that the Admiralty have now been allotted the numbers 801 to 1000 for seaplanes and aeroplanes belonging to the Naval Wing, in view of the probability that the numbers already allotted, viz.: 1 to 200, will all be appropriated during the present year.

The allocation of numbers will, therefore, be as follows:
    1–200   Naval Wing aeroplanes
    201–400 Military Wing aeroplanes
    401–600 C.F.S. aeroplanes
    601–800 Military Wing aeroplanes.'

The second Admiralty allotment was either overlooked or forgotten by the RFC in the Field in the early weeks of the war, for on 30 September, 1914, Brig-Gen Sir David Henderson, O.C. the RFC in the Field, wrote to ADMA:

'I beg to inform you that aeroplanes now arriving, delivered direct by the makers in this country, are being numbered 801 upwards.'

ADMA replied on 10 October, informing Henderson

'. . . that 800 [sic] to 1000 are Naval numbers, and that we are taking numbers from 1600 [sic] upwards for land service. All land machines with numbers 800 and upwards should therefore be altered to 1800 and upwards to 2000.'

Also on 10 October, 1914, a communication sent to the Secretary of the Committee of Imperial Defence intimated that

'. . . the Lords Commissioners of the Admiralty now desire to appropriate the numbers 801 to 1600, for seaplanes and aeroplanes belonging to the Royal Naval Air Service . . . the [Army] Council concur in the above proposal, and, if it meets with the approval of the Air Committee, the allocation of numbers then will be
No.1 to 200 for R.N.A.S. aeroplanes
201 to 400 for Military Wing aeroplanes
401 to 600 for Military Wing aeroplanes
801 to 1600 for R.N.A.S. aeroplanes
1601 to 1800 for Military Wing aeroplanes
1801 to 2000 for aeroplanes delivered direct from the makers to the R.F.C. in the Theatre of War.'

On 19 October, 1914, Sykes informed ADMA that, in accordance with the latter's instruction of 10 October, the following aircraft had been renumbered in France:

1801	Henri [sic] Farman	No.5 Squadron
1802	” ”	” ”
1803	” ”	” ”
1804	” ”	” ”
1805	” ”	No.3 Squadron
1806	” ”	” ”
1807	B.E.2c	No.2 Squadron
1808	Blériot	No.3 Squadron
1809	”	” ”
1810	”	” ”
1811	”	” ”
1812	”	” ”
1813	Henri Farman	No.3 Squadron
1814	” ”	No.5 Squadron
1815	Blériot	No.3 Squadron
1816	”	” ”
1817	Henri Farman	” ”
1818	” ”	No.5 Squadron
1819	Blériot	No.3 Squadron
1820	”	” ”
1821	Henri Farman	” ”
1822	” ”	No.4 Squadron
1823	” ”	” ”

In the administrative background, minutes were exchanged about the allotment of blocks of numbers, and on 6 November Maj Fulton, Chief Inspector, AID, minuted the Secretary to the War Office:

'In accordance with W.O.87/2152(M.A.1) of May 1914; the numbers 601 to 800 are allotted to Military Wing and 801 to 1000 to Naval Wing aeroplanes. As I am now allotting numbers over 700, will you please inform me what numbers I am to proceed with after 800 has been reached.

The series 401 to 600 which is allotted to C.F.S. has not yet been exhausted.

I suggest that that series should be used for all machines, irrespective of whether they are for C.F.S. or R.F.C., and that a further series be allotted for use in a similar manner.

The use of different series for C.F.S. and R.F.C. is confusing, and appears to have no real object.'

In response to this sensible suggestion came vague bureaucratic bleatings about separate log-books being required for CFS and Military Wing aircraft, these being advanced as a reason for retaining separate allocations of serial numbers. The file came to the notice of Lt-Col H. M. Trenchard, who at that time was officer commanding the Military Wing in England. With typical pragmatism he minuted ADMA on 10 November, 1914:

'For the last four months the Central Flying School and Military Wing have not had a different series of numbers in many cases and I am continually handing over machines from M.W. with an M.W. number to the C.F.S. without renumbering and also vice versa, therefore I would request that we all work on C.I.A.'s minute, as it will undoubtedly save much confusion and work here and in the squadrons.'

A week later Mervyn O'Gorman, Superintendent of the Royal Aircraft Factory, supported Trenchard's view. Subsequently an undated minute gave the C.I.A. the following authority:

'One series of numbers will now be used for all machines irrespective of whether they are allotted to C.F.S. or R.F.C. and in this respect it is suggested that the series 401–600 be used up before commencing on the new series allotted in Minute 1.

You are also authorised to allot numbers to machines as soon as any order is placed.

Contractors will then refer to their machines by these numbers and paint them on before delivery.'

Thereafter the allotment of serial numbers rested with the AID, the only notable exceptions being the individual aircraft numbers within the blocks 1801—1900, 5001-5200, A116—315, A6601—6800, B1501—1700, B3451—3650 and B6731—7130 allotted to the RFC in France for locally purchased aircraft. Aircraft Repair Depots in the United Kingdom allotted serial numbers to the aircraft they reconstructed, within the batches B701—900, B4001—4200, B7731—8230, B8831—9030, C3381—C3480, D4961—5000, F321—350, F616—700, F2189—2208, F9396—9420, F9421—9445, F9496—9545, F9573—9622, F9639—9694, F9996—9999 and H8113—8252, while the batch of numbers D7401—7500 were allotted to 'X' Aircraft Depot in Egypt for similar purposes. The allotment of blocks of numbers to Aircraft Supply Depots in France for reconstructed aircraft did not begin until June 1918, but for the record the blocks in question were F5801—6300*, H6843—7342 and J4592– 5091.

*In fact, the numbering of reconstructed aircraft mistakenly continued beyond F6300 at least as far as F6513, and these had to be renumbered in the batch H6843—7342.

Many other serial numbers were allotted to other aircraft rebuilt from salvage or constructed from spares by a wide variety of units, and it appears that control of these was exercised centrally by the AID.

It should be noted that the serial numbers given in each aircraft history in this book represent only numbers known to the author and are therefore not necessarily complete. Where appropriate they include all known numbers allotted after the RFC had been assimilated into the Royal Air Force.

The Official System of Aircraft Nomenclature, 1914

In the pre-1914 period very few aircraft manufacturers anywhere in the world employed any system of designations for their products. This inevitably led to the use of some long and wordy titles in identifying aeroplane types, and the War Office felt a need for a simple, concise and uniform system of nomenclature for the various aircraft then in service with the Military Wing.

An initial proposal of February 1914 was for a very complicated system of letter-plus-number designations, but this was not adopted, perhaps because it was recognised as no workable improvement on the haphazard and often cumbersome names then in use. Evidently further official thought was applied to the matter, for on 10 June, 1914, War Office letter 87/2353 (M.A.2) was sent to the Officer Commanding, RFC (Military Wing). It read:

> 'With reference to previous correspondence on the subject of the adoption for simplicity and uniformity of a definite system of abbreviated nomenclature for aeroplanes which have passed the experimental stage, please note that it has now been decided to adopt for service machines the following nomenclature, which should be brought into use forthwith for all purposes.
>
> *Army Aeroplanes*
> B.E.2—70 Renault to become   R.A.
> B.E.8—80 Gnome   R.B.
> Avro—50 Gnome   R.C.
> Avro—80 Gnome   R.D.
> R.E.1—70 Renault   R.E.
> R.E.5—120 Austro-Daimler   R.F.
> Sopwith two-seater—80 Gnome   R.G.
> Blériot—80 Gnome   R.I.
> Maurice Farman, 1913 type—70 Renault R.K.
> Sopwith Scout—80 Gnome   S.A.
> Henri Farman—80 Gnome   M.A.
> Maurice Farman, 1914 type—70 Renault M.B.
>
> New designs of machines will be allotted new letters as they pass the experimental stage and are finally adopted for service.'

The requirements of this instruction were promulgated in Military Wing Orders on 15 June, 1914, while the RFC was in Concentration Camp at Netheravon, and these designations were duly employed in contemporary documentation.

It seems that in operational practice the system led to confusion, and the dangers that lay in possible misinterpretation had to be recognised. The squadrons had not been quite two weeks in France when Lt-Col W. Sefton Brancker wrote to the O.C. Military Wing on 26 August, 1914:

'Referring to the circular lists issued under War Office number 87/2353, dated 10 June 1914, on the subject of the nomenclature of aeroplanes, it has been decided, owing to the liability of the letters therein laid down for designation of types of aeroplanes to be confused in telegraphic transmission, that, until the termination of the present war, the use of that nomenclature shall be abandoned, and that types of aeroplanes shall be referred to by their vernacular description, as follows:

B.E.2	R.E.5
B.E.2B	Bleriot 50
B.E.2C	Bleriot 80
B.E.8	Maurice Farman, old pattern
B.E.8A	Maurice Farman, Shorthorn
Avro 50	Henri Farman
Avro 80	Sopwith Scout'
R.E.1	

The nomenclature system of War Office letter 87/2353 (M.A.2) was never revived. In view of the brevity of its existence it will not be incorporated in the aircraft histories that appear in this book.

Abbreviations used

A.A.P.	Aircraft Acceptance Park
A.B.C.	All-British [Engine] Co
A.D.	Aircraft Depot
A.D.A.E.	Assistant Director of Aircraft Equipment
A.D.M.A	Assistant Director of Military Aeronautics
AFC	Air Force Cross
A.F.C.	Australian Flying Corps
AID	Aircraft Inspection Department (later Directorate)
A.P.	Aircraft Park
A.R.D.	Aircraft Repair Depot
A.R.S.	Aeroplane Repair Section
A.S.D.	Aeroplane Supply Depot
ASI	Air-speed Indicator
B.A.S.D.	British Aviation Supplies Depot
B.E.	Blériot Experimental
BEF	British Expeditionary Force
B.H.P.	Beardmore Halford Pullinger (engine)
Brig-Gen	Brigadier-General
C.A.S.	Controller of Aeronautical Supplies
CB	Companion, Order of the Bath
CFS	Central Flying School
C.G.E.	Comptroller General of Equipment
C.I.A.	Chief Inspector of Aircraft
CMG	Companion, Order of St Michael and St George
C.O.	Commanding Officer
C.T.D.	Controller of the Technical Department

D.A.D.A.E.	Deputy Assistant Director of Aircraft Equipment
D.A.D.M.A.	Deputy Assistant Director of Military Aeronautics
D.A.E.	Director of Aircraft Equipment
D.A.O.	Director of Air Organisation
D.D.M.A.	Deputy Director of Military Aeronautics
D.F.W.	Deutsche Flugzeug-Werke
D.F.W.	Director of Fortifications and Works
D.G.M.A.	Director-General of Military Aeronautics
DSO	(Companion of the) Distinguished Service Order
E.F.	(British) Expeditionary Force
E.N.V.	A re-arrangement of 'En V' to indicate the V-form configuration of the aero-engines bearing the name.
F.B.A.	Franco-British Aviation (Company)
F.E.	Farman Experimental
Fl Abt	Flieger-Abteilung (Flying Section)
Flt Lt	Flight Lieutenant
ft	foot/feet
gal	gallon(s)
G.O.C.	General Officer Commanding
H.D.	Home Defence
hp	horse-power
HQ	Headquarters
hr	hour(s)
I.C.F.S.	Indian Central Flying School
I.F.	Independent Force
in	inch(es)
KCB	Knight Commander of the Order of the Bath
kg	Kilogram(s)
km	Kilometre(s)
lb	pound(s) (avoirdupois)
Lt	Lieutenant
2/Lt	Second Lieutenant
Lt-Col	Lieutenant-Colonel
m	metre(s)
MA	Military Aeronautics (*i.e.* Directorate of)
Maj	Major
Maj-Gen	Major-General
MC	Military Cross
min	minute(s)
Mk	Mark
mm	millimetre(s)
mph	miles per hour
M.R.A.F.	Marshal of the Royal Air Force

M.W.	Military Wing (of the Royal Flying Corps)
N.E.C.	New Engine Company
N.P.L.	National Physical Laboratory
OBE	Officer of the Order of the British Empire
O.C.	Officer Commanding
P.C.10	Colour specification for the standard finish applied to the upper surfaces of RFC aeroplanes in general. It was dark khaki and resembled the colour of wet mud. Under surfaces were normally clear doped.
RAE	Royal Aircraft Establishment
R.A.F.	Royal Aircraft Factory (up to 31 March, 1918)
RAF	Royal Air Force (from 1 April, 1918)
R.A.S.	Reserve Aeroplane Squadron
R.E.	Reconnaissance Experimental
R.E.	Royal Engineers
R.E.P.	Robert Esnault-Pelterie
RFC	Royal Flying Corps
R.G.A.	Réserve Générale de l'Aviation
R. & M.	Reports & Memoranda (of the Aeronautical Research Council)
RMLI	Royal Marine Light Infantry
RN	Royal Navy
RNAS	Royal Naval Air Service
rpm	revolutions per minute
S.A.D.	Southern Aircraft Depot
S.A.R.D.	Southern Aircraft Repair Depot
S.E.	Santos Experimental (in relation to the S.E.1 only); subsequently Scouting Experimental.
S.F.A.	Service des Fabrications de l'Aviation
sec	second(s)
Sgt-Maj	Sergeant-Major
sq	square
Sqn(s)	Squadron(s)
S.R.A.F.	Superintendent of Royal Aircraft Factory
S.T.Aé	Section Technique de l'Aéronautique
T.D.	Technical Department
T.D.S.	Training Depot Station
T.S.	Training Squadron
V.C.	Victoria Cross
W/V	Wind velocity
yd	yard(s)

Measurements used

In this book Imperial measurement has been used except in the case of aircraft designed to metric standards.

For conversion multiply as:

Feet to metres—by	·3048
Miles to kilometres	1·60932
Metres to feet	3·2808
Kilometres to miles	·62137
Square feet to square metres	·0929
Square metres to square feet	10·764
lb to kg (1–99)	·454
(100–999)	·4536
(1,000–9,999)	·45359
(over 10,000)	·453592
kg to lb (1–99)	2·205
(100–999)	2·2046
(1,000–9,999)	2·20462
(10,000 up)	2·204622

The Aeroplanes of the Military Trials, 1912

The subject of aeroplanes for the British Army's Air Battalion was discussed in the House of Commons on 30 October, 1911. On that occasion Col J. E. B. Seely, the Under-Secretary for War, stated:

'We have at present in the various stages nineteen aeroplanes, but I must admit that one is broken beyond repair, and one is quite out of date. We have been trying out all the different types. We have eleven types, seven of which are biplanes and four monoplanes. We have learnt most useful lessons from these different types of airships, and we are now engaged in testing some of the more speedy monoplanes.

We are arriving at a point when we think we see our way to choose what is the best type, first, for teaching people to fly and, secondly, for the purpose of war, should war unfortunately break out. As soon as the moment for choice comes—and it will come very soon—we propose to purchase an adequate number of aeroplanes, on which a large number of officers who, no doubt, will be forthcoming, will be able to fly.

Army flying is different from civilian flying, and, for war purposes, it is necessary to have a machine for two men, one to steer and the other to observe. Therefore we want a very special type of Army aeroplane. The specifications for the prizes for the Army aeroplane are now practically complete. . . . I hope that before the end of the present year we shall be able to announce the prize which the War Office and the Admiralty propose to offer for an Army and Navy aeroplane.'

Details of an official competition were announced in December 1911 but the Admiralty had no part in them; it was to be a War Office competition, judged by a committee of judges appointed by the War Office. The prizes (which were far from generous) and conditions were defined in the issue of *Flight* dated 23 December, 1911, as follows:

'The prizes to be awarded by the War Office on the recommendation of a Committee, which will judge the tests and will decide whether any machine submitted is to be subjected to any test.
    A.—Prizes open to the world for aeroplanes made in any country—
        1st prize . . . £4,000   2nd prize . . . £2,000
    B.—Prizes open to British subjects for aeroplanes manufactured wholly in Great Britain, except the engines—
        1st prize . . . £1,500
        Two 2nd prizes £1,000 each   Three 3rd prizes . . . £500 each
   No competitor to take more than £5,000. The War Office to reserve the right to vary the proportions of totals under A and B between the various prizes if the merits of the machines warrant it, or to withhold any prize if there is no machine recommended for it by the Testing Committee.

    The War Office to have the option of purchasing for £1,000 any machine awarded a prize.

The owners of 10 machines which are submitted to all the flying tests and are not awarded a prize to receive £100 for each machine tested.

Oil and petrol to be supplied free for the tests.

The place of delivery of aeroplanes entered for the competition will be announced later.

The following conditions are those required to be fulfilled by a military aeroplane:

1. Be delivered in a packing case suitable for transport by rail and not exceeding 32 ft by 9 ft by 9 ft. The case must be fitted with eyebolts to facilitate handling.

2. Carry a live load of 350 lb in addition to its equipment of instruments, &c, with fuel and oil for $4\frac{1}{2}$ hours.

3. Fly for three hours loaded as in Clause 2 and maintain an altitude of 4,500 ft for one hour, the first 1,000 ft being attained at the rate of 200 ft a minute, although a rate of rise of 300 ft per minute is desirable.

4. Attain a speed of not less than 55 mph in a calm loaded as in Clause 2.

5. Plane down to ground in a calm from not more than 1,000 ft with the engine stopped, during which time a horizontal distance of not less than 6,000 ft must be traversed before touching.

6. Rise without damage from long grass, clover, or harrowed land in 100 yards in a calm, loaded as in Clause 2.

7. Land without damage on any cultivated ground, including rough plough, in a calm, loaded as in Clause 2, and pull up within 75 yards of the point at which it first touches the ground when landing on smooth turf in a calm. It must be capable of being steered when running slowly on the ground.

8. Be capable of change from flying trim to road transport trim, and travel either on its own wheels or on a trolley on the road; width not to exceed 10 ft.

9. Provide accommodation for a pilot and observer, and the controls must be capable of use either by pilot or observer.

10. The pilot and observer's view of the country below them to front and flanks must be as open as possible, and they should be shielded from the wind, and able to communicate with one another.

11. All parts of aeroplane must be strictly interchangeable, like parts with one another and with spares from stock.

12. The maker shall accurately supply the following particulars, which will be verified by official test:

    a. The h.p. and the speed given on the bench by the engine in a six hours' run.

    b. The engine weight, complete (general arrangement drawing), and whether air or water cooled.

    c. The intended flying speed.

    d. The gliding angle.

    e. Weight of entire machine.

    f. Fuel consumption per hour at declared h.p.

    g. Oil consumption per hour at declared h.p.

    h. Capacity of tanks.

13. The engine must be capable of being started up by the pilot alone.

14. Other desirable attributes are:

    a. Stand still with engine running without being held. Engine preferably capable of being started from on board.

    b. Effective silencer fitted to engine.

    c. Strain on pilot as small as possible.
    d. Flexibility of speed; to allow of landings and observations being made at slow speeds if required, while reserving a high acceleration for work in strong winds.
    e. Good glider, with a wide range of safe angles of descent, to allow of choice of landing places in case of engine failures.
    f. It is desirable that the time and number of men required for the change from flying trim to road trim, or packed for transport by rail, and *vice versa*, should be small, and these will be considered in judging the machine. The time for changing from road trim and packed condition to flying trim to include up to the moment of leaving the ground in flight, allowance being made for difficulty in starting the engine.
    g. Stability and suitability for use in bad weather, and in a wind averaging 25 miles per hour 30 ft from the ground without undue risk to the pilot. Stability in flight is of great importance.
    h. The packing case for rail transport to be easily dismantled and assembled for use, and when dismantled should occupy a small space for storage.'

    The poverty of the prizes did not deter entries from a number of competitors, including several French, one German, and one Austrian. By 2 July, 1912, a total of 21 entrants had submitted no fewer than 32 highly assorted aeroplanes. These aircraft were allotted consecutive serial numbers from 1 to 32, as follows:

    1 Hanriot monoplane (100 hp Gnome)
    2 Hanriot monoplane (100 hp Gnome)
    3 Vickers No.6 monoplane (70 hp Viale)
    4 Blériot XI-2 monoplane (70 hp Gnome)
    5 Blériot XXI monoplane (70 hp Gnome)
    6 Avro Type G biplane (60 hp Green)
    7 Avro Type G biplane (60 hp ABC)
    8 Breguet U2 biplane (110 hp Canton-Unné)
    9 Breguet U2 biplane (110 hp Canton-Unné)
  10 Coventry Ordnance Works biplane (100 hp Gnome)
  11 Coventry Ordnance Works biplane (110 hp Chenu)
  12 Bristol G.E.2 biplane (100 hp Gnome)
  13 Bristol G.E.2 biplane (70 hp Daimler-Mercedes)
  14 Bristol-Coanda monoplane (80 hp Gnome)
  15 Bristol-Coanda monoplane (80 hp Gnome)
  16 Flanders biplane (100 hp ABC)
  17 Martin-Handasyde monoplane (75 hp Chenu)
  18 Aerial Wheel (50 hp NEC)
  19 Mersey monoplane (45 hp Isaacson)
 *20 (British) Deperdussin monoplane (100 hp Anzani)
 *21 (British) Deperdussin monoplane (100 hp Gnome)
  22 Maurice Farman biplane (70 hp Renault)
  23 D.F.W. Mars monoplane (100 hp Mercedes)
  24 Lohner biplane (120 hp Austro-Daimler)
  25 Harper monoplane (60 hp Green)
  26 (French) Deperdussin monoplane (100 hp Gnome)

*Some contemporary accounts transpose the identities of the two British-built Deperdussins. Those quoted above are the majority presentation.

27 (French) Deperdussin monoplane (100 hp Gnome)
28 Handley Page Type F monoplane (70 hp Gnome)
29 Piggott biplane (35 hp Anzani)
30 Cody IV monoplane (120 hp Austro-Daimler)
31 Cody V biplane (120 hp Austro-Daimler)
32 Borel monoplane (80 hp Gnome)

In the event, eight of the original entrants did not reach Larkhill, and several of those that did took little or no active part in the flying, for varying reasons. Exhaustive reports of the activities of the Trials appeared throughout August 1912 in the contemporary issues of *Flight* and *The Aeroplane*, and readers who require a blow-by-blow account of all that happened are referred to those sources. Here we shall briefly examine the aircraft, which have a place in the history of military aviation in Britain. A motley and unmilitary lot they were – but there speaks hindsight, for in 1912 no one had any very clear idea of what military uses aeroplanes might have, and this unlikely collection of aircraft were all contenders to determine which of them came closest to fulfilling the somewhat naïve specification laid down by the War Office (which obviously had no more idea of the military possibilities of aeroplanes than did the optimistic entrants). In the end, one of the least practical of them all was declared winner.

The contest's title was the Military Aeroplane Competition but it has for so long been popularly known as the Military Trials that that more convenient and compact name will be freely used in this book without apology.

The Aerial Wheel

In its issue of 10 August, 1909, the weekly journal *The Aero* published a letter from George Sturgess under the heading 'The Sturgess Flying Wheel'. This described the advantages of fitting an aeroplane with immense wheels of special design: Sturgess claimed *inter alia* that such large wheels would minimise the risk of sinking into the ground. A diagram accompanying his letter showed a biplane of Wright configuration with the enormous wheels completely encircling the wing cellule; on a fixed circular frame carrying three small wheels an outer peripheral wheel rotated and alone contacted the ground. Two such wheels set 12 ft apart were envisaged.

Sturgess went on to develop his idea, and formed The Aerial Wheel Co. Late in 1910 the correspondence pages of *Flight* (10 December) and *The Aero* (14 December) carried letters from Sturgess in which he again stressed the ability of his invention to enable an aeroplane to land on uneven ground with minimal risk of damage. Both letters were accompanied by a photograph of a model monoplane equipped with two of his wheels; these were seen to go right round and over the wing. Sturgess indicated that the full-size diameter of these wheels was to be no less than 12 ft, and that they were so constructed that each consisted of a periphery that was free to run round a fixed framework.

In November 1911, Henri Salmet's 50 hp Blériot monoplane at Hendon was fitted with a pair of these enormous wheels, each consisting '. . . of a pair of large concentric hoops, separated by wooden roller bearings in such a manner that the exterior hoop is free to revolve over the interior one, which latter is directly connected to the existing springing arrangements of the Blériot. In its action it

The Aerial Wheel photographed some time after the Military Trials.

performs the function of an endless skid . . .' Nothing further seems to have been recorded about this first practical installation of the Sturgess landing wheel.

By 1912 Sturgess had moved from Mablethorpe to Leamington Spa, and on 30 July, 1912, jointly with Ralph Platts of Birmingham, he applied for Patent No. 17649, which sought to protect a specialised form of wing warping that was intended 'to check and neutralize sideslip and slewing.' In view of the date, it seems likely that this warping device would have been incorporated in the Aerial Wheel, the aircraft entered in the Military Trials by the Aerial Wheel Syndicate Ltd, which had been set up by Sturgess and Platts.

The aircraft had reached Larkhill before 31 July, 1912, and was therefore qualified to participate in the Military Trials. Its Trials number was 18. It was reported to be in course of construction on Saturday, 3 August, but such progress as was made with it was in no way publicized, and it rapidly acquired the adjective 'mysterious'. When the Trials had been running for about a week it became known that the Aerial Wheel was to be flown by Cecil Pashley, who arrived at Larkhill on about 10 August in the hope of flying the aircraft. He remained a disappointed man, however, for the Aerial Wheel was never completed at Larkhill, and did not emerge from its shed during the competition.

Not until 12 June, 1913, were photographs of the completed aircraft published in *The Aeroplane*, when it could be seen to be a peculiar canard monoplane powered by a 50 hp NEC and seating its pilot and passenger side by side in the middle of the mainplane. Its landing gear consisted primarily of a single central Sturgess wheel with outboard skids providing lateral support; the NEC engine drove a four-blade tractor airscrew that rotated within the Sturgess wheel. The arrangement of bracing and control wires suggested that the Sturgess and Platts warping system was incorporated, but the only vertical fin area of any kind was provided by a vestigial surface at the midpoint of the foreplane.

It seems highly unlikely that the Aerial Wheel ever flew. Had it done so it could hardly have failed to be extremely dangerous, but it and its designers sank into oblivion and were never heard of again.

The Avro Type G at Larkhill. (*Flight International A381A*)

Avro Type G

The first Avro Type E biplanes (see page 110) for the Military Wing had been delivered before the Military Trials began. The firm's entry for the Trials was obviously related to the Type E but differed remarkably in housing its occupants in a wholly enclosed cabin created by so deepening the fuselage that it filled the entire gap. In view of the importance attached to field of view in the requirements of the Trials this was a remarkable design decision, the more so because there was no directly forward view at all. Landing must have been a hazardous business.

All the flight surfaces on the Type G were those of the Type E, consequently lateral control was again by warping and the base of the rudder was shod to serve as a tailskid. The undercarriage closely resembled that of the Type E.

Two Type G biplanes were put in hand in 1912 and were allotted the Trials numbers 6 and 7. The former was to have the 60 hp Green, the latter a 60 hp A.B.C., but the A.B.C. engine could not be finished in time, consequently No.7 was completed with the Green and packed off to Larkhill without being flown, so great was the urgency. The Trials began on Thursday, 1 August, 1912, the first exercise being appropriately the assembly test. This was won convincingly by the Avro, which was assembled by its pilot, Lt Wilfred Parke, R N, and five men in the astonishing time of $14\frac{1}{2}$ minutes. Parke then flew the aircraft on what may have been its first flight, for its performance was marred by an over-large airscrew. On 7 August he was again flying the Type G, now with a new airscrew, when turbulence forced him to land downwind at considerable speed on rough ground. The Avro turned over and was extensively damaged, but Parke's safety belt saved him from injury.

The wreckage was promptly sent back to the Avro factory and the repaired aircraft was miraculously back at Larkhill a mere nine days later. There can be little doubt that this feat owed something to the available components of No.6. It was flying again on 19 August, when Parke tested it in rain, as if to prove one of the advantages of an enclosed cabin. On 25 August he undertook the 3 hr endurance test and, having completed it, went into a spiral glide at 600 ft that swiftly became one of the earliest identified spins. Parke applied opposite rudder and, despite having his elevators fully up, recovered in time to avoid disaster.

With its power loading of 27·2 lb/hp the Avro's performance overall was too poor for it to stand a chance of winning any of the major prizes in the competition. Its showing in the climbing trial was abysmal, but it bravely submitted to all the trials, for which A. V. Roe & Co were awarded £100 in the final prize list; and it retained the distinction of being the only all-British aircraft to finish the course. The Avro's performance in the trials can be summarised as follows:

Test	Position	Result
Quick assembly	1	14 min 30 sec
3-hours test	Passed	—
Climbing	Failed	105 ft/min
Gliding	3 equal	1 in 6·5
Consumption		
(a) Petrol	1	4 gal/hr
(b) Oil	2	0·5 gal/hr
Calculated range	4	345 miles
Speed		
(a) Maximum	8	61·8 mph
(b) Minimum	4	49·3 mph
Speed-range	7	12·5 mph
Landing	Passed	65 yd run
Quick take-off	Failed	—
Rough weather	Failed	Wind from 16 to 33 mph
Transport	1	Dismantled and re-assembled in 23 min

Avro Type G
60 hp Green
Span 35 ft 3 in; length 28 ft 6 in; height 9 ft 9 in; wing area 335 sq ft.
Empty weight 1,191 lb; loaded weight 1,792 lb.
Maximum speed 61·8 mph; climb to 1,000 ft, 9 min 30 sec.

The Avro Type G undergoing the assembly test at Larkhill. In this it comfortably outstripped all other competitors, being fully assembled in 14½ minutes.

The Blériot XI-2 was at first incorrectly numbered 3 in the Trials series of numbers, but this was removed and the correct identifying number 4 was painted under each mainplane and on the rudder. Its pilot, Perreyon, is in the front cockpit.
(*Peter Liddle's 1914–18 Personal Experience Archives, presently housed within Sunderland Polytechnic*)

The Blériots

The French Blériot company entered two quite different monoplanes in the Military Trials. Both were powered by the 70 hp Gnome engine, and it was initially intended that both should be flown by Gustav Hamel, but there the similarities ended. The Blériots were allotted the Trials numbers 4 and 5, the former being a more or less standard Blériot XI-2, the latter a Blériot XXI with side-by-side seating.

By August 1912, the RFC had in fact had long experience of the Blériot XXI as a type (see page 135) but no Blériot XI of any kind had seen official service at that time. The XI-2 that was the Trials entrant had the widely spaced tandem seating that characterised the earlier Blériot two-seaters, and it had a high-set angular rudder combined with a standard tailplane and elevators. It was also somewhat smaller than the generality of Blériot two-seaters later used by the RFC in some numbers.

Starting up the Blériot XI-2 at Larkhill. As chocks were not in use the aircraft had to be manually restrained. (*RAF Museum*)

8

This Blériot XI-2, No.4 at the Trials, was taken over by the RFC as No.221 and is discussed in that context on page 130. Something is said about the Blériot XXI as a type on pages 135–7. It will therefore suffice to summarise the performance of these two entrants specifically in relation to the Military Trials, in which the aircraft were flown by Edmond Perreyon.

Test	Position		Result	
	XI-2	XXI	XI-2	XXI
Quick assembly	9	10	59 min 57 sec	1 hr 27 min
3-hours test	Passed	Passed	—	—
Climbing	5	6	250 ft/min	235 ft/min
Gliding	Failed	Failed	1 in 5·6	1 in 5·3
Consumption				
(a) Petrol	2	3	5·4 gal/hr	6·3 gal/hr
(b) Oil	6 equal	6 equal	1·7 gal/hr	1·7 gal/hr
Calculated range	9	11	295 miles	252 miles
Speed				
(a) Maximum	9	10	61·1 mph	58·9 mph
(b) Minimum	5	2	52 mph	40 mph
Speed-range	9	2	9·1 mph	18·9 mph
Landing	Passed	Passed	60 yds	45 yds
Quick take-off	Passed	Failed	250 yds	—
Rough weather	Failed	Passed	—	Wind velocity 21–31 mph
Transport	3	5	Dismantled and re-assembled in 44 min	50 min

Perreyon and Lt M. Read in the Blériot XXI, No.5 in the Military Trials.

Considered purely as aeroplanes, especially if one sets aside the results achieved in the more unrealistic of the tests, the two Blériots gave quite good accounts of themselves; as well they might, for they were, relatively speaking, well-tried designs. Because they had submitted to all the tests at Larkhill they were each awarded one of the £100 consolation prizes, and the XI-2 was purchased for the use of the Military Wing, thus becoming the first of many Blériot XI two-seaters to be used by the RFC (*see* page 129).

Blériot XI-2
70 hp Gnome
Span 9·65 m; length 8·33 m; wing area 18·414 sq m.
Empty weight 401 kg; loaded weight 680 kg.
Maximum speed 98 km/h; climb to 1,000 ft, 4 min.

Blériot XXI
70 hp Gnome
Span 11·075 m; length 8·3 m; wing area 27·621 sq m.
Empty weight 389 kg; loaded weight 672 kg.
Maximum speed 95 km/h; climb to 1,000 ft, 4 min 15 sec.

The Borel monoplane entered in the Military Trials was to be similar to this aircraft.

Borel monoplane

Gabriel Borel's interest in aviation began in 1909, and he gave practical expression to it by setting up a flying school at Mourmelon. He built up a team of pilots, one of whom was Léon Morane, and in 1911 the Société anonyme des Aéroplanes Morane-Borel-Saulnier was founded to construct aircraft designed by Raymond Saulnier, who had earlier worked with Louis Blériot. This company was short-lived, however, for Borel separated from it in the autumn of 1911 and went into business as an aircraft constructor on his own.

Borel's earliest products were monoplanes that, in appearance and geometry, owed something to Blériot and Saulnier designs. One of these, powered by an 80 hp Gnome engine, was entered for the British Military Trials of 1912, receiving the identifying number 32. Its designated pilot was one Chambenois.

The Borel monoplane of 1912 was a typical, simple, shoulder-wing aircraft with balanced rudder and elevators, warping wings, and a plain V-strut undercarriage. Its Gnome installation was unusual at that time in having the engine on an overhung mounting. Unfortunately it never had the opportunity of showing its paces, for it was not ready in time and did not in fact leave France. The manufacturers informed the judges by telegram that the monoplane could not be ready until 3 August and therefore could not reach Larkhill until the 5th; this date was too late to be acceptable.

Borel monoplane
80 hp Gnome
Span 10·363 m; length 7·01 m; wing area 15·345 sq m.
Empty weight 333 kg; loaded weight 515 kg.
Speed 113 km/h.

This was the Breguet biplane flown from France to England by W. B. Rhodes Moorhouse on 4 August, 1912. It was the parent Breguet company's entry for the Military Trials and would have been No.9, but was too badly damaged after landing to participate.
(*The late André de Bailliencourt*)

The Breguets

In the French competition for military aeroplanes held late in 1911 Breguet biplanes were very successful, consequently it is not surprising that the British company of Breguet Aeroplanes Ltd entered two aircraft in the Military Trials. It was originally intended that both should be built in the British Breguet factory, but administrative delays in the formation of the new company, coupled with difficulty in finding and equipping a suitable factory, made it impossible to start construction work until 5 July, 1912. To be eligible for the competition aircraft were to be at Larkhill by 31 July, so time was very short.

The company at once began to build one biplane but obtained permission to enter a French-built Breguet as the second. One of these biplanes was to be flown by W. B. Rhodes Moorhouse (later to win immortality as the RFC's first VC of the war) and the other by René Moineau. Rhodes Moorhouse went to France to fly the French-built aircraft to England but was delayed by unfavourable weather. Not until Sunday, 4 August, was he able to set out, and took off from Douai with his wife and a friend as passengers. The Channel was safely crossed despite rainstorms and deteriorating weather, but a landing became essential. This was made, with difficulty, at Bethersden, where the Breguet was blown against some trees and damaged too extensively to have any hope of participating, even belatedly, in the Military Trials.

This aircraft, No.9 in the Trials series of identifying numbers, was a typical Breguet biplane with two-bay bracing applied to a single row of interplane struts, the wings being flexibly built about a single main spar. The fuselage had an aluminium skin overall, and the tail unit was the usual Breguet cruciform all-flying assembly on a single universal joint. Power was provided by a conventional 110 hp Canton-Unné water-cooled radial engine, the twin radiators being disposed as large vertical elements on either side of the forward fuselage. In the Breguet system of designations the aircraft would have been described as a Type U2.

The British-built Breguet, No.8 in the Trials, was in its essential geometry very similar to the unlucky No.9, having a metal-clad fuselage and two-bay wings. Its engine was a 110 hp Canton-Unné, but of the peculiar horizontal type that lay flat

11

The British-built Breguet, No.8 in the Trials, at Larkhill. (RAF Museum)

on its back and drove the airscrew via a right-angle bevel reduction gear having a ratio of 1:1·8. The two radiator units were installed above the engine and on each side of the airscrew shaft. This installation, to which was added a massive four-blade airscrew, gave No.8 a completely different frontal appearance.

Misfortune dogged the British-built Breguet from the outset. It left London on 30 July transported on a trolley hauled by a steam traction engine. Wheel and axle breakages of the transporter occurred en route, but the aircraft eventually reached Larkhill undamaged. It arrived too late to undertake the quick-assembly test, and was reported to be in course of construction by Saturday, 3 August. This Breguet had experienced delays in manufacture owing to material-supply difficulties and had been rushed away from the factory completely untested. Its engine gave trouble, apparently because its bevel gears had not been run in, and the original airscrew was too large. The 'kind of glorified hop' that the aircraft made on 10 August must, therefore, have been its first flight. It had the misfortune to break a wheel on landing.

Towards the end of the Trials Moineau was sufficiently satisfied with the aircraft to be willing to undertake at least some of the tests, but magneto failure thwarted this. Another magneto was sent for from Paris but by the time it arrived and was fitted the competition was declared closed. The unlucky Breguet

No matter from which angle it was viewed, the British Breguet was not a thing of beauty.

therefore took no active part in the Trials, and evidently did nothing of note thereafter. It flew briefly at Hendon early in 1913.

Breguet Type U2
110 hp Salmson (Canton-Unné)
Span 13·65 m; length 8·75 m; wing area 36 sq m.
Empty weight 798 kg; loaded weight 1,098 kg.
Maximum speed 115 km/h.

Breguet No.8
110 hp Salmson (Canton-Unné)
Dimensions as Type U2.
Loaded weight 975 kg.

Dimensions are unconfirmed and U2 weights are typical.

The Bristol biplanes

The largest entry in the Military Trials was that of The British & Colonial Aeroplane Co Ltd, who entered no fewer than four aircraft, two biplanes and two monoplanes. The biplanes were designed by E. C. Gordon England and, as they represented his second design for the firm, were designated Bristol G.E.2. They were side-by-side two-seaters with dual control, and had the Bristol works numbers 103 and 104, the former having a 100 hp Gnome two-row rotary, the latter a 70 hp Daimler-Mercedes engine with chain-drive to the airscrew. In the Military Aeroplane Competition the Bristols Nos.103 and 104 had, respectively, the identifying numbers 12 and 13, and were to be flown by Gordon England and Howard Pixton.

By 10 August, No.13 had been withdrawn owing to poor running by its Daimler engine which, according to C. G. Grey in *The Aeroplane*, was 'so badly balanced

No.12 in the Trials was the Bristol Gordon England biplane with 100 hp Gnome engine. (*Flight International 043*)

The second Gordon England biplane had the 70 hp Daimler-Mercedes engine, which ran so badly that the aircraft had to be withdrawn from the Trials.

that in a long run it would pull itself out of the fuselage'. Gordon England went on flying No.12, but Pixton was left without an aircraft for a time. He later took over one of the Bristol-Coanda monoplanes.

On Saturday, 24 August, Gordon England took up No.12 with full competition load, but the biplane could not be induced to climb. This reluctance to climb may have been attributable to the fact that England's mechanic, in the course of making adjustments in the cockpit, had tied the elevator cables together to prevent elevator movement in the wind. In his haste to get airborne, England had failed to check for freedom of control movement. Nevertheless, the biplane's chance of completing all the other tests was fairly remote; consequently it was brought in and withdrawn from the competition, having participated only in the quick-assembly trials, in which No.12 was put together by four men in 54 min 5 sec, No.13 in 1 hr 32 min. They were thus 8th and 11th respectively in this test.

Bristol biplane No.12
100 hp Gnome
Span 40 ft; length 31 ft; wing area 400 sq ft.
Empty weight 1,474 lb; loaded weight 2,174 lb.
Speed 60 mph.

Bristol biplane No.13
70 hp Daimler-Mercedes
Dimensions as No.12.
Empty weight 1,650 lb; loaded weight 2,350 lb.
Speed 57 mph.

Bristol-Coanda monoplanes

Both of the Bristol-Coanda monoplanes, Nos.14 and 15 in the competition, were taken over by the Military Wing on completion of the Trials. Their history is therefore related on pages 153–6.

Cody IV monoplane

Samuel Franklin Cody was the very epitome of the pioneer aviator-constructor in the early years of the twentieth century, working with great dedication and single-mindedness on the steady development of his biplanes. It was only to be expected that he would enter the Military Aeroplane Competition; not only had he been long associated with the Army Aircraft Factory at Farnborough, but a challenge of the kind presented by the War Office's specification was something that he relished. Although he was virtually a one-man organisation he entered two aircraft: one was the Cody III, the other a monoplane of unusual design that had the designation Cody IV and the Military Trials number 30.

It was a shoulder-wing monoplane that embodied a number of features typical of Cody design, and was distinguished by a tricycle undercarriage. The 120 hp Austro-Daimler engine had originally been fitted to the Etrich monoplane flown by Leutnant H. Bier in the 1911 Circuit of Britain contest, having been bought by Cody after the Etrich crashed. It was mounted low in the fuselage and drove the high-set airscrew via a chain. Pilot and passenger sat side by side behind the engine and in line with the trailing edge of the mainplane.

The aircraft's most curious feature was the unique design of the rear fuselage and tail unit. Four divergent tail-booms of bamboo carried twin rudders and twin elevators in a form of controllable box-kite; all four tail surfaces were all-flying controls, each being kite-shaped and so mounted that it was aerodynamically balanced. Abaft the cockpit, fabric was fitted between the tail-booms in such a way that a pointed cruciform was created. This form of tail unit was the subject of Cody's patent No. 12,348 of 1912; he claimed that it increased stability.

Control was provided by a single central control column, accessible to either occupant of the cockpit, which actuated the elevators in the conventional way and the warping by moving from side to side; rudder control was provided by rotating the column by means of a substantial wheel mounted axially on it. Fore and aft movement was transmitted to the lower elevator by a length of bamboo.

The monoplane was completed in June 1912 and made initial trial flights during that month. Unfortunately its career was brief and the aircraft never left Farnborough. On 8 July, just as Cody was approaching to land on Laffan's Plain, a cow ran into the monoplane's path; the aircraft was wrecked, but Cody was thrown clear and was not badly hurt. The Cody III had crashed five days earlier,

The Cody monoplane was of generous proportions, with a cavernous cockpit and distinctive tail unit. (*Royal Aircraft Establishment 0159; Crown copyright*)

so Cody was left without an aeroplane, yet with typical determination he built another biplane embodying usable parts of Cody III and the Austro-Daimler engine from the monoplane. (see pages 198–200)

Cody was killed on 7 August, 1913, when his last biplane broke up in the air. His workshop and its contents were auctioned by Messrs Kingham and Kingham of Aldershot on 8 September, 1913, and the lots were reported to include 'a complete Cody monoplane'. Presumably this represented the remains of the Cody IV, but the ultimate fate of these relics of the great man remains unrecorded.

Cody IV
120 hp Austro-Daimler
Span 43 ft 6 in; length 37 ft; height 12 ft 6 in; wing area 260 sq ft.
Empty weight 1,400 lb.
Speed approximately 70 mph.

Cody V biplane

As this aircraft was taken over by the Military Wing of the RFC it is described on pages 198–200.

Coventry Ordnance Works biplanes

In 1911 the Coventry Ordnance Works took over the aircraft-construction business of Warwick Wright, and with it the persons of W. O. Manning and Howard Wright. The new company decided to enter the Military Aeroplane Competition with two aircraft designed by Manning and Wright; one was to be powered by a 100 hp Gnome rotary, the other by a 110 hp Chenu engine.

Most of the metal components were made at Coventry, but construction of the airframes was undertaken in the former Warwick Wright workshop under the railway arches at Battersea just south of the Thames. Work proceeded so well that the Gnome-powered biplane was completed in April 1912 and created something of a sensation when it appeared at Brooklands in May. It was an unusual but remarkably workmanlike biplane with single-bay bracing, long extensions on the upper wings, side-by-side seating, and twin fins and rudders. The rudders and elevators were balanced, lateral control was by wing warping, and the only measure of springing in the undercarriage was provided by the deep-section low-pressure tyres. In its early flights the aircraft performed well.

The second biplane was basically similar but was appreciably smaller. Its Chenu engine, being an inline unit, gave the nose a different aspect, and the whole fuselage was slimmer in plan because the pilot and passenger sat in tandem. The tail unit had a single fin and rudder, and was distinguished by flowing fish-like outlines.

In the Military Trials the Gnome-powered aircraft was No.10, its stable-mate No.11. Both were to be flown by T. O. M. Sopwith. No.10 survived transport contretemps to arrive in time to undergo the quick-assembly test, being erected by six men in 1 hr 51 min. It made a trial flight on 1 August and started the 3-hour

The Gnome-powered Coventry Ordnance Works biplane at Larkhill. Behind it is an early Deperdussin monoplane.

flight test next day but had to abandon the attempt because it could not be persuaded to climb. The aircraft was still refusing to climb on 8 August, and next day the efforts of Sopwith and F. P. Raynham failed to cajole it into a climb. Sopwith had to leave the competition before the C.O.W. aircraft could seriously undertake any of the tests, for he was committed to driving the racing motor-boat *Maple Leaf* in the United States. He handed over responsibility for the aircraft to F. P. Raynham, who made a few flights in No.10, but to no avail, and the biplane did not complete any of the tests.

No.11 never had any chance to demonstrate its capabilities, for its Chenu engine gave endless and irremediable trouble. Its magneto drive-shaft sheared repeatedly, and by 9 August it was decided to send part of the crankcase to France to have the magneto drive modified. Two weeks later the retaining lugs for the reduction gear housing broke, and it appears that all thought of getting the aircraft into the air was then abandoned.

The Coventry Ordnance Works' second entry had a refractory 110 hp Chenu engine and a tail unit of great elegance.

17

The Gnome-powered aircraft was fitted with a new set of lengthened three-bay wings some time after the trials and flew successfully. The Chenu-powered biplane was never heard of again, and presumably was scrapped.

C.O.W. biplane No.10
100 hp Gnome
Span 40 ft (upper), 24 ft 8 in (lower); length 33 ft 3 in; wing area 350 sq ft.
Empty weight 1,200 lb; loaded weight 2,000 lb.
Maximum speed 60 mph.

C.O.W. biplane No.11
110 hp Chenu
Span 35 ft (upper), 22 ft (lower); length 31 ft 3 in; wing area 325 sq ft.
Empty weight 1,300 lb; loaded weight 2,100 lb.
Maximum speed 63 mph.

The British-built Deperdussin monoplane with 100 hp Anzani radial engine was No.20 in the Trials. In this photograph it is seen at Hendon. (*Flight International, Deperdussin 2*)

The Deperdussins

By 1912 the Deperdussin *marque* was well-known in the aviation world, and the French parent company had established a British subsidiary. The British Deperdussin Aeroplane Syndicate Ltd became The British Deperdussin Aeroplane Company Ltd in April 1912. In the Military Aeroplane Competition the French and British Deperdussin companies each entered two monoplanes. Both French aircraft were to be powered by the 100 hp Gnome and were to be flown by Maurice Prévost; one of the British-built monoplanes was to have the 100 hp Anzani and be flown by Lt John C. Porte, RN, one of the directors of the British Deperdussin company; the other was to be fitted with a 100 hp Gnome radial engine and flown by Jules Védrines. The British Deperdussins had respectively

Lt John C. Porte, RN, in the pilot's (rear) cockpit of No.20 during the trials. (*RAF Museum*)

the competition numbers 20 and 21, the French monoplanes 26 and 27. In the event, No.27 did not arrive.

The three participating monoplanes looked generally similar, but there were differences. On the French-built monoplane the pilot sat in the front seat, the cabane had three struts on either side, and there were frontal skids on the undercarriage. The British Deperdussins sat their pilots in the rear seat, had two-strut cabanes, and were without frontal undercarriage skids. The Anzani-powered monoplane differed further from the other two by having inversely tapered wings.

In the various tests the French-built Deperdussin outstripped all the other competitors by completing the full set of tests by the evening of Saturday, 10 August. As the table of results indicates, it did well. John Porte flew the Anzani-

This French-built Deperdussin with 100 hp Gnome was No.26 in the Trials.

19

powered British Deperdussin a good deal without managing to reach a significant level of achievement in any test; and the other, Gnome-powered, aircraft had to have a change of pilots when Védrines left for the United States on 15 August to take part in the Gordon Bennett contest. The aircraft was thereafter flown with some success by Gordon Bell.

The judges awarded the £2,000 second prize in the competition to the French-built Deperdussin No.26, and also gave a £500 prize to the British-built Deperdussin No.21, obviously the aircraft flown by Gordon Bell since Porte's Anzani-powered aircraft failed to distinguish itself. The results for the three monoplanes may be tabulated as follows:

Test	French Position	British (Gnome) Position	British (Anzani) Position	French Result	British (Gnome) Result	British (Anzani) Result
Quick assembly	15	18	16	1 hr 56 min	7 hr 15 min	2 hr 55 sec
3-hours test	Passed	Passed	—	Passed	Passed	—
Climbing to 1,000 ft	2	4	8	3 min	3 min 45 sec	4 min 45 sec
Gliding	7	4	—	1 in 5·4 (failed)	1 in 6·2 (passed)	—
Consumption						
(a) Petrol	6	9	—	8·4 gal/h	9·8 gal/h	—
(b) Oil	4	7	—	1·3 gal/h	1·8 gal/h	—
Calculated range	8	7	—	315 miles	320 miles	—
Speed						
(a) Maximum	6	7	—	69·1 mph	68·2 mph	—
(b) Minimum	8	6	—	59 mph	54·6 mph	—
Speed-range	8	6	—	10·1 mph	13·6 mph	—
Landing	7	—	—	73 yd run	—	—
Quick take-off from harrowed field	1	3	—	132 yd run	200 yd run	—
Rough weather	—	—	—	W/V 29-13 mph (average below requirement)	—	—
Transport	6	7	—	Dismantled and re-assembled in 1 hr 44 min	1 hr 54 min	—

This is thought to have been the French-built Deperdussin, No. 26 in the Trials.
(*RAF Museum*)

As recounted on page 215, the two Gnome-powered Deperdussins were taken over by the RFC Military Wing after the Military Trials ended, No.26 becoming 258, No.21 becoming 259. It was on 258 that Captain Patrick Hamilton and Lt Wyness-Stuart were killed on 6 September, 1912; and, as noted on page 218, No.259 apparently did no flying with the RFC.

French Deperdussin
100 hp Gnome
Span 12·37 m; length 6·836 m; wing area 23 sq m.
Empty weight 537 kg; loaded weight 847 kg.
Speed 111 km/h; climb to 1,000 ft, 3 min.

British Deperdussin
100 hp Gnome
Span 39 ft 6 in; length 24 ft 6 in; wing area 236 sq ft.
Empty weight 1,226 lb; loaded weight 2,037 lb.
Speed 68 mph; climb to 1,000 ft, 3 min 45 sec.

British Deperdussin
100 hp Anzani
Span 41 ft 6 in; length 24 ft; wing area 270 sq ft.
Empty weight 1,200 lb; loaded weight 2,000 lb.
Speed 70 mph; climb to 1,000 ft, 4 min 45 sec.

D.F.W. Mars monoplane

In 1912 the Deutsche Flugzeug-Werke was virtually unknown in Britain, possibly because it chose to call its products Mars aircraft; certainly its initials, D.F.W., had no currency at the time, though they were later to become well-known during the First World War.

A Mars monoplane was entered for the Military Aeroplane Competition by Cecil E. Kny. The aircraft concerned was a somewhat ponderous-looking mid-wing monoplane, with a mainplane that appeared to owe a good deal to contemporary Etrich designs. It was a tandem two-seater, with the pilot in the rear cockpit; the power unit was a 100 hp Mercedes engine. In the competition the Mars monoplane was allotted the number 23, and its designated pilot was Leutnant Bier. This was the same Bier whose crashed Etrich monoplane of the Circuit of Britain contest had provided the Austro-Daimler engine that eventually propelled Cody to triumph in the Military Trials.

The Mars might have been a formidable contestant, for it was reported that the monoplane had already attained a speed of 80 mph. However, it never went to Larkhill, for the German Government forbade its participation. Probably no one at Larkhill, or elsewhere, saw anything particularly sinister or significant in that prohibition at the time.

D.F.W. Mars monoplane
100 hp Mercedes
Span 16·154 m; length 13·715 m; wing area 32·655 sq m.
Empty weight 531 kg; loaded weight 848 kg.
Speed 129 km/h; endurance 4 hr.

The Maurice Farman S.7 entered by The Aircraft Manufacturing Co, at Larkhill wearing the Military Trials number 22. (*RAF Museum*)

Maurice Farman biplane

The Aircraft Manufacturing Co Ltd was registered in June 1912, having been created by George Holt Thomas to exploit his rights for the manufacture of Henry and Maurice Farman aeroplanes. In the Military Aeroplane Competition the British company entered a Maurice Farman biplane of the type that became known as the Longhorn in later R F C parlance.

The aircraft concerned was the biplane that had been in use for some months as a demonstrator in the hands of Pierre Verrier; it must, in fact, have been the Farman that Verrier flew impressively at the reception following Claude Grahame-White's wedding on 27 June, 1912. It had not, therefore, been specially designed to compete in the Military Trials but was a standard machine taken from stock. Although the Longhorn later came to be affectionately derided in the R F C and R N A S, in 1912 it was still a relatively new type and, by the standards of that summer, was not unimpressive. In the Competition it had the identifying number 22.

This Farman was representative of the early form of the Longhorn, having the extended upper wings, upswept forward skids, and frontal elevator, that were the outstanding characteristics of the type, together with the short-span tail unit incorporating the kidney-shaped, cambered, lower tailplane, and the tail-booms curving inwards in plan towards the tail unit. The power unit was a fan-cooled 70 hp Renault mounted in total exposure at the stern of the primitive little nacelle in which the crew sat in tandem, the pilot in front.

In the Military Trials the Farman was flown by its regular pilot, Verrier, and underwent all the tests. Its complexity was reflected by its ignominious last place in the quick-assembly tests, its team of five men taking all of 9 hr 29 min to put it all together. A much better performance in the transport test saw it dismantled and re-assembled in a total of 3 hr 6 min. In the flying tests the Farman floated serenely about, creating something of a sensation with its remarkably flat glide. Its wing loading of 2·8 lb/sq ft was the lightest of all the competing aircraft and enabled the aircraft to do outstandingly well in the gliding, landing, and slow-speed tests.

Although the Farman won only one of the £100 consolation prizes for undertaking all of the tests, it was in the future to outstrip all of the other competitors by being widely adopted and built in enormous numbers. At Larkhill in that August of 1912 its universality still lay in the future, and what it achieved in the Military Trials is summarised in the following table.

Test	Position	Result
Quick assembly	19	9 hr 29 min
3-hours test	Passed	Passed
Climbing	9	207 ft/min
Gliding	1	1 in 6·8
Consumption		
(a) petrol	4 equal	7 gal/hr
(b) oil	3	0·73 gal/hr
Calculated range	10	266 miles
Speed		
(a) maximum	11	55·2 mph
(b) minimum	1	37·4 mph
Speed-range	3	17·8 mph
Landing	Passed	64 yd run
Quick take-off	2	140 yd run
Rough weather	—	Wind from 14 to 29 mph
Transport	8	Dismantled and re-assembled in 3 hr 6 min

Maurice Farman No.22
70 hp Renault
Span 15·392 m (upper), 11·278 m (lower); length 12·14 m; wing area 65·1 sq m.
Empty weight 598 kg; loaded weight 876 kg.
Speed 88·8 km/h; climb to 1,000 ft, 4 min 50 sec.

Flanders biplane

At the time of the Military Trials, R.L. Howard-Flanders was already a contractor to the War Office, for the first of four officially ordered Flanders F.4 monoplanes had arrived at Brooklands late in June 1912. His entry for the Military Aeroplane competition was a sensibly designed and well-proportioned biplane with some unusual structural features. It was designed for the 100 hp A.B.C. engine, and was to be flown in the Trials by F. P. Raynham.

The fuselage was of distinctive shape, being very deep at the front and having a form of keel member that geometrically ran all the way back to the sternpost; the aft portion of the fuselage was of triangular cross-section. There were two bays of interplane bracing but no centre-section struts connecting the fuselage and upper mainplane. Short undercarriage legs were attached to the mainspar of the lower wings directly under the forward inboard interplane struts; each wheel had its own half-axle, the inboard end of which was anchored to the fuselage keel. A long frontal skid protected the airscrew against the risk of nosing over on landing or

The Flanders B.2 biplane as originally constructed, waiting for its 100 hp A.B.C. engine. (*Flight International*)

take-off. Tailplane and elevators were typical of Flanders design, and the comma-shaped rudder might have been the prototype of the surfaces that later became so familiar on Avro 500 and 504 aircraft.

Misfortune dogged the Flanders in the Trials. Reports about the aircraft and its engine were at the time somewhat confused and contradictory, but it is clear that the aircraft was delivered to Larkhill before 31 July but without its A.B.C. engine, which was not ready in time. It is uncertain whether the delay was caused by a flawed cam-shaft or a damaged crankcase, but the engine did not reach Larkhill until 12 August, by which time the Flanders had been disqualified following the lodging of an objection against it by another British entrant. It had meantime undergone the quick-assembly test, in which it was placed 6th, having been put together by six men in 40 minutes. The aircraft had had to be hurriedly repaired, for its undercarriage had been damaged on arrival at Larkhill: while being towed backwards behind a car (the method of transport from Brooklands) one wheel struck a post and was wrenched away. This had been repaired by Saturday, 10 August, the day on which the biplane was officially disqualified.

The A.B.C. engine was installed in the Flanders on 13 August, and the aircraft managed to make one straight flight on 22 August. It was obliged to land with camshaft trouble, and was immediately dismantled and packed off to Brooklands. Although the Flanders took no effective part in the Military Trials it outlived the other competitors, doing much flying with a succession of engines, all of lower power than the A.B.C. for which it was designed, and eventually being impressed for service with the RNAS in the early months of the war.

Flanders biplane
100 hp A.B.C.
Span 43 ft (upper), 27 ft (lower); length 31 ft 6 in; wing area 400 sq ft.
Empty weight 1,250 lb; loaded weight 2,000 lb.
Estimated speed 65 mph.

Handley Page Type F

The first aircraft built by Handley Page Ltd were a series of monoplanes with curving sweptback wings based on the early work of José Weiss. The firm's entry in the Military Trials was a monoplane of similar configuration with a deep fuselage having dorsal and ventral fairings; pilot and passenger sat side by side under the cabane structure, and power was provided by a 70 hp Gnome in a semi-enclosed housing. Lateral control was by wing warping, and all the tail surfaces looked rather too small to be adequate.

Presumably the aircraft reached Larkhill in time to be admitted by the judges, but it seems that it had not got there by 31 July, 1912. It did not take part in the quick-assembly test and was only reported to be in course of construction by Saturday, 3 August. Its Trials number was 28 and its pilot was to be Edward Petre, who replaced his brother, Henry Petre, because the latter distrusted the monoplane's crescent wing and wanted to substitute straight-edged sweptback wings.

Erection was a protracted business, for the Type F was said to be 'rapidly approaching completion' on 8 August. It did not emerge to fly until 21 August, when it was reported to have flown well, albeit with 'a long slow regular roll from side to side'. While taxying after a landing, a minor contretemps was caused by a wheel and the frontal skid catching in rough ground, but the aircraft sustained no damage. Next day the Handley Page was out again, intent on completing the 3-hour test, but on landing downwind from a test flight Petre had to swerve first to avoid running into the chains round the Bristol sheds and again to avoid some spectators, at which point a 30 mph gust blew the monoplane over on to one wing, damaging the mainplane and undercarriage.

This effectively put the aircraft out of the Trials, and it had to be taken by road to the Handley Page works at Cricklewood. There it was subsequently fitted with

The Handley Page Type F, No.28 in the Trials, by its shed at Larkhill.
(*Flight International A374*)

new wings and was flown successfully from 9 November until its fatal crash on 15 December, when engine failure and turbulence brought it down near Wembley. Lt Wilfred Parke, who had flown the Avro Type G in the Military Trials, lost his life, as did his unfortunate passenger, A. Arkell Hardwick, the Handley Page works manager.

Handley Page Type F
70 hp Gnome
Span 43 ft 6 in; length 30 ft 2 in; height 10 ft 6 in; wing area 250 sq ft.
Empty weight 850 lb; loaded weight 1,450 lb.
Speed 55 mph.

The Hanriot monoplanes

There had been Hanriot monoplanes since René Hanriot had designed and built his first at Châlons-sur-Marne in 1907; its successors had a grace and elegance that set them apart in their time. In the French military trials of 1911 a Clerget-powered Hanriot monoplane found as one of its competitors a sturdy and sensible Nieuport monoplane designed by M. Pagny, who subsequently joined Hanriot as a designer.

In 1911 the Hanriot company had started a flying school at Brooklands, and in May 1912 a new British company, Hanriot (England) Ltd, set itself up in offices at 412 Moorgate Station Chambers, London E.C. One of the British Hanriot company's first acts was to enter two monoplanes in the Military Aeroplane Competition. These were identical examples of a thoroughly workmanlike design by Pagny and were powered by the 100 hp two-row Gnome rotary engine. They had the Trials numbers 1 and 2, and were to be flown by J. Bielovucic (a Peruvian pilot of note) and Sidney Sippe respectively.

The Hanriot of the Military Trials was a shoulder-wing monoplane with a tapered wing of fairly high aspect ratio and a broad-chord tailplane. Lateral control was by wing warping, the rudder was aerodynamically balanced, and there was no fin. Structure was conventional for the period, but was distinguished by excellent detail design. Pilot and passenger sat in close tandem in a single communal cockpit under the cabane.

No.1 in the Military Trials series of numbers was this Hanriot monoplane.

The other Hanriot entry in the Trials bore the number 2. (Flight International)

The Hanriots made an impressive start in the Trials by being assembled in only 14 min 43 sec and 22 min 58 sec, and went on to return consistently good performances in the flying tests. Bielovucic had the bad luck to have an inlet valve spring break when he was only 18 minutes short of the 3-hour flight, and was obliged to land. He was later able to perform this test successfully on 12 August.

If place marks had had anything to do with the final result, the Hanriot monoplanes would have been well ahead of the rest of the field and the RFC would have had a very good aeroplane. As it was, they were merely awarded £100 each for undergoing all the tests, and were thus among the consolation prize winners. The blatant injustice of this is immediately apparent if the results in the following table are compared with those for the Cody biplane (see page 199).

Test	Position		Result	
	No.1	No.2	No.1	No.2
Quick assembly	2	4	14 min 43 sec	22 min 58 sec
3-hour test	Passed	Passed	Passed	Passed
Climbing	1	2 equal	364 ft/min	333 ft/min
Gliding	2	5 (failed)	1 in 6·6	1 in 5·9
Consumption				
(a) petrol	5 equal	7	8 gal/hr	8·6 gal/hr
(b) oil	9	8	2·4 gal/h	2·1 gal/h
Calculated range	2	3	400 miles	361 miles
Speed				
(a) maximum	2	1	75·2 mph	75·4 mph
(b) minimum	9	10	59·9 mph	66·6 mph
Speed-range	4	10	15·3 mph	8·8 mph
Landing	–	–	120-yd run	119-yd run
Quick take-off	4	–	206-yd run	–
Rough weather	4	2	Wind 25–31 mph	Wind 21–31 mph
Transport	2	–	Dismantled and reassembled in 31 min	–

By the starboard side of the Hanriot's Gnome engine stands J. Bielovucic, by the port side Sidney Sippe. These were the pilots of the two Hanriot monoplanes in the Trials. Second from left is Gordon Bell.

Hanriot monoplane No.1
100 hp Gnome
Span 12·725 m; length 7·315 m; wing area 27·9 sq m.
Empty weight 529 kg; loaded weight 871 kg.
Speed 121 km/h; climb to 1,000 ft, 2 min 45 sec.

Hanriot monoplane No.2
Engine and dimensions as No.1
Empty weight 526 kg; loaded weight 861 kg.
Speed 121·3 km/h; climb to 1,000 ft, 3 mins.

Harper monoplane

The identifying number 25 in the Military Aeroplane Competition was allotted to a monoplane entered by A. Monnier Harper and constructed in the Paddington works of Weston Hurlin & Co. It was reported to be under construction in mid-July 1912, at which time it was also announced by Mr Hurlin of the company that Harper had, in conjunction with the Weston Hurlin firm, opened a seaplane school at Scheveningen in the Netherlands. This venture intended to give exhibitions 'all over Holland', using two Weston Hurlin biplanes that were to be flown by J. R. Duigan.

According to a contemporary report, the Harper monoplane was to be powered by a 60 hp Green engine, and its fuselage was of lozenge-shaped cross-section. Lateral control was to be by wing warping, and the undercarriage was 'a composite structure of bamboo, ash and steel tubing' with wheels and skids of Farman pattern. Side-by-side seating was to be provided for the two occupants.

The Harper monoplane (referred to as the Weston Hurlin monoplane in some accounts) never arrived at Larkhill to participate in the Military Trials. Indeed, confirmation that it was ever completed has yet to be found.

Harper monoplane
60 hp Green
Span 35 ft; length 27 ft.
Estimated speed 60 mph.

The intended Lohner entry would probably have been a biplane of this type.

Lohner biplane

Cecil Kny, who entered the D.F.W. Mars monoplane in the Military Trials, was also the representative in Britain of the Jacob Lohner company of Vienna. A Lohner biplane was entered for the Trials, and according to one contemporary report it was to be the same aircraft on which Leutnant von Blaschke had established new world altitude records at the Vienna meeting on 23 and 29 June, 1912. The biplane was a three-seater, and with all three seats occupied it reached 3,500 metres (11,483 feet); with only one passenger, von Blaschke climbed to 4,260 metres (13,977 ft).

The aircraft was a large, ungainly, three-bay biplane powered by a 120 hp Austro-Daimler engine. The wings were swept back, staggered, and carried on wide-span centre sections; the upper wings had extensions, compound taper, and ailerons. A complex system of struts carried the mainwheels of the undercarriage, and an unusual feature for the time was the use of a large tailwheel in place of a skid. The Lohner was reported to have 'a land brake at the tail to decelerate it quickly on landing'.

Like the D.F.W. Mars monoplane, however, the Lohner biplane had no opportunity to prove itself in the Larkhill trials, for it was forbidden to participate by the Austro-Hungarian Government. Had it taken part in the Trials it might have proved to be formidable, and it was to have been flown by Leutnant von Blaschke. Its Trials number was 24.

Lohner biplane
120 hp Austro-Daimler
Span 16·15 m approximately; length 9·45 m.
Empty weight 699 kg; loaded weight 971 kg.
Speed 113 km/h.

Martin-Handasyde monoplane

As they were already contracted to the War Office to supply a monoplane to the Military Wing, it is not surprising that Martin and Handasyde entered an aircraft in the Military Aeroplane Competition. Its competition number was 17. It is equally not surprising that the entry was another monoplane, its elegance of form and quality of workmanship typical of the products of the partnership. It had the usual slender fuselage, king-posted tapered wing, and characteristic undercarriage of the *marque*, but George Handasyde had forsaken his beloved Antoinette engine, perhaps inevitably, for a 75 hp Chenu. This was a disastrous choice.

The Chenu engine of the handsome Martin-Handasyde monoplane gave quite as much trouble as that of the second Coventry Ordnance Works biplane. Here Gordon Bell waits while efforts to start the Chenu are made. (*Flight International 0173*)

The Martin-Handasyde made its first flight at Brooklands on Friday, 19 July, 1912, when it was reported to have flown 'remarkably well with a passenger'. It was at Larkhill in time for the start of the Military Trials but its Chenu engine was immediately in trouble, for it sheared a bevel gear in the magneto drive, and its pilot, Gordon Bell, discovered instant frustration. Repairs enabled the monoplane to essay the rough-weather test on 6 August, when the wind was reported to be gusting from 16 to 34 mph. Bell made another successful flight next day, but on 8 August sudden engine failure immediately after take off brought the Martin-Handasyde down safely but spectacularly. Apparently the magneto drive had failed again, whereupon Martin and Handasyde set about the complete dismantling and virtual rebuilding of the engine. As on the 100 hp Chenu fitted to the second Coventry Ordnance Works biplane, the magneto drive was in this process drastically modified to be re-installed at right angles to its original position. Despite these strenuous endeavours the engine still gave trouble, for the reduction gear proved refractory. Although it was at one stage reported that the partners were giving up the Chenu as a bad job and replacing it with an unspecified Renault, this did not happen.

Thus when the trials ended the handsome Martin-Handasyde had undertaken only two of the tests: the quick-assembly test, in which it was in 12th place with an assembly time of 1 hr 33 min; and the rough-weather test, in which it was one of only five aircraft officially considered to have passed. It deserved better, and with a less troublesome engine might well have done better, although it should be observed that it proved to be considerably overweight, turning the scale at 1,671 lb empty, whereas the estimated weight had been only 1,250 lb. After the Trials it was re-engined, and by 27 September, 1912, was being flown with a 65 hp Antoinette.

Martin-Handasyde monoplane
75 hp Chenu
Span 42 ft; length 38 ft; wing area 310 sq ft.
Empty weight 1,671 lb.
Estimated speed 65 mph.

Mersey monoplane

The company known as Planes Ltd was formed in 1909 to exploit an aircraft configuration created by W. P. Thompson. A biplane embodying his ideas was built in 1909 by Handley Page Ltd, a good deal of the work being done by Thompson's assistant, Robert C. Fenwick, who doggedly rebuilt the aircraft in 1910 after it had been extensively damaged when a gale blew down part of the Handley Page factory at Barking. Fenwick took the reconstructed biplane to Freshfield, Lancashire, where he flew it successfully enough to gain his pilot's certificate on 29 November, 1910.

In the following year, Fenwick, in collaboration with Sydney T. Swaby, designed a monoplane that was completed in late 1911 and was flown at Freshfield by Fenwick. The aircraft had therefore been designed before the specifications for the Military Trials were announced. A certain amount of secrecy surrounded the early flights of this monoplane, but if flew with some regularity in the spring of 1912.

Precisely what happened in May 1912 is not absolutely clear, but it seems that Fenwick and Swaby independently set up the Mersey Aeroplane Company, bought the monoplane from Planes Ltd, and remained at Freshfield as licensees and tenants of that company. Passengers flown in the monoplane at that time

The extraordinary Mersey monoplane, on which Robert Fenwick met his death at Larkhill on 13 August, 1912. (*Flight International A382*)

included Mr Isaacson, maker of the aircraft's engine, and Mr Thompson, who was then reported to be over 70 years of age.

The original Mersey monoplane was extensively damaged in a crash in the summer of 1912, when it alighted on some quicksand and turned over at speed. Fenwick was uninjured and set about rebuilding the aircraft, for the Military Trials were approaching, and such was the partners' faith in the monoplane that they had entered it in the competition. In the reconstruction, longer wings were fitted and were rigged without the dihedral used on the original aircraft; the areas of the tail surfaces were increased and the tail-booms shortened; only the engine and nacelle remained in their original form.

The reconstructed monoplane was a bizarre device with perilously exiguous bracing. It was a pusher, the 45 hp Isaacson radial engine being mounted reversed on the nose of the short nacelle, driving a two-blade pusher propeller via a long extension-shaft that passed between the side-by-side seats. A reasonably sturdy undercarriage provided an anchorage for the flying wires, but the landing wires were attached to two individual struts placed centrally. Of these the rear, raked, strut provided an attachment point for the upper wires bracing the tail unit. This was connected to the mainplane by only two tail-booms of steel tubing, the whole composing a terrifyingly inadequate structure.

In the Military Aeroplane Competition the only test in which the Mersey participated was that for quick assembly, but it was well down the list in 17th place, for it took three men 4 hr 25 min to assemble it.

The monoplane made a good flight on Friday, 9 August, with Fenwick at the controls, Swaby in the passenger's seat, and fuel for six hours in the tank. On landing, the rubber shock absorbers in the undercarriage were damaged but it was flying again by 11 August. Tuesday, 13 August, was windy and cold, but at 6.06 p.m., when the wind had momentarily dropped, Fenwick took off and flew towards Stonehenge. Precisely what happened is nowhere clearly recorded, but when the monoplane was at about 200 ft it was seen to be in difficulties; it dived for about 50 ft, recovered momentarily, then plunged vertically to the ground. Fenwick was killed.

The Royal Aero Club enquiry attributed the accident to turbulence and the Mersey's instability, though what the Club's Public Safety and Accidents Investigation Committee could possibly have known about the aircraft's stability or lack of it is by no means clear.

Mersey monoplane
45 hp Isaacson
Span 35 ft; length 34 ft; wing area 220 sq ft.
Speed 55 mph; endurance 6 hr.

Piggott biplane

Smallest of all the contestants in the Military Trials was the Piggott biplane, a far cry from the earlier Piggott pusher biplane of 1910 and the extraordinary enclosed monoplane of 1911. In its general configuration the Military Trials biplane looked ahead of its time, being a compact tractor aircraft with a simple V-strut

The Piggott biplane, No.29 in the Trials, did not succeed in flying at Larkhill.

undercarriage. The equal-span wings were of narrow chord and had only a single row of interplane struts, divided into three bays on each side. Upper and lower wings were attached at the levels of the corresponding longerons; pilot and passenger sat in tandem, their heads above the upper wing. Power was provided, inadequately, by a 35 hp Anzani three-cylinder radial.

Numbered 29 in the Competition, the Piggott was to be flown by S. C. Parr. At Larkhill it did not emerge until Tuesday, 13 August, when its engine was run. Not until 21 August was any attempt made to move the aircraft under its own power, but while taxying it ran into a ground obstruction and heeled over on to its starboard wings, causing damage to the structure. By Sunday, 25 August, the Piggott was declared out of the competition, having taken part in none of the tests. With its low power and small wing area it is doubtful whether it could possibly have flown with two men aboard, and it is equally doubtful whether it was tried seriously at Larkhill or elsewhere.

Piggott biplane
35 hp Anzani
Span 25 ft 6 in; length 17 ft 6 in; wing area 100 sq ft.
Empty weight 300 lb; loaded 700 lb.
Estimated speed 55 mph.

Vickers No.6 monoplane

The first aeroplanes built by the great Vickers company were monoplanes based on contemporary R.E.P. designs. Sixth of the series was a shapely shoulder-wing monoplane with side-by-side seating for two and a 70 hp Viale radial engine. It was the Vickers entry in the Military Aeroplane Competition; as such it had the Trials number 3 and was flown at Larkhill by Leslie F. McDonald.

Steel was extensively used in the structure of the fuselage, which was wider than it was deep in order to seat the two occupants in reasonable comfort. Taper towards the tail was pronounced; the tailplane faired gradually forward into the fuselage sides, and the elevators and elegant rudder were balanced; there was no fixed fin. The mainplane's landing wires were attached to a cabane structure consisting of two upright and parallel inverted-V struts; lateral control was by wing warping. In the undercarriage the wheels were independently sprung by arranging each main leg to slide on an upright of the fuselage against rubber springs in tension, and there was a substantial central skid.

A gallant attempt was made to fly the Vickers monoplane to Larkhill for the Trials, for McDonald and Challenger, a member of Vickers' design team, set out to fly from Bognor on 31 July, 1912, only to lose their way in bad visibility and come down at Brockenhurst, whence the monoplane was towed to Larkhill on its own wheels behind a car.

The Vickers No.6 monoplane, photographed at Larkhill wearing its Trials number 3 on the rudder. (*Flight International 0115*)

The Vickers monoplane started the Trials by taking seventh place in the quick-assembly test: it was put together by six men in 53 min 30 sec. Its first flight followed assembly and was successful; so was its flight on 3 August, when McDonald took it up to 1,000 ft for gliding tests. The aircraft's first trouble occurred on 7 August, when excessive pressure in the fuel tank caused petrol to squirt into McDonald's eyes and he was obliged to come down. The monoplane was flying again next day, but on 9 August the engine was behaving erratically, making take-off hazardous; it continued to give trouble on 12 August. Nevertheless, McDonald made a start on the 3 hr test two days later, only be forced to land by an increasing wind after 2 hours 28 min.

Although the Vickers monoplane did a fair amount of flying during the Trials it essayed only one other flying test, the so-called 'rough weather' exercise. It was successfully flown in a wind varying from 14 to 29 mph, but this gave a mean wind velocity of only 21·5 mph, too low to qualify in a test that specified an average wind of 25 mph. The aircraft was therefore among the also-rans. Some time after the Military Trials the No.6 monoplane was re-engined with a 70 hp Gnome, when it had modified cabane struts. The design provided the basis for the generally similar Vickers No.8 of 1913, last of the early series of Vickers monoplanes.

Vickers No.6 monoplane
70 hp Viale

Span 34 ft; wing area 220 sq ft.

The Aeroplanes of the Royal Flying Corps (Military Wing)

Aeronautical Syndicate Valkyries

Horatio Barber was one of the true pioneers of aviation in Britain. Inspired by the European flights of the Wright brothers in 1908 and the Paris Aero Salon of that year, he took occupancy of one of the railway arches in Battersea that year. Early in 1909 he moved to Larkhill on Salisbury Plain, formed the Aeronautical Syndicate Ltd, and began to build aircraft. From the beginning of his aeronautical activities Barber was convinced that aircraft would have a key role to play in any future war, and all his subsequent designs were seen by him as potential reconnaissance aircraft.

By March 1910 successful flights were being made by a tail-first pusher monoplane, and by September the fifth of his experimental designs, named Valkyrie I, was flying. This new aircraft was also a canard monoplane but had several individualistic features, notably a fixed forward plane with separate frontal elevator well below it and twin rudders behind the mainplane. The 35 hp Green engine was mounted on the wide-span centre section and drove a propeller that rotated just ahead of that surface, while the outer wing panels had marked dihedral, and carried ailerons to provide lateral control.

In September 1910 the Aeronautical Syndicate became, in company with Louis Blériot's School, the first occupants of the row of simple sheds erected on the London Aerodrome Company's newly created flying field at Hendon. The Valkyrie was airborne at Hendon on 1 October, 1910, even though the aerodrome

The Valkyrie Monoplane with 50 hp Gnome engine that was exhibited at the 1911 Olympia Aero Show. The aircraft on which Lt Cammell met his death was probably similar.

was not considered to be fully prepared. Late in October a three-seat Valkyrie with a 60 hp Green was undergoing trials, and a series of experimental flights with the original single-seater had led to some modifications. The three-seater was initially known as Valkyrie II, and it was reported that two new monoplanes similar to Valkyrie I were being built. Presumably these were the small monoplanes that emerged as Valkyries III and IV, the latter having a 30-35 hp Green engine.

The line continued into 1911, and by April a change in nomenclature came when alphabetical type identification was introduced. In general, the Valkyries were sub-divided into Type A (single-seat, 35 hp Green), Type B (single or two-seat, 50 hp Gnome) and Type C (three-seat, 60 hp Green). By May 1911 five Valkyries were in use at Hendon, and the type was offered for sale at prices that started at £280; apparently eleven had been built up to that time.

On 12 May, 1911, an elaborate demonstration was organised by the Parliamentary Aerial Defence Committee and presented at Hendon. Its objective was to impress on Members of Parliament and senior officers of the Admiralty and War Office that aviation in general and aeroplanes in particular possessed military potentialities that the nation could no longer afford to ignore. The event was a considerable success, but the flying was dominated by French aircraft (Farman biplanes and Blériot monoplanes), a fact that must have rankled with Horatio Barber.

For reasons that have never been explained, a letter had been sent by the organisers of the demonstration to the Aeronautical Syndicate, telling them that they must not fly during the demonstration except by invitation, and at the same time stating that it was not possible to extend any such invitation to the Syndicate. Apparently no reason was given for this pronouncement, which excluded the Valkyries from participating, and had the additional effect of substantially reducing the representation of British aircraft in what was, at the time and in the circumstances, an event of great significance and importance. As far as is known, only the Aeronautical Syndicate was thus stigmatised.

If Barber knew why or by whom he was debarred from flying on 12 May, 1911, he gave no public or overt sign of it. Instead of lamenting or accusing, he presented to the nation four Valkyrie monoplanes, suggesting only that, as the structure of the Valkyrie lent itself readily to the fitting of a combination of floats

Horatio Barber on the Valkyrie Type B, No.5 (50 hp Gnome), at Hendon.

and wheels, two of the aircraft might be allotted to the Royal Navy. Additionally, he offered his services as a designer, constructor and pilot to the Government as far as his time might permit.

Of the four Valkyries thus donated to the Government two were single-seaters, one having a 30 hp Green engine, the other what was described as a 40–50 hp Green. One of the two-seaters had a 60 hp Green; the other, which was newly built, was intended to have a 50 hp Gnome.

Although contemporary reports stated that Barber's gift and his suggestion had been accepted with alacrity by 'the authorities', the subsequent histories of these presentation aircraft remain shrouded in mystery. Any record of Naval use of a Valkyrie has yet to be found, but the War Office apparently decided to collect at least one of theirs, namely the brand-new aircraft intended for the 50 hp Gnome. Barber had personally tested this Valkyrie in August 1911, subsequently removing his own Gnome engine from it. Early in September Lt R. A. Cammell, R.E., a pilot of the Air Battalion, brought a 50 hp Gnome to Hendon for installation in the Valkyrie. The engine was fitted, with considerable difficulty, by Air Battalion mechanics.

Cammell was quite an experienced pilot, having flown the Air Battalion's Blériot XII occasionally and his own Blériot XXI extensively. On Sunday, 17 September, 1911, he took off in the Valkyrie, apparently with the intention of spending half an hour familiarising himself with it before flying it to Farnborough. It was said that Cammell liked to perform steep turns in his Blériot, and that he made similar turns in the Valkyrie, despite his complete lack of familiarity with it. While gliding down he made a steeply banked turn to port but obviously failed to maintain adequate air speed, for the aircraft 'side-slipped to the ground' from a height of 30 ft. Cammell died of his injuries on the way to hospital.

This fatality may well have inspired official antipathy towards the Valkyries. It is not known with certainty whether a Valkyrie was numbered among the nineteen unidentified aeroplanes that Col J. E. B. Seely admitted to being in the possession of the War Office at the end of October 1911. Four of the nineteen were reported to be monoplanes, and at that date these could have been Cammell's Blériot XXI B2, the Nieuport monoplane B4, the Deperdussin B5—and one other.

Whatever the truth of the situation may have been, it seems that none of the donated Valkyries was flown by any pilot of the Air Battalion or of the Royal Navy. Barber continued his research and experimental work for a few months, but early in April 1912 he withdrew from active aviation owing to increasing costs. The aircraft and stock of spares and materials belonging to the Aeronautical Syndicate were auctioned on 24 April. In 1913 Barber presented the Britannia Trophy for the most meritorious achievement by a British pilot, and during the 1914–18 war he served in the RFC and RAF as an instructor, in which capacity he invented several ingenious aids to training. His country, to which he gave so much, did not recognise or acknowledge in even the most modest way anything that he had done.

Valkyrie Type B
50 hp Gnome

Span 31 ft; length 26 ft; wing area 168 sq ft.
Empty weight 550 lb.
Maximum speed 70 mph.

An early appearance of the D.H.1 prototype at Hendon. The air-brakes originally fitted can be seen as small wings at the lower ends of the forward centre-section struts. The Deperdussin monoplane behind was No.885 of the RNAS unit at Hendon.
(*Flight International 0192*)

Airco D.H.1

The British manufacturing rights for the Henry and Maurice Farman aircraft were acquired in 1911 by George Holt Thomas. In the spring of 1912 he created the Aircraft Company Ltd as the means of exploiting these rights: its first advertisements appeared early that April. In its issue of 15 June, 1912, the journal *Flight* recorded the registration of The Aircraft Manufacturing Company Ltd as a new company with a capital of £14,700. In spite of its full title, the company continued to advertise simply as the Aircraft Company for some years, and was generally known as Airco; the latter name was formally adopted in October 1918.

For two years, Holt Thomas was content for the firm to function as an agent and licensee for the Farman designs. In 1914, however, Geoffrey de Havilland suggested to him that his firm should undertake original design, and in May of that year Holt Thomas engaged de Havilland to create and control a design department. De Havilland had worked for some $3\frac{1}{2}$ years as a designer and test pilot at the Royal Aircraft Factory, Farnborough, and had a wide range of experience in design and practical flying. One of his outstanding achievements was the design of the B.E. series of biplanes and he brought with him an unusual range of knowledge and experience of meeting contemporary requirements for military aircraft.

War was declared before de Havilland could produce an aircraft design of any kind for his new employers. He had been a Lieutenant in the RFC Reserve since 1912, and reported for duty on 4 August, 1914. This earned him a posting to Montrose, but good sense eventually prevailed and about three months later he returned to The Aircraft Manufacturing Company's design office. He set about designing the first of the series of aircraft that were to bear his name; significantly, these were almost invariably recorded and referred to as de Havilland types rather than Aircraft or Airco types.

The first de Havilland design built by the Aircraft Company was designated D.H.1. Geoffrey de Havilland has recorded that he had started to design a tractor biplane but was obliged to abandon it when the War Office made it clear that what was wanted was a pusher giving its gunner an unobstructed field of fire. Thus the aircraft that emerged late in January 1915 was a two-seat pusher biplane powered by a 70 hp Renault. According to more-or-less contemporary reports, it was intended to have a more powerful engine, apparently the 100 hp Green, which was also the designed power unit of the Royal Aircraft Factory F.E.2a. At that time, the modest output of the Green Engine Company could not hope to meet the requirements of more than one manufacturer, and only forty-two engines of the 100 hp model were ordered from Mirlees, Bickerton & Day Ltd. It seems that the Royal Aircraft Factory claimed the entire output for the F.E.2a, for the D.H.1 had to be redesigned to have the very different 70 hp Renault.

Despite the reduced power, the D.H.1 still had a good performance and created a considerable impression with its early flights at Hendon. In its earliest form, it was fitted with air-brakes on the nacelle sides; these looked like small wings and could be rotated about a spanwise axis to be normal to the airstream. They were soon discarded, and the aircraft was no longer fitted with them when it was tested at Farnborough. The undercarriage was a simple structure but incorporated coil springs and compact oleo dampers at the apices of the V-struts.

The prototype did a good deal of flying at Hendon, frequently in the hands of its designer, during the early months of 1915. It was officially adopted with the serial number 4220 and an order for 49 production D.H.1s was placed in the spring. By then, however, The Aircraft Manufacturing Company had heavy production commitments for Henry and Maurice Farman biplanes, and Geoffrey de Havilland was totally absorbed in the design of the D.H.2, for which there was an urgent need. Responsibility for the production order was therefore transferred by subletting the contract to Savages Ltd, a Kings Lynn firm that made fairground equipment and knew nothing of aircraft construction. This delayed production, and not until 7 November, 1915, was No. 4600 at Farnborough for inspection and allocation. By then the F.E.2a had been in operational use for some months, despite the irony of having to replace the 100 hp Green by the 120 hp Beardmore, and had given a good account of itself. With nearly twice the power of the D.H.1 it naturally commended itself more to the RFC. By the end of 1915 five D.H.1s had been issued to training units.

The production D.H.1s had modified front cockpits, simple rubber shock-cord for springing in the undercarriage, and horizontal exhaust pipes with forward outlet. A few aircraft of the first production batch were fitted with the 120 hp Beardmore and therefore became directly comparable with the F.E.2a and F.E.2b. In this form, the aircraft was designated D.H.1A.

Production D.H.1, A1649. (*RAF Museum*)

D.H.1A in operational use with No.14 Squadron, Palestine.

A series of comparative trials was conducted at CFS in May 1916; in these the D.H.1A No. 4605 was matched against the F.E.2b No. 6337. On every significant count the D.H.1 was markedly superior to the F.E.2b: it was appreciably faster on the level, climbed faster, carried more fuel, gave its gunner a better field of fire, and was more manoeuvrable. With the F.E.2b then well established in production, a change of type was not practicable, but the result of these trials probably encouraged the RFC to send the D.H.1A to the Middle East for use as an escort fighter. The type began to arrive in Egypt in the middle of 1916 and saw limited operational use with No. 14 Squadron in Palestine. On 2 August, 1916, one of these D.H.1As shot down an Aviatik two-seater near Salmana. The type was still in use in March 1917, for on 5 March a D.H.1A of No. 14 Squadron was brought down and its crew made prisoners.

Only six D.H.1As went to the Middle East but twenty-four D.H.1s or 1As were allotted to Home Defence units. A second production batch of D.H.1s (A1611—1660) was ordered and again the contract was sublet to Savages. It appears that all were built but only 73 in all were issued to RFC units. The D.H.1 was used by a number of training units and, apparently, saw some service as a night-flying trainer. It was an aircraft that, in its day, deserved better.

D.H.1 and D.H.1A

70 hp Renault (D.H.1), 120 hp Austro-Daimler (D.H.1A)
Two-seat fighter-reconnaissance biplane.
Span 41 ft; length 28 ft 11¾ in; height 11 ft 2 in; wing area 362·25 sq ft.
Empty weight 1,356 lb (D.H.1), 1,610 lb (D.H.1A); loaded weight 2,044 lb (D.H.1), 2,340 lb (D.H.1A).
D.H.1; maximum speed at 3,500 ft, 80 mph; climb to 3,500 ft—11 min 15 sec.
D.H.1A: maximum speed at sea level 90 mph, at 6,000 ft—90 mph, at 10,000 ft—86 mph; climb to 6,000 ft—13 min 10 sec, to 10,000 ft—27 min 30 sec; service ceiling 13,500 ft.
Armament: One 0·303-in Lewis machine-gun on No.4 Mark I rising pillar mounting.

Manufacturers: Airco; Savages.

Service use: Palestine—No.14 Sqn.
Training duties: CFS, Upavon; Reserve Sqns 6, 10, 19, 35, 46, 59; No.200 Depot Sqn. *Other Units*: Orfordness.

Known serials: 4220, 4600—4648, A1611—1660. *Rebuilds*: A5211, A9911, B3969.

Airco D.H.2

Following the completion and flying of the D.H.1 prototype Geoffrey de Havilland sketched out the basic design of a twin-engined tractor biplane of practical appearance. His earliest note on this design is dated 12 February, 1915, but it seems that he abandoned it to design a single-seat scout. Notes dated 4 March, 1915, are accompanied by a sketch of a single-seat tractor biplane with heavily staggered wings; span and chord were to be 28 ft and 4 ft 9 in respectively, the loaded weight 1,090 lb.

This tractor scout design was not developed in any detail, for by 9 March, 1915, de Havilland was setting down the basic calculations for the design of a pusher scout to be powered by a 100 hp Gnome Monosoupape, with reserve consideration given to the 80 hp Gnome. This type was to have the same span and chord as the projected tractor, and its estimated weight was 1,250 lb.

In view of its ability to carry a gun, the pusher was proceeded with, and was completed in May 1915 with the designation D.H.2. It was a compact little two-bay biplane of typical nacelle-and-tailbooms configuration, giving its pilot an unobstructed field of view in most forward directions. The original rudder was a plain surface totally devoid of any balance area, and generally resembled that of the D.H.1. Indeed, the whole design was somewhat like a scaled-down D.H.1.

The D.H.2 prototype made its first flight on 1 June, 1915. Its engine ran badly on that occasion, and it was obvious that the aircraft was tail heavy. The tailplane was adjusted but this did not cure the tail heaviness. A further flight on 2 June produced a speed of about 80 mph and a climb to 3,500 ft in the undistinguished time of 5 minutes. On 3 June the D.H.2 went back to the factory to have the nacelle moved forward 4 inches, a gun mounting installed, and a new and larger

The prototype D.H.2, 4732, after capture by II Marine Feldfliegerabteilung at Moorseele. A fair amount of repair work was done by the Germans before this photograph was taken, for the D.H.2 sustained damage on being brought down. (*Alex Imrie*)

rudder fitted. The structure of the tail was apparently lightened further to relieve the tail heaviness.

With these modifications incorporated, the D.H.2 was flying again by July 1915. Its original gun mounting permitted some movement of the single Lewis gun and was mounted on the port side of the nacelle. The aircraft was allotted the official serial number 4732 and went to France for operational evaluation on 26 July, 1915. Apparently it went straight to No.5 Squadron, but its operational career lasted only a bare two weeks, for it was shot down on 9 August and fell behind the German lines; its pilot, Capt R. Maxwell-Pike, died of his wounds.

Although the aircraft turned over on landing and was damaged, it was carefully rebuilt by the Germans (even to the extent of marking it anew with its RFC number 4732, the original having been 'souvenired'), and presumably was flown by them. Its acquisition at this time was perhaps of greater significance than its German captors could know, for their Fokker E I had then been operational for only a few weeks and had yet fully to make its mark as one of the significant aerial weapons of the war. Thus, the Germans may not have recognised in the D.H.2 a potential Fokker-beater. It is necessary to make this point because it has been alleged many times that the D.H.2 was designed specifically to counter the Fokker. In fact, the first two operational Fokker E Is did not reach the front until a few days after the D.H.2 made its first flight on 1 June, 1915.

Production of the D.H.2 was ordered promptly, probably about the time when the prototype went to France, for the initial order was obviously for 50 aircraft, being the prototype 4732 and 49 production D.H.2s, 5335—5383. These production serial numbers seem not to have been used, however, for the earliest recorded production aircraft were from the batch of 100 numbered 5916—6015 that were ordered under Contract No.87/A/36 about mid-September 1915.

An unusual view of the D.H.2 No.7850 at No.2 Aircraft Depot, Candas. The ailerons are at full deflection, and the rubber-bungee return cords on the upper ailerons can be seen clearly.

No.5925 was a true veteran of war, having been with No.24 Squadron in February 1916 and serving with the squadron for over a year. It was flown back to England on 22 May, 1917, and is here seen at Brooklands with traces of No.24 Squadron's characteristic Flight markings of red and white stripes on the outer interplane struts. (*RAF Museum*)

Curiously, in at least one official document the aircraft are listed as 'De Havilland (2a) Scouts', a designation that appeared again against the later batch 7842—7941.

The first production D.H.2s appeared in November 1915. On 24 November, Nos.5917 and 5918 were at Hendon, allotted to the RFC in France, but a change of allocation on 28 November sent them to CFS instead. Probably the first to go to France was 5919, which was flown over by Lt Breeze on 8 January, 1916; No.5920 followed two days later. The production D.H.2 had a modified nacelle with, at first, the Mark I gun mounting which was officially described as 'Rising pillar, balanced by rubber and locking knob; no arm. Windscreen on gun. No clamping handles'. This gun mounting was installed centrally in the nacelle nose. The fuel system was also modified by the addition of a gravity tank that was at first fitted under the port upper wing but was subsequently carried above the wing or centre section.

When the earliest D.H.2s went to France it was still the practice for single-seat scouts to be allotted in ones and twos to the two-seater squadrons; thus 5919 and 5916 went to No.18 Squadron, 5917 and 5920 to No.5 Squadron, and 5918 to No.11 Squadron. On 7 February, 1916, the first homogeneous single-seat fighter squadron ever formed went to war. This was No.24 Squadron, RFC, which was equipped throughout with D.H.2s; and it was followed in March and May by Squadrons 29 and 32, similarly equipped.

In its early operational days, the D.H.2 was not popular, its poor speed-range and sensitive controls demanding careful handling. A number of spinning accidents occurred, but in No.24 Squadron, Maj L. G. Hawker, VC, DSO, and Lt S. E. Cowan did much to foster confidence in the aircraft. In combat, the gun mounting proved to be highly unsatisfactory, and eventually a Mark II mounting was evolved and introduced. This had clamping handles for elevation and traverse, but the real solution lay in fixing the gun to fire forwards, thereafter aiming the aircraft at the target. Spare ammunition drums were carried in external racks on the cockpit sides, but by October 1916 No.24 Squadron had devised a means of carrying three 97-round drums inside the cockpit. This practice was adopted for the type, but doubtless arrangements varied between squadrons and individual aircraft.

A few D.H.2s were armed with two Lewis guns, but this installation seems to have been short-lived. On 30 September, 1916, No.5956 was tested for an installation of Le Prieur rockets, and one or two aircraft were thus equipped. Both attempts to augment the armament must have seriously impaired the D.H.2's meagre performance. One D.H.2 of No.29 Squadron was experimentally armed with a 0·45-in Maxim gun, but a report on this installation has yet to be found.

The Monosoupape engine gave much trouble, which is probably why Brig-Gen Brooke-Popham wrote to the Director of Aircraft Equipment on 19 September, 1916:

> 'Will you please say whether a 110 hp Le Rhône has yet been fitted to a De Havilland Scout? If so, it is requested that a report of its performance may be sent out at an early date, and that one of these machines may be sent out to this country for trial as soon as possible.
> Messrs. Allens will presumably be turning out the 110-hp Le Rhône in fair numbers by January, and it is important that as soon as engines are available, there shall be no delay in replacing the Monosoupapes in the De Havillands by 110 hp Le Rhônes.'

In response to Brooke-Popham's request, there went to France on 26 October, 1916, a D.H.2 fitted with a 110 hp Le Rhône. This aircraft at first had no number, but on 29 October No.1 Aircraft Depot allotted it the serial A305. At least two other D.H.2s were fitted with the Le Rhône engine; these were A2538 and A2594, which were tested at No.2 Aircraft Depot in November 1916. The test reports on these aircraft were dated 23 and 27 November, and both were signed by Capt Vernon Busby, who later was to lose his life in the crash of the first Handley Page V/1500 prototype.

No.7851, Aircraft C1 of No.32 Squadron, had a spanwise balance cable between the upper ailerons. This D.H.2 was sent to the Expeditionary Force in June 1916 and apparently served with No.32 Squadron until 7 January, 1917, when 2/Lt E. G. S. Wagner was reported missing.

A few D.H.2s were armed with Le Prieur rockets. One such was 7862, which was first allotted to the RFC in France on 15 June, 1916, and was later used by No.32 Squadron.

Despite repeated adjustments to rigging and engine, the performance of the D.H.2 with the Le Rhône was disappointing and handling seemed to be adversely affected. The proposal thus to re-engine the aircraft was immediately abandoned, and it is known that A305 and A2538 were later fitted with Monosoupape engines.

The Le Rhône installation necessitated a small reduction in the chord of the mainplanes inboard of the tail-booms, but it seems that a number of wings were modified in this way. The disparity in chord caused some puzzlement in squadrons for a time.

An earlier alternative power unit may have been the 110 hp Clerget 9Z, for No.5994 was said to have had such an engine in June 1916.

As early as 16 April, 1916, the D.H.2's lack of performance was emphasised in comparative trials in which one was flown against a Nieuport Scout, presumably a Nieuport 16. The D.H.2 was 10 to 12 mph slower at 8,000 ft and took about $9\frac{1}{2}$ minutes longer than the Nieuport to climb to that altitude.

Despite its several shortcomings, the D.H.2 gave a gallant account of itself in combat and registered a large number of victories. A well-deserved VC was awarded to Maj L. W. B. Rees of No. 32 Squadron for the action on 1 July, 1916, when, flying 6015, he attacked a German formation of ten bombers and put them to flight. Other distinguished pilots who flew the type included Flt Sgt (later Maj) J. T. B. McCudden.

In France the withdrawal of the D.H.2 began in March 1917, but not until June was replacement complete. A few remained operational in Macedonia and Palestine for a time but eventually the D.H.2 was flown only by training and experimental units and it is doubtful whether any outlived A5058 which was still flying at Turnberry in January 1919. Although serial numbers were allotted for 501 production D.H.2s, official statistics aver that only 400 were distributed to units of the RFC, 266 of these allocations being to the RFC in France, 32 to the Middle East Brigade and two to Home Defence units.

D.H.2
100 hp Gnome Monosoupape
110 hp Le Rhône 9J
110 hp Clerget 9Z

Single-seat fighting scout.
Span 28 ft 3 in; length 25 ft 2½ in; height 9 ft 6½ in; wing area 249 sq ft.
Empty weight 943 lb (Monosoupape), 1,004 lb (Le Rhône); loaded weight 1,441 lb (Monosoupape), 1,547 lb (Le Rhône).
With Monosoupape: maximum speed at sea level 93 mph, at 5,000 ft—90 mph, at 11,000 ft—73·5 mph; climb to 6,000 ft—11 min, to 10,000 ft—24 min 45 sec; service ceiling 14,000 ft; endurance 2¾ hr.
With Le Rhône: maximum speed at sea level 92 mph, at 5,000 ft—85 mph, at 11,000 ft—72 mph; climb to 6,000 ft—12 min, to 10,000 ft—31 min; endurance 3 hr.
Armament: One 0·303-in Lewis machine-gun on No.4 Mark I, later No.4 Mark II mounting; some D.H.2s of No.24 Sqn had two Lewis guns. On a few aircraft six Le Prieur rockets were carried.

Manufacturer: Airco.

Service use: Western Front: Sqns 5, 11, 18, 24, 29 and 32. *Macedonia*: A Flt, No.47 Squadron; RFC/RNAS Composite Fighting Squadron. *Palestine*: Sqns 14 and 111; 'X' Flt at Aqaba. *Training Duties*: Reserve Sqns 6, 10, 13, 20 and 24. *Middle East*: No.22 Training Squadron. *Other Units*: School of Military Aeronautics. Oxford; Schools of Aerial Gunnery, Hythe and Loch Doon; No.8 School of Aeronautics, Cheltenham; No.40 Sqn, Gosport.

Serials: 4732, 5335—5383 (not built), 5916—6015, 7842—7941,
A305, A2533—2632, A4764—4813, A4988—5087,
B1389
Rebuild—B8824.

Airco D.H.3

As noted in the history of the D.H.2, Geoffrey de Havilland had roughed out the general arrangment of a twin tractor biplane as early as 12 February, 1915. On 25 May, 1915, he began work on the design of a twin pusher biplane having an estimated empty weight 'with 2 Austro-Daimler or similar engines' of 3,200 lb and a loaded weight, including fuel for 6 hours, of 4,550 lb. His preliminary calculations included values for the use of two 100 hp Gnome engines. By 27 July, 1915, de Havilland had progressed to considering two 140 hp R.A.F.4a engines as the power installation and had come round to visualising the aircraft as having its fuselage underslung about the spars of the lower wings. A loaded weight of 7,000 lb was now estimated, with an endurance of 5 hours; various spans and chords were considered, the apparent optimum dimensions being 65 ft by 9 ft, giving a wing area of about 1,200 sq ft.

The design was pursued and, modified and refined in various ways, emerged in the spring of 1916 as the D.H.3. It was a handsome equal-span three-seat biplane of unusual appearance, powered by two Beardmore (Austro-Daimler) engines of 120 hp each, mounted between the mainplanes and driving two-blade pusher propellers by short extension shafts that placed the airscrews' plane of rotation behind the trailing edges of the wing. The mainplanes could be folded to conserve hangar space. In the tail unit there appeared for the first time the characteristically

The D.H.3 as it first appeared.

elegant profile of fin and rudder that was to typify succeeding generations of D.H. aircraft; the large elevators had horn balances. The aircraft was intended to be a long-range bomber capable of permitting a measure of strategic bombing.

The D.H.3 wore no serial number when it first appeared, but the numbers 7744—7745 were officially allotted for two prototypes on 7 February, 1916, under Contract No.87/A/337. The aircraft was at Upavon for official trials in mid-May 1916 and sustained damage necessitating repairs to a wing and the replacement of a propeller and extension shaft. Evidently this delayed the testing of the aircraft, for the trials reported on were not conducted until 2 and 9 July. The report attributed quite a fair performance to the D.H.3, and the assessment of its stability and controllability was commendatory. Clearly, the aircraft was greatly superior to the contemporary and somewhat similar F.E.4.

Despite the F.E.4's poor performance, a contract for 100 was given in July 1916, whereas it was not until September that Contract No.87/A/744 was let with The Aircraft Manufacturing Company for no more than fifty D.H.3s; the serial

The D.H.3 at Central Flying School, fitted with tall exhaust stacks. Behind at left is the first F.E.4.

numbers A5088—5137 were allocated on 29 September. It is perhaps significant that this was two and half months after the first contract for the single-engined D.H.4 was given to the company.

Orders for the 120 hp Beardmore engine totalled only 400, whereas the eventual total of 160 hp Beardmores was no less than 2,937. Many of these were ordered relatively late to sustain the extended production of the F.E.2b; nevertheless 1916 deliveries of the 160 hp engine totalled 317, and in the following year 933 were made. There must have seemed to be every advantage in equipping the production D.H.3 with the more powerful engine. This is the possible explanation of how the D.H.3 prototype 7744 came to be fitted with two 160 hp Beardmore engines driving four-blade propellers without extension shafts. The wing trailing edges were cut away to allow the propellers to rotate freely, and the chord of the ailerons was increased. The balance area of the rudder was also enlarged in somewhat makeshift fashion, and a large vertical fin surface was at one time mounted between the propellers, its purpose being obscure. Both gunners' cockpits were provided with two bracket-type mountings for Lewis guns.

The D.H.3 numbered 7744, with two 160 hp Beardmore engines driving four-blade propellers on standard shafts, necessitating the cutting away of part of the trailing portion of the upper wing. A vertical fin surface was fitted between the wings in line with the propellers.

This form of the type was designated D.H.3A but little is known about it. Presumably it was tested in some way, but any test report has yet to be found. Production began, and some work was done on A5088. The contract was abandoned before the first production aircraft was completed because (so it has been said) official policy changed to favour only single-engine aircraft. This may have been a decision forced by the problems of aero-engine production and a feeling that, in the circumstances, twin-engine types were a form of extravagance. The current production of the excellent D.H.4 was probably a further reason for abandoning the D.H.3, but it meant that the RFC was thereby deprived of its first long-range bomber. Nevertheless, the experience was not wholly wasted, for the D.H.3 eventually provided the design basis for the D.H.10 and 10A that would have equipped squadrons of the Independent Force, RAF, had the war continued into 1919.

The D.H.3 was sent to the Southern Aircraft Repair Depot at Farnborough with instructions that its engines were to be removed. Apparently, no specific instructions relating to the airframe were given, but the SARD dismantled the aircraft and it was struck off charge.

D.H.3 and D.H.3A
Two 120 hp Beardmore (D.H.3)
Two 160 hp Beardmore (D.H.3A)

Three-seat long-range bomber.

Span 60 ft 10 in; length 36 ft 10 in; height 14 ft 6 in; wing area 793 sq ft (D.H.3), 770 sq ft (D.H.3A).

Empty weight 3,980 lb (D.H.3); loaded weight 5,810 lb (D.H.3), 5,776 lb (D.H.3A). Loaded weight including 8 hr fuel but no bomb load.

Maximum speed at sea level 95·1 mph, at 9,500 ft—88 mph; climb to 5,000 ft—16 min 12 sec, to 10,000 ft—58 min; endurance 8 hr. Performance figures for D.H.3.

Armament: At least two 0·303-in Lewis machine-guns. Bomb load presumably varied in proportion to fuel load.

Manufacturer: Airco.

Service use: Prototype flown at CFS.

Serials: 7744—7745, A5088—5137.

Airco D.H.4

The D.H.4 was one of the truly great aircraft that were designed and built during the 1914–18 war, yet its history was, even in a time when aero-engine development and production were critical, remarkably affected by the availability of the great variety of engines fitted to it. In design terms the D.H.4 marked an immense advance on the D.H.1 and D.H.2, being a handsome two-bay tractor biplane. It was initially designed for the 160 hp Beardmore engine, but the first prototype was completed in time to take advantage of the new B.H.P. engine that, when tested in June 1916, delivered more than 200 hp. The B.H.P. was, like the Beardmore, a water-cooled six-cylinder upright inline engine, but its dimensions gave the first D.H.4 a nose that was both longer and higher than originally intended.

With Geoffrey de Havilland at the controls the D.H.4 made its first flight at Hendon in mid-August 1916. The initial excellent impression was confirmed by CFS Testing Flight, who evaluated the aircraft between 21 September and 12 October, 1916: their report was highly commendatory. From CFS the prototype flew direct to France on Sunday, 15 October, 1916; two days later, Geoffrey de Havilland wrote to Trenchard, advising him of a number of modifications that were to be incorporated in later D.H.4s, and indicating that the second prototype was expected to fly by the following week-end, powered by a 250 hp Rolls-Royce engine. The first D.H.4 flew back to England from No.2 A.D. on 18 October, piloted by Capt R.H. Mayo.

When the second prototype emerged it incorporated at least some of the modifications listed by Geoffrey de Havilland in his letter of 17 October. The engine had been lowered by 3 in to improve forward view; the mainplanes had

The first prototype D.H.4 at Central Flying School while undergoing its official handling and performance trials, September–October 1916. (*T. Heffernan*)

been moved forward 4 in, in consequence of which the positioning and geometry of the centre-section struts had been suitably altered; and the design of the undercarriage V-struts had been revised.

The RFC in France obviously liked the D.H.4, seeing it primarily as a fighter-reconnaissance type, as was indicated by Trenchard when he wrote to the Director of Air Organisation (at that time Brancker) on 16 October, 1916:

> 'Captain Mayo is returning to England with the de Havilland 4 machine with Hulford [*sic*] engine.
> We have suggested that the chief function of this machine should be a reconnaissance fighter, but should be able to carry the new 200-lb or 2—112-lb bombs if necessary.'

Brancker's reply of 25 October deserves to be quoted *in extenso*: not only does it reveal interesting forward thinking, but it provides more evidence that the higher command of the RFC was perfectly capable of recognising a potentially valuable

The second prototype D.H.4 with 250 hp Rolls-Royce engine, photographed at Hendon.

aircraft when it saw one and was not exclusively bound to the products, especially the B.E.2 series, of the Royal Aircraft Factory as was so frequently alleged by Farnborough's clamorous critics.

'It is now necessary to come to some decision as to the exact role of the de Havilland 4. It was designed as a bomber, and it is intended to equip it, if employed as such, with a gyroscopic sight fitted in the rear seat and used by the passenger looking through the floor.

Very accurate results have been obtained lately with this sight in the hands of the passenger, his pilot being directed over the target by an automatic indicator.

If this is agreed to, it would be possible to standardise on R.E.8 (first with the R.A.F.4a and eventually with the Hispano-Suiza), and on the de Havilland 4 (first with the Rolls-Royce and eventually with the B.H.P. engine), for all work other than that of fighting pure and simple and of long range bombing.

I would be glad of your opinion on this suggested policy which would eventually eliminate all B.E.2s, R.E.7s, Armstrong Whitworth 160 hp Beardmore, and Martinsydes from the programme sent over to you under this office No.87/Aeros/597(A.O.1), dated 17th October 1916.'

The second prototype D.H.4 with modified coaming about the rear cockpit, and fitted with fore and aft mountings for the Lewis gun, as described in the text.
(*Sir Robin Rowell, from Peter Liddle's 1914–18 Personal Experience Archives, presently housed within Sunderland Polytechnic*)

On 29 October Brooke-Popham replied to Brancker, agreeing fully with his proposals and forward thinking, and asking that arrangements should be made for seven D.H.4 squadrons to be provided for use in France. Nine days earlier Trenchard had replied to Geoffrey de Havilland, and had opined:

'As a reconnaissance fighter I think it will be a first rate machine, but I do not think it is entirely suitable for bomb dropping.

For a large machine it is extremely handy to fly. It is quick on its turns, with very sensitive fore and aft controls and has a very large range of speed.'

It was well that the RFC liked the D.H.4, for the first production contract, No.87/A/496, had been placed as early as 11 July, 1916, more than two months

A2152, a production D.H.4 of the initial Airco-built batch, photographed at Farnborough, where it was first recorded on 19 February, 1917. It went to a Reserve/Training squadron (possibly No.44) at Harlaxton but was wrecked at Spittlegate in August 1917, when Sgt Dunville lost his life in the crash.

before the first prototype was tested at CFS. The serial numbers A2125—A2174 were allotted on 13 July for the first 50 aircraft, and about three months later a further 690 D.H.4s were ordered from The Aircraft Manufacturing Co as A7401—A8090, by far the largest order placed by the RFC for any type of aircraft up to that time. The implications in terms of aero-engine supply were serious.

The D.H.4's high performance gave rise to doubts, in mid-November 1916, about the observer's ability to use a ring-type mounting satisfactorily at high air speeds, for the RFC in France had specifically asked for the observer's gun to be carried on a Scarff mounting. This concern led to trials of a two-mounting installation with high protective fairings; the forward mounting was a fixed pillar, the rear very similar to the Anderson mounting used on the F.E.2b. This arrangement was made on the second prototype, which was then tested at Orfordness, but was not adopted for production D.H.4s.

Early in January 1917 it was forecast that thirty D.H.4s would be produced that month, 45 in February and 100 in March. Of these the first 60 were to be built for Rolls-Royce engines, followed by 30 for the R.A.F.3a, whereafter production airframes for these two engines would come out in equal numbers. It was further intended to produce 50 aircraft in May 1917 for the B.H.P. engine. The adoption of so many types of engine, although inevitable, implied production and spares problems.

At no time were there ever enough Rolls-Royce engines of the types that were later named Eagle and Falcon, because officialdom in 1916 prevented the Rolls-Royce company from equipping a new factory, refused the company permission to build a special repair shop that would have relieved pressure on the production factory, and in July 1917 declined a Rolls-Royce proposal that would have provided 2,000 Eagle engines between June 1918 and February 1919. Additionally, procrastination by the Treasury frustrated an ambitious scheme for Rolls-Royce components to be made in the USA.

The brilliant 250 hp Rolls-Royce engine had originally been designed at the request of the Admiralty Air Department to be a suitable power unit for seaplanes and flying-boats, and the earliest deliveries had been made late in 1915.

At a remarkably early date the War Office secured a number of Series I and Series III engines (later named Eagles I and III) for installation in the first batch of F.E.2ds, followed by more Series III and Series IV (Eagles III and IV) engines for later batches; later still, some of the F.E.2ds built by Boulton & Paul had Rolls-Royce 275 hp Series I and II engines (Eagles V and VI). All were engines that had been built as right-hand tractors (at that time the various Series of the 250 hp engine were built alternately as right-hand and left-hand engines), and by the end of 1916 a total of seventy-five F.E.2ds built at the Royal Aircraft Factory had been fitted with these engines; at that time a total of 394 Rolls-Royce 250 hp engines of Series I to IV had been delivered.

The D.H.4 A7624 was completed with a Rolls-Royce Eagle V engine and was at Hendon Aircraft Acceptance Park when it was allotted to the Expeditionary Force on 3 September, 1917. It was used operationally by No.55 Squadron and is here seen as aircraft M of that unit. Later it saw service with No.99 Squadron, RAF, on mail-carrying duties.
(*RAF Museum*)

Enough left-hand engines were available to enable most of the D.H.4s of the first production batch to be powered by the earlier marks of Eagle, and deliveries of the aircraft began in January 1917. First unit to receive the new type was No.55 Squadron, then mobilising at Lilbourne. Much to the puzzlement of RFC Headquarters the D.H.4s supplied to No.55 Squadron were not fitted with bomb racks, for it had been decided that the Rolls-Royce version should be employed as a fighter-reconnaissance type and that the versions with the R.A.F.3a and B.H.P. engines should be bombers. It was conceded that if it proved essential to use the Rolls-Royce D.H.4 as a bomber this could be done provided the camera, wireless, R.L. Tube, signal pistol, and 19 gallons of petrol were removed (the Rolls-Royce engine was 300 lb heavier than the B.H.P.).

To sustain D.H.4 production while awaiting quantity deliveries of the B.H.P. engine, recourse was had to Rolls-Royce engines intended for F.E.2ds. As late as May 1917 The Aircraft Manufacturing Co were told that they could expect to receive 20 to 30 of the 275 hp Rolls-Royce (Eagle V to VII) engines, followed by some eighty F.E.2d engines. These were of various Marks and output, and in at least some cases had modified sumps that necessitated modifications to the D.H.4 airframe. Additionally, the different Marks had to have separately designed

A7446 was the first D.H.4 to incorporate modifications that enabled any Mark of Rolls-Royce engine to be installed in the airframe. It is here seen at Martlesham Heath in August 1917, fitted with an Eagle VIII. As recorded in the text, its operational service was with the RNAS and RAF. (*T. Heffernan*)

cowlings, oil pipes and airscrews. A hastily summoned conference on 23 May, 1917, decided that available F.E.2d Eagles I and III would be brought up to Eagle IV standard, and corresponding Eagle Vs to Eagle VI standard, to provide 104 Eagle IVs and 57 Eagle VIs for D.H.4s. Further conversions of standard Eagles were expected to produce 59 Eagle IVs and 55 Eagle VIs, but these groups comprised mixtures of right-hand and left-hand engines. There is a small amount of evidence, however, that suggests that these figures of conversions were not achieved.

This may have been because by June 1917 Airco had rationalised matters by so modifying the D.H.4 fuselage that it could accommodate any type of Rolls-Royce engine. The first aircraft to have this form of fuselage was A7446, which arrived at Martlesham Heath on 9 August, 1917, fitted with an early example of the 375 hp Rolls-Royce Eagle VIII. It was exhaustively tested, returned to its makers on 17 September, and was subsequently lent to the RNAS. It flew to Dunkerque on 24 October and subsequently served with No.2 (Naval) Sqn and No.202 Sqn RAF. Its allocation to the RNAS was formally recorded on 14 March, 1918, and it survived at least until 25 October, 1918, when it was at No.4 Aircraft Supply Depot.

Rolls-Royce D.H.4 with the taller Mk II undercarriage that became standard.

The more powerful engines drove larger airscrews, consequently a taller undercarriage was designed and standardised in order to provide adequate ground clearance. Another important modification of earlier date raised the observer's Scarff ring mounting significantly above the position it occupied on the earliest production aircraft, thus increasing its effectiveness in action; at the same time the rear top decking was simplified and became flat-topped and flat-sided.

As early as January 1917 it was intended that the first RFC squadron to have the R.A.F.3a-powered D.H.4 would be No.63, but this did not happen. The earliest known installation of an R.A.F.3a was made in A2168, which was officially tested at Martlesham Heath in March 1917. This variant began to equip No.18 Sqn from 26 June, 1917, onwards, and the squadron had many engine difficulties.

A7701 had the 200 hp R.A.F.3a engine. It was photographed at Martlesham Heath, where it arrived for trials on 10 November, 1917. It was extensively tested with various modifications of the power installation before flying to the Isle of Grain on 11 January, 1918.
(*T. Heffernan*)

By that time enough Rolls-Royce D.H.4s had been completed to permit the re-equipment of No.57 Sqn and to start the same process in No.25 Sqn. In that June RFC Headquarters were planning to replace the F.E.2bs of No.22 Sqn with B.H.P.-powered D.H.4s, but production difficulties and development problems frustrated deliveries of this variant for some months and No.22 Sqn received Bristol Fighters in July 1917.

Much of the B.H.P. delivery delays arose from the Siddeley-Deasy redesign of the engine into the Puma. Although the original B.H.P. was built by the Galloway Engineering Co, as the Adriatic, with deliveries starting in April 1917, production was slow and only 43 engines were delivered in 1917. First deliveries of Pumas began in July 1917, but The Aircraft Manufacturing Co immediately found that the redesigning process had so altered the original B.H.P. dimensions that the Pumas would not fit the D.H.4 airframes for which they were intended. The essential modifications delayed production, and the situation was not improved by an official decision of September that the Galloway engine was not to be used in France*. By then the RNAS were waiting for deliveries of B.H.P.-powered D.H.4s, having handed over fourteen Westland-built D.H.4s to the RFC in June, all with Rolls-Royce** engines, and the RFC were anxious to re-equip No.27 Sqn with B.H.P. D.H.4s by October 1917 at latest. The process began on 29 September, when A7677 joined the squadron, but not until late December was the last Martinsyde Elephant replaced by a D.H.4. In fact A7677 reached No.27 Sqn before the results of official trials of the B.H.P. version of the D.H.4 were available: B9458 with a Puma reached Martlesham on 22 September, A7671 (Galloway) not until 3 October, 1917. To add a touch of irony to the situation, the prototype D.H.9 A7559 (Puma) arrived at Martlesham only two days later than A7671.

Despite all the production problems of the D.H.4 and its engines it was agreed in the spring of 1917 to supply fifty D.H.4s to the Russian Government. These were to be fitted with 260 hp Fiat A.12 engines that were to be supplied by the Russian authorities, and the initial Fiat installation was made in A7532, which arrived at Martlesham Heath for official trials on 30 June, 1917.

German night bombing attacks on London began on 2 September, 1917, and the British Cabinet was quickly convinced of the need to bomb Germany in retaliation. From the Cabinet meeting of 2 October, which was attended by Trenchard, came the decision to set up the 41st Wing under Lt-Col C. L. N. Newall; one of the first two squadrons to join the new Wing was No.55, which took its D.H.4s to Ochey. With winter so close at hand the Russian D.H.4s could not then be delivered, whatever the outcome of the Bolshevik revolution, and the 50 aircraft would be a welcome addition to the RFC's bombing strength. The British Government therefore asked Russia to relinquish her fifty D.H.4s on the understanding that they would be replaced by 75 in the spring of 1918, and the Russians agreed readily.

This agreement had obviously been reached in advance of the Cabinet meeting of 2 October, for on that same day the Progress and Allocation Committee discussed the supply of the necessary Fiat engines, delivery of which against RNAS orders was then only beginning. It emerged that ten engines had already been handed over to the RFC. The installation of the engines in the D.H.4s was to be effected at the Southern Aeroplane Repair Depot, Farnborough. Although the task was viewed with urgency its completion was delayed by lack of minor components and airscrews.

It is known that only forty D.H.4s (C4501—4540) were assembled with Fiat engines, the balance of ten airframes being delivered as spares. All 40 Fiat-powered D.H.4s were officially allotted to the Expeditionary Force in France on 19 October, 1917, and it is known that 18 of them went to France in January 1918. There the Fiat-powered D.H.4s were flown operationally by No.49 Sqn in the spring of 1918. At least twelve of these D.H.4s were used by the squadron, though a record dated 15 March, 1918, suggests that seven of the 18 were dismantled on

---

*Nevertheless, on 24 September, A7671 and A7677, both fitted with the Galloway engine, were allotted to the RFC in France, and other Galloway-powered D.H.4s were similarly allotted. Although A7671 was re-allotted four days later to the Technical Department for evaluation at Martlesham Heath, A7677 was used operationally by No.27 Sqn, as were A7706 and A7710.

**But B3987 (ex N6380), which replaced B3968 (ex N6007) on 25 July, 1917, had a R.A.F.3a.

A7671, one of the relatively few D.H.4s to have the Galloway-built B.H.P. (Adriatic) engine, was at Hendon Aircraft Acceptance Park on 24 September, 1917, when it was officially allotted to the E.F. for the re-equipment of No.27 Squadron. On 28 September it was re-allotted to the Controller of the Technical Department and was sent to Martlesham Heath on 3 October for evaluation. Its engine was borrowed and temporarily fitted to the D.H.9 prototype A7559 but was returned to A7671 by 20 October. Thereafter this D.H.4 was used in a variety of tests, and on 28 December it returned to Hendon to have a high-compression Siddeley Puma fitted. With this new engine it arrived back at Martlesham Heath on 25 February, 1918, but stayed only until 7 March, when it went back to Hendon.

A7671 crashed on 8 May and was written off on 4 July, 1918. (*T. Heffernan*)

B3957 was one of the group of Westland-built D.H.4s that were transferred from the RNAS to the RFC; it was originally N5986 and had a 275 hp Rolls-Royce engine. It retained the twin Vickers guns and characteristic rear cockpit of the early Westland-built aircraft. This D.H.4 was first allotted to the Expeditionary Force on 12 June, 1917, and was used by No.55 Squadron. In January 1919 it was still on the strength of No.57 Squadron, RAF.

One of the forty Fiat-powered D.H.4s, with bomb ribs under the lower wings and a rack for a 230-lb bomb under the fuselage.

that date and sent to No.1 A.D. to be used as spares. C4526 was still on the strength of the BEF on 30 June, 1918, and as late as 10 October C4521 was on the strength of the Independent Force, while C4523 returned to France four days later to join the Independent Force (I.F.). Whether they still had their Fiat engines at those relatively late dates is not known.

It was with the Rolls-Royce engine that the D.H.4 enjoyed the greatest success and, despite the initial official misgivings as to its use as a bomber, it was in its

The standard Eagle VIII installation in a D.H.4. (*RAF Museum*)

Rolls-Royce form that it is best remembered as a high-performance day bomber. From the beginning of its operational career with No.55 Sqn the Rolls-Royce powered D.H.4 proved its superiority; for a time it was able to operate without fighter escort, as if to foreshadow the achievements of its brilliant descendant of the next war, the Mosquito. Throughout the battles of 1917 a substantial part of the RFC's bombing and photographic-reconnaissance effort was sustained by the three D.H.4 squadrons, Nos.55, 57 and 25, all of which flew Eagle-powered aircraft.

For all that, it seems that the RFC received very few of the ultimate and finest operational version of the D.H.4, that with the Eagle VIII engine. Late in September 1917 it was official policy that all Eagle VIIIs were to go to the RNAS, doubtless with use in Felixstowe flying-boats and Handley Page bombers in view. At the time the RNAS was starting to hand over to the RFC the first of 36 Eagles of earlier Marks for use in D.H.4s. By mid-October 1917, however, the RNAS wanted to put twenty Eagle VIIIs into D.H.4s that were to be fitted with large cameras for special photographic-reconnaissance work, and this was regarded as of sufficient importance to justify some interruption of normal D.H.4 output to expedite the building of these 20 aircraft. A few of these Eagle VIII D.H.4s were allotted to the RFC, one of the first being A7788, allotted to the RFC in France on 9 November, 1917. It had been preceded by two D.H.4s, A7652 and A7657, that were originally allocated to the RNAS but were instead allotted without engines to the RFC in France on 24 October, 1917; both later saw operational use with No.25 Sqn and then had Eagle VIIIs. Unfortunately, delays in deliveries of suitable radiators retarded effective deliveries of Eagle VIII D.H.4s until late January 1918.

Taken very soon after the formation of the Royal Air Force, this photograph illustrates an operational Puma-powered D.H.4 of No.27 Squadron at Ruisseauville in the spring of 1918. The aircraft, here seen bombed up for a mission, must almost certainly have joined the squadron before the RFC ceased to exist.

The D.H.9 was expected to supersede the D.H.4, and indeed the production batches B7581—7680, C1151—1450, and C2151—2230, were originally ordered as D.H.4s but were amended to call for D.H.9s. In service, however, the D.H.9 was dogged by the deficiencies of the Siddeley Puma engine, and its performance was generally inadequate for the conditions of 1918. In January 1918 two contracts for a total of 130 D.H.4s were placed, and the Royal Air Force found itself obliged to order 100 more in June and a further 100 in September. For these belated orders the RAF had to go to contractors who had not previously built the D.H.4, and these late D.H.4s all had the Rolls-Royce Eagle engines. The Puma and R.A.F.3a variants of the D.H.4 were declared obsolete for all purposes when A.M.O. 896/1919 was issued on 7 August, 1919.

D.H.4

200 hp B.H.P.; 230 hp Siddeley Puma; 230 hp Galloway Adriatic; 200 hp R.A.F.3a; 260 hp Fiat A.12; 250 hp Rolls-Royce Marks I, II, III and IV (Eagles I, II, III and IV); 275 hp Rolls-Royce Marks I, II and III (Eagles V, VI and VII); 375 hp Rolls-Royce Eagle VIII.

Two-seat day bomber and fighter-reconnaissance.

Span 42 ft 4$\frac{5}{8}$ in; length 30 ft 2$\frac{3}{16}$ in (Eagle VI and VII), 30 ft 8 in (Puma and Adriatic); height 10 ft 1$\frac{3}{4}$ in (Eagle VI and VII); wing area 434 sq ft.

Weights and Performance:

	Puma	Adriatic	R.A.F.3a	Fiat	Eagle III	Eagle VIII
	4 ×112 lb bombs	4 ×112 lb bombs	without bombs	2 ×230 lb bombs	without bombs	without bombs
Weight empty, lb	2,197	2,209	2,304	2,306	2,303	2,387
Weight loaded, lb	3,610	3,641	3,340	3,822	3,313	3,472
Maximum speed, mph at						
10,000 ft	106	104·5	117·5	106·5	113	113·5
15,000 ft	—	—	110·5	—	102·5	126
Climb to						
10,000 ft	24 min 36 sec	24 min 55 sec	14 min 15 sec	26 min 40 sec	16 min 25 sec	9 min 00 sec
15,000 ft	— —	— —	29 min 20 sec	— —	36 min 40 sec	16 min 30 sec
Service ceiling, ft	13,500	13,500	17,500	14,000	16,000	22,000
Endurance, hr	—	—	4	4$\frac{1}{2}$	3$\frac{1}{2}$	3$\frac{3}{4}$

Armament: One 0·303 in Vickers machine-gun (two on some Westland-built aircraft), one 0·303-in Lewis machine-gun; bomb load two 230-lb or four 112-lb bombs or equivalent.

Manufacturers: Airco, F. W. Berwick, Vulcan Motor & Engineering, Westland. Later RAF contracts to Glendower Aircraft and Palladium Autocars. Glendower order for F2633—2732 was for D.H.4s less mainplanes and ailerons.

Service use: Western Front: Sqns 18, 25, 27, 49, 55 and 57. *Home Defence*: No.51 Sqn. *Mesopotamia*: Sqns 30 and 72. *Training duties*: Training Sqns 9, 18, 19, 26, 31, 44, 46, 51, 52, 61 and 75. *Training Depot Stations*: Nos.2, 6 and 7. *Other units*: No.83 Squadron working up; No.4 (Auxiliary) School of Aerial Gunnery, Marske; Schools of Aerial Navigation and Bomb Dropping Nos.1 and 4; School of Photography, Farnborough.

Serials: A2125—2174; 7401—8090*.

B394 briefly allotted to 3697, but allotment to Expeditionary Force cancelled and aircraft retained by RNAS. 1482; 2051—2150; 3955—3968 Westland-built for RNAS but delivered direct to RFC; 3968 cancelled and replaced by 3987. 5451—5550; 9434—9439

*A8090 cancelled and skeleton airframe sent to Farnborough, but apparently completed later and used by No.57 Sqn.
A8059, A8063, A8065, A8066, A8067 and A8079 fitted with Eagle VIII engine and transferred to separate contract; replaced by D9231—9236. A7559 completed as D.H.9 under separate contract.

Westland-built for RNAS but delivered to RFC; 9456, 9458, 9460, 9461 and 9470 Westland-built for RNAS but transferred to RFC; 9476—9500.
C4501—4540.
D1751—1775; 8351—8430; 9231—9280.
E4624—4628.
F1551—1552—to replace D8408 and D9231; F2633—2732; 5699—5798; 7597—7598.
H5290; 5894—5939; 8263.
Rebuilds: B774—776, 882, 884, 7747, 7764, 7812, 7866, 7878, 7891, 7910, 7911, 7925, 7933, 7938—7942, 7950, 7964—7967, 7969, 7982, 7986, 7987, 7991, 9471, 9951, 9994.
F5825—5833, 5835, 5837, 5842, 5846, 6001 (ex A8034), 6002 (ex A8021), 6003 (ex A8047), 6059, 6070 (ex B884), 6096, 6099, 6103, 6104, 6114, 6115, 6119, 6133, 6139, 6165, 6167, 6168, 6187, 6207 (ex A8028), 6209 (ex D9263), 6212 (ex A7713) 6214 (ex A7507), 6215 (ex A7626), 6222 (cx A8016), 6234, 6253, 9511.
H6858 (ex A7964), 6859 (ex D8427), 6873 (ex F5826), 6881 (ex A7820), 6882 (ex A8088), 6885 (ex D8378), 6887 (ex B7911), 7118—7120 (F6511—6513 wrongly used and replaced), 7123, 7124 (ex D8383), 7125 (ex A8029), 7147 (ex D8377), 7148 (ex D8382).

Airco D.H.5

To the fighter pilots of the 1914–18 war, view from the cockpit was of great importance, and aircraft designers inevitably had to have regard to this factor when creating new fighters. This occasionally led to the designing of some bizarre aircraft. Whatever the performance shortcomings of the D.H.2 may have been, its pusher configuration gave its pilot a superlative field of view in almost all forward directions. Geoffrey de Havilland's next design for a single-seat fighting scout was a radical endeavour to combine the superior performance of the tractor biplane with the good forward field of view of the pusher.

The new tractor scout had the type number D.H.5 and the prototype was completed in the autumn of 1916. In appearance it was strikingly unorthodox, for marked negative stagger on the mainplanes was employed in order to place the pilot ahead of the upper mainplane. The cockpit was so situated that the pilot's head was in line with the leading edge of the upper wing: forwards and upwards he

The D.H.5 prototype as it first appeared, photographed at Hendon.

The prototype D.H.5, now wearing its serial number A5172, photographed at No.2 A.D., Candas, where it arrived from No.1 A.D. on 28 October, 1916. It was flown back to Farnborough on 16 November, but was back at No.1 A.D. in April 1917, doubtless much modified. It saw brief operational use with No.24 Squadron before returning to No.2 A.D., damaged, on 3 June, 1917.

had a clear field of view; rearwards and upwards he had none at all, for the upper centre section cut off all hope of outlook in that direction. In overall design, the D.H.5 was a neat and compact single-bay biplane; on the prototype the fuselage had flat sides and fabric-covered built-up flank fairings behind the engine cowling. The fin and rudder looked distinctly small, and the rudder was horn-balanced.

Although the prototype was unnumbered when it first appeared at Hendon, it was quickly allotted A5172, and with that identity it was flown to No.1 Aircraft Depot on 26 October, 1916, by Maj Mills. It was apparently allotted to the 4th Brigade next day, but on 28 October it flew to No.2 Aircraft Depot at Candas. It may well have gone to a squadron (and No.24 would seem to be the most likely); certainly it must have undergone some testing before it was flown back to Farnborough by Flt Sgt Piercey; it arrived there on 16 November.

The prototype D.H.5 A5172 fitted with the elevating gun mounting requested by the RFC in France; the fin and rudder had been revised in shape and size.
(*Imperial War Museum Q68244*)

Apparently the D.H.5 was regarded as satisfactory, for on 17 November Brig-Gen Brooke-Popham wrote to the Director of Aircraft Equipment: 'Reference the D.H.5 it is not considered that any alterations require to be made to this machine before fitting the gun. This gun should be mounted on the machine to fire through the propeller and be movable in a vertical plane so as to give an elevation of from 0 to 45 degrees. Movement in a horizontal plane is not necessary. Arrangements to be made for 500 rounds of ammunition in a disintegrating belt . . . An improvement which should be considered in future types of this machine is doing away with the centre section entirely so as to allow the pilot a good view to the rear'. In its reply the War Office indicated that the gun and mounting would be fitted in two to three weeks, and went on: 'It is proposed to do away with the centre section of the top plane as an experiment in the third or fourth machine of this type'.

A single Vickers gun offset to port on an elevating mounting was installed in A5172, and at about the same time an enlarged rudder of more pleasing outline was fitted. In this form, the D.H.5 was tested at CFS on 9 December, 1916, and received a generally favourable report. This seemed to be regarded as good enough for production to be initiated, and on 5 January, 1917, George Holt Thomas of The Aircraft Manufacturing Company forecast that 20 would be built in February, 40 in March and 60 in April. Serial numbers under the two initial contracts were allotted on 13 January: A9163—9362 were to be built by the Airco concern under Contract 87/A/1286 (but A9362 was subsequently cancelled because A5172 was covered by this contract), and A9363—9562 were ordered

A9507, a Darracq-built D.H.5, carried the presentation inscription *Christchurch Overseas Club*. It was first allotted to the RFC in France on 10 October, 1917, and was used by No.64 Squadron, in whose markings it is seen here. It survived its operational career and was flown back to England on 15 February, 1918. (*RAF Museum*)

from The Darracq Motor Engineering Company under 87/A/1358. These were the largest single orders placed for single-seat fighters up to that time and were followed early in February by a contract for fifty (B331—380) ordered from The British Caudron Company. There can be no doubt that RFC Headquarters staff were anxious to have delivery of the production D.H.5s as early as possible.

On 9 January, 1917, the A.D.A.E. informed Maj-Gen Trenchard that it was hoped to deliver twenty-four D.H.5s to France by 15 March; no doubt this was based on Holt Thomas's forecast of production. Unfortunately, production delays occurred, and late in February it was expected that the first two D.H.5s would reach France by the middle of March.

When the production D.H.5s appeared they differed noticeably from A5172. To the basic fuselage structure of longerons and spacers was applied a superstructure of formers and stringers to create a smooth blending of the circular engine cowling into a rear fuselage of octagonal cross-section. An external gravity fuel tank was mounted on the starboard upper wing and the elevating gun mounting was abandoned. The rudder was a plain surface without horn-balance. On the earliest production machines the aileron-return control was provided by rubber bungee cords on the upper ailerons.

While the D.H.5 was in production, the Sopwith-Kauper interrupter gear was adopted as the standard mechanism to be used on rotary-powered aircraft in the RFC, but quantity production could not start early enough to provide for the arming of the D.H.5. The Aircraft Manufacturing Company had themselves designed an interrupter gear that would have been fitted if deliveries of it had not been delayed. It was therefore decided late in March 1917 to equip the production D.H.5 with the Constantinesco hydraulic synchronising mechanism until the Sopwith-Kauper gear was available, and to use the Airco gear if any hold-up in Constantinesco deliveries occurred. How many, if any, D.H.5s had the Airco gear is not known.

A further delay in deliveries occurred because it was found that excessive vibration was experienced in flight. By 17 April five different D.H.5s, all fitted with French-made 110 hp Le Rhône engines, had been tested and all were 'found to vibrate so badly that the instruments were affected and the main engine controls could not be secured in position on their quadrant'. Engine changes produced little improvement, and it was believed that the vibration was largely attributable to the design of the engine bearer plate, which was unusually wide and was therefore thought to accentuate engine vibration. The design of this bearer plate was revised and strengthened, and presumably this modification

The D.H.5 A9186 was experimentally armed with a Vickers gun mounted at an upward angle of 45 deg. It was tested at Martlesham Heath in July 1917 and went to Orfordness on 20 July. Although it returned to Martlesham Heath on 8 September, it went back to Orfordness on 24 October, 1917. (*Gordon Kinsey*)

A9377, seen here at Brooklands, had a cut-away engine cowling when photographed. This Darracq-built D.H.5 was at Hendon on 31 May, 1917, when its initial allocation to the Training Brigade was cancelled and it was allotted to the Expeditionary Force. On 11 June it was again allotted for training duties and was subsequently used by No.40 Training Squadron at Croydon and by No.62 Training Squadron, Dover. It crashed on 25 September, 1917, killing 2/Lt H. M. Lee. (*RAF Museum*)

made the type operationally acceptable. From August 1917, however, A9403 was tested at Farnborough and Orfordness with a ply-covered fuselage and the Lloyd Lott jettisonable fuel tank; and while at Farnborough the vibration from which it suffered made it impossible to evaluate the aircraft.

The prototype D.H.5 A5172 was back in France in April 1917. On 20 April it crashed at No.1 A.D. while being flown by Capt Ainslie, but it was repaired and was flying again on 30 April. It was subjected to performance tests on 1 May, when it proved to have a speed of 113 mph at ground level and climbed to 10,000 ft in 12½ min. A5172 subsequently went to No.24 Squadron and was returned damaged to No.2 A.D. on 3 June, 1917.

No.24 Squadron had received its first D.H.5 on 1 May, 1917, its second on the following day, but deliveries were slow, for on 7 June both No.24 and No.32 Squadrons still had a mixture of D.H.2s and D.H.5s. The latter unit had also begun to re-equip with the D.H.5 in May. By 12 June both squadrons were out of action, their aircraft rendered impotent by troubles experienced with the Constantinesco gear. They were not alone in their difficulties, for all squadrons using the Constantinesco were similarly afflicted at that time.

After successful experiments by No.32 Squadron had demonstrated the advantages of fitting a spanwise inter-aileron balance cable instead of the rubber bungee cords, this was standardised for the D.H.5 by September 1917. After some indecision, it seems that this important modification was introduced from A9290, A9462 and B4902 in their respective production batches; the Aircraft Depots had been instructed as early as 1 August to modify all D.H.5s overhauled in their Aeroplane Repair sections. Another modification made about this time was the application of stiffening ribs to the engine cowling.

The D.H.5 never made a name for itself as a dog-fighter. Many pilots distrusted its unorthodox layout and it acquired a reputation, very largely undeserved it seems, of being dangerous and tricky to land. Yet as early as 13 May, 1917, Maj

M. H. B. Nethersole, then commanding No.65 Squadron which was evidently using some D.H.5s in its working-up period at Wyton, was able to write:

'The machine is extremely simple to fly, easy in landing, and light in control; it is very quick on turns but is so stable that it has been found impossible so far to get the machine out of control, e.g. in a spin, etc. It loops easily and well. The greatest disadvantage from an Active Service point of view is the impossibility of seeing anything behind and above'.

No.24 Squadron's historian lamented the fact that with the D.H.5 the squadron was unable to claim combat victories on the scale that it had enjoyed with the D.H.2. Other units had the same experience, but in the First Battle of Ypres the D.H.5 found its métier in ground attack and thereafter gave valuable but costly service in that role. Last of the squadrons to take the field equipped with the D.H.5 was No.64, and its pre-operational training included much low flying. Replacement of the D.H.5 began in October 1917, when No.41 Squadron began to re-equip with S.E.5as, and by the end of January 1918 none was left with operational units. Many were flown back to England in February and March 1918, where they served for a time with training units. Some of those used on training duties had the 100 hp Gnome Monosoupape engine.

D.H.5
110 hp Le Rhône 9J
110 hp Clerget 9Z
100 hp Gnome Monosoupape

Single-seat fighting scout.
Span 25 ft 8 in; length 22 ft; height 9 ft 1½ in; wing area 212·1 sq ft.
Empty weight 1,010 lb; loaded weight 1,492 lb. Weights with Le Rhône.
Maximum speed at 10,000 ft, 102 mph, at 15,000 ft, 89 mph; climb to 6,500 ft, 6 min 55 sec, to 10,000 ft, 12 min 25 sec, to 15,000 ft, 27 min 30 sec; service ceiling 16,000 ft; endurance 2¾ hr. All performance figures with Le Rhône.
Armament: One 0·303-in Vickers machine-gun; four 25-lb bombs.

Manufacturers: Airco, British Caudron, Darracq, and Marsh, Jones & Cribb.

Service use: *Western Front*: Sqns 24, 32, 41, 64 and 68 (Australian). *Training Duties*: Reserve/Training Sqns 30, 40, 43, 45, 55, 62, 63; No. 3 Training Depot Station; Sqns working-up—Nos.64, 65, 68.

Serials: A5172, A9163—9361, A9363—9562, B331—380, B4901—5000, *Rebuilds*: B897, B7750, B7753, B7762, B7775, B9468, B9942.

Airco D.H.6

In June 1916 the RFC had 26 squadrons in France with No.70 Squadron's arrival impending. It was at that time that Field Marshal Sir Douglas Haig requested that the number of RFC squadrons in France should be increased to 56 by the spring of 1917; in doing so, he recognised that such an increase would demand a very large number of pilots and observers. This perturbed the War Office, and initially only half of the proposed expansion was provisionally approved. Haig wrote to the War Office on 1 November, 1916, asking whether his proposals for expansion were approved, and on 15 November he was told that approval had been given for the creation of the additional units to the total of 56.

A9576 was a standard D.H.6 built by The Grahame-White Aviation Co. It is believed that the aircraft was with No.13 Training Squadron, Yatesbury, when this photograph was taken.

The implications for the training units were substantial, and after much study and debate the Army Council at its meeting of 12 December, 1916, formally approved the expansion of the RFC to a total of 106 operational squadrons and 95 reserve (*i.e.* training) squadrons. To meet the need for training aircraft that this enormous development implied, Geoffrey de Havilland designed a simple two-seat tractor biplane designated D.H.6, in which ease of manufacture, maintenance and repair were primary considerations. The square-cut wing panels were interchangeable as between upper and lower, and their heavily cambered aerofoil was chosen in the belief that it would give good low-speed performance and handling qualities. The fuselage of the prototype had a long communal cockpit for instructor and pupil, behind which there was a rounded top decking. The tail unit incorporated a typically elegant de Havilland fin and horn-balanced rudder, but the tailplane was as uncompromisingly square-cut as the mainplanes. Power was provided by a 90 hp R.A.F.1a engine.

Serial numbers were allotted for two prototypes, A5175—5176, on 18 October, 1916, but it is uncertain whether both were built. Production was ordered and in January 1917 a batch of 200 was ordered from The Grahame-White Aviation Co; the serial numbers A9563—9762 were allotted on 13 January. Doubtless Grahame-White was chosen because the design firm was heavily committed to production of the D.H.4 and D.H.5. In April 1917 a further 500 (B2601—3100) were ordered from The Aircraft Manufacturing Company.

Further orders were not placed until the autumn of 1917, for serious difficulties arose over the timber that was apparently forced upon contractors. The D.H.6, in common with most contemporary types of aircraft, had been designed to have most of its structure made of spruce, which was obtained from America. Following the entry of the USA into the war in 1917, shipment of spruce to Britain virtually ceased and a substitute had to be found. This was undoubtedly made more difficult by the fact that in 1917 the total imports of timber of all kinds fell from 6,300,000 tons to 2,800,000 tons; consequently, the Ministry of Munitions may have had very little choice but to allocate to aircraft constructors quantities of swamp cypress, which had not previously been used for aircraft construction.

The tables of timber strength and other factors then available to the aircraft

B2612, Airco-built, illustrates the full chord and deep camber of the mainplanes, together with the original rudder and elevators of the unmodified D.H.6. In fact B2612 was not used by the RFC but was transferred to the RNAS and saw service at RNAS Chingford. (*RAF Museum*)

From the same production batch as B2612, this D.H.6, B2840, was one of those modified to improve the aircraft's handling and safety. Comparison with the photograph of B2612 illustrates the reduction in chord and under-camber of the mainplanes, the marked negative stagger, and the narrow-chord rudder and elevators of this modified D.H.6. (*RAE*)

industry did not include values for swamp cypress. The cypress supplied to Grahame-White was wet, unseasoned and short-grained, and Claude Grahame-White had a sample length tested in his materials laboratory. This showed that the wood was dangerously useless, for it would warp and shrink so quickly that accurate construction would be impossible. Grahame-White's repeated protests to official quarters were brushed aside or ignored, and he was obliged to go on using the swamp cypress in his production of D.H.6s. Several months passed before the timber was condemned by the Technical Department, possibly following the structural testing of A9605 at Farnborough in September 1917.

Production of the D.H.6 increased spectacularly towards the end of 1917. Between September and 2 November no fewer than 1,700 were ordered from seven contractors, and this was reflected in the output of 1,099 D.H.6s in the first three months of 1918: the total for all of 1917 had been only 612. These aircraft were issued to most training units in the United Kingdom and Egypt; a few went to the Australian Central Flying School at Point Cook, and some were transferred to the RNAS.

The production aircraft differed from the prototype and were rendered even more angular by the use of a plain triangular fin and parallelogram-shaped rudder with four straight edges. Even the rounded top decking of the fuselage was replaced by an angular superstructure.

In the air the D.H.6 was normally viceless; indeed, some instructors considered it to be too docile. However, accidents occurred when aerobatics were attempted, for it proved to be difficult to pull the aircraft out of dives at speeds over 100 mph. Investigations at the Royal Aircraft Factory proved that in a power-on dive the force required to hold the D.H.6 in a dive increased rapidly with the speed, but if the pilot slackened his pull on the stick for fear of overstraining the airframe the aircraft would dive even more steeply. Farnborough continued the investigation by cutting off the forward 4 inches of the chord of the mainplanes, thus significantly reducing the camber of the aerofoil; additionally, the upper wing was rigged with 10·2 in of negative stagger, the elevator chord was reduced by 12 in, and the tailplane was rigged with its incidence reduced by 4 deg. Various combinations of aerofoil, stagger and incidence were tested and eventually the D.H.6 was considered to be safe though longitudinally unstable.

The essential investigative work had been done and the basic solution determined by March 1918, and the various contractors were asked to state the point of their production at which the necessary modifications could be incorporated on D.H.6s leaving the production lines. Inevitably, responses varied: The Aircraft Manufacturing Co claimed that incorporating the modifications would interfere with their D.H.9 and D.H.9A programme; Ransomes, Sims & Jefferies expected to incorporate the 10-in negative stagger from C7363 onwards but objected that the modifications to the leading edges, rudder and elevators would delay their F.E.2b production. The Grahame-White Aviation Co thought they

B3010, when this photograph was taken, bore night-flying national markings. It is known that this D.H.6 was at one time used by No.2 Wireless School, Penshurst.

The only D.H.6 to be built in Canada had a 90 hp Curtiss OX-5 engine. *(K. M. Molson)*

could introduce the stagger from 1 April, 1918; The Gloucestershire Aircraft Co could do this on C9471—9485 but not the structural modifications; Harland & Wolff's contract was to be completed by 30 March and no alterations could be made; The Kingsbury Aviation Co were confident that they could make all the modifications from 1 April; Morgan & Co could rig the stagger from C6636; Savages would make the structural modifications from C6867, the negative stagger from C6868. Such was the position reported to the Progress and Allocation Committee on 29 March, 1918, but the general introduction of the various modifications came after the formation of the Royal Air Force.

The RFC's use of the D.H.6 was as widepread as it was unspectacular, and the aircraft was satisfactory enough once its structural problems were resolved. It was something of a joke, enjoying perhaps the most extensive range of nicknames earned by any aircraft of the period. Of these, the more printable were The Clutching Hand, The Sky Hook, The Crab, The Clockwork Mouse, The Chummy Hearse and, from the Australians, The Dung Hunter.

The sole Canadian-built D.H.6, with national and unit markings applied, photographed at Leaside. *(K. M. Molson)*

Some of the later production D.H.6s had the 90 hp Curtiss OX-5 engine, an installation that changed the nose contours of the aircraft, for it was a water-cooled engine that required a flat frontal radiator. Later still, when Curtiss engines became scarce, the 80 hp Renault was fitted. Shortly before the RAF was formed, the RNAS acquired more D.H.6s for oversea patrols on anti-submarine duty, with which task the type continued after the formation of the RAF.

Something of the history of the RFC's training aircraft and units in Canada is related on pages 211–215. Although the Curtiss JN-3 was adopted as the basic type to be built in Canada, it was apparently thought prudent to have the D.H.6 in reserve as a kind of insurance against failure of the JN-3. One example of the D.H.6 powered by a Curtiss OX-5 engine was therefore built by Canadian Aeroplanes Ltd; its control system was that of the JN-3 in all essential respects. The solitary Canadian-built D.H.6 was tested by Brig-Gen C. G. Hoare, officer commanding the RFC in Canada, who had personally directed its construction. It evidently went to a training unit at some time, but production in Canada did not ensue.

D.H.6
90 hp R.A.F.1a
90 hp Curtiss OX-5
80 hp Renault

Two-seat trainer.

Span 35 ft 11¼ in; length 27 ft 3½ in; height 10 ft 9½ in; wing area 436·3 sq ft.

Empty weight 1,460 lb; loaded weight 2,027 lb. Weights with R.A.F.1a.

Maximum speed at 6,500 ft, 66 mph; climb to 6,500 ft, 29 min. Performance with R.A.F.1a.

Armament: One 0·303-in Lewis machine-gun on rear cockpit. Bomb load of about 100 lb could be carried on anti-submarine patrol.

Manufacturers: Airco; Grahame-White; Gloucestershire Aircraft; Harland & Wolff; Kingsbury Aviation; Morgan; Ransomes, Sims & Jefferies; Savages; Canadian Aeroplanes.

Service use: Training duties: Reserve/Training Sqns 4, 8, 11, 13, 15, 16, 17, 20, 21, 23, 24, 25, 29, 31, 35, 39, 42, 44, 46, 48, 50, 51, 52, 53, 54, 59, 61, 64, 66, 68, 69, 190, 191, 193, 194; No.200 Night Training Squadron. Australian Flying Corps Training Sqns 5 and 7. Training Depot Stations Nos.1, 3, 5, 8, 9, 11, 16, 20, 35, 40. *Other units:* Service Sqns 64, 97, 110, 121; in 1918, seventy-one D.H.6s were issued to Home Defence squadrons, including Nos.76 and 77. No.2 School of Aerial Fighting; Artillery Observation School, Almaza.

Serials: A5175—5176, 9563—9762.
  B2601—3100, 9031—9130.
  C1951—2150, 5126—5275, 5451—5750, 6501—6700, 6801—6900, 7201—7600, 7601—7900, 9336—9485.
  D951—1000, 8581—8780.
  F3346—3441.
Rebuilds: B8789, B8790, B8799, B8800, B9472, B9992,
  C3506, C4281, C9994.

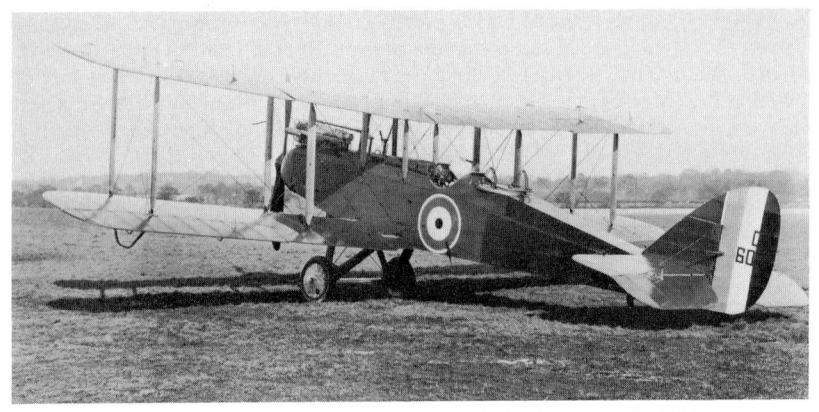

The first production D.H.9, C6051, photographed at Hendon. (*RAF Museum*)

Airco D.H.9

The D.H.9's history began in January 1917, when the Air Board asked the internal-combustion engine sub-committee of the Advisory Committee for Aeronautics to consider the 200 hp B.H.P., 200 hp Sunbeam Saracen, 200 hp Hispano-Suiza and 200 hp Sunbeam Arab engines and recommend which was most suitable for mass production. The sub-committee recommended the Arab, then untried, and additionally regarded the B.H.P. as superior to the Saracen. A total engine output of 2,000 engines per month was the aim at the time, and both the Arab and the B.H.P. were ordered in large numbers, 2,000 of the latter being ordered from The Siddeley-Deasy Motor Car Co of Coventry, and a more modest number (apparently only 60 initially) from the Galloway Engineering Co of Dumfries.

In the design of the B.H.P. engine Frank Halford had introduced some structural features of the contemporary Hispano-Suiza design, which he greatly admired. He employed an aluminium monobloc, into which each cylinder was screwed as a closed steel liner threaded over its full length; the cylinder heads were of cast iron and the water jackets of sheet steel. Ostensibly to render the engine more suitable for mass production, Siddeley-Deasy modified it extensively; J. D. Siddeley was himself responsible for introducing cast aluminium cylinder heads into which open steel liners were screwed on short threads; the water jackets were of aluminium. The revised design was named Siddeley Puma, and its weight came out at 645 lb, or 45 lb lighter than the original B.H.P. design.

It took some time for the RFC to realise that the original B.H.P. (produced as the Galloway Adriatic, though seldom recorded as such) and the Puma were two distinct and different engines that were not in any way interchangeable. The Aircraft Manufacturing Co soon found that the two engines were dimensionally different, but this disparity was insignificant when compared with the magnitude of the production difficulties that retarded output of the Puma. By July 1917, when it was expected that 100 engines would be made per month, it was found that 90 per cent of the aluminium cylinder blocks were defective. When that difficulty had been overcome trouble was experienced with the burning out of exhaust valves, this further setback delaying the attainment of full output until the spring of 1918. Yet the eventual overall total of Pumas ordered was no less than 11,500, whereas the corresponding total of Galloway Adriatics was no more than 560 and deliveries totalled only 94 by December 1918; by that month 4,288 Pumas had been delivered.

The numerous shortcomings of the Puma had not manifested themselves at the time when it was decided to build the D.H.9 in large quantities. Following the German daylight bombing attack on London of 13 June, 1917, it was decided on 21 June to increase the operational strength of the RFC from 108 to 200 squadrons, the majority of the additional units to be bombing squadrons. Because the D.H.4 was of full production status at the time a further 700 were ordered on 28 June, but when the Air Board met on 23 July the Controller of the Technical Department produced drawings of an extensively modified D.H.4 with the new type number D.H.9. It was claimed that the new type would have a speed of 112 mph at 10,000 ft fully loaded, combined with longer range. At the Air Board's 119th Meeting on 27 July, 1917, 'Sir William Weir stated that as a result of a conference with General Pitcher and Capt de Havilland it had been decided to adopt the de Havilland 9 design for the 700 bombers to be ordered. This would involve a loss of 3 or 4 weeks now but would place us ultimately in a much better position owing to the superiority of the new design. Drawings would be ready in four weeks and production in bulk could begin in six weeks. The board approved.'

The prototype D.H.9 was created by modifying the D.H.4 A7559 and was apparently in existence by the end of August 1917. Martlesham Heath at first expected it to arrive there about the middle of September but it did not do so until 5 October. Within a week it had had an engine change, its original Siddeley-made Puma engine being replaced by the Galloway-made Adriatic from the D.H.4 A7671, and it was with the latter engine that its recorded performance and climbing tests were flown. Ominously, the speed at 10,000 ft without bombs was only 110·5 mph.

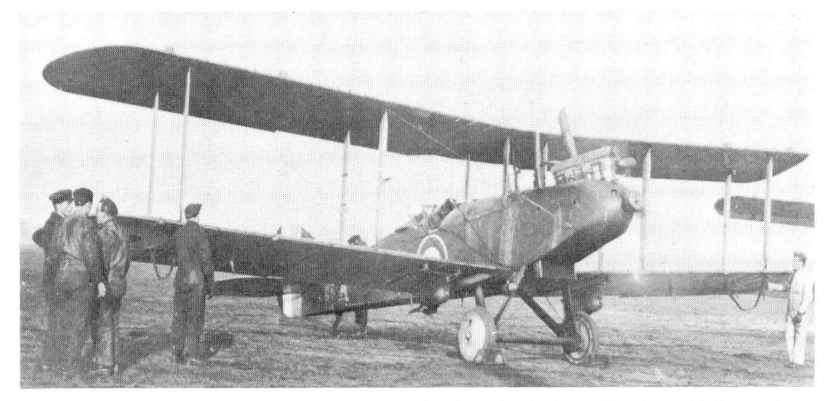

This D.H.9 is thought to be C6053, which arrived at Martlesham Heath on 22 December, 1917, as one of several very early production aircraft used for evaluation and development trials. Here it is seen carrying two 112-lb bombs and with a clear-view cut-out at the root of the lower wing, in which configuration C6053 was being tested by 5 January, 1918. This D.H.9 was exhaustively tested before going to Orfordness on 20 April, 1918.
(*Gordon Kinsey*)

C6109 was one of the first production D.H.9s to be allotted to the RFC in France, being one of five allotted on 27 February, 1918. It went to No.27 Squadron, with which unit of the RAF it is seen here, and was lost on 16 June, 1918, when 2/Lt H. Wild and Sgt E. Scott were killed in action.

In the D.H.9 the pilot's cockpit was moved aft to abut that of the observer, and immediately in front of the pilot was an internal bomb compartment that could hold up to fourteen 25-lb Cooper bombs stowed vertically. Between the bomb cell and the engine were the main petrol tanks of 34 and 28 gallons capacity. The six-cylinder upright engine was only partly cowled and its underslung radiator was retractable into the underside of the fuselage as the means of regulating cooling. The mainplanes, rear fuselage, undercarriage and complete tail unit were identical with those of the D.H.4.

Martlesham devoted much time and effort to testing A7559. It was sent to Orfordness, presumably for armament evaluation, returned to Martlesham on 24 October, and went back to Orfordness on 28 November. Meanwhile, the first production D.H.9, C6051, had arrived at Martlesham on 6 November and had gone to Orfordness three days later, returning to Martlesham on 28 November. It went to France on 18 December, and a succinct critique on the aircraft was drawn up next day by the indefatigable Brooke-Popham. On 20 December he informed the 12th Wing that C6051 was to be issued to No.27 Sqn for evaluation and report.

The report that Martlesham had rendered on A7559 was no better than lukewarm. About C6051 Brooke-Popham was neutrally factual, confining his criticisms to aspects of equipment and design. Absence of enthusiasm from his report would have been wholly understandable, for early in November Geoffrey de Havilland had unofficially told Trenchard (possibly following initial trials at Martlesham) that the D.H.9's performance would be poorer than that of the D.H.4 with the 275 hp Rolls-Royce, and that it would be unable to fly in formation at 15,000–16,000 ft with full bomb load.

On 14 November, 1917, Field Marshal Sir Douglas Haig asked that the orders for the D.H.9 should be reduced to no more than the total needed to equip and maintain fifteen squadrons, for the type would be outclassed as a day bomber by June 1918. Thereafter the D.H.9 might have some use as a night bomber but could only be expected to be useful as a day bomber if it could be fitted with the Rolls-Royce Eagle or Liberty engine. On 16 November Trenchard wrote to Maj-Gen J. M. Salmond, then Director-General of Military Aeronautics (D.G.M.A.):

'I do not know who is responsible for deciding upon the D.H.9, but I should have thought that no-one would imagine we should be able to carry out long distance bombing raids by day next year with machines inferior in performance to those we use for this purpose at present. I consider the situation critical and I think every endeavour should be made at once to produce a machine with a performance equal at least to the existing D.H.4 (275 hp Rolls-Royce) and to press on with the output with the utmost energy. I am strongly of opinion that unless something is done at once we shall be in a very serious situation next year with regard to this long-distance day bombing.'

Salmond conveyed Trenchard's views to the Air Board urgently, but this drew only a statement from Sir William Weir that the choice lay between the D.H.9 with the Puma engine or nothing at all. Trenchard attended the Air Board meeting of 28 November and reiterated his objections.

This D.H.9, B7651 built by Westland, was allotted from the Yeovil works to Brooklands Aircraft Acceptance Park on 28 March, 1918. It is here seen as aircraft 1 of the Wireless Experimental Establishment, Biggin Hill, and also saw service with No.99 Squadron, RAF.

As early as mid-September 1917 the use of the Fiat A.12 engine as an alternative to the Puma was being pursued, and on 18 September action to allocate a Fiat engine to Airco was initiated. By early November it had apparently been decided that the Fiat installation was to be made by Westland Aircraft, but on 13 November it was again agreed to allocate a Fiat engine to Airco. Short Brothers had been instructed to fit Fiat engines to their batch of one hundred D.H.9s (D2776—2875) to be built under Contract No. A.S. 34886 of 19 November, 1917, presumably in expectation of deliveries of the 1,000 Fiat engines that had, late in August, been ordered for use in the D.H.9.

By 28 December, 1917, Westland had asked for the immediate allotment of a Fiat engine for installation in a D.H.9. A total of nine Fiats was reported to be available by 2 January, 1918, but confirmation that any went to Westland has yet to be found. On 8 January, 1918, it was officially stated that the Puma was to be the power unit of the RFC's D.H.9s and the Fiat was regarded as only a stopgap in the event of failure of deliveries of the Puma. By 5 February Short Brothers had been informed that their D.H.9s were to have the Puma engine instead of the

Fiat, and it seems that very few D.H.9s had the Italian engine. Although the Fiat had the same basic configuration as the Puma its dry weight was 910 lb as against the 645 lb of the Siddeley engine.

The second Airco-built aircraft, C6052, had the first Fiat installation. It arrived at Martlesham Heath from Orfordness on 8 January, 1918, and underwent performance tests before departing for Orfordness on 12 February. It had spent a week there during its testing period, and at some time while at Orfordness it was fitted with navigation lights and with an experimental device, the Static Head Turn Indicator, that was intended to facilitate instrument flying. This Indicator was standardised for the D.H.9A, D.H.10, Handley Page O/400 and V/1500, and Vickers Vimy in 1919.

C6052 had the first installation of the Fiat A.12 engine and is here seen at Martlesham Heath. This D.H.9 went first to Orfordness on 4 January, 1918, and arrived at Martlesham on 8 January. Its testing was conducted at both places, and it finally left Martlesham for Orfordness on 12 February. (*R. C. B. Ashworth*)

Initially, output of production D.H.9s was slow. By 25 January, 1918, only Airco had delivered any D.H.9s, and the total was a mere five; the situation was made worse by a strike of all metal workers in London. G. & J. Weir's output was described as 'very disappointing', no more than three or four aircraft being hoped for by the end of January, and Short Brothers were expected to deliver only one or two. Expectations rose slightly a few days later, and it was hoped that 40 would be delivered in February and 100 in March.

Indicative of the official determination to evaluate the D.H.9 as thoroughly as possible is the decision to allocate the first six aircraft to the Technical Department. Early in December it was decided to send the seventh D.H.9 to the USA to assist in the preparations for American production of the type. Additionally, on 3 January, 1918, it was agreed to allocate one D.H.9 to the French Government. Martlesham received C6053 on 22 December, 1917, C6060 (recorded as 'first production machine') on 29 January, 1918, and the first Weir-built aircraft, C1151, on 28 February.

Although 336 D.H.9s had been completed by the end of March 1918, it seems

D.H.9 C6052 fitted with navigation lights and, at each upper wingtip, equipment that was part of the installation of the Static Head Turn Indicator.

that none saw operational use with any squadron of the RFC. New RFC squadrons had begun to form with the type early in 1918 and official allotments of D.H.9s to the RFC in France apparently began on 26 February, 1918, when D5554 and D5559 were allotted from Hendon Aircraft Acceptance Park, followed next day by the allotment of C6093, C6109, C6110, C6114 and C6098. By 31 March, 1918, a total of thirty-three D.H.9s had been formally allotted to the RFC in France, but seven of these had been re-allotted to No.99 Sqn mobilising at Old Sarum, and two were diverted to the Training Division.

With Royal Air Force and Independent Force day-bomber squadrons the D.H.9 saw extensive but, on the whole, undistinguished service until the war ended. Although it was intended that the D.H.9 should replace the D.H.4 it never did and, as noted elsewhere, production of the D.H.4 had to be revived in 1918 to provide at least some of the day-bomber squadrons with aircraft having a reasonable hope of survival. It was planned to have more than 30 squadrons equipped with the D.H.9, a figure that was not quite achieved, although the D.H.9 saw service in virtually every theatre of war before the Armistice.

D.H.9
230 hp Siddeley Puma
230 hp Galloway Adriatic
260 hp Fiat A.12

Two-seat day bomber.
Span 42 ft 4⅝ in; length 30 ft 6 in; height 11 ft 2 in; wing area 434 sq ft.
Weights and Performance:

	Puma, with two 230 lb bombs	Adriatic, without bombs	Fiat, without bombs
Weight, empty, lb	2,203	2,193	2,460
Weight loaded, lb	3,669	3,283	3,600
Maximum speed, mph			
at 10,000 ft	111·5	110·5	117·5
at 15,000 ft	97·5	102	107·5
Climb to 10,000 ft, min and sec	20 5	19 55	16
to 15,000 ft, min and sec	45	42 25	32 20
Service ceiling, ft	15,500	16,000	17,500
Endurance, hr	4½	4½	—

Armament: One 0·303-in Vickers machine-gun, one 0·303-in Lewis machine-gun. Two 230-lb or four 112-lb bombs, or equivalent weight of smaller bombs.

Manufacturers: Airco; F. W. Berwick; Cubitt; Mann, Egerton; National Aircraft Factory No.2; Short Brothers; Vulcan Motor & Engineering; Waring & Gillow (and Wells Aviation under sub-contract); G & J. Weir; Westland; Whitehead Aircraft.
Later contracts after 1 April, 1918, with The Alliance Aeroplane Co and National Aircraft Factory No.1.

Service use: Training duties—Sqns 98, 99 and 103 while working up to operational status.

Serials: A7559,
B7581—7680 (B7664 became D.H.9A), 9331—9430,
C1151—1450, 2151—2230, 6051—6350 (C6122 and C6350 became D.H.9A prototypes),
D451—950, 1001—1500, 1651—1750, 2776—2875, 2876—3275, 5551—5850, 7201—7300 (D7215, D7216 and D7251—7300 cancelled; D7209 dismantled for spares), 7301—7400 (D7331—7400 cancelled under original contract; D7331—7380 re-ordered under new contract), 9800—9899,
E601—700, 5435 (replacement for C6350), 5436 (replacement for C6122), 8857—9056,
F1—300, 1101—1300, 1767—1866,
H3196—3395 (cancelled), 4216—4315, 4316 (cancelled), 4320—4369 (cancelled), 5541—5890, 7563—7612, 7913—8112, 9113—9412.

Rebuilds: B7886, B7945,
F5841, F5845, F5847, F5849, F5850, F5864, F6055, F6057, F6066, F6068, F6072, F6073, F6074, F6098, F6112, F6113, F6125, F6141, F6172, F6183, F6196, F6205, F6213.
H7075, H7201 (ex E8932), H7202 (ex E8938), H7203 (ex E8983).

Note: Although F6055 and F6172 have for some time been reported as rebuilt D.H.9s, the log book of Lt C.E.F. Arthur, who for a time in 1918 was a test pilot at No.2 A.S.D., records these two as rebuilt Sopwith F.1 Camels.

Airco D.H.9A

The initiation of the process that led to the evolution of the D.H.9A might be traced back to the 256th Meeting of the Progress and Allocation Committee on 21 December, 1917, when

> 'Major Gray put forward a request from D.S. for a Mark VIII Eagle Rolls-Royce to be allocated to the Aircraft Co. for D.H.9.
> It was queried whether this engine was not intended for a D.H.4. Major Gray said he would have this confirmed.
> As it was understood that whether the machine was a D.H.9 or D.H.4 it was going to the Navy, Lt Cdr Andrews agreed to supply an engine for either machine.'

Next day the Committee received the following request from Sir William Weir, Controller of Aeronautical Supplies in the Ministry of Munitions:

> 'I understand that the Technical Department desire the installation of an Eagle engine in the D.H.9 machine. I consider this necessary, as towards the middle of next year we may have to put such engines in D.H.9 and the earlier

the performance is made and the production drawings prepared, the better. Please arrange for an engine to be allotted to the Aircraft Manufacturing Co. for this experiment.'

Reaction was instantaneous, for on that same day, 22 December, 1917, the Eagle VIII No.8/Eagle/250 W.D.33024 was allotted to the Technical Department and sent immediately to The Aircraft Manufacturing Co for installation in a D.H.9. Historically, it might be observed that this decisive request by Sir William Weir went some way to offsetting his retort of a few weeks earlier to Trenchard's views on the undesirability of employing the D.H.9.

The airframe in which that Eagle VIII was installed was that of C6350, which should have been the last D.H.9 of the batch C6051—C6350 ordered from The Aircraft Manufacturing Co. The installation did not differ significantly from that of the late marks of Eagle in the D.H.4, but on C6350 the engine was slightly farther forward. The entire rear fuselage, tail unit and undercarriage were those

C6350, the Eagle-powered D.H.9A, photographed at Martlesham Heath, where it arrived for testing on 23 February, 1918. (*Alex Revell*)

fitted as standard to the D.H.4 and D.H.9, but the mainplanes and ailerons were of greater span and chord than those of the D.H.4 and D.H.9, thus providing 52·73 sq ft of additional area, an increase of 12 per cent. The two main petrol tanks in the fuselage held a total of 65 gallons, supplemented by a 6-gallon gravity tank in the centre section. An oil tank holding 4 gallons was carried under the front of the crankcase.

The aircraft had been completed by mid-February 1918 and flew from Hendon to Martlesham Heath for official trials on 23 February. At that time it was still recorded as a D.H.9 with larger planes, but on 9 March Martlesham recorded it as the D.H.9A. Its performance tests were over by 16 March; gun trials followed, and C6350 left Martlesham before the end of that month.

Of the Rolls-Royce Eagle VIII engine the first official Statement of Engines dated 15 March, 1918, said:

'The whole of the Long Distance Night Bombing Programme is dependent on this engine, as also is the Large Flying Boat Programme, and it is of the utmost importance that as large an output as possible both of new and repaired engines should be maintained, and provision of spares on a heavy scale is essential.'

Of the American Liberty 12 engine the same Statement said:

> 'Very large deliveries of this engine are promised from June onwards, and these will be utilised in D.H.9s and D.H.10s, making it one of the most important engines.'

The story of the creation of the Liberty engine is a piece of history in itself that needs no repetition here. Jesse G. Vincent, one of its co-designers, is on record as saying:

> 'Although the [US] government asked for five 12-cylinder engines and five 8-cylinder engines on a test run, the project to supply the Allies with an engine that could be mass-produced quite possibly would have died had not England promptly cabled an order for 1,000 of them as soon as the endurance test on the first 12-cylinder model was completed. Actually, the order to build production Liberty engines for the Army and Navy was not definitely given until September 1917. The first production-built motor was sent to Washington on Thanksgiving Day that year.'*

(See the book *Flight—a pictorial History of Aviation*, p.77, published by the Editors of *Year, Inc*, Los Angeles, 1953.)

With heavy demands for the Eagle VIII in prospect it was natural for Britain to see in the Liberty a potential alternative. Not only was the American engine in the same power range as the Eagle, but it was of broadly similar configuration, was of slightly smaller overall dimensions, and was about 100 lb lighter**. As all available Eagle VIIIs would be wanted for the O/400, Vimy and Felixstowe F.2A/F.3/F.5 programmes, it was inevitable that the Liberty 12 should be considered as the power unit of the D.H.9A and D.H.10.

*It should be noted that this statement of Vincent's is not included in *The Liberty Engine 1918–1942* by Lt-Col Philip S. Dickey III (Smithsonian Annals of Flight, Vol. 1, No. 3, Washington D.C, 1968).

**The Eagle measured 72·5 by 42·6 by 44·5 in and weighed 926 lb dry; the Liberty's dimensions were 69·4 by 26·9 by 43·7 in, its dry weight being 825 lb.

The first Liberty-powered D.H.9A was C6122, which did not reach Martlesham until 15 May, 1918. (*RAF Museum P10342*)

The first Liberty 12 successfully completed its initial 50-hr test run at 1.30 a.m. on 25 August, 1917. Similar 50-hr runs of development engines between September 1917 and February 1918 led to numerous modifications, and these inevitably retarded production. By January 1918 Britain had asked for a total of 3,000 Liberty engines; the USA agreed to supply 11 per cent of total production. The first Liberty 12 engines to arrive in Britain came in March 1918, a total of ten being delivered in that month.

The need to adapt the D.H.9A to take the Liberty 12 occurred just when the Air Board were pressing Airco to expedite the design work on the larger and more complex D.H.10, itself eventually to have twin Liberty engines. Such a combination of tasks was beyond the small drawing office available to Geoffrey de Havilland, and it was decided to make the Westland Aircraft Works responsible for the design of the Liberty installation and the preparation of production drawings. To assist the Westland company, John Johnston, one of Airco's draughtsmen, was sent to Yeovil: he had been doing some preliminary work on adapting the design for the Liberty. Presumably design drawings of the engine had been sent to Britain in advance of deliveries of the production engines.

B7664 has long been wrongly regarded as a D.H.9A prototype, but as explained in the text it was a trial combination of an Eagle VIII engine and a production-type D.H.9A airframe that was made when the supply of Liberty engines ceased in July 1918. Here the aircraft is seen at Yeovil, rigged with the increased stagger that was introduced in early August 1918. B7664 went to Martlesham Heath on 17 August, 1918.

But no Liberty-powered D.H.9A flew before the RFC ceased to exist. The first installation of the American engine was made in C6122, which made its first flight on 19 April, 1918; Martlesham Heath optimistically expected it to go there next day, but it was not until 15 May that C6122 reached Martlesham. Thereafter it was exhaustively tested, and was still pursuing its test programme when it was joined by an early Westland-built production D.H.9A, F966, on 1 August, 1918.

In C6122 petrol and oil tankage was greatly increased. The two fuselage tanks held a total of 100 gallons, the gravity tank 7 gallons; and a 17-gallon oil tank was installed ahead of the fuel tanks.

All major development and operational flying of the D.H.9A was done by or for the Royal Air Force and Independent Force, and by the end of 1918 no fewer than 885 D.H.9As had been officially accepted.

A word needs to be said about B7664, a Westland-built D.H.9A that had a Rolls-Royce Eagle VIII engine. Although two additional D.H.9s, E5435 and E5436, were built as replacements for C6350 and C6122 respectively, the original Eagle and Liberty prototype D.H.9As, no such compensation was made for the conversion of B7664, which had been ordered as a D.H.9. Indeed, the official listing of allocations of serial numbers makes no separate or specific mention of this use of B7664, whereas the use and replacement of C6350 and C6122 were appropriately recorded. It seems possible that B7664 was the subject of Contract No.A.S.14119/18, which was given to Westland for a single D.H.9A on 27 September, 1918. All of this suggests that the conversion of B7664 was not envisaged before the adoption of C6350 and C6122 as D.H.9A prototypes.

Deliveries of Liberty engines to Britain stopped abruptly in July 1918, by which time a total of 1,050 had been received. Subsequently Mr Ryan, the US Under-Secretary for Aviation, informed Winston Churchill, British Minister of Munitions, that the US Admiralty claimed absolute priority for Liberty engines. The Eagle-powered B7664 did not go to Martlesham Heath until 17 August, 1918: it was intended that it should then undergo performance tests with bomb load for comparison with the Liberty-powered D.H.9A. The results of these trials were recorded in report No.M.227 of September 1918.

Surviving photographs of B7664 at Yeovil show that its engine installation closely resembled that of the Liberty in production aircraft and, remarkably, that it was rigged with stagger increased by 4 in, a modification introduced on production D.H.9As early in August 1918.

This combination of events and facts strongly suggests that B7664 was not, as has long been believed, a pre-production prototype, but a hasty conversion to adapt the production D.H.9A airframe to accept the Eagle VIII engine as an insurance against cessation of deliveries of Liberty engines, although there could have been no hope of producing sufficient Eagle VIII engines to satisfy the combined needs of O/400, V/1500, Vimy, F.2A, F.3, F.5, D.H.9A and D.H.10C production. Perhaps some alleviation of the situation was hoped for in the intention that later D.H.9As would have the 500 hp Galloway Atlantic, and that the first squadron of Atlantic-powered D.H.9As would join the Independent Force in January 1919, a second in March and a third in May; but in August 1918 that was pure optimism. Nevertheless, 72 Atlantic engines were delivered in 1918 and it was intended to order 1,000 of them for use in bombers. In the event only one installation in a modified D.H.9A airframe was made, the resulting combination being designated D.H.15 and named Gazelle.

D.H.9A

375 hp Rolls-Royce Eagle VIII, 400 hp Liberty 12
Two-seat bomber-reconnaissance.
Span 45 ft 10⅝ in; length 30 ft 3 in; height 11 ft 4 in; wing area 486·73 sq ft.
Weights and Performance with two 230-lb bombs:

	C6350 Eagle VIII	C6122 Liberty	B7664 Eagle VIII
Weight empty, lb	2,705	2,800	2,832
Weight loaded, lb	4,223	4,645	4,733
Maximum speed, mph			
at 10,000 ft	118	114·5	111·5
at 15,000 ft	104·5	106	—

Climb to 10,000 ft min sec	15 36	15 48	24 36
to 15,000 ft, min sec	33 42	33 0	—
Service ceiling, ft	16,000	16,500	14,500
Endurance, hr	—	$5\frac{3}{4}$	—

Armament: One 0·303-in Vickers machine-gun and one 0·303-in Lewis machine-gun; bomb load of up to 660 lb.

Manufacturers: Airco; F. W. Berwick; Mann, Egerton; Vulcan Motor & Engineering; Westland; Whitehead Aircraft.

Service use: None with RFC. In 1918 RAF Sqns 18, 99, 110 and 205 had the D.H.9A before the Armistice; Sqns 155 and 156 were mobilising with D.H.9As in October 1918; and it was intended to equip Nos.49, 123 and 133 with the type. It was further intended that by 1 June, 1919, there would be six D.H.9A squadrons with the E.F., seven D.H.9A (Liberty) and three D.H.9A (Atlantic) squadrons with the I.F.

Serials: B7664,
C6122, 6350;
E701—1100; 8407—8806; 9657—9756; 9857—9956;
F951—1100; 1603—1652; 2733—2902;
H1—200; 3396—3545; 3546—3795;
J401—450; 551—600; 5192—5491. In the postwar years further serial numbers were allotted to later production and reconstructed D.H.9As.
Wartime rebuilds—F9515 (ex F963), H7107, H7204 (ex F2747).

The first D.H.10 prototype, C8658, with its Puma engines installed as pushers.

Airco D.H.10 Amiens

Early in July 1917 the Technical Committee reported that it had been examining the question of modifying the D.H.3 to meet the then-new official Specification A.2.b. This called for a two-crew day bomber that might be single-engined or twin-engined, capable of carrying bombs and their racks to a weight of 500 lb, two guns and ammunition, and fuel for 500 miles at full speed at 15,000 ft. Speed with bomb load at 15,000 ft was to be at least 110 mph, the ceiling not less than 19,000 ft.

83

The Technical Committee considered that the adaptation of the D.H.3 to meet these exacting (for 1917) requirements '. . . could be done by using the 200 hp B.H.P. engine, and changing the wing section, together with some slight further redesign'. Possibly the Committee was mindful of the fact that experiments were at that time being undertaken at the Royal Aircraft Factory with a so-called Supercharger variant of the 200 hp B.H.P. engine that was described as giving 260 hp and held out the hope of improved performance in the future.

This development of the D.H.3 was the principal subject of British Requisition No.144, and on 4 August, 1917, in advance of a formal contract, an order was placed with Airco for three aircraft described as 'De Hav. 3 (designed to take B.H.P. engine)'. Apparently this original order was intended to be the subject of Contract No.A.S.22046. Only about six weeks earlier, on 18 June, 1917, Contract No.A.S.15292 had made official provision for the supply of the aircraft that came to be known as the Avro 529A; at that time it was officially described as 'A.D. experimental fighter No.3695 twin engine'. However, in the second week of July 1917 it was reported that the Technical Committee had '. . . decided that the 200 hp B.H.P. is to be fitted in the twin-engined Avro, which will be fitted as a bomber to meet A.2.b Specification'.

A week later the Committee reported that it had been examining the test results obtained with 'the Avro twin-engined Fighter [presumably the Avro 529] . . . in order to estimate its suitability for conversion to A.2.b type'. The Avro 529A hardly had a chance to prove whether it could satisfy Specification A.2.b: it arrived at Martlesham Heath for official trials on 29 October, 1917, powered by two Galloway-built B.H.P. engines, but its rudder broke in the air on 7 November and the aircraft was wrecked.

Obviously Geoffrey de Havilland undertook design work on the D.H.3 development in the late summer and autumn of 1917, and what he evolved was a new type with the designation D.H.10. British Requisition No.221 dated 18 October, 1917, called specifically for four experimental 'de Hav. 10' aircraft, and on 30 October, in advance of formal Contract No.A.S.31576, an order was placed for four prototypes. The superseded Contract No.A.S.22046 was cancelled during the week ending 24 November; serial numbers for three of the D.H.10s (C8658—8660) were allotted on 18 December, 1917, followed by the fourth (C4283) on 31 December.

What Geoffrey de Havilland had designed was a handsome three-seat biplane, similar in general appearance to the D.H.3, and powered by two 230 hp Siddeley Puma engines installed as pushers. As the six D.H.7s ordered on 29 May had been cancelled on 5 December, some of the load on the Airco drawing office had been lightened, and work on the D.H.10 proceeded so well that by 23 January, 1918, the first fuselage had been erected, the wings were nearly finished, and the engines were in place on their bearers. Delivery of the necessary Rafwires for the interplane bracing was held up, but it was hoped that the aircraft would be flying by about 7 February. Strength calculations were in hand, as were further calculations with additional engine data to assess the D.H.10's potential for fulfilling other operational requirements.

The first D.H.10, C8658, made its first flight at Hendon on 4 March, 1918. By that time the official name Amiens had been bestowed on the type; the second prototype, which was to have two Rolls-Royce Eagle VIIIs, was progressing well and had been designated Amiens Mk II. At that stage it was intended that the third prototype would have two Pumas, and it had been agreed on 4 March that two such engines should be delivered to Airco for the third D.H.10.

The second prototype D.H.10, C8659, had two Rolls-Royce Eagle VIII engines installed as tractors.

Considerable importance was attached to the D.H.10, for the initial contract had been increased to 20 aircraft by absorbing Contract No. A.S.31576 into a new contract, 35a/509/C.385, under which the additional serial numbers F1867—1882 were allotted on 27 April, 1918. By 6 March, however, the D.H.10 was apparently regarded primarily as a long-range escort fighter to protect bomber formations (A.F. Type IVa), and was under consideration as being likely, with alternative engines, to meet the requirements of Air Force Types VI, VIII and IX*. Two weeks later the design had been accepted for production as a long-range day bomber (A.F. Type VIII) with two Liberty 12 engines, and it was then intended to fit these to the third prototype, which was designated Amiens Mk III.

The period during which the first prototype underwent its maker's trials extended over 1 April, 1918, consequently the RFC had ceased to exist a week before C8658 went to Martlesham Heath on 7 April, 1918, for official testing. The aircraft was tested both as a bomber, with a crew of three, and as a Home Defence fighter with only the pilot and rear gunner aboard. Armament of the latter version consisted of only two machine-guns, one of which was to be wielded by the pilot, and it seems likely that the nosewheels were fitted as a safeguard against damage in nocturnal landings.

With the Puma engines, C8658 was quite seriously underpowered, and its performance fell short of the requirements of Specification A.2.b, even though its representative bomb load was less than the specified 600 lb (but this was 100 lb more than Specification A.2.b. had originally demanded). Presumably because its rate of climb and ceiling in its Home Defence form were quite good the Amiens I was sent straight from Martlesham to No.51 (Home Defence) Squadron at Marham on 4 May, 1918.

It was on that date that Martlesham first reported that it was expecting C8659, the Amiens II, in its turn. This second D.H.10 was still being modified at Hendon

---

*The various Air Force Type categories developed from a list that was in existence in March 1918. Those quoted here were as follows: Type IVa: Long-distance reconnaissance fighter; Type VI: Short-distance day bomber; Type VIII: Long-distance day bomber; Type IX: Night flying Home Defence machine.

Third prototype D.H.10, C8660, with Liberty engines in bulky nacelles and square-cut wingtips. This aircraft arrived at Martlesham Heath, where this photograph was taken, on 24 June, 1918. (*Imperial War Museum Q67541*)

two weeks later, but it seems that it never went to Martlesham, for on 1 June, 1918, Martlesham omitted it from the weekly reports, substituting C8660 which, powered by two Liberty 12 engines, was expected to do its maker's trials in the following week. Perhaps it was considered pointless to evaluate or develop C8659 in view of the unlikelihood of sufficient Eagle engines being available. Assuming that it was decided instead to concentrate on the use of the powerful Liberty, the abandonment of the Amiens II was, at the time, sensible.

C8660 reached Martlesham on 24 June and was extensively flown on test. Various propellers were tried and by 20 July, 1918, high-compression pistons had been fitted to the engines. A crash brought trials to a stop, but C8660 had been joined by the fourth prototype, C4283, on 28 July. This final prototype represented the production form of the design in such details as wing and aileron planform and in the mounting of the tail unit. Both Liberty-powered prototypes had provision for a Lewis gun firing under the tail; C4283 had high-compression engines. When repaired, C8660 went to Orfordness on 16 August, 1918, and on the following day the first D.H.10A, F1869, arrived at Martlesham Heath. This last variant had the engines mounted directly on the lower wing instead of being in the mid-gap position employed on the four prototypes.

The fourth D.H.10 prototype, C4283, arrived at Martlesham on 28 July, 1918. It had modified nacelles and raked wingtips.

With F1869 sustaining the test programme, C4283 left Martlesham on 29 August with the declared intention of its joining the Independent Force. However, its arrival at No.3 Aircraft Depot was not recorded until 18 September, actually three days later than F1867, described as a 'special machine,' being flown there by Capt B. C. Hucks. A third D.H.10, F1868, got to France on 10 November. The first two aircraft went to No.104 Sqn of the Independent Force at Azelot, and on 10 November Capt Ewart Garland flew F1867 to bomb Sarrebourg aerodrome on what was almost certainly the only D.H.10 operation of the war.

A further D.H.10 variant was built in small numbers. This was the D.H.10C, which was powered by two Rolls-Royce Eagle VIII engines; these, as on the D.H.10A, were mounted directly on the lower mainplane. It seems likely that this change of power unit was dictated by the cessation of Liberty deliveries in July 1918, just as that event led to the installation of the Eagle VIII in the D.H.9A B7664. Very few D.H.10Cs were built.

The D.H.10 was used in various experiments in armament and aerodynamics for a few years, and its postwar RAF squadron service was limited and obscure, being mostly spent with No.97 Sqn (later renumbered as No.60) in India and with No.216 Sqn in Egypt.

D.H.10

C8658—two 230 hp Siddeley Puma; C8659—two 360 hp Rolls-Royce Eagle VIII; C8660—two 396 hp Liberty 12; C4283 and production D.H.10 and 10A—two 405 hp Liberty 12; D.H.10C—two 375 hp Rolls-Royce Eagle VIII.

Three-seat day bomber or escort fighter; two-seat Home Defence fighter.

Span 62 ft 9 in (C8658—8660), 65 ft 6 in (C4283 and production); length 38 ft 10¼ in (C8658 and C8659), 39 ft 6 in (C8660), 39 ft 7$\frac{7}{16}$ in (C4283 and production); height 14 ft 6 in (with 900 ×200 wheels), 15 ft (with 1,100 ×220 wheels); wing area 787·sq ft (C8658), 834·8 sq ft (C8659), 837·4 sq ft (C4283 and production).

Weights and Performance:

	C8658		C8660	C4283
	As bomber	As Home Defence aircraft		
Weight empty, lb	5,004	5,004	5,600	5,585
Weight loaded, lb	6,950	5,814	9,000	9,000
Maximum speed, mph				
at 10,000 ft	100·5	—	—	112·5
at 15,000 ft	89·5	—	—	106
Climb to				
10,000 ft, min sec	20 55	15 10	18 30	16 5
15,000 ft, min sec	50 30	29 50	—	34 20
Service ceiling, ft	15,000	18,000	15,000	16,500
Endurance, hr	3½	—	—	5¾

Armament: Bomb load of up to six 230-lb bombs or equivalent; three 0·303-in Lewis machine-guns, one in bow cockpit, one in rear cockpit, and one firing aft under the tail; double-yoked Lewis guns could be fitted to nose and dorsal positions.

Manufacturers: Airco; Alliance Aeroplane; Birmingham Carriage Co; Daimler; Mann, Egerton; National Aircraft Factory No.2; Siddeley-Deasy.

Service use: None in RFC. First prototype went to No.51 (Home Defence) Sqn, RAF, and two went to No.104 Sqn of the Independent Force, RAF. It was intended to have, by 1 June, 1919, two D.H.10 squadrons with the Expeditionary Force, eight squadrons with the Independent Force, and three squadrons based in the British Isles for anti-submarine patrol. It is probable that it was intended that one of the Independent Force squadrons would be No.104, and it is known that the first two of the new D.H.10 squadrons for the Independent Force would have been Nos.121 and 122.

Serials: C4283; 8658—8660;
E5437—5636; 6037—6136; 7837—7986; 9057—9206;
F351—550; 1867—1882; 7147—7346; 8421—8495;
H2746—2945.

A.R.1

The A.R.1 was a French two-seater that was designed in 1916 by Capitaine G. Lepère of the Section Technique de l'Aéronautique. It was intended to be a replacement for the outdated Farman F.40 range of pushers, initially using the same 160 hp Renault engine installed in many of the Farmans. The fact that the Lepère design was a tractor was proclaimed in its designation, the initial letters signifying Avant-Renault.

At the time the director of the S.T.Aé was Col Dorand, who before the war had designed an inelegant tractor biplane with pronounced negative stagger on the mainplanes. This form of rigging was also adopted on the A.R.1, and it seems that the two connections with Dorand sufficed to ensure that for many years the A.R.1 was regarded erroneously as a Dorand type. Later versions of the design were powered by the 190 hp Renault 8Gd or 240 hp Lorraine 8A, the latter occasioning the revised designation ARL 1.A2. A further operational variant with wing area reduced to 45 sq m and some refinement was designated ARL 2.A2, having the 240 hp Lorraine 8Bb or 190 hp Renault 8 Gd/Gdx. Neither was an aircraft of conspicuously good performance.

The A.R. underwent its first official trials on 24 September, 1916, and, with a speed unmatched in the case of any other production type in France. orders were placed for 650 A.R. aircraft on 4 October. The type's acceptance trials were not held until 18 October but evidently the official mind was made up and production went ahead. The A.R. was issued to a considerable number of escadrilles, but was never regarded as better than mediocre. By 1 January, 1917, the total on order had risen to 1,435, despite the fact that no A.R.1 had been evaluated at the Front.

More were ordered in mid-February 1917—ironically, these were to be built by Farman—and it was alleged that orders for Moranes, Breguets, Sopwiths (1½ Strutters) and Letords were passed over in favour of the official design.

By 1 February, 1917, there were 216 A.R. two-seaters with escadrilles and aircraft parks and 80 in reserve. In its operational use the A.R.1 was not confined to the Western Front, seeing some use on the Italian Front and in Macedonia. In this last-named theatre of war it is possible that units were only partially equipped, one such being Escadrille BR2, which had a few A.R.1s on its strength.

It was in Macedonia that the Royal Flying Corps had some very limited experience of the A.R. type. The German cruisers *Goeben* and *Breslau* had long lain inactive at Constantinople, but on 20 January, 1918, the entire Allied naval and air presence in that part of the Mediterranean was galvanised into action by

Typical A.R.1 two-seater, possibly of Escadrille 122.

the intimation that the German warships had passed out of the Dardanelles at 5 a.m., obviously intent on attacking the British monitors *Raglan* and *M.28* in Kusu Bay, Imbros, and bombarding Mudros. With six or seven well-placed salvos the monitors were destroyed, and the German cruisers turned towards Mudros, despite the fact that *Goeben* had already struck a mine. Aircraft from Imbros so harassed the two warships that *Breslau* sailed into a minefield and had her stern shattered by a mine. Despite salvage attempts by *Goeben* the other cruiser had to be left to her fate. She struck more mines and sank, while *Goeben* again set course resolutely for Mudros. She struck a mine, turned back and struck another. RNAS Sopwith Baby seaplanes and D.H.4s made bombing attacks and the *Goeben* went aground south of Nagara.

In subsequent bombing attacks on the grounded ship the RNAS was augmented by the RFC, which took over the naval air station at Stavros and sent aircraft to Mudros to take part in the bombing. Most of the RFC contribution comprised five B.E.12s, the bomb load of which was not impressive. The Navy asked whether an aeroplane capable of carrying a 450-lb depth charge could be provided. None of those available to the RFC could do so, and the RNAS had not replaced its solitary Handley Page O/100 that had been lost on 30 September, 1917. An approach to the French aviation units in the area brought the prompt offer of an A.R.1 provided the RFC supplied a pilot. The choice fell on Lt W. J. Buchanan of No.17 Sqn, who flew the A.R.1 to Mudros on 28 January, 1918.

This A.R.1 in fact made no attack on the *Goeben*. During bad weather between 25 and 29 January the German cruiser succeeded in refloating herself, having suffered no significant damage from the 15 tons of light bombs aimed at her by RNAS and RFC aircraft. Lt Buchanan flew the borrowed A.R.1 back to Salonika on 29 January, and presumably it was returned to its escadrille.

A.R.1 data overleaf.

A.R.1
190 hp Renault*

Two-seat reconnaissance-bomber
Span 13·2 m; length 9·3 m; height 3·1 m; wing area 50 sq m.
Empty weight 810 kg; loaded weight 1,250 kg.
Maximum speed at 2,000 m—152 km/h, at 3,000 m—147 km/h, at 4,000 m—141 km/h; climb to 2,000 m—13 min, to 3,000 m—22 min 20 sec, to 4,000 m—39 min; ceiling 5,500 m; endurance 3 hr.
Armament: One fixed 7·7 mm Vickers machine-gun; one, occasionally two, 7·7 mm Lewis machine-guns. Normal bomb load four 120 mm bombs.

Manufacturers: Various French constructors. The maker of the RFC aircraft is not known.

Service use: One A.R.1 briefly lent to RFC group supporting RNAS from Mudros, piloted by Lt W. J. Buchanan of No.17 Sqn, RFC.

Serial: No British serial allotted.

*Engine assumed in view of load proposed.

Armstrong Whitworth F.K.2

On the outbreak of war in August 1914 the War Office and Admiralty ordered substantial numbers of B.E.2cs from many contractors. One of these was the Armstrong Whitworth company, from whom some fifty B.E.s were ordered. By the standards of the time the B.E.2c was considered difficult to produce, and Armstrong Whitworth undertook to design an aeroplane that could be more easily manufactured but would be equally efficient. Remarkably, the firm were allowed to do this, and early in 1915 Frederick Koolhoven designed a two-seater powered initially by the 70 hp Renault air-cooled V-8 engine.

The prototype emerged as a two-bay, equal-span biplane with straight raked wingtips, tailplane and elevator of unusually long span, a vertical tail assembly of low aspect ratio and Germanic appearance, and the pilot in the rear seat, as on the B.E.2c. The two cockpits were separated by a rounded top decking, and large external levers actuated the elevators. Behind the Renault engine the exhaust

The first Koolhoven-designed Armstrong Whitworth two-seater, with Renault engine, pilot in rear seat, and original flight surfaces. (*Thijs Postma*)

manifolds were united centrally and led upwards in a tall central stack that exhausted above the centre section. The only marking was a small Union Flag low down on the rudder, surmounted by A.W. in large letters.

The designation of this aircraft is uncertain. According to the late R. B. C. Noorduyn, an early collaborator with Koolhoven, the first Armstrong Whitworth two-seater was the F.K.2, but that may have applied to the aircraft of the officially ordered batch of seven, for what appeared to be an otherwise unnumbered intermediate aircraft with an enlarged horn-balanced rudder, revised elevator controls, roundels and rudder stripes, bore the marking A.W.1 on its fin.

An order was given under Contract No.94/A/103 for seven aircraft numbered 5328—5334, and these must, almost certainly, have been designated F.K.2. They bore a general resemblance to the B.E.2c, a likeness that was perhaps accentuated by the use of a standard B.E.2c tail unit in place of the original low-aspect-ratio fin and rudder and the long-span horizontal surfaces. The fuselage differed little from that of the prototype, the pilot again occupying the rear seat, but the forward actuating mechanism of the elevators was mounted internally. The planform of the wingtips was much modified, and it is possible that a new aerofoil section was used. Doubtless the detail design of the F.K.2 differed materially

Early Armstrong Whitworth with redesigned tail unit, internal control runs to the elevators, and the marking A-W 1 on the fin. (*Thijs Postma*)

from that of the Royal Aircraft Factory design, but it is questionable whether the Armstrong Whitworth type was in fact simpler to produce than the B.E.2c, for it inherited from the prototype an oleo undercarriage of original design and its engine was more elaborately cowled.

In July 1915, No. 5328 was delivered to Farnborough and was recorded as being there on the 13th of that month. Deliveries of the other aircraft of the batch were made in the following weeks: it is known that 5331 arrived at Farnborough from the makers on 13 August, 1915, and No. 5333 on 28 August.

Only one F.K.2 went to France. This was 5332, which had been tested at CFS on 31 August, 1915. It went to the 1st Aircraft Park at St-Omer on 8 September, 1915. On arrival it was tested by that great pilot Norman Spratt and by Maj J. H.

Armstrong Whitworth F.K.2 in what is thought to be the original form of the design, with B.E.2c fin and rudder. This aircraft was used by No.7 Reserve Squadron, Netheravon.

Becke, neither of whom evinced any noticeable enthusiasm for the F.K.2. Its aileron control was adversely criticised and it was thought likely that the aircraft would be difficult to fly in rough weather; Brig-Gen Trenchard regarded its fuel capacity of 28 gallons as inadequate. The F.K.2 was not found to be superior to the B.E.2c; Trenchard sensibly did not want the complication of having to provide stores for a further type and recommended against its use in France. No.5332 returned to England on 12 September.

When tested in France this F.K.2 had a 90 hp R.A.F.1a engine, and at some stage it acquired a modified rudder with a large overhanging horn balance, similar to the rudder of the aircraft that had been marked A.W.1. Presumably the latter modification was made in an attempt to improve the aircraft's directional control.

No.5332 was, as noted in the text, the only F.K.2 to go to France, arriving at St-Omer on 8 September, 1915. At that time it had a 90 hp R.A.F.1a engine, and is here seen at Gosforth with a horn-balanced rudder.

It is not known whether these modifications were incorporated in other F.K.2s. The type's service was apparently confined to training units. No.5329 was allotted to No.4 Reserve Aeroplane Squadron on 20 August, 1915; No.5333 was allotted to CFS on 31 August; No.5334 was used by No.7 Reserve Squadron at Netheravon; and 5328 and 5330 were with No.42 Squadron at Patchway in May 1916. No.5330 is also known to have been at Netheravon at some time.

Armstrong Whitworth F.K.2
70 hp Renault or 90 hp R.A.F.1a

Two-seat reconnaissance biplane.
Span 40 ft; wing area 457 sq ft.
Indicated speed at 300 ft, 74 mph; at 6,000 ft, 68 mph.

Service use: Used for training at CFS; No.4 Reserve Aeroplane Squadron, Northolt; No.7 Reserve Aeroplane Squadron, Netheravon; No.42 Sqn forming at Patchway.

Serial numbers: 5328—5334

When tested at CFS in the spring of 1916, No.5552 carried a dummy Lewis gun and had a non-standard rudder and small spinner. Its CFS trials were flown on 10 May, 1916.
(*T. Heffernan*)

Armstrong Whitworth F.K.3

The memorandum in which Brig-Gen Trenchard advised against the supplying of the F.K.2 to units of the RFC in the Field is dated 10 September, 1915, but it appears that the War Office had already committed itself to quantity production of the type. In addition to the F.K.2s numbered 5328—5334, a further 93 aircraft (5504—5553, 5614 and 6186—6227) were ordered under the same Contract, No.94/A/103, and it is known that the serial numbers 6186—6227 were allotted on 21 September, 1915.

These later aircraft were of a revised design to which the new designation F.K.3 was applied. An entirely new tail unit incorporating a balanced rudder and tapered adjustable tailplane was fitted; the crew positions were reversed to place the pilot in front, and there was emergency dual control for the observer; there was no top decking to separate the cockpits. Two sliding mountings for Lewis guns could be fitted to the observer's cockpit.

F.K.3s of No.47 Squadron at Yanesh in March 1917. (*The late Wg Cdr W. A. E. Featherstone*)

No.6191 was a typical trainer F.K.3, possibly of No.36 Training Squadron, Montrose.

The F.K.3 A1505 was built by Hewlett & Blondeau and is here seen on the aerodrome at Bogton that served the School of Aerial Gunnery, Loch Doon. It is known to have been used, more individualistically marked, by another training unit.

On the majority of the production aircraft the standard power unit was the 90 hp R.A.F.1a engine, but a few F.K.3s, mostly of the initial batch, had the 120 hp Beardmore, for which more than one form of installation existed. In *The Aeroplane* of 20 June, 1917, there appeared a thoroughly confused and confusing article by Charles L. Freeston, in which what purported to be a history of the Armstrong Whitworth company's early aircraft products was related. This alleged that

'In the early part of 1915 a serious shortage of the 90 hp R.A.F. engine arose, and, like many other manufacturers, the Armstrong-Whitworth firm were brought to a standstill accordingly. They had no fewer than a hundred machines ready and awaiting engines, but all had to be hung from the roof, three or four deep.

At the request of the War Office a large water-cooled Beardmore engine was sent to Newcastle to be fitted, if possible, into an Armstrong-Whitworth machine in place of the 90 hp R.A.F, for which the machine was designed.'

An Armstrong Whitworth F.K.3 with a Scarff ring mounting for the trainee observer; on the mounting is a Thornton-Pickard Mark IIIH camera gun. (*RAF Museum*)

That, like much that had gone before, is utter nonsense. In the early part of 1915 no production R.A.F.1a engines were available, for deliveries from Daimler did not begin until March 1915. As far as the F.K.3 was concerned this did not matter, as no aircraft had even been ordered, let alone completed. Moreover, deliveries of the 120 hp Beardmore were then so slow that there could have been barely enough to meet the needs of the R.E.5s and F.E.2as in service, and those of the F.E.2bs in prospect: certainly none would have been diverted for F.K.3s.

Even if one charitably assumes that '1915' was a misprint for '1916', the circumstances still seem somewhat improbable. By early 1916, over four hundred R.A.F.1as had been delivered (by Daimler, Siddeley-Deasy and Austin, who by the end of 1915 had delivered 385, 25, and 18, respectively: Wolseley deliveries did not start until March 1916, Lanchester not until June 1916), whereas the corresponding total of 120 hp Beardmores was only 249 by the end of 1915. This was hardly a situation in which the operationally desirable Beardmore would be pressed on Armstrong Whitworth in preference to the more numerous R.A.F.1a. Yet, as noted below, it was in March and May 1916 that Beardmore-powered F.K.3s were officially tested, consequently such installations could only have been made at about that time, certainly not a full year earlier.

The highly suspect article in *The Aeroplane* went on:

'Twelve of these [*i.e.*, Beardmore-powered] machines had been made when a notification was received that the supply of 90 hp R.A.F. engines was once more normal, and the arrangement of the pilot in front and observer behind having proved exceedingly popular, the firm were requested to alter all the 90 hp type machines, which was done.'

The installation of the 120 hp Beardmore engine in the F.K.3 varied considerably. In 5504, seen here, the engine was partly exposed and had a tall exhaust stack, while a large aerofoil-section gravity tank was incorporated in the centre section; and (*right*) A close-up of the engine installation, radiator block and gravity tank on F.K.3 5528. (*Peter Halliday*)

That, too, is nonsense, but one has to remember that the journal concerned was edited by C. G. Grey, whose hatred of the Royal Aircraft Factory was so unassuageable that he was not one to let a trifling consideration of truth frustrate an opportunity for implying any shortcoming of a Royal Aircraft Factory design, such as the R.A.F.1a engine.

What seems more likely, by 1916 when the Beardmore installations were actually made, is that operational employment for the F.K.3 may have been foreseen and, quite naturally and simply, more power was wanted. Equally probably, the Beardmore may have been subsequently withheld from the F.K.3

On F.K.3 5528, seen here with standard wings, the Beardmore was more neatly cowled and drove a two-blade airscrew, while the gravity tank in the centre section was of streamline form. (*Peter Halliday*)

because it was wanted for F.E.2bs and Martinsydes that were needed on the Western Front.

Beardmore-powered F.K.3s were tested at CFS in March 1916, and on 5 May, 1916, CFS Experimental Flight tested No.5528, which on that occasion not only had the Beardmore with a 41-gallon fuel tank beneath the upper wing but was fitted with lengthened wings having a span of 42 ft 5 in. It might be noted that this unusual F.K.3, identified in official documents simply as an Armstrong Whitworth two-seater, was responsible for a long-held erroneous belief that there was an early version of the F.K.8 with the 120 hp Beardmore.

The Beardmore engine of F.K.3 6218 was fully enclosed, and the cowling presumably housed an internal radiator. (*D. F. Woodford, via Harald Penrose*)

In F.K.3 6210 an experimental installation of a 140 hp R.A.F.4a engine was made, its fuel system incorporating an underwing gravity tank that might have been borrowed from a B.E.12. The pilot had no forward view. (*T. Heffernan*)

This version of the F.K.3 earned a complimentary report and was regarded as 'undoubtedly superior' to the B.E.2c. Nevertheless, the narrative of the test report on No.5528 ended with the observation: 'All machines of this type are now being fitted with 90 hp R.A.F.1a'. On 10 May, No.5552, fitted with the R.A.F. engine, was tested at CFS and also won praise for its excellent handling qualities, the observer's wide field of fire, and the oleo undercarriage. Later in May, No.5519 was tested with a 105 hp R.A.F.1b engine, which gave endless trouble and was not adopted. It is uncertain whether any of the Beardmore-powered aircraft saw operational use, but 6218 is believed to have been with No.47 Sqn at some time. Three Beardmore-powered F.K.3s, Nos.5504, 5505 and 5508, were sent to the Middle East, but what they did there remains unknown.

It seems that the only F.K.3s to see operational service were those that were used by No.47 Sqn in Macedonia. The squadron was formed at Beverley and received some F.K.3s while there. It arrived at Salonika on 19 September, 1916, and subsequently operated from Yanesh at Hajdarli with a variety of types of aircraft. Its operational R.A.F.-powered F.K.3s are known to have included 6199, 6212 and 6213. In the conditions prevailing in the Macedonian theatre of war No.47 Sqn was obliged to use its Armstrong Whitworths as general-purpose aircraft for bombing, artillery co-operation and contact patrols. On some bombing missions the F.K.3's were flown without observers, and this may have encouraged attempts to fit fixed guns that could be fired more or less forwards. The type lingered on in dwindling numbers in Macedonia until the war ended: on 31 October, 1918, three were still on charge at Salonika.

The F.K.3's good flying qualities made it suitable for use as a trainer, and further contracts were given for the production of more aircraft. In June 1916 fifty (A1461—1510) were ordered from Armstrong Whitworth, dual control being specifically mentioned; this contract was sublet to Hewlett & Blondeau, who later received a supplementary order for a further fifty (A8091—8140) under the same contract. Despite the ubiquitous Avro 504 variants so widely used as trainers, a batch of three hundred F.K.3s (B9501—9800) was ordered from Hewlett &

Blondeau in August 1917, but it is doubtful whether all of these were delivered.

The F.K.3 trainers saw quite widespread use at home and in Egypt. On them, pilots were taught to fly, observers to shoot and take photographs. Some of the aircraft used as gunnery trainers had Scarff ring mountings for the trainee observers' Lewis guns.

F.K.3

120 hp Beardmore or 90 hp R.A.F.1a

Two-seat reconnaissance, general purpose or training aircraft.

Span (standard) 40 ft 0⅝ in; length 29 ft; height 11 ft 10¾ in; wing area 442 sq ft.

Beardmore: Empty weight 1,682 lb; loaded weight 2,447 lb.

R.A.F.1a: Empty weight 1,386 lb; loaded weight 2,056 lb.

Performance:

	Beardmore engine		5552 with	5519 with
	Standard F.K.3	5528 with long wings	R.A.F.1a engine	R.A.F.1b engine
Maximum speed, mph				
at ground level	84	90·7	88·9	92·8
at 8,000 ft	—	83·5	81	—
at 10,000 ft	—	—	—	80
Climb to				
5,000 ft, min sec	13 30	13 30	19	8 5
10,000 ft, min sec	—	35	48 56	23 30
Service ceiling, ft	—	12,000	12,000	13,000
Endurance, hr	—	3	3	2½

Armament: One or occasionally two 0·303-in Lewis machine-guns and load of bombs.

Manufacturers: Armstrong Whitworth; Hewlett & Blondeau.

Service use: Macedonia—No.47 Sqn. *Training units* —Reserve/Training Sqns 9, 15, 17, 18, 20, 31, 35, 36, 42, 43, 46, 50, 51 and 53; Central Flying School; No.39 Training Depot Station; Sqns working up—Nos.52, 58, 110 and 121; Schools of Aerial Gunnery at Hythe, Marske, Turnberry and Loch Doon; School of Aerial Gunnery, Egypt; No.5 Fighting School, Heliopolis; School of Photography.

Serials: Production—5504—5553, 5614, 6186—6227. A1461—1510, A1967 (reported to have 70 hp Renault engine), A8091—8140. B9501—9800.

Rebuilds: A9972, B8827, B9968, F4219.

The Armstrong Whitworth F.K.7, A411, photographed at Gosforth.

Armstrong Whitworth F.K.7 and F.K.8

While the 120 hp Beardmore-powered Armstrong Whitworth F.K.3 No.5528 was being tested at CFS, another and more powerful two-seater designed by Frederick Koolhoven was nearing completion at Gosforth. This was an entirely new design, powered by a 160 hp Beardmore, intended to undertake the same duties as the Royal Aircraft Factory R.E.8. The new Armstrong Whitworth two-seater emerged in the early summer of 1916, wearing the serial number A411; it seems virtually certain that it was designated F.K.7 by its makers.

It was a sturdy, sensible aircraft, larger than the standard F.K.3 but clearly derived from the earlier type. The undercarriage incorporated oleo shock absorbers, there was a Scarff ring mounting for the observer's Lewis gun but no weapon for the pilot, and the rudder had a long, pointed horn-balance area. In this form A411 left Newcastle on the evening of 16 June, 1916, piloted by Capt E. D. Horsfall, who flew the aircraft to CFS. He arrived there next day, well satisfied with A411's handling qualities, and the aircraft was immediately put through the range of official trials.

The CFS Testing Flight's report on A411's flying qualities was generally good. Some features of construction and equipment were criticised, but all were remediable. Koolhoven's initial thoughts on forward-firing armament were to install a Vickers gun in the bottom of the fuselage, where it would have been totally inaccessible to the pilot: remarkably, CFS passed no comment on this outlandish proposal. The F.K.7 was sent to the Royal Aircraft Factory for structural testing. There it was subjected to the usual static loading tests then practised and was tested to destruction on 1 July, 1916.

At that time only the initial batch of fifty R.E.8s was on order. With the R.A.F. aircraft still an unknown quantity, it is perhaps not surprising that a similar number of Armstrong Whitworth two-seaters should be ordered. The serial numbers A729—777 were allotted early in June 1916, and by a remarkable effort on the part of the Armstrong Whitworth factory the first production aircraft, A729, was flying in August; the first R.E.8 was ready at Farnborough only on 13 September. On 21 August, 1916, the Assistant Director of Aeronautical Equipment informed RFC Headquarters '. . . . relative to the allotment of Armstrong Whitworth biplane A729 . . . . this machine will be delayed from despatch to France for a short while . . . . it will be despatched to Orfordness for trials prior to its being flown to the E.F.' On 25 August, for unexplained reasons, the batch of forty-nine numbers A729—777 was abandoned and replaced by fifty as A2683—2732, and it was as A2683 that the first production aircraft was officially allotted to the RFC in France on that date, although it was supposed to go first to Farnborough from Orfordness. Whether it made that detour is uncertain, but it is known that A2683 went to France for evaluation on 17 September, 1916, direct from the experimental armament unit at Orfordness.

The production aircraft were designated F.K.8 and resembled A411 in all principal features; in particular A2683 and several of the earliest production aircraft had the pointed horn balance on the rudder. A fixed Vickers gun was mounted within the fuselage, to port of and slightly behind the engine. Its

The earliest production F.K.8s had a long, pointed balance area on the rudder, as seen here on A2692. This aircraft was used by No.35 Squadron.

synchronising mechanism was designed by Armstrong Whitworth, but development of the gun gear had not been completed by the time A2683 went to France. A weapon installation employing the Arsiad interrupter gear existed and may have been used on some production F.K.8's.

Some modifications were made, and others recommended, by the RFC in France, but the F.K.8's handling qualities generally found favour. Enough production aircraft were available to equip No.35 Sqn, which went to France on 24 January, 1917, thereafter being obliged to resolve the early operational problems that arose in the Field. These were mostly created by the undercarriage, tailskid and radiators. Rough usage on operational aerodromes overtaxed the oleos; tailskid breakages were infuriatingly frequent; and the original form of tubes used in the radiators furred up quickly. It seems, too, that the Armstrong Whitworth gun synchronising mechanism was unsatisfactory, for it was replaced by the Constantinesco gear from early April 1917.

101

Typical F.K.8 with original nose cowling and oleo undercarriage. Allotments of aircraft in the Sanderson-built batch C3507—3706 began late in September 1917, and C3541 must have been allotted to the Training Division. In September 1918 it was with the Artillery and Infantry Co-operation School, Worthy Down.

At least four of these early production F.K.8s had the original long and pointed horn-balance area on the rudder, but this was quickly shortened and modified following the criticism that the pointed balance could be too easily jammed by combat damage. The shape of the fin was correspondingly modified.

The undercarriage problems were resolved following an instruction from Brig-Gen Brooke-Popham dated 30 April, 1917. This required No.1 A.D. at St-Omer to fit a standard Bristol Fighter undercarriage to an F.K.8; this was done on B201 and was an instant success. On 12 June, 1917, both A.D.s were instructed to fit the

From an early date the RFC wanted F.K.8s to have an exhaust stack led over the upper wing to eliminate the heat distortion of the pilot's view. As their request apparently went unheeded B233 was modified in the Field in the manner shown; the modifications were complete by 19 July, 1917, and on 22 July the installation was tested by No.10 Squadron, who reported very favourably on it. Nevertheless, it was not adopted, and it was some time before a long horizontal pipe was standardised.
(*Public Record Office: AIR 1/1066/204/5/1605*)

V-type undercarriage to all Armstrong Whitworths; this was still regarded as experimental at that time. Even this led in five weeks to a shortage of Bristol Fighter undercarriages and the practice had to be discontinued for a time. It is known to have been revived by May 1918, and numbers of F.K.8s had Bristol Fighter undercarriages.

Various other modifications to the wings, gunner's seat and exhaust system were tried, and new-section tubes were introduced in the characteristic inverted-V radiators. On later aircraft a redesigned nose cowling was introduced and remained standard, but it appears that the small flank radiators and long exhaust pipe were introduced on an experimental variant of the F.K.8 that appeared in December 1917. This aircraft had no official serial number and remained the private property of Armstrong Whitworth; whether it ever had a separate type designation is unknown. It went to France and was allotted to No.2 Sqn, who reported favourably on the new radiators, exhaust manifold, and special V-strut undercarriage. By 13 February, 1918, this experimental biplane had flown some

C8636, seen here fully bombed up in RAF service, had the Bristol Fighter V-type undercarriage. At one time this F.K.8 was used by No.110 Training Squadron, Sedgeford.

10 hr, including operational missions, and in the following month was reported to be with No.5 Sqn.

Despite its functional problems the F.K.8 gave good service and was well liked. Production burgeoned: 201 additional aircraft were ordered under the original contract, and subsequent RFC orders called for 600 F.K.8s from Armstrong Whitworth and 400 from Angus Sanderson. More orders were later placed for the Royal Air Force.

As further F.K.8s became available more squadrons were equipped with the type. It was decided in April 1917 to re-equip No.2 Sqn with Armstrong Whitworths, and the unit had F.K.8s when the Battle of Messines began on 7 June, 1917. In July, No.10 Sqn was re-equipped with the Armstrong Whitworth, as was No.8 Sqn next month; on 20 November, 1917, No.82 Sqn went to France equipped with F.K.8s. The Big Ack-W also saw service in Macedonia with Sqns No.17 and 47, and in Palestine with No.142 Sqn. Individual aircraft gave limited service on Home Defence duties in the United Kingdom, and one such aircraft, B247 of No.50 (HD) Sqn crewed by 2nd Lts F. A. D. Grace and G. Murray, shot down a Gotha into the sea on 7 July, 1917.

In operational service in France and elsewhere the F.K.8 proved to be effective, sturdy and dependable. It was also versatile, being used as a reconnais-

F.K.8 B3312 was used by a school of photography and had the revised nose cowling.

sance, artillery-spotting, ground-attack, contact-patrol, and day and night bombing aircraft with success throughout the battles of 1917 and 1918. During the strenuous fighting associated with the German offensive of March 1918, B5773 of No.2 Sqn was the vehicle of one of the war's most gallant combats. Its pilot, 2nd Lt A. A. McLeod, was awarded the Victoria Cross for the action of 27 March in which he was wounded five times and his observer, Lt A. W. Hammond, MC, six times. Although attacked by eight Fokker Dr Is, McLeod so handled his F.K.8 that Hammond was able to shoot down three of the triplanes; then, with the two-seater's petrol tank ablaze, McLeod climbed on to the port lower wing and retained control of the aircraft to side-slip the flames away from the fuselage until it came down in No-Man's Land. Despite his own wounds he then extricated

This is believed to be the unnumbered experimental variant of the F.K.8 that, although the property of the Armstrong Whitworth company, went to France and was flown operationally by No.2 Squadron, RFC. The aircraft had a purpose-designed V-strut undercarriage, long exhaust pipe and small flank-mounted radiators, all of which were favourably reported on by No.2 Squadron. In March 1918 the experimental Armstrong Whitworth was with No.5 Squadron. (*Vickers Ltd*)

Hammond from the wreckage and placed him in comparative safety before collapsing from exhaustion and loss of blood.

That was not the only action in which the F.K.8 proved to be a redoubtable fighter, and it was popular with its crews. To the end, however, only five squadrons of the type were operational in France, and it remains one of the more obscure aircraft of the war.

Its standard power unit remained unchanged, but experimental installations of other engines were made. The 140 hp R.A.F.4a was tried out in A2725 and B215, the 200 hp R.A.F.4d in B214, and the 150 hp Lorraine-Dietrich in A2696.

F.K.7 and F.K.8
160 hp Beardmore

Two-seat reconnaissance-bomber.

Span 43 ft 6 in; length 31 ft; height 11 ft; wing area 540 sq ft.
Empty weight 1,916 lb; loaded weight, without bombs, 2,811 lb.
Maximum speed at sea level 98·4 mph, at 5,000 ft—97 mph, at 10,000 ft—88 mph; climb to 5,000 ft 11 min, to 10,000 ft 27 min 50 sec; service ceiling 13,000 ft; endurance 3 hr.

Armament: One fixed 0·303-in Vickers machine-gun and one Scarff-mounted 0·303-in Lewis machine-gun; bomb load of eight 25-lb Cooper bombs or equivalent.

Manufacturers: Armstrong Whitworth, Angus Sanderson.

Service use: *Western Front*—Sqns 2, 8, 10, 35 and 82; Headquarters Communication Sqn. *Home Defence*: Sqns 36, 39 and 50. *Macedonia*: No.17 (part) and No.47 (part). *Palestine*—No.142 Sqn (part). *Training duties*: Training Sqns 3, 15, 31, 39, 50, 57, 61, 110 and 127; No.1 Training Depot Station. *Squadrons working up*— Nos.58 and 94. *Other Units*: Wireless Testing Park, Biggin Hill; No.1 (Auxiliary) School of Aerial Gunnery, Hythe; Artillery and Infantry Co-operation School, Worthy Down; No.1 Wireless School.

Serials: *Production*—A411, 729—777 (surviving records suggest that A729 was renumbered A2683 and this batch of numbers abandoned), 2683—2732, 9980—9999.
B201—330, 3301—3400, 5751—5850.
C3507—3706, 8401—8651.
D5001—5200.
E8807—8856.
F3442—3491, 4221—4270, 7347—7546.
H4425—4724.
Rebuilt aircraft: B856, 870, 871, 1497, 4014, 4017, 4018, 4049, 4061, 4080, 4081, 4108, 4120, 4142, 4145, 4146, 4147, 4150, 4151, 4154, 4160, 4161, 4163, 4165, 4166, 4169, 4170, 4174—4177, 4179, 4188, 4190, 4194, 4195, 4198, 4200, 8068.
C4550, 9997.
F617, 618, 623, 625, 626, 632, 633, 634, 638, 642, 646, 647, 649—651, 5801, 5803—5808, 6124, 6137, 6159—6163.
H7156 (ex F7419), 7157 (ex F7539).

The Armstrong Whitworth F.K.9 as it first appeared, with small, tapered ailerons, no dihedral and small tail unit.

Armstrong Whitworth F.K.9 and F.K.10

Frederick Koolhoven had tried his hand at the design of multiplanes before the F.K.8 appeared, for he created two remarkable three-seat triplanes early in 1916. These did not proceed beyond the prototype stage; indeed, the first seems never to have flown, and the second was incapable of doing anything warlike.

The year 1916 was to see the appearance of a triplane that was dramatically successful and a joy to fly. This was the little Sopwith triplane, a truly inspired design that had a considerable impact on the aerial war when it first went into action. How much the Sopwith triplane inspired Koolhoven's bizarre quadruplane two-seater of 1916 will probably never be known, but its structure and appearance suggested a strong Sopwith influence. The first aircraft to appear had the designation F.K.9.

It was an ugly, ill-proportioned little thing, powered by a 110 hp Clerget. The mainplanes had no dihedral and the minuscule tapered ailerons were inset from the wingtips. Pilot and observer were widely separated in the shallow fuselage and there seemed to be no provision for armament. The undercarriage was an insecure-looking affair in which a single main leg on each side was cross-braced fore and aft to the fuselage. All the tail surfaces looked perilously inadequate.

The F.K.9 first flew on 24 September, 1916, piloted by an RFC officer and with a Clerget engine lent by the War Office. It was then the property of Armstrong Whitworth, and it was reported that most of the parts of a second aircraft had been made. Slight damage sustained on the first landing was repaired and the F.K.9 flew again on 28 September with Flt Lt Peter Legh of the RNAS at the controls. At the end of the landing run the aircraft ran into Wing Cdr Alec Ogilvie of the Admiralty Air Department, who had witnessed both flights and was obviously impressed by the F.K.9's performance. Ogilvie was slightly injured and the F.K.9 damaged, but enough flying had been done to indicate that, as might have been expected, lateral control and stability were deficient, and the design had to be drastically revised. New wings with enlarged ailerons had to be made, and were given a small dihedral angle; rudder and fin area were enlarged; and the track of the undercarriage was increased.

Ogilvie displayed considerable enthusiasm for the quadruplane, and discussed with Frederick Koolhoven the possibility of adapting the design to have the 150 hp or 200 hp Hispano-Suiza and to build it in both patrol two-seater and single-seat bomber versions. Not surprisingly, Koolhoven was confident that, particularly with the 200 hp engine, 'a very fine machine' could be made. At the time, the Armstrong Whitworth company were exclusively War Office contractors, and in a letter to Wing Captain C. L. Lambe dated 26 September, 1916, Ogilvie indicated that the firm's directors were 'scared of disobeying the War Office'. In fact, the likelihood of either the War Office or the Admiralty being able to spare Hispano-Suiza engines for any version of the quadruplane was remote. Next day, 27 September, Ogilvie wrote to Lambe again, developing somewhat the possibility of using the geared 200 hp Hispano-Suiza with a machine-gun firing through the hollow airscrew shaft, and reporting that Arm-

The modified F.K.9 at Central Flying School, with new wings, enlarged ailerons, dihedral and redesigned tail unit.

strong Whitworth had that day given to Hewlett & Blondeau an order for three quadruplanes for the Admiralty. It is doubtful whether this order progressed very far. Possibly part of it was subsequently transferred to The Phoenix Dynamo Manufacturing Co, of Bradford, who early in December 1916 were given an order for two F.K.10 quadruplanes with the Admiralty serial numbers N511 and N512; while N514, a third F.K.10 for the Admiralty, was built by Armstrong Whitworth.*

---

*It has for long been believed that the four aircraft N511—514 were all originally ordered as Armstrong Whitworth quadruplanes, but this is incorrect. N513 was an Armstrong Whitworth two-seat reconnaissance biplane powered by a 200 hp Sunbeam and originally ordered late in December 1916; it was apparently delivered in June 1918, but was rejected after trials. In August 1918 the serial number H4424 was allotted to this biplane, but was soon cancelled in favour of the original identity N513. In at least one official record H4424 is listed as a 'Sunbeam two-seater', presumably from erroneous association with its engine; and this in turn has inevitably increased the confusion.

An F.K.9 incorporating the new wings and the other modifications went to CFS to be tested in November 1916; this may have been the second aircraft mentioned by Ogilvie, completion of which had originally been expected for 3 October. Although the aircraft was light on the controls and manoeuvrable, performance proved to be abysmal; the cockpit was too small and cramped to allow the pilot to make full use of the controls; and the fuselage had to be trued up anew after every landing, so inadequate was the structure.

Someone somewhere in the War Office must have been persuaded that the quadruplane had some potential that had not been detected during the CFS trials, for Contract No.87/A/1457 for a batch of 50 aircraft, apparently for the RFC, was given to Angus Sanderson & Co in December 1916; the serial numbers A8950—8999 were allotted on 30 December. On 25 November the numbers A5212—5213 had been allotted for two prototypes ordered from Armstrong Whitworth.

The two aircraft to which these last serial numbers apparently belonged were of a completely revised design. The fuselage was deeper and had faired sides; the entire tail unit had new profiles and there was no fixed tailplane; only the single-strut undercarriage and plank-type interplane struts recalled the original design. There was a fixed Vickers gun for the pilot and a Lewis gun on a pillar mounting for the observer. This revised design was given the new designation F.K.10.

An Armstrong Whitworth built F.K.10, powered by a 130 hp Clerget engine, was tested at Martlesham Heath in March 1917. Although described as 'light to handle and easy to manoeuvre' it returned performance figures appreciably worse than those of the F.K.9; and its undercarriage, although sketchily reinforced, was still found to be unacceptably weak. The test aircraft was almost certainly N514, the only A.W.-built F.K.10 to be ordered for the RNAS, but the results were valid enough for the RFC to reach the sensible (indeed, essential) decision to cancel the production order, and this was done on 27 March, 1917. The serial numbers A8950—8999 were re-allotted for other aircraft.

Production-type F.K.10 with 110 hp Le Rhône engine. (*Vickers Ltd*)

F.K.10 with 110 hp Clerget engine and cowling over barrel casing of the Vickers gun. This may have been A5213.

Evidently Sanderson production had already got under way and five production F.K.10s had been built. These were consigned to Orfordness to serve as static damage-assessment targets for gunfire, the allocation being made on 21 July, 1917. On 3 August the five aircraft were allotted the new serial numbers B3996—4000 under the original contract number, presumably as an administrative action to provide identities for writing-off purposes, but at least one of the F.K.10s is known to have been marked with a number from this batch.

F.K.9 and F.K.10
F.K.9—110 hp Clerget 9Z
F.K.10—110 hp Le Rhône 9J or 130 hp Clerget 9B

Two-seat fighter-reconnaissance quadruplane.

Span 27 ft 10 in; length 22 ft 3 in; height 11 ft 6 in; wing area 390·4 sq ft. All figures for F.K.10.*

Empty weight 1,226 lb (F.K.9), 1,236 lb (F.K.10); loaded weight 2,038 lb (F.K.9), 2,019 lb (F.K.10).

F.K.9—maximum speed at 6,500 ft—94 mph, at 10,000 ft—87·5 mph; climb to 6,500 ft—14 min 20 sec, to 10,000 ft—25 min; service ceiling 13,000 ft; endurance 3 hr.

F.K.10—maximum speed at 3,000 ft—95 mph, at 6,500 ft—84 mph, at 10,000 ft—74 mph; climb to 6,500 ft—15 min 50 sec, to 10,000 ft—37 min 15 sec; service ceiling 10,000 ft; endurance 2¼ hr.

Armament: One fixed 0·303-in Vickers machine-gun and one 0·303-in Lewis machine-gun on pillar mounting.

Manufacturers: Armstrong Whitworth, Angus Sanderson.

Serials: A5212—5213; A8950—8999, cancelled and numbers re-allocated; B3996—4000.

*These dimensions evidently came from the makers shortly after the aircraft had existed. Other published figures for F.K.9 and F.K.10 dimensions appear to have been derived from the notoriously inaccurate measurements given in the Martlesham Heath reports of the time and are not to be trusted.

Avro Types E and Es

The basic design of the Avro Type E was first expressed in a biplane built in November 1911 to the order of John R. Duigan. A slightly modified development with a 60 hp E.N.V. engine first flew at Brooklands on 14 March, 1912, and a further redesign the next month produced an aircraft with a slightly wider fuselage, a 50 hp Gnome and a revised tail unit in which the rudder was shod to serve as a tailskid and sprung against vertical travel for shock-absorption.

This variant proved to have a good performance and impressed the official observers when it underwent official trials at Farnborough on 9 May. The War Office negotiated its purchase and ordered two more, which were delivered to the RFC on 5 June and 22 July. These three aircraft were allotted to CFS and eventually received the serial numbers 404—406. As training aircraft they gave

Avro Type E No.404 with original rudder, running up at Central Flying School. The instructor in the rear cockpit is Maj J. D. B. Fulton; the West Highland terrier behind the Avro belonged to Lt A. M. Longmore, RN.

satisfactory service, and four more were ordered late in 1912. These were numbered 430, 432, 433, and 448, and were delivered to CFS between 18 March and 29 May, 1913.

A further order was placed by the War Office in January 1913. This was for five single-seaters which, on delivery in the early summer of 1913, were initially allocated to No.3 Squadron with the serial numbers 285 and 288—291. In the RFC Military Wing the single-seaters were known as the Avro Type Es to distinguish them from the two-seat Type E. When the Avro company introduced its numerical system of type designations the basic two-seater became the Avro 500, the single-seater the Avro 502, but there is nothing to suggest that these type numbers found any currency in the RFC.

As far as can now be determined, the Avro Type Es biplanes were delivered with the distinctive comma-shaped rudder that replaced the original sprung surface, and had separate tailskids mounted on a downward extension of the rudder post. No.285 was fitted with ailerons, in place of the original wing-warping, in February 1914, and the other four Type Es biplanes were returned to

110

Avro Type E No.406 with the later rudder of comma form. (*RAF Museum*)

the Avro works to be overhauled, re-covered and fitted with ailerons at a cost of £155 each. Of the Type E two-seaters, it is known that 430, 433 and 448 were at Avros in August 1914 having ailerons fitted, and the comma rudder was also introduced, notably on 406, which then had constant-chord ailerons.

The single-seaters were transferred to No.5 Squadron by January 1914 and subsequently went to CFS, where 285, 288, 289 and 291 were reported in September 1914. No.285 also saw some service with No.1 Reserve Aeroplane Squadron at Farnborough.

At least two Avro 500 two-seaters were flown by civil owners before the war, both having comma rudders and inversely tapered ailerons. One of these, owned by J. Laurence Hall, was a familiar sight at Hendon in 1914. It was impressed for the RFC at Shoreham on the outbreak of war with the military identity of

Avro Type Es No.288 photographed at Larkhill. When the numerical series of Avro type numbers was introduced, the Type E was allotted 500, the Type Es 502. (*RAF Museum*)

No.491*; its purchase price was £600 and it was delivered to Farnborough by Hall personally. This impressed Avro was intended for CFS and apparently went there. Its stay at Upavon may have been brief, for it was apparently sent to No.7 Squadron. It survived well into 1915, latterly incomplete or damaged, for on 25 November, 1915, the remains of 491 and 448 were inspected with a view to assembling one complete Avro from the two. This was deemed impossible, however, and authority to write off was requested. This was approved on 14 December.

Apart from the Avro 500 sold to Portugal in 1912, two others were supplied to the Naval Wing and numbered 41 and 150. The latter had the comma rudder and constant-chord ailerons.

There was no operational employment for the Type E or Type Es, and it is doubtful whether any of the RFC's Avros survived into 1916. Longest-lived of the type was the RNAS's No.939, which was still at Chingford in June 1917.

Avro Types E and Es
50 hp Gnome

Type E—two-seat trainer; Type Es—single-seat reconnaissance and training aircraft.
Span 36 ft; length 29 ft; height 9 ft 9 in; wing area 330 sq ft.
Type E—empty weight 900 lb; loaded weight 1,300 lb.
Maximum speed 61 mph; initial rate of climb 440 ft/min.

Manufacturer: Avro.

Service use: Type E—CFS, Upavon; No.4 Reserve Aeroplane Sqn, Northolt; No.7 Sqn.
Type Es—Sqns 3 and 5; CFS; No.1 Reserve Aeroplane Sqn, Farnborough.

Serials: Type E—404—406, 430, 432, 433, 448, 491.
Type Es—285, 288—291.

Avro 504

The first Avro 504 made its public début at Hendon on 20 September, 1913, and created something of a sensation in the aviation world. It underwent several modifications before mid-November 1913 and on the 24th of that month returned impressive performance figures under test at Farnborough.

Twelve were ordered by the War Office on 1 April, 1914, under Contract No.A2367. The first, numbered 376, was delivered on 12 June, only to be tested to destruction at Farnborough. 390, 397 and 398 soon followed and these three were all used by No.5 Squadron, and it appears that 397 went to that unit on 1 August, 1914. No.5 Squadron was the first RFC unit to use the Avro 504, and its three examples went with it to France on 15 August. First to fall in action was 390 on 22 August while being flown by Lts V. Waterfall and C. G. G. Bayly, and it seems likely that this unfortunate Avro provided the Germans with the first positive confirmation that British forces were in the field opposing them.

*It has hitherto been believed that Hall's Avro 500 was impressed by the RNAS as No. 939, but this is not so. The RNAS aircraft, first recorded at Hendon on 10 October, 1914, may have been C. F. Lan Davis's Avro, which had earlier been reported (on 6 August, 1914) as being 'a complete wreck' at Hendon.

Avro 504 No.637 of No.5 Squadron at Abeele undergoing maintenance, literally in the Field. This Avro arrived at Farnborough by rail on 17 August, 1914, and went to join No.5 Squadron in France some time before 3 November, 1914. It remained with the squadron until 16 July, 1915, when it was returned to England. (*RAF Museum*)

In the RFC, operational Avro 504s were never numerous. A production batch of fifty aircraft (750—799) had been quickly ordered, the first 44 being delivered as standard two-seaters, the last six (794—799) as single-seaters designated Avro 504D; these are mentioned later. Deliveries of the two-seaters began early in February 1915, many of the aircraft going to operational units in France. Most of the early production Avros had the 80 hp Gnome engine, but a few installations of 80 hp Le Rhônes or 80 hp Clergets were made. These aircraft were modestly supplemented by a few transferred from the RNAS: the first of these were originally 1005—1008, renumbered 2857—2860, all of which were on the strength of No.1 Squadron when it went to France on 7 March, 1915. They were accompanied by 752—755 and four B.E.8s. In May 1915 further Avro 504s

No.753 was one of the earliest batch-produced Avro 504s for the RFC. It went to No.1 Squadron in France on 7 March, 1915, but had the misfortune to come down at St Kruis in the Netherlands on 21 March. Interned by the Dutch, it was officially struck off charge on 23 March, but was purchased by the Luchtvaart Afdeling on 27 October, 1915. In Dutch service it was numbered LA-13, and survived until 1918, when it was given the new identity A-21.

113

(1020—1025) were transferred from the RNAS to the RFC and renumbered 4221—4225 and 4255. Of these, 4223 was used by No.1 Squadron, 4225 by No.5.

The Avro 504 was not particularly well suited to the tasks that the RFC was compelled to impose on it. Like the contemporary B.Es, it was flown from the rear seat, consequently the observer was hemmed in by the upper wing and its centre-section struts and wires. When combat demands necessitated the installation of a gun he was no more able to make effective use of the weapon than could the observer of any B.E. One of the earliest armed Avro 504s was Lt Louis Strange's aircraft in No.5 Squadron (wrongly recollected by him as No.383, which was in fact a B.E.2a of No.2 Squadron), in which a Lewis gun was carried on an arrangement that consisted essentially of a steel tube, a length of rope and a pulley. Primitive though it was, this 'mounting' sufficed to make Strange's Avro a popular vehicle among No.5 Squadron's observers.

Apart from some inaccurate references in Louis Strange's book *Recollections of an Airman* virtually nothing has been recorded about the actions or activities of the RFC's operational Avro 504s in France. The type survived on the Western Front for a surprisingly long time. No.773 was withdrawn from No.1 Squadron on 2 October, 1915, and, on 12 October, Nos.758, 769, 784 and 2858 followed it to the Aircraft Park. From No.5 Squadron No.4225 returned to the Park on 11 October, No.783 two days later.

Those Avros that remained in Britain had proved to be useful training aircraft, and subsequent production for the RFC consisted almost entirely of dual-control

A5921 was an Avro 504A of 'A' Flight, No.13 Reserve Squadron, Dover.

trainers. The first separate version was designated Avro 504A; it had shortened ailerons and broad-chord interplane struts but retained the 80 hp Gnome, and, at least on the earlier production aircraft, the tailskid was attached directly to the base of the rudder. Many hundreds of 504As were built, and the type saw largescale use with training squadrons throughout the war.

The Avro 504B and 504C were RNAS variants, the latter being a long-range single-seater, some of which were armed for anti-airship patrols. As an RFC counterpart, the Avro 504D was also a single-seater, having the shorter ailerons and comma rudder common to all RFC variants. In contrast, it had the cockpit sides cut below the level of the upper longerons, a feature otherwise found only on RNAS Avros. No reference to operational employment of any Avro 504D has yet been found, and it seems that most of the six examples went to training units. Delivery of the six was effected during August 1915.

114

Only six Avro 504D single-seaters were built in August 1915. The aircraft illustrated is 796, first recorded in the AID shed at Farnborough on 18 August, 1915. It went first to No.1 Reserve Aeroplane Squadron on 28 August, 1915, and by January 1916 was with No.11 Reserve Squadron at Northolt.

A much-modified Avro 504A converted into a single-seat fighter with overwing Lewis gun, 100 hp Gnome Monosoupape engine, and V-strut undercarriage as used on the Avro 521 and some of the single-seat night-fighter 504Ks.

Avro 504J B3103, sometime of No.3 Training Depot Station, Lopcombe Corner, had a distinctively simplified undercarriage.

A well-known Avro 504J of the School of Special Flying, Gosport.

Avro 504J with large-diameter engine cowling. The aircraft is known to be B3165, an Avro of 'A' Flight of the School of Special Flying.

As the war went on and production expanded, other contractors undertook the manufacture of the Avro. Its performance was enhanced by the installation of the 100 hp Gnome Monosoupape engine, in which form the type was designated Avro 504J. The more powerful engine was installed in production aircraft as and when supplies permitted, consequently production batches were mixed, including 504As and 504Js. Some 504Js had enlarged engine cowlings that produced a distinct hump above the line of the top decking. There were other variants in profusion, for the Avro was nothing if not adaptable. Undercarriages were modified, fuel systems varied, and one aircraft was rigged with Camel-like dihedral on the lower wing, the upper being flat; this Avro had a four-blade airscrew, a V-strut undercarriage and cut-away lower wing roots. It may have had some connection with the development of the single-seat night-fighter 504K mentioned later.

When the Avro 504J was adopted by Maj R. R. Smith-Barry as the standard training aircraft of the School of Special Flying at Gosport, which was formed in August 1917, its immortality was assured. Thereafter it became universally adopted in the RFC and production had to be expanded. This promptly created an engine-supply problem, for production of the Monosoupape was running

This modified Avro 504J (100 hp Gnome Monosoupape) had increased dihedral on the lower mainplanes, none on the upper, and was fitted with a four-blade airscrew and a V-strut undercarriage. It may have belonged to No.198 Training Squadron. The original of this photograph was dated 10 December, 1917.

down. Recalling all surplus rotary engines from all RFC units produced a mixed bag of 130 hp Clergets, 110 hp Le Rhônes and 80 hp Le Rhônes. The idea of making these engines interchangeable in the Avro airframe has been attributed to Smith-Barry himself, and indeed the first installation of a Clerget engine was made in B3157 by the engineering staff at Gosport. This carried the engine on an overhung mounting with an open-fronted cowling, and H. E. Broadsmith of Avro went on to evolve a similar installation using special bearer plates that permitted the use of any suitable rotary engine.

All Avros with the overhung mountings were designated 504K, and again the modifications were introduced into the production lines as resources and circumstances allowed. On 5 February, 1918, it was officially expected that, in their respective batches, the following aircraft would be completed as 504Ks, whereafter subsequent production would be of that sub-type: B8665, D1625, D1967, D4451, D5851, D6276, D7051, D7501 and D8781. In reality this did not happen, at least in some cases, for it is known that many Avro-built aircraft in the batch D7501—7800 were completed as 504Js.

Various experimental modifications of Avros were made at Gosport. B4222 was one of several 504Js that were flown there with single-bay wings of reduced area. (*K. M. Molson*)

Gosport evolved other experimental variations of the 504J and 504K, including a set of short-span wings with single-bay bracing. A few Avros had the single-bay wings, but Gosport's other experiments were not developed further.

Early in 1918 it was decided to equip the northern Home Defence squadrons with Avro 504Ks powered by the 110 hp Le Rhône engine. These were flown as single-seaters, usually with the front cockpit faired over, and could reach a ceiling of 18,000 ft. Armament consisted of a Lewis gun on an overwing Foster mounting, and flare brackets were fitted under the lower wings. Some of the Home Defence single-seaters had V-strut undercarriages similar to that of the unsuccessful Avro 521.

It seems that the Avro 504K Home Defence fighters did not equip squadrons until some months after the RFC had ceased to exist, but it seems likely that some preliminary experiments had been conducted with a Monosoupape-powered Avro that had begun life as a 504A; it had the V-strut undercarriage.

But it is as a trainer that the Avro will always be remembered with respect and great affection. It had been decided to standardise the 504K as the sole training type in February 1918 and the process of replacing all other trainers had begun before 1 April, 1918. The Avro lived on long after that.

One of the most historic British aircraft of all time was B3157, here seen at Gosport after being modified to become the first Avro 504K. Its engine was a 130 hp Clerget.

Avro 504

504—80 hp Gnome, 80 hp Le Rhône or 80 hp Clerget
504D—80 hp Gnome
504J—100 hp Gnome Monosoupape
504K—130 hp Clerget, 110 hp Le Rhône or 100 hp Gnome Monosoupape
Two-seat reconnaissance; two-seat trainer; single-seat Home Defence fighter.
Span 36 ft; length 29 ft 5 in; height 10 ft 5 in; wing area 330 sq ft.
Empty weight 924 lb (504), 1,050 lb (504A Le Rhône), 1,100 lb (504K Monosoupape), 1,231 lb (504K Le Rhône); loaded weight 1,574 lb (504), 1,700 lb (504A Le Rhône), 1,800 lb (504 Monosoupape), 1,829 lb (504K Le Rhône).
Maximum speed at sea level 82 mph (504), 95 mph (504K Le Rhône); maximum speed at 6,500 ft—62 mph (504A Le Rhône), 82 mph (504K Monosoupape); maximum speed at 8,000 ft, 87 mph (504K Le Rhône). Climb to 3,500 ft—9 min 30 sec (504A Le Rhône), 5 min (504K Le Rhône), to 6,500 ft—25 min (504A Le Rhône), to 10,000 ft—16 min (504K Le Rhône).

Armament: Appropriate variants had one 0·303-in Lewis machine-gun. Several small bombs as necessary.

Manufacturers: Avro, Blériot & Spad, Eastbourne Aviation, Humber, Parnall, S. E. Saunders, Sunbeam, Harland & Wolff, Sage, Brush, Henderson Scottish, Hewlett & Blondeau, Grahame-White.

A Parnall-built Avro 504K, E3273, after conversion to a single-seat Home Defence fighter. The front cockpit was faired over and a Lewis gun on a Foster mounting fitted on the centre section. This aircraft retained the standard undercarriage, but many of the H.D. Avros had the V-strut undercarriage. E3273 was used by No.77 (H.D.) Squadron at Penston, East Lothian.

Service use: *Western Front*—Sqns 1, 3 and 5 (Avros 504 and 504A). *Training duties*: Avro 504D used by No.12 Sqn, Reserve Sqns 5, 9 and 11, and by No.24 Training Sqn. Avro 504, 504A, 504J and 504K—Reserve (Aeroplane) Sqns, later Training Sqns, 1, 2, 3, 4, 5, 6, 7, 8, 10, 11, 12, 13, 14, 15, 16, 17, 18, 19, 20, 22, 23, 24, 28, 30, 31, 34, 36, 37, 38, 39, 40, 42, 43, 44, 45, 50, 52, 53, 54, 55, 56, 57, 58, 60, 62, 63, 64, 65, 67, 72, 73, 74, 89, 194, 195, 196 and 198; Night Training Sqns 186, 198 and 200. *Squadrons working up*: Nos. 9, 15, 23, 28, 29, 43, 46, 49, 51, 52, 54, 58, 62, 65, 71 (A.F.C.), 72, 73, 74, 80, 81, 83, 84, 85, 86, 87, 88 and 94.

Training Depot Stations: Nos.1, 2, 3, 4 and 7. Schools of Aerial Fighting at Ayr (No.1) and Heliopolis. No.1 School of Aerial Navigation and Bomb Dropping, Stonehenge; CFS, Upavon; No.1 School of Special Flying, Gosport; No. 2 School of Special Flying, Redcar; Wireless Section, Brooklands; No.2 Wireless School, Penshurst.

Serials: 376, 390, 397, 398, 637, 638, 652, 665, 683, 692, 715, 716, 750—799, 2857—2860 (ex 1005—1008), 2890—2939, 4020—4069, 4221—4225 and 4255 (ex 1020—1025), 4737—4786, 7446—7455, 7716—7740, 7943—7992.
 A412—461, 462—511, 512—561, 1970—2019, 2633—2682, 3355—3404, 5900—5949, 8501—8600, 9763—9812, 9975—9977.
 B382—385 (ex N6018—6021), 389—392 (ex N6022—6025), 395—396 (ex N6026—6027), 901—1000, 1390—1394 (ex N6028, N6029 and N6650—6652), 1397—1398 (ex N6653—6654), 3101—3250, 3251—3300, 4201—4400.
 C551—750, 4301—4500, 5751—6050.
 D1—200, 1601—1650, 1976—2125, 4361—4560, 5451—5550, 5851—5950, 6201—6250, 6251—6400, 7051—7200, 7501—7800, 8251—8300, 8781—9080, 9281—9380.
 E301—600, 1601—1900, 2901—3050, 3051—3150, 3254—3403, 3404—3903, 4104—4303, 4324—4373, 6737—6786, 9207—9506.
 F2233—2332, 2533—2632, 8696—8845, 8846—8945, 9572 (ex A8591), 9696—9745, 9746—9845.
 H201—350, 1896—2145, 2146—2645, 2946—3195, 5140—5239, 5240—5289, 6543—6842, 7413—7562, 9513—9812, 9813—9912.
 J731—1230, 2142—2241, 3992—4091, 5092—5191, 5492—5591.
Rebuilds: A8966, B8792, B8796, B8797, B8814, B8818, B8819, B9452, B9901, B9932, B9939, B9958, B9959, B9995, C3505, C4290, C4299, C4549, C8652, C9986, C9998, C9999, E9984, E9989, E9991—9995, E9999, F2188, F2209, F2231, F2232, F4171—4174, F4217, F5799, F5800, F9546, F9547, F9625, F9633, F9636, H8254, H8255, H8257, H8289, H8290.

Avro 521

The Avro 521 was possibly the most uninspired and unsatisfactory aeroplane to leave the Avro factory during the 1914–18 war. It seemed to have been built in the hope that someone would think of something for it to do, for it was intended from the outset to test it with three different sets of wings: as the Avro 521 with single-bay wings of 30 ft span, the Avro 521A with 42 ft wings, and the Avro 521B with standard Avro 504 wings spanning 36 ft.

Design work was done during the autumn of 1915 and the Avro 521 was completed by January 1916; its 30 ft span, single-bay mainplanes were referred to as 'Scout wings' in documents of the time. Its derivation from the Avro 504 was obvious, and the airframe embodied some 504 parts. The cockpits of the Avro 521 were widely separated and the centre-section struts were acutely angled, presumably to facilitate ingress to the front cockpit.

The prototype had the number 1811 painted on its fuselage at one stage, but this cannot have been an officially allocated serial number, for the true and original No.1811 had been a Blériot monoplane delivered to No.3 Squadron as early as 10 September, 1914, and struck off charge on 10 June, 1915. The first Avro 521 was flown at Alexandra Park, Manchester, by F. P. Raynham, apparently in January 1916, probably about the middle of the month. A batch of 25 aircraft, to be numbered 7520—7544, were ordered before the prototype flew: the serial numbers were allocated on 30 December, 1915, under Contract No.87/A/234.

The Avro 521 prototype at Farnborough, January 1916.
(*RAE; Crown copyright*)

By 20 January, 1916, the prototype was at Farnborough, and a week later it was recorded there as 'Avro Scout 1811'. How long it stayed at Farnborough is not known, but it had gone to CFS some time before 3 March, 1916, for it had been tested there by that date. The pilot was Flt Cdr Dalrymple Clark RN and the only known CFS report on the aircraft is worth putting on record:

'In this machine the passenger is seated behind. Owing to the machine being delivered here out of balance — and no further adjustment of the tail being possible — it has only been possible to do a speed test with a passenger. A full service load behind would have made the machine almost unmanageable. The machine, as she stands, is quite unsuitable for service requirements. There is no provision for gun mountings, and even if provided, it would be almost impossible to use them owing to the dual control gear behind interfering with the observer's movements. The pilot's rudder bar is so close to a cross member of the fuselage that the pilot's feet continually jam between this and the rudder bar.

The top of the fuselage is fitted with a tin streamline fitting to the pilot's and passenger's heads [*sic*], making the view ahead very bad. The instruments are badly placed and almost in darkness. The machine, being out of balance, "hunts" at full speed. It is recommended that this machine be returned to the makers and be correctly balanced before further trials are proceeded with.'

It is not known how far production of the twenty-five Avro 521s went, nor what their configuration was. Avro drawing No. A 69, dated primarily 23 March, 1916, secondarily 4 July, 1916, depicts a modified single-bay Avro 521 with a rotating circular gun mounting on the rear cockpit, reduced gap, fixed fin and plain rudder, but without the head fairings behind and in front of the cockpits. A further drawing, A 76, dated 1 June and 27 June, 1916, delineates the Avro 521B, which was to have the fuselage, undercarriage and tail unit as in drawing A 69 but fitted with standard Avro 504 two-bay wings and interplane struts.

An Avro 521 was again awaited by CFS in the latter half of May 1916, at which time it was reported 'Machine climbing unsatisfactorily in preliminary trials at

Manchester'. The use of the adjective 'preliminary' at that date is puzzling, unless a production aircraft was meant. The expected aircraft was at CFS that summer of 1916 and one of the few who flew it was Lt H. H. Balfour (later Lord Balfour of Inchrye, PC, MC). He found that 'it had an unpleasant habit of trying to plunge its nose earthwards directly you tried to make a right-hand turn', and specifically warned Lt W. H. Stuart Garnett, one of the scientist pilots of the CFS Testing Flight, not to try to fly the Avro 521. Garnett chose to disregard this advice on 21 September, 1916, essayed a right-hand turn at 1,500 ft, and failed to recover from the spin into which his aircraft promptly fell. This tragedy evidently marked the end of the road for the Avro 521, and no production aircraft entered service.

Avro 521A with three-bay wings of 42 ft span. (*RAF Museum*)

It is known that at least one aircraft was completed in the Avro 521A configuration, its 42-ft wings having three-bay interplane bracing. Presumably it was flown, but it can only have been assessed as useless for any military application.

It seems that production of fuselages at least may have gone ahead to some extent, for what appeared to be an engineless Avro 521 fuselage without head fairings about the cockpits was to be seen in a shed belonging to the Murray Dutton Aircraft Flying School at Queensferry near Chester. Moreover, the great similarity of the fuselage and centre-section bracing of the Avro 504E to those of the Avro 521 may have been more than coincidence, and it is possible that the ten Avro 504Es delivered to the RNAS may have had fuselages originally designed for the Avro 521s.

Avro 521
110 hp Clerget 9Z
Probably intended to be a fighter-reconnaissance two-seater.
Span 30 ft; length 28 ft 2 in; height 9 ft 10 in; wing area 266 sq ft.
Empty weight 1,150 lb; loaded weight 1,995 lb.
Maximum speed 94·6 mph at sea level; climb to 6,000 ft, 14 min; endurance 4½ hr.
Armament: It was intended to fit one 0·303-in Lewis machine-gun on the rear cockpit.

Manufacturer: Avro.

Service use: Flown experimentally at Farnborough and at CFS, Upavon.

Serials: 1811 (evidently applied in error and subsequently removed), 7520—7544.

Beatty-Wright biplanes

George W. Beatty was an American pilot who made something of a name for himself in the USA before coming to Hendon in the summer of 1913. He marked his début on Thursday 10 July by flying his Wright biplane with great panache, making banked turns that approached the vertical. He had apparently built the Wright himself; it was a two-seater and was powered by a 50 hp Gyro rotary engine, an American power unit. Beatty had come to Britain primarily to demonstrate the Gyro engine. It evidently performed well by the standards of the time, but the aircraft in which it was fitted was reported by *The Aeroplane* to be 'of terrifyingly primitive construction'.

In the latter half of 1913 Handley Page Ltd tried hard to convince the aviation world that they had a flying school at Hendon. By the early spring of 1914 their advertisements proclaimed: 'Pupils at the Handley Page Flying School are trained by Mr George W. Beatty, at first on slow speed dual control biplanes. They complete their tuition and take their Certificates upon Handley Page Monoplanes'. (The plurality, at least of the H. P. Monoplane, was a kind of poetic licence, for only the Type E was in fact available). This Handley Page advertisement appeared regularly, even after *Flight* of 4 April, 1914, carried an advertisement for the Beatty School of Flying, in which its proprietor offered 'Tuition on Handley Page Monoplane and Wright Biplanes'. Not until 17 July, 1914, did the Handley Page advertisement resolve this confusion, but then, in an announcement side by side with the Beatty School's advertisement, Handley Page Ltd stated boldly 'Tuition at the Beatty School of Flying, Hendon'. The historical fact of the matter was that George Beatty had opened his school on 16 February, 1914, using three differently-powered Wright biplanes (50 hp and 60 hp Gyro and 40 hp Wright engines) and the Handley Page Type E. He was reported to have simultaneous charge of the Handley Page Flying School.

Beatty remained faithful to the Wright biplane long after everyone else had sensibly discarded it, and he built a series of Wright-type biplanes to keep his

An early Beatty-Wright biplane that apparently retained the original (and highly unnatural) Wright form of controls.

Beatty-Wright biplane with 35 hp Wright engine and dual wheel controls on Deperdussin-type rocking bridge. (*RAF Museum*)

Edouard Baumann at the controls of a Wright-powered Beatty-Wright biplane. This photograph clearly illustrates the peculiar 'blinker' surfaces at the top of the intermediate forward interplane struts, together with the forward vertical surfaces formed on the landing skids. (*RAF Museum*)

Beatty pusher biplane of 1916, basically the wings, tail-booms and tail unit of the Beatty-Wright combined with a rudimentary nacelle carrying a conventional pusher-mounted Gnome engine with a single direct-driven propeller.

flying school supplied with aircraft. The earliest of these resembled the Wright Type B, having no forward elevator but a flimsy horizontal control surface tremulously mounted immediately behind the biplane rudder. Beatty was obliged to recognise the unnatural nature of the original Wright pilot's controls and substituted a somewhat ponderous pivoted frame of steel tubing on which were mounted two substantial control wheels that operated the wing warping. For unknown reasons he introduced small rectangular fin surfaces fitted at the top of the forward interplane struts immediately outboard of the propeller supports on each side.

Nacelle design varied considerably on the Beatty pushers and it is now impossible to determine the order of their appearance. This was one of the simplest forms. (*R. C. Bowyer*)

On the earliest and most primitive Beatty-Wrights, instructor and pupil sat side by side in total exposure on the leading edge of the lower wing. In this respect, these aircraft were decidedly more advanced than most contemporary types used as trainers, in which the pupil usually had to stand close behind the instructor and place his hands over those of the instructor on the controls in order, so it was hoped, to gain some feel of the flying controls.

Inevitably, the war brought many more pupils to all the schools at Hendon, necessitating increases in Beatty's fleet of aircraft and staff of instructors. These included, at various times, W. Roche-Kelly, C. B. Prodger, P. A. Johnston, G. Virgilio, R. W. Kenworthy, A. E. Mitchell, L. L. King, H. Fawcett and H. Sykes. Several dual-control Beatty-Wrights were in use, one of them having, in 1916, a 50 hp four-cylinder engine designed by Beatty himself.

For more advanced pupils Beatty created in 1916 a single-seat biplane of singular appearance. This had rather short wings of Wright form but extensively cut away to allow the single pusher propeller to revolve; the Wright-pattern biplane rudder and single elevator were carried on convergent tail-booms, and the 'blinker' surfaces on the interplane struts favoured by Beatty were fitted. At least four different forms of nacelle existed, but all variants appeared to have in common a 50 hp Gnome as power unit, driving a single pusher propeller direct. This was a great improvement on the terrifyingly precarious chain-driven propellers of the two-seaters. The following passage appeared in *Flight* of 13 April, 1916:

> 'A short time ago the Beatty-Wright biplane fitted with an enclosed nacelle made her debut under none too favourable weather conditions, and so well did she behave that during the next few days her services were much sought after by "ticketers', who one and all agreed that she was very easy to handle and that the enclosed nacelle afforded much greater comfort than the exposed position of the usual Wrights. In addition to her popularity at the Beatty School the new machine is quite an attraction at Hendon, where her appearance is a never failing source of merriment. More than one of those who have seen her in the air . . . have described her to me as the funniest thing that ever happened . . .'.

This Beatty pusher had in its nacelle a more elaborate structure of transverse formers.

Funny or not, the nacelle single-seater seemed to work so well that Beatty built several of them.

Although no Beatty-Wright of any kind was ever officially taken on strength by any RFC unit, it seems that the aircraft of the Beatty School of Flying, together with at least some of those of the Grahame-White School and various Caudrons of other civil flying schools, were regarded as being attached to the RFC 18th Wing School of Instruction that was formed at Hendon on 22 September, 1916. The relationship of these assorted aircraft to the RFC was sufficient to merit their inclusion in official casualty reports, which indicate that some of the Beatty-Wrights bore some purely local identifying numbers. Recorded crashes involving the Beatty biplanes were:

11 September, 1917. Beatty Pusher No.2 (50 hp Gnome No.286).
    This aircraft dived into the ground from 150 ft; the pilot, 2nd Lt H. E. Busby, was killed.

Second Lt H. J. W. Roberts with a Beatty pusher biplane at Hendon, presumably on or about 27 July, 1917, when he was granted his Royal Aero Club pilot's certificate, No.4999.

14 October, 1917. Beatty Pusher No.5 (50 hp Gnome No.250).
4 November, 1917. Beatty Pusher (50 hp Gnome No.307).
  This aircraft dived, turned on its back, and the pilot, 2nd Lt D. C. Bispham, was thrown out and killed. The aircraft crashed at Cricklewood.
18 November, 1917. Beatty Pusher (50 hp Gnome No.286).
  Crashed at Cricklewood.
23 December, 1917. Beatty Pusher No.203 (50 hp Gnome No.286).

Most elaborate of the known forms of Beatty nacelle, this aircraft had a radial arrangement of longitudinal formers in the nose. (*R. C. Bowyer*)

These reports demonstrate that the quaint little Beatty pushers remained in use at least until the end of 1917. They also suggest that the Gnome engine No.286 was remarkably durable, for it was involved in three of these five recorded crashes.

Beatty-Wright biplanes
50 hp Gyro, 50 hp Gnome, 50 hp Beatty (two-seater)
50 hp Gnome (single-seat pusher)
Beatty-Wright—two-seat (side-by-side) primary trainer; Beatty Pusher—single-seat trainer.

Manufacturer: Beatty School of Flying.

Service use: Training duties, 18th Wing School of Instruction, Hendon.

As noted in the text, No.219 was the single-seat Blériot XI that was presented to the War Office by the International Correspondence Schools on 28 January, 1913. It was the first 50 hp Blériot XI to see service with the RFC Military Wing.

Blériot XI

Louis Blériot made many varied attempts to build and fly a practical aircraft before he won immortality by coaxing his little Type XI monoplane across the English Channel on 25 July, 1909. The monoplane provided the design basis of later, larger, and more powerful developments, all of which were known as Blériots Type XI, although they had little in common with the cross-Channel aircraft. The stronger developments, especially the Blériot XI-2 two-seater, found favour with a number of the infant air services of the world.

Officers of the RFC made the acquaintance of the Blériot XI type at an early date, apparently having two monoplanes on Salisbury Plain in 1912. In the last days of the Air Battalion during late March, *Flight* recorded in its issue of 30 March: 'Captain Fulton's old Blériot with 25 hp Anzani engine is also being re-erected and tuned up.' On 8 June the same journal reported: 'The two-seater Blériot has arrived back from the Royal Aircraft Factory, where the planes have been re-covered. Captain Loraine will be the pilot.' That must have been a reference to the Blériot XXI (see pages 135–7), but Loraine was reported to have

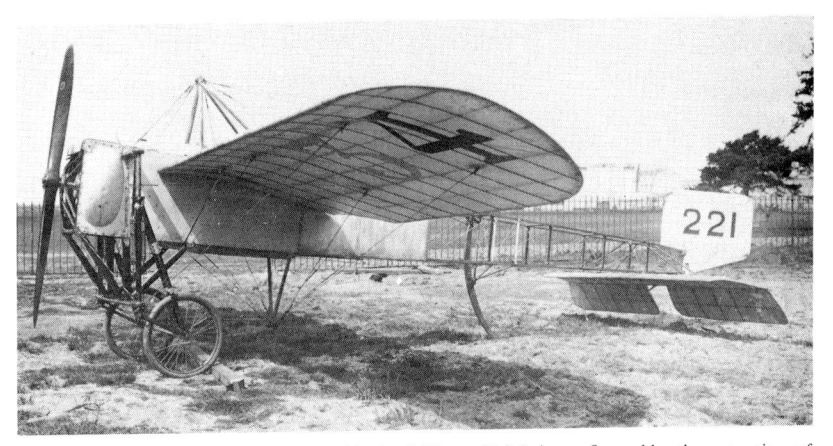

That 221 was the Blériot XI-2 entered in the Military Trials is confirmed by the retention of its Trials markings under the wings. It is here seen at Farnborough in June 1913.
(*RAE 32; Crown copyright*)

flown the Anzani-Blériot on 14 June. It appears that this Blériot was never officially assimilated into the RFC and never had a serial number.

The variants used by the RFC were, in general, those powered by the 50 hp, 70 hp and 80 hp Gnome rotary engines; the 50 hp monoplanes were single-seaters, the others two-seaters. The first 50 hp Blériot XI to join the RFC was No.219 which, in company with the XI-2 No.221, was at Farnborough and eagerly awaited by the RFC on 12 April, 1913; they were identified by their serial numbers on 19 April. Official correspondence of the time identifies 219 as 'the I.C.S. 50 Gnome Blériot', confirming that it was the aircraft presented to the War Office by the International Correspondence Schools on 28 January, 1913. It was the aircraft on which Robert Slack had, in 1912, toured Britain under the sponsorship of the I.C.S., whose students had, by subscription, purchased the Blériot. In the course of his tour Slack covered 1,700 miles without mishap.

No.293 was a 50 hp single-seat Blériot XI that was used briefly by No.3 Squadron in the autumn of 1913. Allotted to the Military Wing on 13 September, it was tested to destruction at the Royal Aircraft Factory on 6 November, 1913.

B. C. Hucks about to take off in his looping Blériot XI, which had roundels painted on the upper surfaces of its wings to provide visible evidence of inversion. This was the aircraft that became 630 on impressment into the RFC after the outbreak of war.

Although this was a testimonial to the aircraft, it signified that the Blériot was far from new when it was acquired by the RFC.

No.221 was a two-seater with the angular form of rudder used on some Blériot XI-2s at that time; it was the aircraft that had participated in the Military Trials of 1912 as No.4. It was taken on charge in September 1912 and may have been the un-numbered 70 hp Blériot that was on the strength of No.3 Squadron at Netheravon late in 1912 and early in 1913. Certainly both Nos.219 and 221 were flying with No.3 Squadron in the summer of 1913, being the only Blériot XIs in the RFC at that time. The two-seater was used until 31 October, 1913, when engine failure followed by rudder jamming led to a crash while being flown by Capt A. G. Fox. The pilot was unhurt but the Blériot was eventually struck off charge on 20 December, 1913.

The single-seater, No.219, was longer lived, surviving at least until August 1914, when it was transferred to CFS without an engine. It was finally destroyed there in December 1914.

These two aircraft must have created quite a favourable impression, for further Blériots of both 50 hp and 80 hp were ordered for the RFC from late 1913. At least nine 80 hp and four 50 hp Blériots were acquired by the RFC in the prewar months. These later aircraft had the more typical Blériot rudder; most had the

No.292 of the RFC was an 80 hp two-seat Blériot that originally had the Blériot works number 791. It was with the RFC Military Wing in the autumn of 1913, and on 31 July, 1914, was with 'A' Flight of No.3 Squadron. It went to France with the Squadron in August 1914 but was wrecked on 10 September. (*RAF Museum*)

two seats arranged in close tandem, but it is known that 292 at least had the earlier more widely spaced disposition of the seats.

On the outbreak of war both the RFC and the RNAS swiftly rounded up virtually all and any aircraft that offered even a remote prospect of being in some way useful in service. The RFC's haul included one 80 hp Blériot purchased from Mme Salmet for £700 and numbered 626, together with three Blériots that belonged to that great British demonstration pilot B. C. Hucks. One of this trio, impressed at Bournemouth, was an 80 hp aircraft and became No.619. It was allocated to none other than 2nd Lt B. C. Hucks, RFC (Reserve Squadron). The other two were 50 hp monoplanes and became Nos.621 and 630 in the RFC, the latter being recorded as Hucks's looping aircraft. It seems that it did nothing of note. Neither, for that matter, did No.621, which was allotted to Capt Fox in August 1914 but was struck off charge on that 25th of that month. By 17 September it had been dismantled for spares at Jersey Brow.

Some of the Blériots and personnel of No.3 Squadron near Amiens in the early weeks of the war. The nearest aircraft is the Blériot XI-2 No.810, not yet renumbered as 1810; it was delivered to No.3 Squadron on 9 September, 1914, finally left the unit on 16 October, and was struck off charge on 25 October, 1914. Behind 810 is the Blériot Parasol 616. (*RAF Museum*)

Other assorted Blériot singletons were acquired in the first few weeks of war, as the allotted serial numbers indicate; and 608, 619 and 626 were taken overseas with the RFC when it went to join the BEF in France. The only Blériot XIs to go with an operational unit were the five (292, 296, 388, 389 and 473) taken to France by No.3 Squadron together with the Blériot Parasol 616. All were 80 hp two-seaters.

With the inevitable attrition of war at close hand, more Blériots were ordered for the RFC. Three 80 hp and two 50 hp monoplanes, respectively 570—572 and 573—574, were ordered from the English-based firm of Blériot Aeronautics, while the RFC acquired others direct from the parent Blériot works in France. The first of these were originally numbered 808—812, 815—816 and 819—820 but were renumbered 1808—1812, 1815—1816 and 1819—1820 when it was realised that 801—1600 had been allotted for RNAS aircraft. All went to No.3 Squadron.

It was on a Blériot XI-2 of this famous unit that the RFC's first reconnaissance of the war was flown on 19 August, 1914, by Capt P. B. Joubert de la Ferté. He carried no observer, but was accompanied by a B.E.2a of No.4 Squadron flown by Lt G. W. Mapplebeck.

Active service conditions proved to be too tough for the lightly-built Blériots, open-air picketing leading to loss of fabric tautness and deterioration of the

No.7652 was a late-production Blériot XI-2 that was used for training purposes. It was first reported with the AID at Farnborough on 14 May, 1916. (*The late Peter W. Moss*)

aircraft's basically marginal performance. The Type XI carried on in No.3 Squadron, and in smaller numbers with Squadrons No.6, 9 and 16, in the spring of 1915. Last to go was No.1811 of No.3 Squadron, which was struck off charge on 10 June, 1915.

Thereafter the RFC's Blériots saw a good deal of use as trainers, and the later batches ordered from Blériot Aeronautics were intended solely for that purpose. The Blériot XI-2 was not ideally suited to training functions, but at the time the most desirable qualities of a sound trainer had not been identified and instructors and pupils had to do the best they could with the available equipment. Deliveries of the last batch (7645—7664) did not begin until mid-May 1916, and these Blériots remained in use for some time. The type was finally declared obsolete in the autumn of 1918, by which time few can have existed outside ground schools.

Blériot XI

50 hp or 80 hp Gnome on production aircraft. A few individual Blériots of slightly differing versions had the 28 hp or 45 hp Anzani, or 80 hp Le Rhône

Type XI—single-seat military monoplane; Type XI-2—two-seat reconnaissance.

Single-seater: Span 8·9 m; length 7·8 m; height 2·5 m; wing area 15 sq m.

Two-seater: Span 10·35 m; length 8·4m; height 2·5 m; wing area 19 sq m.

Empty weight 265 kg (single-seater), 335 kg (two-seater); loaded weight 415 kg (single-seater), 585 kg (two-seater).

Maximum speed 100 km/h (single-seater), 120 km/h (two-seater); climb to 500 m—5 min (single-seater), 6 min (two-seater); endurance 3 hr (single-seater), 3½ hr (two-seater).

Armament: Makeshift or purely portable disposition of pistols and/or rifle or carbine.

Manufacturers: Blériot Aéronautique (France), Blériot Aeronautics (United Kingdom).

Service use: Prewar—No.3 Sqn. *Western Front*: Sqns 3, 6, 9 and 16.
Training duties: Reserve Aeroplane Sqns 1, 2, 3 and 4. *Squadrons working up*: No.13.

Serials: It is now impossible to be certain that all of the following Blériots were of Type XI or XI-2; a few may have been Parasols (*see* page 141). *With 50 hp Gnome*: 219, 293, 297, 298, 323, 573—574, 608, 621, 630, 673, 2852—2854, 4650—4661. *With 70/80 hp Gnome*: 221 (ex 4 of Military Trials), 260, 271, 292, 296, 374, 375, 388, 389, 473, 570—572, 618, 619, 626, 647, 662, 681, 706, 1808—1812, 1815—1816, 1819—1820, 1825, 1828, 1831—1834, 1836—1838, 1842, 1847, 1850, 2338, 2869, 5604, 5605, 7645—7664. *With unknown engine*: 711, 714, 721, 723, 726, 732, 4296. *Others*: 671 (28 hp Anzani), 672 (45 hp Anzani), 694 (80 hp Le Rhône).

Blériot XII

A certain amount of mystery surrounds the Blériot XII monoplane that saw brief service with the Royal Engineers late in 1910. The Blériot XII was of unusual design, having its mainplane mounted on top of the deep fuselage frame, in which the pilot and passenger sat immediately behind the engine; the airscrew shaft was roughly level with the mainplane and was driven by a chain. It was an ill-proportioned and uninspired design, and its flying qualities proved to be unsatisfactory.

The first Blériot XII was built in May 1909 and first flew on 21 May. It participated in the Rheims meeting held that August, by which time its tail unit had been completely revised and simplified; its engine was a 50 hp E.N.V. One who was there and admired the Blériot XII was Claude Grahame-White; seeking out Louis Blériot, he arranged to buy the aircraft on completion of the Rheims meeting. Unfortunately the Type XII crashed while being flown by Blériot and was destroyed by fire, but Blériot promised Grahame-White the next Type XII to be built. During its construction Grahame-White worked on it in the Blériot works and eventually took delivery of the completed monoplane at Issy-les-Moulineaux on 7 November, 1909. Its power unit was a 60 hp E.N.V. engine, and he named the aircraft *White Eagle*.

Although Grahame-White had had no flying training of any kind he succeeded in making a number of successful flights at Issy, but the area available there was restricted. He therefore went south to Pau and the aircraft arrived at the Blériot aerodrome there on 24 November. Despite a good flight on 28 November the monoplane was damaged in a forced landing while being flown by Blériot with

Claude Grahame-White helps pull along his Blériot XII *White Eagle*, which was later flown by pilots of the Royal Engineers. This photograph was taken at Pau on 28 November, 1909.

Grahame-White as passenger. It appears that this accident convinced Blériot that the Type XII was unstable, and further that he offered to replace it by two of the more conventional Type XI monoplanes.

Precisely what happened to *White Eagle* thereafter is not known. If, as reported, Blériot did in fact recognise that the Type XII was dangerously unstable, and if he took it away from Pau, there appears to be no good reason why he should repair and preserve it. Equally, there is likewise no reason why he should build a further example of a type he is said to have regarded as unsatisfactory. Nevertheless, a second-hand Blériot XII with a 60 hp E.N.V. engine was available for purchase in June 1910, when it was bought and presented to the British War Office by the Duke of Westminster and Col Laycock, a director of the E.N.V. Engine Company.

This Blériot had to be collected from France, and Lt R. A. Cammell, RE, was instructed to go to the Blériot aerodrome at Etampes, familiarise himself with the construction of the monoplane, learn to fly it, and bring it back to England by rail. With all this in prospect Cammell went to Étampes on 28 June, 1910, and set about learning to fly on Blériot XI monoplanes. The Blériot XII was flown once or twice, but its engine was badly out of tune and the flights were not regarded as successful. The aircraft sustained damage in a minor mishap on 1 July and repairs had to be put in hand. This, coupled with a lack of flying instructors at Étampes, left Cammell with a few days of enforced idleness, but he put these to good use by going to the semaine d'aviation at Rheims and compiling reports on the aircraft he saw there.

The Blériot XII was officially accepted on 3 August, 1910, and was subsequently taken to Amesbury by rail, thence to Larkhill. It did not start to fly in England until 9 October, and its earliest flights proved disappointing. Remarkably, these activities and the fact of War Office ownership escaped mention in the aviation press for a surprisingly long time. The Blériot's designated pilot was Lt Cammell, and on Sunday 27 November, 1910, he brought out the aircraft and flew eight circuits each about 4 miles in circumference before landing with a broken wire. With this repaired, he set off for Aldershot at a height of 1,000 ft, but strong winds forced him to land at North Waltham.

A somewhat garbled report of this flight appeared in *The Aero* (then edited by C. G. Grey) in its issue of 4 January, 1911. What is significant is that this report unequivocally identified the aircraft as 'the old Blériot type XII *White Eagle*'. It went on to regret 'that so capable and versatile an aviator as Captain Felton [*sic*] should risk his valuable life on a machine which is so dangerous, owing to its low centre of gravity, that the type has been condemned and given up entirely by its designer.' This report drew a related letter from Capt J. D. B. Fulton, published on 18 January, 1911, in which he reiterated the name 'Blériot Type XII *White Eagle*' without comment or amendment. Later, when C.G. Grey was editing the newly-launched weekly *The Aeroplane*, he repeated in the issue of 27 July, 1911, that the Army's Blériot XII had earlier been Grahame-White's *White Eagle*.

On 29 December, 1910, Mervyn O'Gorman, the Superintendent of the Balloon Factory, wrote to the War Office that he had 'the Blériot here to repair,' and seemed to have considered offering the Type XII to Blériot in part exchange for a Type XI. Repaired the aircraft never was, though, as noted on page 614, about February/March 1911 O'Gorman listed the Blériot XII among the aeroplanes then at Farnborough, and the Army Estimates published on 25 February, 1911, included it in the five aeroplanes then 'available for Army work'. There is no reason to suppose that the Blériot did not then exist, but the date of its final

134

demise is not known. Apparently it had acquired the nickname of *The Man-Killer* in the Royal Engineers, and it seems virtually certain that its engine was used to power the S.E.1 that O'Gorman recorded as being nearly completed on 26 April, 1911.

No identity has ever been attributed to the War Office's Blériot XII, but it is possible that the number B1 may have been allotted to it, perhaps only theoretically or retrospectively.

Blériot XII
60 hp E.N.V. Type F

Two-seat (side-by-side) monoplane.
Span 9·5 m; length 8·5 m; wing area 22 sq m.
Loaded weight 620 kg.
Speed 96 km/h.

Manufacturer: Blériot Aéronautique.

Service use: Royal Engineers, Larkhill.

Serial: The Blériot XII may have been regarded as B1, but this is unconfirmed.

Blériot XXI

The Blériot company produced a considerable number of variations on the successful Type XI theme, including several types that had attenuated rear fuselages and sweeping curves to the tailplane leading edges, that surface being made to blend gradually into the lines of the fuselage. One such monoplane was a side-by-side two-seater powered by a 70 hp Gnome rotary engine and designated Blériot XXI. The first example to be seen in Britain was exhibited at the 1911 Olympia Aero Exhibition, which opened on 24 March.

A Blériot of this exotic type was purchased by Lt R. A. Cammell, RE, and figured among the primitive aircraft used by that small but historic band of true pioneers, who operated from their camp at Larkhill on Salisbury Plain. In a note on those remote days at the very birth of military aviation in Britain,

The Air Battalion's Blériot XXI, which successively had the official identities of B2 and 251.

G. B. Cockburn wrote: 'Lieutenant Cammell was flying a Blériot of weird and wonderful type, his own property.' In some surviving official documents this aircraft is described as a Blériot XXII, yet it was to all external appearances indistinguishable from the Blériot XXI, and it seems likely that the type number was wrongly recorded.

This Blériot XXI was flying with the Air Battalion before May of 1911 was out, its third officially recorded flight being logged on 28 May. Experiments in the installation of an early Box compass were conducted, and on 2 June, 1911, the machine was flown with a 100 ft wireless aerial carried on a drum and led back to the tail through rings; it also served as a test vehicle for a Dines air-speed indicator, a Baird altimeter, and Heath and Clift compasses. Experiments in dropping messages on 7 July, 1911, were regarded as successful.

In mid-July Lt Cammell, with Lt Reynolds as his passenger, made a long flight under the terms of the Mortimer Singer prize competition for Army and Navy pilots. The flight lasted 2¾ hr and covered 112 miles. Cammell entered and flew the Blériot in the 1911 Circuit of Britain contest; he got as far as Wakefield, but his engine burst a cylinder and he was compelled to withdraw. The aircraft was damaged at Hendon on 31 August, when variable winds swept Cammell and his Blériot against the door of an aeroplane shed. Repairs were needed, but Cammell never flew the aircraft again for he was killed at Hendon on 17 September, 1911, while flying a Valkyrie monoplane.

The Blériot XXI re-emerged on 6 December, 1911, making a fleetingly brief flight with Lt D. G. Conner at the controls; unfortunately the monoplane ran into a ditch and broke its undercarriage, and it did not fly again until 11 January, 1912. It flew little and by 18 February was reported to be 'much out of adjustment', to rectify which it was sent to the Army Aircraft Factory at Farnborough for overhaul.

At some time the Blériot XXI was quietly taken over officially by the Air Battalion, and when the Battalion's aeroplanes were given identifying numbers it became B2. It returned from Farnborough about 3 June and was flown once or twice but evidently remained unsatisfactory, for it was tested on 27 July by Perreyon, who advised that it should be used for local flying only and that a factory overhaul would be necessary if cross-country flying were to be required.

The Blériot XXI at Farnborough. (*RAE: Crown copyright*)

Lt R. A. Cammell, R.E., with a passenger in his Blériot XXI, which was taken over by the Air Battalion after his death on 17 September; 1911. (*RAE 0156; Crown copyright*)

It did not go to the Blériot works, but was on the strength of No.3 Squadron, RFC, with the new identity 251 late in October 1912, by which date the War Office ban on monoplanes was in force. At that time its engine was out for overhaul, and by November the aircraft itself was reported to be in need of overhaul to make it suitable for training duties. By the beginning of February 1913 a decision to reconstruct had been taken, but with the intention of 'rebuilding' the Blériot as a rotary-powered B.E. of the B.E.3/4 configuration. Indeed, in its weekly return of 12 April, 1913, No.3 Squadron listed No.251 as 'B.E. 70 Gnome: to be converted into B.E. type'. By 30 April the Blériot XXI had gone to Farnborough for its reincarnation. The reconstruction was abandoned in December 1913 and No.251 did not re-emerge from the Royal Aircraft Factory.

Blériot XXI
70 hp Gnome

Two-seat (side-by-side) monoplane.
Span 11 m; length 8·24 m; wing area 25 sq m.
Empty weight 330 kg; loaded weight 630 kg.
Speed 95 km/h.

Manufacturer: Blériot Aéronautique.

Service use: No.2 (Aeroplane) Company, Air Battalion, Royal Engineers; No.3 Sqn, RFC.

Serial: Originally B2, renumbered 251.

The Blériot Parasol that was exhibited in the Olympia Aero Show in March 1914. It may have been the aircraft that was impressed on the outbreak of war to become No.616 in the RFC. (*Flight International*)

Blériot Parasol monoplane

As recorded on page 289, the Morane-Saulnier G-19 parasol monoplane, later designated Type L, was designed in August 1913, was flying by October, and was exhibited at the Paris Salon in December. It was closely followed, early in 1914, by a Blériot monoplane of similar configuration: the British aeronautical world had an opportunity of comparing the two French parasol monoplanes at the 1914 Olympia Aero Show held in March of that year, for an example of each type was exhibited. The Blériot Parasol was a single-seater, powered by an 80 hp Gnome rotary engine.

The Blériot company described its new creation as providing 'visibilité totale', presumably because the mainplane was placed about level with the pilot's eyes, where it interfered least with his field of view. This specific configuration was the invention of Lieutenant Gouin of the French Aviation militaire, consequently the Blériot Parasol monoplane was sometimes referred to as the Blériot-Gouin. In reality, of course, the totality of the visibility conferred on the pilot by the parasol arrangement was mostly downwards. Nevertheless, this quality held instant appeal to the military aviators of the time, and on 27 March, 1914, the Officer Commanding the Military Wing of the Royal Flying Corps forwarded to the Director-General of Military Aeronautics a request made by Capt A. G. Fox of the RFC for permission to fly the new Blériot. Permission was granted on 2 April, subject to the proviso that no expense to the public would be incurred.

Evidently some delay occurred before Fox actually flew the aircraft, for his report is dated 4 June, 1914. The Parasol may well have been the example displayed at Olympia. Fox flew it for only 15 minutes, not without difficulty, but

Blériot Parasol 577 at Farnborough, a photograph dated 4 February, 1915. On that date 577 went to No. 9 Squadron and returned to the Aircraft Park on 12 March, 1915, only to be struck off charge on 15 March. *(RAE 700; Crown copyright)*

found that it 'seemed to fly quite nicely'. At that time the tailplane and elevator were those fitted to the 50 hp Blériot XI; Fox quite sensibly thought that the larger surfaces of the 80 hp aircraft might provide improved control. He found the view from the cockpit excellent and recommended that an example of the type be purchased for further trial subject to modification to provide more leg room. This recommendation was supported by the O.C. when he forwarded Fox's report to the Director General of Military Aeronautics on 16 June.

No.2861 was one of two Parasols built by Blériot Aeronautics; it was with the AID at Farnborough on 8 April, 1915, and for a time was used by No.4 Reserve Aeroplane Squadron at Northolt. This photograph of 2861 was taken at Gosport, possibly at a later date; the aircraft has an external fairing beside the cockpit that may have housed a camera.

Events evidently overtook consideration of this matter, and among the aeroplanes that were impressed for the RFC on the outbreak of war was one Blériot Parasol. This might well have been the aircraft that Fox had tested. On 7 August, 1914, it was flown by E. L. Gower from Brooklands to Farnborough, where its official number 616 was allotted. The price paid to the Blériot company for it was £1,200, and it was initially issued to Lt E. L. Conran. It was taken to France on 14 August, 1914, by No.3 Squadron, which used it until 16 March, 1915. It then went to the Aircraft Park and was re-issued to No.5 Squadron five days later; on 5 May, No.616 went back to the 1st Aircraft Park and was struck off on 25 May.

While with No.3 Squadron, No.616 was cared for by Air Mechanic J. T. B. McCudden, who was later to win immortality in an outstanding career as a fighting pilot. No.616 retained the small tailplane and elevator of the 50 hp Blériot XI until, in McCudden's own words, '. . . . on this flight Mr Conran's Parasol was badly shot about, necessitating the fitting of a new airscrew and tail-plane. The tail-plane of this Blériot was of the 50 hp Blériot type, but as we did not have one to spare we fitted the 80 type, which is a good deal larger. However, it behaved quite well.' McCudden also recorded that one morning Lt Conran 'left the aerodrome on the Blériot Parasol to bomb Laon railway station, with the following loads of bombs: Sixteen hand grenades, two shrapnel bombs, each one in a rack on the outside of the fuselage, out of which the pilot had to throw them by hand, and a Mélinite bomb tied on to an upper fuselage longeron with string, so that when he wanted to unload this bomb he had to cut the string first with a knife and then push the bomb overboard.'

As far as can now be determined, a further fourteen Blériot Parasols were built for the RFC by Blériot Aeronautics in England; they were numbered 575—586 and 2861—2862. Deliveries of the main batch began in mid-December 1914 and continued until the end of March 1915. No.575 went to the Aircraft Park at St-Omer on 6 January, 1915, but crashed on arrival; it was struck off on that day. Nos.576—579 went to No.9 Squadron on various dates and served until 12 March, when 577 went to the Aircraft Park, the others to No.5 Squadron. In operational terms all were outlived by No.616, but the last to be struck off the strength of the RFC in the field was No.576, which was deleted on 1 June, 1915.

The Blériot Parasol thus preceded the Type XI into retirement by a few days only, never having been numerous enough to make any lasting impression on the war's history. The fact that it was a single-seater was probably against it: it totally lacked the performance necessary to become a successful high-speed scouting aircraft, which is perhaps how its designer conceived it. McCudden has left another vivid impression of No.616 when, fitted with a signalling lamp, it was flown by Lt A. S. Barratt (later Air Chief Marshal Sir Arthur Barratt, KCB, CMG, MC) of 'B' Flight, No. 3 Squadron: 'Mr Barratt [sic] did about forty hours flying on this machine, directing artillery fire, and how he managed to fly the machine and Morse to the battery by means of this lamp single-handed, still remains a mystery to me.'

Most of the other Blériot Parasols of the RFC were used for training after the type's withdrawal from operational duties, surviving at least until the autumn of 1915.

Blériot Parasol
80 hp Gnome

Single-seat general-purpose monoplane.
Span 8·95 m; length 7·8 m; wing area 18 sq m.
Empty weight 280 kg; loaded weight 420 kg.
Speed 110 km/h approximately.
Armament: Makeshift arrangements for carrying assorted bombs.

Manufacturers: Blériot Aéronautique (France), Blériot Aeronautics (United Kingdom).

Service use: *Western Front*—Sqns 3, 5 and 9. *Training duties*: Reserve Aeroplane Sqns 2, 4 and 13; 4th Wing, Netheravon.

Serials: 575—586, 616, 2861—2862.

The Breguet L1 that became B3 in the Air Battalion was at first referred to as a 'Cruiser-type aeroplane'. This photograph, probably taken shortly after its arrival at Farnborough, shows the aircraft in an early (possibly its original) configuration with three-blade airscrew. (*RAE 0376; Crown copyright*)

The Breguet biplanes

By 1910 the great French pioneer constructor-pilot Louis Breguet had evolved a remarkable form of biplane. It was characterised by wings of unusual flexibility and by the extensive use of steel tubing in the airframe. The mainplanes had a single steel-tube spar on which the ribs were mounted with a certain amount of rotational freedom, thus giving the wings more than the normal amount of flexibility. Interplane bracing consisted of a single row of struts along the spars and the entire structure looked perilously inadequate. Breguet insisted, for a

Close-up of the power installation and undercarriage of the Breguet L1 as it first appeared at Farnborough. (*RAE 0162; Crown copyright*)

The Breguet L1 at a later date, with four-blade airscrew and numbered B3.

time, on calling his aircraft 'double monoplanes', but they were, in every essential sense, biplanes.

Breguet biplanes appeared in a remarkable profusion of shape, size and variation, and were fitted with many widely differing power units. The engines alone determined the only more-or-less consistent form of designations applied to these idiosyncratic aircraft. An initial letter indicated the type of power unit installed (D = Dansette-Gillet, G = Gnome, L = Renault, R = R.E.P., Ch = Chenu, U = Salmson, etc), and a numeral related approximately to the power of the engine. Thus G1 signified a 50 hp Gnome, G2 a 70 hp Gnome, G3 an 80 hp Gnome, U2 a 130 hp Salmson (Canton-Unné); and so on.

The purchase of examples of Breguet, Nieuport, Deperdussin and Sommer aircraft for the use of the Air Battalion was officially recommended by Capt J. D. B. Fulton in a memorandum dated 8 September, 1911. This was followed remarkably quickly by a War Office purchase of a Breguet L1, and Lt G. B. Hynes was sent to Douai to learn to fly it. It arrived in England in October 1911 and was flown on speed trials at Farnborough by Geoffrey de Havilland on 14 October, when it attained a speed of 49 mph. This Breguet had

The end of B3. The wreckage of the aircraft after its crash on 7 March, 1913, the rudder still bearing its original number B3 although the Breguet was by then 202. (*RAF Museum*)

two interplane struts on each side, yet its basic rigging was that of a single-bay biplane. It had the typical Breguet cruciform 'all-flying' tail unit mounted precariously on a single universal joint. The tricycle undercarriage had a short wheelbase with steerable nosewheel.

This Breguet biplane acquired the identifying number B3 and did a good deal of flying in the hands of Lt Hynes. It was of this aircraft and of this time that G. B. Cockburn recalled: 'It was a most unwholesome beast, with flexible wings, steel spars, and wheel control.' Sir Walter Raleigh, having thus quoted Cockburn in Volume I of *The War in the Air*, went on to add: 'It required enormous strength to steer it, and was perseveringly and valorously flown by Lieutenant Hynes.' One surviving photograph suggests that on at least one occasion Hynes took up three passengers in the unwholesome beast.

Breguet G3 No.210. (*RAE*)

On 1 April, 1912, the Breguet was dismantled, allegedly to demonstrate how quickly it could be prepared for movement by road, but was soon flying again. It was overhauled in the Royal Aircraft Factory in June, flying again at Farnborough on 20 June, whereafter B3 flew frequently from that historic field until it was returned to the British Breguet works at the end of August. After an extensive overhaul it was returned to Farnborough on 7 March, 1913, being flown there from Hendon, but was totally wrecked on landing. Although still marked B3 it had, during its absence, been allotted the number 202 in the new and sensibly simplified system of serial numbers that was introduced in the autumn of 1912.

The RFC had several other Breguet biplanes of assorted types in the prewar period. Four more were ordered in the financial year that ended on 31 March, 1913, and in August 1912 two Breguet G3s joined the Military Wing of the RFC; numbered 210 and 211, they were initially allotted to No.2 Squadron. Both were powered by the 100 hp two-row Gnome rotary. Early in October these two biplanes were joined by two Breguet L2s, which were numbered 212 and 213. These four Breguets constituted the equipment of 'A' Flight of No.2 Squadron in November 1912; it was reported that No.211 was being tested by Geoffrey de Havilland on 6 November and was handed over to the RFC on 8 November, but

Breguet G3 No.211 was generally similar to 210.
(*Peter Liddle's 1914–18 Personal Experience Archives, presently housed within Sunderland Polytechnic*)

this was probably a consequence of repair or overhaul rather than the initial handing over. The four Breguets 210—213 did not stay with No.2 Squadron for long: on 21 December, 1912, all four were transferred to No.4 Squadron.

A gale that swept Farnborough on 22 March, 1913, caused the collapse of the sheds housing Nos.210 and 213, and both were extensively damaged, 'A' Flight of No.4 Squadron resourcefully undertook the repair of these aircraft, and No.210 at least remained with the squadron until December 1913. In August 1913, when No.211 had been in service for a full year, the RFC asked the Royal Aircraft Factory to inspect the aircraft and report on its condition. It was duly inspected on 3 September and was given an excellent report, which is perhaps not surprising in view of the fact that it had been fitted with new mainplanes and

Breguet No.212 was originally an L2 powered by a 70 hp Renault engine, but in April 1913 it was re-engined with an 85 hp Salmson (Canton-Unné) water-cooled radial. This photograph depicts the aircraft in its second form.

The Breguet L2 No.313 retained its original Renault-powered configuration. (*RAE 1; Crown copyright*)

elevators only on 18 August, and the rudder was a recently fitted replacement.

No.212 was modified in service by having its Renault engine replaced by an 85 hp Salmson (Canton-Unné) water-cooled radial in April 1913. In this form it had a large shallow spinner, but its performance was considered to be too poor and it did little flying. When finally struck off charge its total flying time was only 1 hr 30 min.

A further seven Breguet biplanes were ordered for the RFC Military Wing but all were not delivered. Two that did reach the RFC were allotted the official identities of 310 and 312, the former built by the British branch of the Breguet firm and powered by a 110 hp Salmson, the latter French-built with an 85 hp Salmson and large spinner on the airscrew. The precise identity of No.310 is uncertain, but

No.213 in unhappier circumstances, after the gale of 22 March, 1913, had caused the collapse of its shed. (*RAF Museum*)

it might have been the Breguet seen in the photograph below, which had a 110 hp Salmson engine, a wholly metal-clad fuselage and a four-wheel undercarriage supplemented by a tall rear skid directly under the pilot's cockpit. No.310, identified as such, was flying with No.4 Squadron at Netheravon on 30 June; No.312 also went to No.4 Squadron and was flying with that unit on 20 August, 1913. Both 310 and 312 were still on No.4 Squadron's strength on 19 December, 1913, at which date their respective totals of flying time were 7 hr 35 min and 5 hr 49 min. In that month, however, Nos.211, 212, 310 and 312 were all struck off charge, and 202 and 213 were deleted in January 1914.

This might have been the Breguet biplane that was allotted the number 310. The photograph was taken at Farnborough and bears the date 6 August, 1913. *(RAE 152; Crown copyright)*

Breguet Biplanes
B3 (202)—60 hp Renault, later 85 hp Salmson
210 and 211—100 hp Gnome
212—70 hp Renault, later 85 hp Salmson
213—70 hp Renault
310—110 hp Salmson
312—85 hp Salmson

Two-seat military biplane; G3 could operate as a three-seater.
G3: Span 13·65 m; length 8·75 m; wing area 36 sq m.
G3: Empty weight 550 kg; loaded weight 950 kg.
G3: Maximum speed 110 km/h.

Manufacturers: Breguet, British Breguet.

Service use: No.202 flown by No.2 Sqn; Nos.210—213 by Nos.2 and 4 Sqns; Nos.310 and 312 by No.4 Sqn.

Serials: B3(202), 210—213, 310, 312.

Bristol Boxkite F4 of the Air Battalion at Larkhill.

Bristol Boxkite

In its dimensions, geometry and aerodynamics the Bristol* biplane of July 1910 was an unabashed copy of the contemporary Henry Farman III. It was popularly known as the Bristol Boxkite (later, in the RNAS, it was designated Bristol School Biplane) and substantial numbers of this primitive aircraft were built and flown with, by the standards of the time, considerable success. Various engines were fitted.

The War Office was in no hurry to order aeroplanes for the Air Battalion but eventually, on 14 March, 1911, placed the first British contract for a quantity of aircraft for military purposes: this was for four Boxkites with 50 hp Gnome engines. The first to be delivered had the Bristol works sequence number 37 and was handed over to the Air Battalion at Larkhill on 18 May, 1911. On 22 May, Bristol No.40 was supplied, being flown on that day, without upper-wing extensions, by Collyns Pizey; two days later it was flown with extensions fitted, and was then accepted by Capt J. D. B. Fulton. Its instrumentation consisted of a Hicks aneroid, an Elliott speedometer, a Clift compass, two Adams gradometers and a watch; its total purchase price was £849 9s 0d.

With these deliveries made, the War Office asked for its next two Boxkites to be powered by the 60 hp Renault engine. Inevitably, some delay in delivery was occasioned by the necessary modifications, and Bristol No.39 was not delivered

---

*On 19 February, 1910, Sir George White founded The Bristol Aeroplane Co Ltd, The Bristol Aviation Co Ltd, The British and Colonial Aeroplane Co Ltd and The British and Colonial Aviation Co Ltd. Trading was begun by The British and Colonial Aeroplane Co, the others having only a nominal capital of £100 each.

148

until 9 July. It had the Renault engine and a skeletal nacelle, and was apparently followed by a second Renault-powered Boxkite on 2 August; this second aircraft cannot now be positively identified.*

A second batch of four Bristol Boxkites had been ordered by the War Office, two to be delivered with 50 hp Gnome engines, the other two as engineless airframes. This batch probably included Bristol Nos.41, 48 and 49; one was being tested at Larkhill on 6 August, 1911, and deliveries ran up to mid-August.

The Air Battalion gave numbers to only six of these eight Boxkites; these were F4—F9 inclusive, of which F4 is known to have been Bristol No.37, F5 formerly Bristol No.40, and F6 the Renault-powered No.39. In service the Boxkites saw brief quasi-operational use in the Army manoeuvres of August 1911 but subsequently were used solely as trainers. As early as 8 September, 1911, the Commandant of the Air Battalion wrote to the Director of Fortifications and Works a paper in which he suggested types of aircraft that might be purchased for the Air Battalion in 1912. Of the Bristol Boxkite he wrote:

'It must be recognised that the Bristol biplanes do not fulfil all the conditions required of military aeroplanes. As beginners' machines they are admirable, and none could have served better to allow of the Air Battalion getting experience in flying, maintenance and organisation during the first

Bristol Boxkite, Works Sequence No.39, with 60 hp Renault and rudimentary nacelle, photographed with Henri Jullerot at the controls. (*Bristol Aeroplane Co*)

year of its existence. But they have not the speed which is at present necessary to allow aeroplanes to fly in any but the finest weather, and the Air Battalion pilots now possess skill and experience enough to allow of their being put with advantage into aeroplanes of a more advanced type than the Bristol.'

*This Renault Boxkite was previously believed to be No.42, but it is now known that that Bristol sequence number belonged to the first Bristol Type T flown by Tabuteau in the 1911 Circuit of Europe, which in turn had been wrongly regarded as No.45.

The Air Battalion's Bristol Boxkite F4 photographed at Southall on 3 July, 1911, while being flown from Farnborough to Hendon by Lts B. H. Barrington-Kennett and H. R. P. Reynolds, whose objective was to meet the participants in the Circuit of Europe contest. The officers flew back to Farnborough on 5 July. The aircraft bears the words 'Air Battn' on the rudder and on a special pennant carried on the port forward elevator outrigger. This is probably the earliest recorded instance of a British (or perhaps any) military flying Service having its name inscribed on its aircraft.

And so the RFC's Boxkites served as trainers, mostly at CFS, though F7 was on the strength of No.3 Squadron for a time in 1912. F8 was then at CFS, the two seemingly being the only active Boxkites in the Military Wing. When serial numbers were introduced both must have been at CFS, for F7 became 408 and F8 became 407. F5 had done a fair amount of flying, its total airborne time by May 1912 being 79 hr 12 min. It went to the Royal Aircraft Factory for overhaul during the week ending 10 August, 1912, but apparently did not assume a numerical identity. The careers of 407 and 408 apparently ended in December 1912, when the War Office decreed that they were to fly no more.

With the impressment of virtually everything flyable on the outbreak of war the RFC acquired four later Boxkites. These were the aircraft having the Bristol works sequence numbers 202, 203, 226 and 347; they had been the training aircraft of the Brooklands flying school and were impressed by Capt F. V. Holt in mid-August 1914. The British and Colonial Aeroplane Company received £200

Another photograph of a Boxkite at Larkhill. The officer at the controls appears to be Lt B. H. Barrington-Kennett, the first adjutant of the RFC (*RAF Museum*)

each for 202 and 203, £350 each for 226 and 347. For good measure Capt Holt also bought the Bristol P.B.8 pusher for a mere £100, but it seemed to find no employment in the RFC.

Known RFC identities for three of these Boxkites were 640, 641 and 657; the fourth may have been 718, which was reported as having been erected by No.1 Squadron on 7 November, 1914, and transferred to No.2 Reserve Aeroplane Squadron at Brooklands a week later. Certainly the fourth of the impressed Boxkites was reported to be anonymously serviceable with No.2 RAS on 12 December, 1914, and 718 is known to have survived with that unit until 20 February, 1915, when it crashed; it was written off on 28 February. It outlived the other three, for 640 was struck off on 29 September, 1914, after a brief sojourn with No.1 Squadron; while 641 and 657 also went to No.1 Squadron and lasted until 14 October and 16 October respectively.

The Boxkite remained in use with the RNAS until early 1916 and continued in production for that Service well into 1915. These late-production Boxkites for the RNAS (8442—8453 and 8562—8573) have hitherto been wrongly identified as Bristol T.B.8s.

Bristol Boxkite
50 hp Gnome, 60 hp Renault

Two-seat trainer.

Span 46 ft 6 in (with extensions), 34 ft 6 in (without extensions); length 38 ft 6 in; height 11 ft 10 in; wing area 527 sq ft (with extensions), 457 sq ft (without extensions).

Empty weight 900 lb; loaded weight 1,150 lb. Weights with wing extensions.

Speed 40 mph.

Manufacturer: British and Colonial.

Service use: No.2 (Aeroplane) Company, Air Battalion, Royal Engineers; Sqns 1 and 3, RFC; CFS; No.2 Reserve Aeroplane Sqn.

Serials: F4—9, 407 (ex F8), 408 (ex F7), 640, 641, 657, 718.

Bristol-Prier monoplane

The British and Colonial Aeroplane Company built a number of monoplanes designed by Pierre Prier, who had joined the company's staff in June 1911. These were built in various configurations, the first having the Bristol works sequence number 46.

On 5 September, 1911, Prier demonstrated one of the earlier monoplanes before Capt J. D. B. Fulton of the Air Battalion. This flight did nothing to inspire confidence in the aircraft's stability or controllability, and in concluding his report to the Officer Commanding the Air Battalion, Capt Fulton wrote: 'I consider that the machine in its present state of development is completely unsuitable'. Although Fulton's report was forwarded to the Director of Fortifications and Works it did not deter the War Office from ordering a Prier monoplane for the Air Battalion in January 1912.

This monoplane was delivered to the Air Battalion at Larkhill on 17 February, 1912, at a purchase price of £850 and was given the official identity B6. It made a six-minute flight next day with Prier as pilot, but seems not to have been flown by

The earlier of the RFC's two Bristol-Prier monoplanes came to the Air Battalion as B6 and was renumbered 256 on assimilation into the RFC.

an Air Battalion pilot until 17 March, when Lt H. R. P. Reynolds was airborne for eight minutes. He flew it again for 37 minutes on 26 April, but the engine failed during his landing approach and the aircraft crashed. It was returned to the British and Colonial works for repair, and was back at Larkhill by 20 June. Misfortune, again in the form of engine failure, struck the hapless B6 on that day while on its test flight; the monoplane landed heavily and overturned, sustaining extensive damage. Again it went back to Filton for repair.

The circumstances of both crashes suggested catastrophically inadequate elevator control, and this may be why the aircraft was extensively modified to incorporate the improvements introduced by Capt Bertram Dickson into the basic Prier design. Thus, when B6 went back to Larkhill it had a lengthened fuselage, and a revised horizontal tail was fitted. This consisted of a fixed tailplane and plain elevator in place of the low-aspect-ratio all-flying surface of the original Prier design.

On 17 September, 1912, the RFC took on charge a second Bristol-Prier monoplane, which was numbered 261. Owing to the imposition of the monoplane ban it did little flying.
(*RAE 46; Crown copyright*)

On 22 August the rebuilt monoplane made its first test flight in its new form. It was flown by Capt C. R. W. Allen and Lt C. A. Bettington, and was reported to have handled and climbed well. At this time it received the new RFC serial number 256, being a No.3 Squadron aircraft; its designated pilot was Capt Allen.

A second Prier monoplane, Bristol No.91, was taken on charge by the RFC on 17 September, 1912; it was allotted the serial number 261 and was allocated to No.3 Squadron. Neither of the Prier monoplanes had done much flying when the ban on the flying of monoplanes by the Military Wing was imposed in September. Both were still on the strength of No.3 Squadron on 21 December, 1912, but by 28 March, 1913, they were included in an official list of monoplanes 'unallotted to squadrons'. In fact they were in the Royal Aircraft Factory, temporarily on charge of the RFC Flying Depot. Both were struck off charge on 5 August, 1913, their total flying time being 8 hr 33 min in the case of No.256 and 4 hr 38 min for No.261.

Bristol-Prier monoplane
50 hp Gnome

Two-seater.
Original form: Span 32 ft 9½ in; length 23 ft 7½ in; wing area 200 sq ft.
Original form: Empty weight 670 lb; loaded weight 1,200 lb.
Original form: Speed 60 mph.

Manufacturer: British and Colonial.

Service use: No.2 (Aeroplane) Company, Air Battalion, Royal Engineers; No.3 Sqn, RFC.

Serials: B6 (renumbered 256) and 261.

Bristol-Coanda monoplane

Henri Coanda, a Rumanian engineer and scientist of great ability and versatility, had designed several aircraft of startling originality before he joined The British and Colonial Aeroplane Company in January 1912. He had done a good deal of work on the design of a two-seat military monoplane when the conditions for the entries in the British Military Aeroplane Competition were announced. As there was little time to design anything more specifically intended to meet the trial requirements of the competition, the Bristol company decided to enter two monoplanes to the Coanda design, together with the two Gordon England biplanes.

The monoplanes had the Bristol works sequence numbers 105 and 106 and emerged as clean shoulder-wing aircraft in which care and ingenuity had been exercised to reduce drag. Both were powered by the 80 hp Gnome rotary engine and had a four-wheel undercarriage on which the aircraft sat tail-high when at rest; long rearward extensions of the longitudinals in the undercarriage prevented the tail from striking the ground. Structure was generally conventional, with wing warping for lateral control.

In the Military Trials the Coanda monoplanes were given the identifying numbers 14 and 15 and were flown respectively by Harry Busteed and James Valentine. The latter withdrew as pilot on 9 August and was replaced by Howard

Bristol-Coanda monoplane, No.14 in the Military Trials, in its original form without fin.

No.14 with fin added. In the cockpit is Harry Busteed.

The second Bristol-Coanda entered in the Military Trials had the number 15. In the rear cockpit is Howard Pixton.

Pixton, who had been left aircraftless when the Daimler-powered Bristol biplane had to pull out of the Trials.

The Bristol-Coanda monoplanes did well in the Trials; in particular, Howard Pixton created a considerable impression by flying No.15 in a wind gusting from 17 to 44 mph. Busteed also made a fine flight in a strong wind on No.14. His aircraft had first appeared with a dangerously inadequate-looking rudder, and this was soon supplemented by a curved fixed fin. Directional control was still unsatisfactory, however, and by 9 August the aircraft had been fitted with an enlarged rudder without fin.

The respective showings of the two Bristol-Coanda monoplanes in the Military Trials are summarised in the following table:

Test	Position		Result	
	No.14	No.15	No.14	No.15
Quick assembly	3	5	17 min 52 sec	23 min 35 sec
3-hours test	Passed	Passed	—	—
Climbing	10	7	200 ft/min	218 ft/min
Gliding	3	—	1 in 6·5	—
Consumption				
(a) Petrol	5	4	8 gal/hr	7 gal/hr
(b) Oil	6	5	1·7 gal/hr	1·5 gal/hr
Calculated range	6	1	328 miles	420 miles
Speed				
(a) Maximum	5	3	70·5 mph	72·9 mph
(b) Minimum	11	7	68·3 mph	58·1 mph
Speed-range	11	5	2·2 mph	14·8 mph
Landing	Passed	—	—	—
Quick take-off	Failed	Failed	—	—
Rough weather	—	1	—	Average W/V 30·5 mph
Transport	Did not participate	Did not participate	—	—

In the secondary list of prizes open to British subjects for British aeroplanes the two monoplanes were awarded two of the three third prizes of £500 each, the second prize having been withheld. Within days of the end of the Trials the two monoplanes were acquired by the RFC, being taken on charge as 263 and 262 respectively. They were allocated to No.3 Squadron, but tragedy prevented them from flying much in the RFC. On 10 September Lts E. Hotchkiss and C. A. Bettington were flying 263 from Larkhill to Cambridge, and the monoplane was seen to start a descent at 2,000 ft over Port Meadow, Oxford. At 500 ft the descent became a steep dive; at 200 ft fabric tore off the starboard wing and the aircraft plummeted to the ground; both officers were killed. The subsequent enquiry established that the crash had occurred because of the accidental opening of the quick-release catch that secured the anchorage of the flying wires. The catch had been fitted in order to facilitate the dismantling of the aircraft in the tests of the Military Trials.

This disaster came only four days after the fatal crash of the ex-Trials Deperdussin No.258 and led to the well-known ban on the flying of monoplanes by the Military Wing of the RFC. Inevitably this applied to the surviving Coanda monoplane No.262, which was deprived of its engine at an early date and was

Trials No.15 became 262 in the RFC. It is here seen engineless at Farnborough in March 1913, awaiting scrapping. (*RAE*)

transferred to Farnborough in March 1913. In company with all the remaining monoplanes of No.3 Squadron it had been vouchered to the Flying Depot there, but it appears that it never flew again.

Bristol-Coanda
80 hp Gnome

Two-seat reconnaissance monoplane.
Span 40 ft; length 28 ft 3 in; height 7 ft; wing area 242 sq ft.
No.14: Empty weight 1,144 lb; loaded weight 1,839 lb.
No.15: Empty weight 1,159 lb; loaded weight 1,871 lb.
Maximum speed 73 mph; best climb to 1,000 ft, 5 min.

Manufacturer: British and Colonial.

Service use: No.3 Sqn.

Serials: Military Trials Nos.14 and 15 became 263 and 262.

Bristol T.B.8

It has been said that the remarkable performance of the B.E.2 flown hors concours at the Military Trials inspired Henri Coanda to design a biplane. Late in 1912 he began work on the design of a two-seat tractor biplane to be powered by either the 70 hp Renault or the 90 hp Daimler-Mercedes engine. The resulting aircraft for the Renault was designated Bristol B.R.7. Seven were built, having fuselage, undercarriage and tail unit of typical Coanda design.

When the Admiralty ordered a landplane derived from the contemporary Coanda monoplane, the fuselage of Bristol No.121 (originally built as a military monoplane) was fitted with a set of biplane wings similar to those of the B.R.7, and thus became the prototype Bristol T.B.8. A number of similar conversions were made, and further examples were built as biplanes, including one prescient aircraft that had a neat and ingenious rotary bomb carrier holding twelve 10 lb bombs.

Six of the converted aircraft were delivered to Rumania, and a few T.B.8s went to the RNAS. The only aircraft of the type to be purchased by a private owner was a biplane conversion of Bristol No.143. It was acquired early in July 1914 by R. P. Creagh, but his ownership of it was short-lived, for it was impressed for the RFC on the outbreak of war. Creagh had no financial grounds for complaint, for the price paid by the Service (£1,000) was £300 more than he had paid for it. The RFC gave it the serial number 634, and recorded it as having an 80 hp Gnome, although in May it had had an 80 hp Clerget. Others impressed were given the serial numbers 614, 615 and 620; of these 615 is said in one document to have had a Clerget engine; consequently the possibility of some confusion over the numbering of Creagh's T.B.8 cannot be excluded.

The AID at Farnborough had a poor opinion of the T.B.8s, rejecting 634 as unsafe on 16 August, 614 and 620 three days later. These last two were taken over by the RNAS as 948 and 917 respectively and had reached Eastchurch and Dunkerque, again respectively, by October. On 8 October, No.634 was still at Farnborough and its eventual fate is unknown.

Mr R. P. Creagh's Bristol T.B.8, photographed at Brooklands in the summer of 1914. It was bought by the RFC for £1,000 at the outbreak of war and became No.634. (*Flight International*)

A Bristol T.B.8 at Farnborough, probably one of the impressed aircraft awaiting its official number. (*RAE; Crown copyright*)

One of the only production batch of T.B.8s, photographed at Farnborough while still bearing its original identity of 698. (*Bristol Aeroplane Co*)

Certain modifications to the design were made by Frank Barnwell and a dozen of the modified version were ordered by the War Office. On these aircraft, ailerons replaced wing warping, the mainplanes were staggered, and an open-fronted full-circular cowling was fitted to the engine; the petrol capacity was 25 gallons. Apparently, the serial numbers 691—702 were allotted to this batch, and the first aircraft was delivered on 26 September, 1914. Six more were delivered to Farnborough between that date and 17 October, whereafter the entire dozen was diverted to the RNAS to be renumbered 1216—1227.

Thus the RFC never made any use of the Bristol T.B.8, but those of the RNAS were employed in various ways at a number of stations between Dundee and Dunkerque. The RFC probably did not miss much: the RNAS General Memorandum No.21 of 1915 warned:

> 'The attention of Officers is called to the fact that the 80 hp Gnome Bristol Tractors Nos.1216 to 1227 are machines which were purchased for use as practice machines for advanced pupils. They are not suitable for flying loaded full up with petrol and passenger, and must not be used thus loaded, except under exceptional circumstances by skilled pilots who understand that the machines are in an overloaded condition.'

It should be noted that fewer T.B.8s were built than has been hitherto believed. Nos.1216—1227 (the erstwhile 691—702) were the last to be produced, for the batches of RNAS aircraft numbered 8442—8453 and 8562—8573 are now known to have been Boxkites.

Bristol T.B.8
80 hp Gnome, 80 hp Clerget, 80 hp Le Rhône

Two-seat tractor biplane.
Span 37 ft 8 in; length 29 ft 3 in; wing area 450 sq ft.
Empty weight 970 lb; loaded weight 1,665 lb.
Maximum speed 75 mph; climb to 3,000 ft, 11 min; endurance 5 hr.

Manufacturer: British and Colonial.

Service use: None in RFC.

Serials: 614 (renumbered 948 on transfer to RNAS), 615, 620 (renumbered 917), 634, 691—702 (renumbered 1216—1227).

Bristol G.B.75

Six Bristol T.B.8s, all conversions from Coanda monoplanes, were delivered to Henri Coanda's native Rumania and inevitably came to the attention of Prince Cantacuzene. The prince had led the Rumanian contingent whose members had learned to fly at the Bristol school at Larkhill in 1912. He was therefore well acquainted with the products and activities of British and Colonial, and doubtless took pride in the designs of his countryman, Coanda.

Prince Cantacuzene placed an order with the firm for an improved development of the T.B.8, for which he provided the power unit, a 75 hp Gnome Monosoupape. The aircraft built to meet this order had the Bristol works sequence number 223 and was designated G.B.75. It was exhibited on the British and Colonial company's stand at Olympia in March 1914, when it could be seen to differ appreciably from the T.B.8. Much thought had evidently been given to minimising drag, for the engine was wholly enclosed, its airscrew having a large shallow spinner with radial louvres optimistically provided for the admission of cooling air. The portly fuselage had full-length dorsal and belly fairings, there was a fixed fin and conventional rudder, and a four-wheel undercarriage was provided.

In practice, the refinement of the fuselage and engine cowling did not work; as might surely have been foreseen, the engine overheated. This quickly became evident when the G.B.75 was flown at Larkhill on 7 April, 1914; the aircraft also proved to be nose-heavy. An open-fronted cowling replaced the enclosed installation and the stagger was increased; after these modifications the G.B.75 flew again on 28 April.

For unexplained reasons, Prince Cantacuzene's order for the G.B.75 was cancelled. The British War Office was interested in the aircraft, which was eventually delivered to Farnborough on 2 August, 1914. Its official identity provides a puzzle for historians, for it was marked with the number 601, which was

The Bristol G.B.75 in its original form, as exhibited at the 1914 Olympia Show. (*Bristol Aeroplane Co*)

Bristol G.B.75 after delivery to Farnborough, where it was marked as 601. By that time the spinner had been removed and the wings given slight stagger. (*RAE; Crown copyright*)

incontrovertibly the true identity of an experimental B.E. that eventually became a B.E.2c and was clearly recorded as the aircraft on which Edward Busk was killed on 5 November, 1914. Bristol company records apparently indicate that the G.B.75's true identity was 610 — but that number, too, has been associated with a different aircraft, a Maurice Farman Longhorn of No.4 Squadron.

It was proposed to send the G.B.75 to France with the RFC but it never went, not even when, on 22 August, 1914, the hard-pressed RFC in the Field requested that any spare aeroplanes should be flown over to France. The five that were scraped together did not include the G.B.75. It is not known whether the aircraft still had its 75 hp Monosoupape engine at that time. If it did, that might account for its retention in England, for the early seven-cylinder Monosoupape was a troublesome piece of machinery, and spares for it would have been a problem. Whatever the reason, the G.B.75 was never heard of again.

Bristol G.B.75
75 hp Gnome Monosoupape

Two-seat tractor biplane.
Span 37 ft 8 in; wing area 420 sq ft.
Empty weight 970 lb; loaded weight 1,650 lb.
Speed 80 mph; endurance 5 hr.

Manufacturer: British and Colonial.

Service use: None known.

Serial: Uncertain; 601 painted on aircraft but this duplicated the number of a B.E. then in existence.

Bristol Scouts B, C and D

The original Bristol Scout that went to Larkhill on 23 February, 1914, had been designed by Frank Barnwell as a sporting single-seater. It was flown as such in the prewar months, and was purchased by Lord Carbery in June 1914. It underwent various modifications early in its existence and its main configuration had been settled by the time Lord Carbery acquired it. It was lost in the Channel on 11 July, 1914, when he was flying it in the London—Paris—London race.

Two further Bristol Scouts were under construction when Lord Carbery bought the prototype. The outbreak of war prevented them being flown as sporting aircraft; instead they were delivered to the RFC at Farnborough on 21 and 23 August, 1914. Both had the 80 hp Gnome engine and, allotted the official serial numbers 644 and 648, were sent to France in packing cases on 20 September. On 28 September they were issued to, respectively, Squadrons No.5 and 3, with which units their usual pilots were, again respectively, Maj J. F. A. Higgins and Lt R. Cholmondeley.

One of the two Bristol Scout Bs after delivery to Farnborough, August 1914. (*RAE 545; Crown copyright*)

Attempts were made to arm these two Scouts. No.644 was equipped with a fixed Lee-Enfield rifle at an absurd angle, a Mauser pistol and five rifle grenades carried as hand-dropped bombs. In No.3 Sqn, No.648 was armed with two fixed rifles, one either side of the fuselage angled outwards to miss the airscrew. No.644 was wrecked on 12 November, and its surviving parts were used to maintain 648 which had a remarkably long operational career. This Scout was returned to the Aircraft Park from No.3 Sqn on 25 November, 1914, and was issued to No.4 Sqn on 23 March, 1915. It stayed with No.4 until 18 September, when it was transferred to No.8 Sqn; it was finally struck off charge on 4 October, 1915.

Long before No.648's career ended other Bristol Scouts were in the field. These were designated Bristol Scout C, their two predecessors being named Scout B.

The first production batch of twelve had been ordered by the War Office as early as 5 November, 1914, with the Admiralty following suit two days later. This led to an inter-Service argument over priority in delivery, which was resolved by delivering the first production Scout C (No.1243) to the RNAS and the following twelve to the RFC as 1602—1613. Deliveries to the RFC began on 23 April, 1915, and the first RFC Scout C, No.1602, went direct to No.1 Sqn in France on 15 May. No.1603 was delivered to the Aircraft Park on 2 June and went to No.5 Sqn next day. By 27 June seven Scout Cs had reached France, including 1613, delivered to the Aircraft Park on 12 June. By 30 June, 1602 was with No.1 Sqn, 648 with No.4, 1603 with No.5, 1611 with No.6 Sqn, 1606 with No.7, and 1610 with No.8. All the Scout Cs of that first RFC batch had the 80 hp Gnome engine. It was on No.1611, armed with an obliquely aligned Lewis gun, that Capt L. G. Hawker drove down three enemy two-seaters in the course of a single patrol. For this action he was awarded the Victoria Cross, becoming the first man to win this decoration for an action in aerial combat.

The armament of Scout B No.644 while with No.5 Squadron consisted of an obliquely-mounted Lee-Enfield rifle with stock removed, a Mauser pistol in external holster, and five rifle grenades in an external rack. (*Flight International*)

In the summer of 1915 the Gnome engine became scarce and alternatives had to be found. The first Scout C to have the 80 hp Le Rhône was No.4668; others of that batch initially fitted with that engine were 4671—4678 and 4687, while 4679—4686 and 4688—4699 were delivered without engines. Other modifications were introduced progressively. As early as 27 March, 1915, RFC HQ had recommended that the oil tank should be moved to a position ahead of the cockpit instead of being behind the pilot. This was not acted upon until production was well advanced, but the later Scout Cs incorporated this modification. From 4685

No.1611, the Bristol Scout C of No.6 Squadron on which Capt Lanoe G. Hawker fought the action of 25 July, 1915, for which he was awarded the first Victoria Cross for action in aerial combat. The Bristol had gone to the 1st Air Park on 12 June, 1915, to No.6 Squadron on 26 June, and to No.12 Squadron on 6 November, 1915. (*Lt-Col T. M. Hawker*)

onwards increased dihedral was rigged on the mainplanes; No.5297 introduced new tail surfaces, the medium-size tailplane and elevators, and the medium-size rudder; from 5298 onwards production Scouts had the Scout D type engine mounting plate and cowling; and 5311 and 5312 had Scout D struts and wing-anchorage fittings together with 'cut-away lower wings'—a reference to the creation of a gap between wings and fuselage to provide some downward view.

The Scout D was a modest revision of the design made by Frank Barnwell in the autumn of 1915. This had as basic standard the medium-size tail surfaces, and in its interplane bracing streamline-section Rafwires replaced the stranded steel cables of the earlier variant. Later Scout Ds had a still larger rudder and some had revised wings with shorter ailerons and the underwing skids moved outboard. Engine cowling design varied, and there were three different types.

The Scout remained in service throughout the period when every RFC Squadron had one or two single-seat scouts on its strength, for the day of the

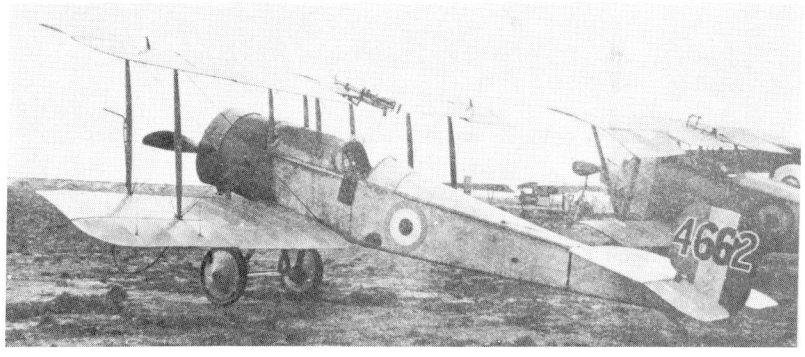

No.4662 was a Scout C that had been delivered to Farnborough by 17 July, 1915. It was with No.12 Squadron from 4 November, 1915, until 18 January, 1916, and it was struck off charge at No.1 Aircraft Depot on 21 January, 1916.
(*Sir Edmund Neville, from Peter Liddle's 1914–18 Personal Experience Archives, presently housed within Sunderland Polytechnic*)

Scout C No.5319 with obliquely-mounted Lewis gun on the starboard side. (*T. Heffernan*)

single-seat fighter squadron did not dawn until 7 February, 1916, when No.24 Sqn flew its D.H.2s to France. The Bristol was never able to make much of a name for itself owing to lack of effective armament. Installations varied, but the basic standard became a Lewis gun mounted on the centre section and firing forward over the airscrew. This could be supplemented in various ways: for example, in December 1915 No.4662 of No.12 Sqn had two Lewis guns, one in the overwing position, the other on the starboard side of the fuselage.

Even the modest overwing Lewis installation created problems. On 8 December, 1915, the O.C. No.4 Sqn reported that Lt Woodhouse had experienced disquieting instability on the squadron's Bristol (which must have been 4678) when the gun was carried. He also found the aircraft excessively fatiguing to fly in rough conditions. This experience does not seem to have been general, however,

Weatherbeaten Bristol Scout C, possibly of No.12 Squadron, with overwing Lewis gun.

164

and No.4 Sqn received no recorded sympathy or recommended remedy for its problems.

What might have marked the beginning of a new era for the Bristol Scout was the despatch to France on 25 March, 1916, of No.5313. This Scout C was armed with a fixed Vickers gun regulated by the first operational example of the Vickers-Challenger interrupter gear. This aircraft went to No.12 Sqn but evidently went later to No.11 Sqn with which unit it was flown by Lt Albert Ball in a combat on 15 May, when he succeeded in driving down an Albatros two-seater out of control. The Bristol crashed next day and was struck off charge. Other Scouts of its production batch, including 5312, 5314 and 5315, had the synchronised Vickers, as did several others of later batches. A similar Vickers-gun installation using the Arsiad interrupter gear was at least designed and may have been fitted to some aircraft.

Vickers gun with Challenger interrupter gear on Bristol Scout D, A1761. (*W. Evans*)

On 28 June, 1916, the Board of Admiralty agreed to turn over sixty aircraft to the RFC in view of that Service's urgent need for more aircraft in France for the impending Battle of the Somme. Of these, 20 were to be Bristol Scout Ds, to be transferred without engines direct from the British and Colonial works. The aircraft concerned were 8981—9000, but 8994—8999 were directed to Mudros, leaving only fourteen aircraft. The RFC serial numbers A2376—2380 and A3006—3020 were allotted for the full twenty Bristols, but it is uncertain how many, if any, of the RNAS Scouts ever wore any of these RFC numbers, for the aircraft were returned to the RNAS and several are known to have served into 1918 with their original identities.

Although the Bristol Scouts were never very numerous, the type saw operational use with almost every squadron of the RFC in France. It might not have been particularly popular in the Field, for on 19 February, 1916, Lt-Col Ludlow-Hewitt, Officer Commanding the III Wing, wrote to RFC Headquarters:

> 'There are very few pilots in this Wing who have learnt to fly the Bristol Scout, owing apparently to an exaggerated idea of the difficulty of flying it, combined with disinclination on the part of Squadron Commanders to run the risk of having their Bristol Scouts broken.'

Ludlow-Hewitt sought to remedy this by giving orders that four pilots per squadron were to be taught to fly the Bristols.

In terms of operational statistics the Bristol Scout showed up badly in comparison with its contemporaries in France. In the autumn of 1915 the average flying life of a Bristol Scout was only 21 hr 43 min whereas that of the B.E.2c was 81 hr 56 min. By the spring of 1916 the average life of Bristols struck off through deterioration was only 28 hr 16 min (B.E.2c 127 hr 30 min); the corresponding figure for aircraft struck off through all causes was a mere 22 hr 45 min (B.E.2c 71 hr 44 min). Further proof that the delicate little Bristol was difficult for the poorly instructed pilots of the time lies in a memorandum dated 22 May, 1916, to the four Brigades from Brig-Gen Brooke-Popham:

> 'The number of Bristol Scouts wrecked by the School at No.1 A.D. has been so large recently as to cause a dangerous shortage in machines and 80 hp Le Rhône engines.'

By the end of July 1916 only six Bristol Scouts were with the squadrons in France, all being fitted with Vickers guns and interrupter gears. RFC Headquarters indicated that they would ask for only six Bristol Scouts in their next monthly demand; seven were finally allotted back to storage a week later. In fact, on 11 August RFC Headquarters instructed No.1 AD: 'Bristol Scouts will not be issued as Service Machines in future.'

That marked the end of the little Bristol's operational career in France, but as late as 2 February, 1917, A1754—1759 were allocated to the RFC in France. This was probably for training purposes, for it is known that A1759 was later on the strength of the Scout School at No.2 AD, Candas. The type remained active in other theatres of war, notably in Palestine, Macedonia and Mesopotamia, where No.63 Sqn still employed the Scout as late as November 1917. Elsewhere it saw

Scout D A1752 in immaculate condition with a training unit in England.

Very much an instructor's pet, this Scout D had a Morane-like spinner. (*R. C. Bowyer*) much use as a training aircraft at home and in Egypt. Although the Scout did nothing of note in the field of Home Defence a few were allocated to HD squadrons. Two, 5570 and 5571, were allotted to No.47 Sqn in June 1916 for this purpose but were transferred to No.33 Sqn later that month. The type was also used by No.39 Sqn.

One or two experimental Scouts existed, fitted with the 110 hp Clerget or Le Rhône, some having very large shallow spinners. Early in 1916, 5303 was tested with various airscrews, including one that had Morane-Saulnier bullet deflectors and spinner. The RNAS experimented in various ways with its Bristol Scouts, for the type saw extensive and varied service with the naval flying service.

Bristol Scouts B, C and D
80 hp Gnome, 80 hp Le Rhône 9C

Single-seat scout.
Span 24 ft 7 in; length 20 ft 8 in; height 8 ft 6 in; wing area 198 sq ft.
Empty weight 760 lb; loaded weight 1,200 lb (Scout C), 1,250 lb (Scout D).
Scout D with 80 hp Le Rhône: Maximum speed at sea level 92·7 mph, at 5,000 ft—90·5 mph, at 10,000 ft—86·5 mph; climb to 5,000 ft—7 min, to 10,000 ft—21 min 20 sec; service ceiling 15,500 ft; endurance 2 hr.
Armament: Normally one 0·303-in Lewis machine-gun above the centre-section, but various side-mounted installations were tried, and the Lewis could be supplemented by a second, or by another weapon. Later Scout Cs and Ds had one fixed 0·303-in Vickers machine-gun regulated by Vickers-Challenger or Arsiad interrupter gear.

Manufacturer: British and Colonial.

Service use: *Western Front*—Sqns 1, 2, 3, 4, 5, 6, 7, 8, 9, 10, 11, 12, 13, 15, 16, 18, 21, 24 and 25. *Home Defence*—Sqns 33, 39 and 47. *Macedonia*—No.47 Sqn. *Mesopotamia*—Sqns 30 and 63. *Palestine*—Sqns 14, 67 (Australian) and 111. *Training duties*—Reserve/Training Sqns 1, 2, 6, 10, 11, 15, 16, 18, 19, 20, 22, 23, 30, 40, 43 and 55; No.2 Training Depot Station; Sqns working up—Nos.15, 17, 46, 56, 65 and 81.

Serials: Scout B—644 and 648. Scout C—1602—1613, 4662—4699, 5291—5327. Scout D—5554—5603, 7028—7057, A1742—1791. A2376—2380 allotted for 8981—8985 to be transferred from RNAS. A3006—3020: allotted for further transfers, to include 8986—8993, from RNAS. Confirmation that these transfers took place has yet to be found.
Rebuilt aircraft: A5209, B763, B813.

The first of the two Bristol S.2As, photographed at Central Flying School, whither the aircraft had gone more probably for evaluation than for employment. (*T. Heffernan*)

Bristol S.2A

The Bristol S.2A was designed to meet an Admiralty requirement for a two-seat fighter. Two prototypes, to be numbered 3692—3693, were ordered, and were still expected by the RNAS in March 1916. The War Office evidently learned about the existence of the design and saw in it an aircraft worth evaluating as a trainer. They, too, ordered two prototypes, for which the serial numbers 7836—7837 were allotted as early as 10 March, 1916.

In the event, only the War Office prototypes were built. The RNAS had by that time probably realised that the Sopwith 1½ Strutter was superior to the Bristol two-

The second S.2A, 7837, in service as a trainer. (*W. Evans*)

seater; certainly the Admiralty abandoned the Bristol in favour of the Sopwith.

What the War Office got was a squat little single-bay biplane, obviously related to the Scout C and D, but seating its two occupants side by side in the wide fuselage. Such a seating arrangment in a two-seat fighter was so nonsensical that it is a wonder that the Admiralty ever gave it serious consideration. In a trainer, however, the side-by-side seats made much sense, and the use of the 110 hp Clerget promised pupils a foretaste of the engine-handling techniques and performance they could expect to encounter in contemporary operational aircraft.

The first of the two S.2As was completed in May 1916 and was nicknamed 'Tubby' in the Bristol works. It was delivered to CFS at Upavon on 11 June and was followed a few weeks later by 7837, which was completed in June and delivered to CFS on 30 July, 1916. The S.2A proved to be quite tractable with a fair performance, but the almost completely enclosed installation of the Clerget created problems for a time. This may explain why one of the aircraft was later seen at Gosport with a 100 hp Gnome Monosoupape engine in a cutaway and fretted cowling.

Bristol S.2A
110 hp Clerget 9Z, 100 hp Gnome Monosoupape

Two-seat trainer.
Span 28 ft 2 in; length 21 ft 3 in; height 10 ft.
Loaded weight 1,400 lb.
Maximum speed 95 mph.
Armament: None.

Manufacturer: British and Colonial.

Service use: CFS; possibly also used at Rendcombe and Gosport.

Serials: 7836—7837.

Bristol F.2A

In the autumn of 1915 Royal Flying Corps Headquarters submitted specifications for a corps reconnaissance and artillery-spotting two-seater to supersede the B.E.2c; not surprisingly, in view of the depredations of the Fokker monoplane fighters, these required the new aircraft to be capable of defending itself. To meet the requirement the Royal Aircraft Factory designed the R.E.8 and Frank Barnwell designed the Bristol R.2A. Both designs were drawn up in March 1916, consequently Barnwell was unlikely to have been in any way influenced by the R.E.8 design.

The R.2A design was for a two-bay, equal-span biplane powered by a 120 hp Beardmore engine. The crew were to sit close together in tandem, the rear fuselage tapering markedly in depth towards the tailplane to give the gunner a good field of fire; the pilot's gun was to be a fixed forward-firing Lewis, and the observer's Lewis was to be on an Etévé rotating mounting. To minimise the extent to which the upper wing obscured the pilot's field of vision the fuselage was placed in a mid-gap position, bringing the pilot's eyes close to the level of the upper wing.

The first Bristol F.2A, A3303, in its original form with flank radiators and adapted B.E.2c wings.

Barnwell must have known that this aircraft would be underpowered with the 120 hp Beardmore, and in May 1916 he redesigned it to have the 150 hp Hispano-Suiza. The revised design was designated R.2B and had unequal-span wings with strut-braced extensions; the fuselage, tail unit, undercarriage and crew disposition were substantially similar to those of the R.2A.

A further redesign to take the 190 hp Rolls-Royce Mark I engine reverted to the two-bay equal-span wing arrangement but the fuselage was revised. Again it was in mid-gap and the crew were accommodated close together, but the downward curve of the upper longerons was increased and ran down to a horizontal knife-edge on the rear spar of the adjustable tailplane. The engine installation was ugly and clumsy, embodying two tall flank radiators at the rear of the engine. Again an Etévé mounting was envisaged for the observer's Lewis gun, but in the Rolls-Royce aircraft the pilot had a fixed Vickers gun mounted on the centre line and passing through a special tunnel in the forward petrol tank.

This revised design was designated Bristol F.2A, signifying a change of intended role from pure reconnaissance to fighter-reconnaissance. Two prototypes and fifty production F.2As, numbered A3303—3354, were ordered under Contract No.87/A/552 in August 1916. The first prototype, A3303, was built in

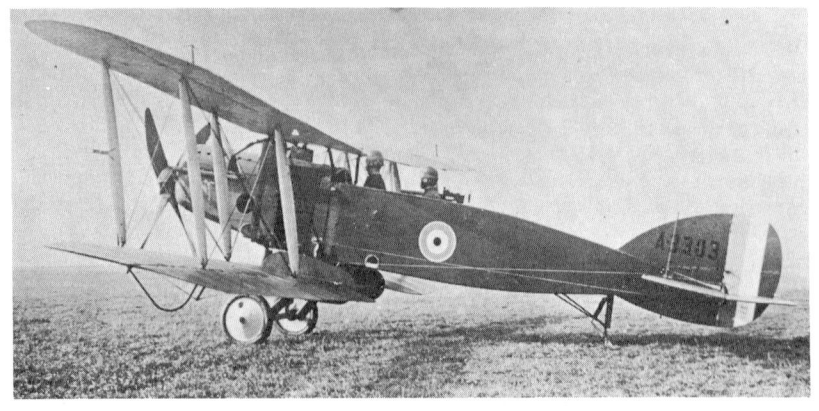

A3303 at Central Flying School in September 1916.

close conformity with Barnwell's layout, but appeared without the proposed top decking on the rear fuselage; no form of gun mounting was provided on the observer's cockpit. The wing panels appeared to be B.E.2c components with modified trailing edges at their inboard ends. This first F.2A made its initial flight on 9 September, 1916, and flew to Upavon for official trials on 21 September. As might have been expected, the radiators obscured the pilot's forward view excessively and a new installation of a single frontal radiator had to be designed. With its new engine installation A3303 was back at CFS for further tests by 16 October and won a favourable report despite rather indifferent performance. A number of minor modifications were called for.

The second prototype, A3304, was fitted with a 150 hp Hispano-Suiza engine, which from the outset had a circular frontal radiator. Apart from its rudder-mounted tailskid it was otherwise similar to A3303, having B.E.-pattern mainplanes, an open lower centre-section frame, and endplates on the inboard ends of the lower wing panels. A3303 went to Orfordness for further evaluation, by which time it had a Scarff ring mounting on the rear cockpit, and saw action on 7 July,

A3303 with frontal radiator and the end-plates removed from the root ends of the lower wings.

1917, when it was one of the 90-odd British aircraft that went up to attack the Gotha formation that bombed London that day. Its pilot was Lt F. D. Holder, the observer Lt F. W. Musson, both of whom fired on the enemy, but they were unable to close the German formation and had to abandon pursuit for lack of fuel.

Before 1916 was out a further 200 Bristol two-seaters were ordered as an addition to the original contract. These were to be numbered A7101—7300 and were delivered as Bristol F.2Bs. The first deliveries of production F.2As began late in 1916, and ten had been handed over by the end of the year. All production F.2As were powered by the Rolls-Royce engine; they had new wings with blunt raked tips and modified engine cowlings. From a very early stage the Bristol two-seater became known as the Bristol Fighter and so, appropriately, it remained.

In July and August 1916 No.48 Squadron was at Rendcombe, equipped with B.E.12s and apparently expecting to go to France with that type. During August, however, its B.E.12s were progressively sent to France for the use of Squadrons No.19 and 21. Eventually, No.48 was equipped with the Bristol F.2A and went to

The second Bristol F.2A, A3304, with 150 hp Hispano-Suiza engine, at Filton. (*Bristol Aeroplane Co*)

France on 8 March, 1917. To achieve maximum surprise, the operational use of the Bristol and the D.H.4 was deferred until the Battle of Arras started, hence it was not until 5 April that the first offensive patrol of No.48 Squadron crossed the lines. It was led by A3337, flown by Capt W. Leefe Robinson, VC, with 2nd Lt E. D. Warburton as his observer, but in a combat with five Albatros D IIIs of Jagdstaffel 11 led by Manfred von Richthofen the Bristols tried to fight in the orthodox two-seater manner, manoeuvring to give their observers favourable shots. These tactics proved to be disastrous: Leefe Robinson's Bristol fell; A3340 and A3343 were shot down by Richthofen as his 35th and 36th victories; and A3320 also went down. A3330 was lost on 8 April, and three days later four Albatros attacked four Bristols and shot down A3318, A3323 and A3338. Other losses occurred, but fortunately some pilots began to fly the F.2A as if it were a single-seat fighter, concentrating on using the front gun and leaving the observer to protect the tail. These tactics proved successful, and the Bristol Fighter thereafter went from strength to strength.

Production F.2A, A3345, photographed at No.2 A.D., Candas; the photograph is dated 23 March, 1917. This aircraft was first allotted to the E.F. on 5 March and was flown back to England on 12 June, 1917.

In No.48 Squadron the F.2A began to be replaced by the F.2B in May 1917, and a number of the earlier type were flown back to England in June and July. One, A3325, was briefly attached to 'B' Flight of No.8 Squadron but returned to England on 13 July, 1917. At home the remaining F.2As saw continuing service with a few training squadrons, at Orfordness Experimental Station, and with the Wireless Experimental Establishment at Biggin Hill.

Bristol F.2A
190 hp Rolls-Royce Mark I (Falcon I); second prototype had 150 hp Hispano-Suiza 8A
Two-seat fighter-reconnaissance.
Span 39 ft 3 in; length 25 ft 9 in; height 9 ft 4 in; wing area 389 sq ft.
Empty weight 1,727 lb; loaded weight 2,667 lb.
A3303 at 2,753 lb loaded weight with frontal radiator: Maximum speed at sea level 110 mph, at 5,000 ft—106 mph, at 15,000 ft—96 mph; climb to 5,000 ft—5 min 25 sec, to 10,000 ft—14 min 30 sec, to 15,000 ft—31 min; service ceiling 16,000 ft; endurance 3¼ hr.
Armament: One fixed 0·303-in Vickers machine-gun and one 0·303-in Scarff-mounted Lewis machine-gun.

Manufacturer: British and Colonial.

Service use: Western Front—No.48 Sqn; one aircraft attached to No.8 Sqn. *Training duties*—Training Sqns 35 and 38. *Other units*—Orfordness Experimental Station; Wireless Experimental Establishment, Biggin Hill.

Serials: Prototypes A3303—3304; production A3305—3354.

Bristol F.2B

The second prototype Bristol F.2A, A3304, underwent official performance trials at Central Flying School on 4 and 6 December, 1916, by which time it had been modified. The pilot's forward view was improved by sloping the upper longerons downwards from the back of the pilot's cockpit to the level of the engine bearers. This allowed the forward top decking to taper upwards and also permitted installation of a larger upper tank and ammunition box for the Vickers gun. A lower centre-section plane was created by building out and covering the lower wing-attachment frame.

Early Bristol F.2B, A7107 with 190 hp Rolls-Royce Mark I (Falcon I) engine. First allotted to the RFC with the E.F. on 25 April, 1917, A7107 was used by No.48 Squadron and was flown back to England on 10 August, 1917. It is here seen as aircraft 6 of the Wireless Experimental Establishment, Biggin Hill.

These modifications were incorporated in all production Bristol Fighters from A7101, the modified aircraft being designated Bristol F.2B. The first 150 aircraft retained the 190 hp Rolls-Royce Mark I (later named Falcon I) in a relatively close-fitting cowling through which the water-conducting pipes of the cooling system partly protruded. When the more powerful 220 hp Rolls-Royce Mark II (Falcon II) became available it was installed from about A7221, the engine cowling being modified to enclose all the plumbing. The Falcon III followed the Falcon II, and the first Falcon III installation was made in A7177. Quantity deliveries of Falcon IIIs began late in 1917 and were fitted to the later F.2Bs of the batch B1101—1350, but the engine could only be fitted as it was available and a number of Bristols of this batch still had the Falcon II; indeed a few (*e.g.*, B1134, 1157) even had the Falcon I.

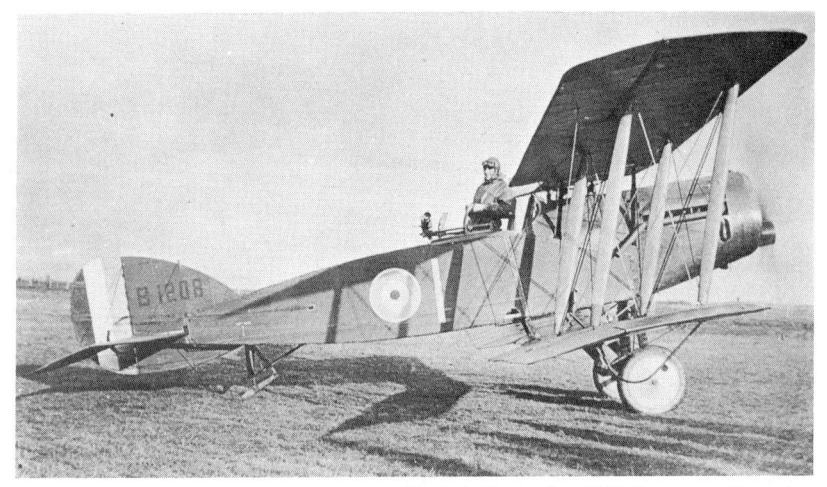

B1208 was a Bristol F.2B that was completed early in December 1917. It was not allotted to France and may therefore have had a Falcon II engine. In this photograph it was with a training unit; a camera gun was fitted to the Scarff ring.

Changes in the design of the horizontal tail surfaces were made in the summer of 1917. In July 1917 some aircraft had F.2B tailplanes modified to take F.2A elevators, but on 10 August it was recorded that 'the first Bristol Fighter machine to be fitted with the new type tailplanes will be No.1240' [*i.e.*, B1240]. These new tail surfaces were slightly smaller than those of the F.2A and early F.2B but were of higher aspect ratio.

The F.2B replaced the F.2A in No.48 Squadron, and other units were re-equipped or formed as the new Bristols became available. Between May and August 1917, Squadrons No.11, 20 and 22 replaced their F.E two-seaters with Bristols, and in No.11 Squadron the type found its most skilful exponent in Lt A. E. McKeever, a Canadian officer who, with his regular observer Sgt (later Lt) L. F. Powell, shot down many enemy aircraft before the end of 1917, and were the victors in several outstanding combats.

An official decision, taken in July 1917, to standardise the Bristol F.2B for all fighter-reconnaissance squadrons led to large additional orders, a substantial expansion of production by contractors additional to the parent Bristol company, and a crisis in engine availability. Although the output of Bristol Fighters increased progressively quarter by quarter, that of Falcon engines lagged further and further behind production of airframes. Alternatives had to be sought, but nothing comparable to the Rolls-Royce engine was available. The difficulties threatened to become greater on 13 September, 1917, when Maj-Gen Sefton Brancker, justifiably impressed by the performance of the Martinsyde R.G. and the outstanding potential of the brilliant Martinsyde F.3, proposed to recommend that the Falcon engine 'should be taken out of Bristol Fighters and put into Martinsyde Scouts.'

Having made this proposal at the 171st meeting of the Progress and Allocation Committee, Brancker went on to enquire how the trial of a Bristol Fighter with a Hispano-Suiza engine was progressing. Gen D. le G. Pitcher replied that there

B1201 at Martlesham, fitted with a 200 hp Hispano-Suiza engine. This Bristol arrived at Martlesham on 31 December, 1917, for performance trials. When these and associated trials were over, B1201's engine was borrowed to keep the Dolphin C3778 flying, but this Bristol eventually went to Farnborough on 28 February, 1918, and was subsequently used to test a 200 hp R.A.F.4d engine. (*RAF Museum*)

had been trouble in obtaining a 200 hp Hispano-Suiza for the Bristol and that it was intended to try a 200 hp Sunbeam Arab in the aircraft. On 2 October, 1917, two Arab engines were available, and it was decided to send one to the Royal Aircraft Factory for installation in an S.E.5a, the other to British and Colonial for a Bristol Fighter.

Much greater urgency attached to the Hispano-Suiza installation at that time, for it was officially expected that of the seventy Bristol Fighters, delivery of which was expected in October, thirty would be fitted with the Hispano-Suiza engine; a further thirty-six, similarly powered, were expected in November. On 3 October, 1917, it was decided that No.62 Squadron would be equipped with Hispano-powered Bristols and that the other four Bristol squadrons should be maintained with Falcon engines. No.62 Squadron was supposed to receive its first aircraft on 20 October, but on 17 October the British and Colonial company had only six suitably adapted F.2B airframes, five Hispano-Suiza engines, but no radiators for them. Indeed no Hispano-powered Bristol had flown at that time.

Inevitably, it was decided on 18 October that No.62 Squadron would after all have to have Falcon-powered Bristols; it was agreed that the available Hispano-Suizas should be installed in aircraft and turned over to the Technical Department for testing. Some delay occurred while the Bristol company redesigned the radiator, and not until the second week of December 1917 did Martlesham Heath begin to expect B1200, fitted with the Hispano-Suiza. Labour troubles at Filton delayed this Bristol; it did not go to Martlesham but was replaced by B1201 with the same engine, which arrived there on 31 December; and on 2 January, 1918, B1119 with the 200 hp Hispano-Suiza was tested in France at No.1 Aeroplane Supply Depot.

Other alternative engines for the Bristol F.2B were considered over the same period. Despite the fact that it had been decided some time before 2 October,

1917, that the 200 hp R.A.F.4d engine was not to be adopted, one was being installed in a Bristol Fighter in October 1917. This combination was not developed, but almost contemporary with the initial Hispano-Suiza installation was the F.2B B7781, which had a 230 hp B.H.P. engine; it was subjected to simple performance trials at No.1 A.S.D. on 10 and 13 January, 1918. This unusual Bristol had a frontal radiator, a gravity tank above the centre section and an externally mounted Vickers gun. It was extensively modified in France, the gun being brought within the cowling, the large exhaust manifold turned upwards to discharge above the wing, and the external gravity tank replaced by two new tanks carried within the centre section. On 27 January, 1918, B1206, also powered by a 230 hp B.H.P., arrived at Martlesham Heath for trials. By mid-February it had been fitted with the engine out of the Sopwith Rhino X7 and its testing extended over several weeks. On 3 April, 1918, B1206 left Martlesham for Hendon.

The B.H.P.-powered Bristol was seen as a possible escort for bombers, but its performance was poor, the pilot's vital forward view was much obstructed by the upright engine and exhaust manifold, and its manoeuvrability was impaired. On 26 February, 1918, Maj-Gen W. S. Brancker, Comptroller-General of Equipment, wrote to RFC Headquarters to the effect that the B.H.P. Bristol was not considered good enough for use as an escort fighter and would not be going into production. The installations of the Siddeley Puma that were made in Bristol F.2Bs later in 1918 were separate developments.

It had been recognised that the Bristol F.2B with the 200 hp Hispano-Suiza could not be expected to have as good a performance as the Falcon-powered version, and the variant was intended to replace the R.E.8 and Armstrong Whitworth F.K.8 in the corps-reconnaissance squadrons. However, late in 1917 the Hispano-Suiza was scarce, troublesome, and sorely needed to power the S.E.5a and Dolphin, then in largescale production. A change of power unit for the Bristol was announced in a War Office letter dated 19 December, 1917, wherein it was stated that the standard engine for the corps-reconnaissance Bristol was to be the Sunbeam Arab. This decision took RFC Headquarters by surprise: they quickly pointed out that the RFC in the Field had no experience

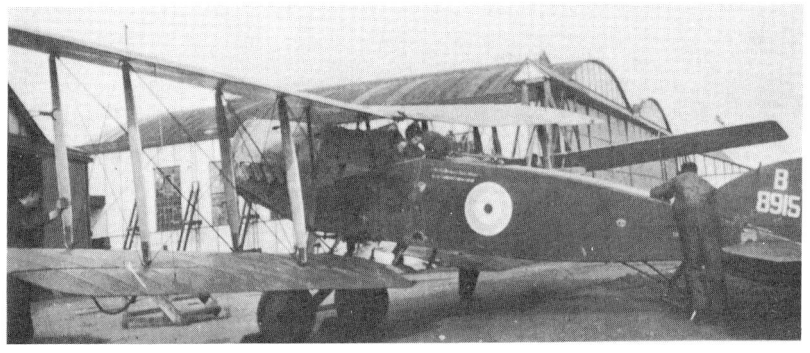

The earliest trial installations of the Sunbeam Arab engine in Bristol F.2Bs were made by Western Aircraft Repair Depot in B8914 and B8915, apparently using a slightly modified Falcon housing. B8915 was initially allotted to the E.F. on 9 January, 1918, but was re-allotted to the C.T.D. on 26 January. It arrived at Martlesham Heath for trials on 8 February and was flown extensively until 27 June, 1918, when it was sent for storage at Coventry. (*RAF Museum*)

whatever of the Arab engine and asked that a specimen aircraft fitted with the new engine should be sent to France.

In response to this justified protest, an Arab engine was sent to France and two Bristol Fighters were hurriedly fitted with Arabs at Western Aircraft Repair Depot, Yate. Why this belated haste was necessary is not clear, for the first Arab had been delivered in May 1917 and the initial decision to fit it to the Bristol F.2B was taken on 7 August, yet nothing was done to implement it then and the Hispano-Suiza was tried instead. Indeed, two Arab engines had been sent to British and Colonial on 1 and 2 October, 1917, but it seems that no urgent steps were taken at Filton to fit the Sunbeam engines to F.2B airframes. The anxiety of RFC Headquarters was increased by the fact that their forward planning called for the re-equipment of corps squadrons with Bristol Fighters to start in March 1918.

The first two Arab-powered Bristols were B8914 and B8915, both Aircraft Repair Depot rebuilds. B8914 had reached France by 25 January, 1918, and on 28 January went to No.11 Squadron for evaluation. Headquarters called for weekly reports on the aircraft, but it was detached to No.12 Squadron, an R.E.8 unit, on 3 February, 1918. During the delivery flight from No.11 to No.12 Squadron five exhaust stubs fell off, an omen of the troubles that lay ahead. On the same day a Falcon Bristol was attached to No.15 Squadron, being one of several aircraft that had been withdrawn from the four Bristol Fighter squadrons for this purpose. Two were allotted to No.35 Squadron on 4 February. B8915 went to Martlesham Heath for trials on 8 February, 1918, and was joined there on 1 March by B1204, which had a geared Arab I.

So hastily had B8914 and B8915 been modified that the Rolls-Royce engine bearers and Falcon radiator were retained. Six more Bristols, C4657—4659 and C4662—4664, were fitted with Arab engines by No.3 (Western) Aircraft Repair Depot and were allotted to the RFC in France in February 1918. These also retained Rolls-Royce engine bearers and had an alarmingly unsatisfactory radiator installation. Apparently C4663 went to No.12 Squadron also, and that unit experienced innumerable problems with the Arab engine. Eventually vibration was established as the cause of many of the difficulties, a finding that was confirmed by Martlesham Heath, where B1204 and B8915 were extensively tested with the Arab.

These problems led to protracted development of the Arab, its radiators and its mountings, until a suitable Lanchester-designed installation was evolved by July 1918. This meant that the RFC was unable to re-equip any of its corps squadrons in March 1918 as had been intended, and in mid-June there was a possibility that the Arab Bristol would be abandoned. This did not happen, but the war was nearly over by the time the Royal Air Force began to receive Arab Bristols in quantity. Before the RFC merged with the RNAS to form the RAF on 1 April, 1918, further large contracts for Bristol Fighters were given to several firms of contractors, and for most of these contracts the Arab engine was the specified power unit. With only one or two exceptions Bristol-built F.2Bs had the Rolls-Royce Falcon.

The Bristol F.2B has always been represented as an aircraft of great structural strength, but there was a period, early in 1918, when much trouble was experienced with buckling compression struts in the lower mainplanes. These had apparently been modified at some time during the production of the type: whereas the struts had been of solid spruce on the F.2A, the F.2B had built-up compression members of ash and plywood. As soon as the problem was identified, arrangements were made for subsequent F.2Bs to have solid spruce compression struts, and those aircraft that had the built-up struts were fitted with local reinforcement. As a temporary expedient ash strips were glued and screwed on externally, but a permanent scheme was also devised, and by 10 February strengthening was required on the second, third and fourth compression ribs.

Two more squadrons, Nos.62 and 88, were to fly the Bristol in France, but only No.62 got there before the RFC lost its separate identity. The redoubtable two-seater fought and flew with great distinction until the end, not only in France but in Palestine with Squadrons No.67 and 111; while in Italy in March 1918 'Z' Flight attached to No.34 Squadron provided the nucleus from which No.139 Squadron, RAF, developed. The Bristol also saw Home Defence service in the RFC with No.39 Squadron from September 1917 and subsequently with four other RAF units in 1918.

When No.62 Squadron arrived in France on 23 January, 1918, its Bristol F.2Bs included C4619, then flown by Lt Hampton and 2/AM F. Smith. When photographed here its crew was Capt W. E. Staton and Lt J. R. Gordon, and the aircraft survived its operational career to be flown back to England on 28 July, 1918. (*D. R. Neate*)

Bristol F.2B
190 hp Rolls-Royce Falcon I, 220 hp Rolls-Royce Falcon II
275 hp Rolls-Royce Falcon III, 200 hp Hispano-Suiza
230 hp B.H.P., 200 hp Sunbeam Arab

Two-seat fighter-reconnaissance.

Span 39 ft 3 in; length 25 ft 10 in (Falcon III); height 9 ft 9 in; wing area 405·6 sq ft.

Empty weight 1,934 lb (Falcon III), 1,886 lb (Arab); loaded weight 2,779 lb (Falcon III), 2,804 lb (Arab).

With Falcon III: Maximum speed at 10,000 ft—113 mph, at 15,000 ft—105 mph; climb to 10,000 ft—11 min 15 sec, to 15,000 ft—21 min 20 sec; service ceiling 20,000 ft; endurance 3 hr.

With Arab: Maximum speed at 10,000 ft—104 mph, at 15,000 ft—94 mph; climb to 10,000 ft—14 min 25 sec, to 15,000 ft—29 min 45 sec; service ceiling 17,000 ft.

Armament: One fixed 0·303-in Vickers machine-gun; one 0·303-in Lewis machine-gun, or double-yoked pair, on Scarff ring mounting; up to twelve 25-lb Cooper bombs.

Manufacturers: British and Colonial, Armstrong Whitworth, Gloucestershire Aircraft, Marshall Sons & Co, National Aircraft Factory No.3, Angus Sanderson, Standard Motor.

Service use: *Western Front*—Squadrons 11, 20, 22, 48, 62 and 88; detachments with Sqns 4, 8, 10, 12, 15, 16 and 35. *Home Defence*—No.39 Sqn; later also with Nos.33, 36, 76 and 141. *Palestine*—No.67 (Australian) Sqn, No.111 Sqn. *Italy*—'Z' Flight. *Training duties*: Training Sqns 2, 7, 8, 31, 35, 38 and 59; Training Depot Stations No.21, 44 and 45; No.1 School of Aerial Fighting, Ayr. *Other units*: Wireless Experimental Establishment, Biggin Hill; Wireless Testing Park.

Serials: A7101—7300,
B1101—1350,
C751—1050, 4601—4800, 4801—4900, 9836— 9985,
D2126—2625, 2626—2775, 7801—8100,
E1901—2150, 2151—2650, 2651— 2900, 5179—5428, 9507—9656,
F4271—4970, 5074—5173,
H834—1083, 1240—1739, 3796—3995, 5940—6539,
J1231—1730. 1731—1930, 2292—2391.
Rebuilds: B808, B817, B878, B883, B7763, B7781, B7947, B8914, B8915, B8918, B8919, B8923, B8925, B8928, B8937, B8938, B8941, B8947, B8948, C4548, F5809—F5817, F5819—5824, F5995, F5997—5999, F6040—6043, F6094, F6101, F6116, F6121, F6131, F6195, F6206, F6208, F6217, F6235, F9516 (ex E2460), F9549, F9616, F9638, H6893 (ex C9896), H7060—7065, H7069 (ex B1209), H7070 (ex B1234), H7171—7175 (ex A7256, B1307, C842, C4746, C4817), H7196 (ex C868), H7197 (ex C983), H7198 (ex D8065), H7291—7293.

Bristol M.1A, 1B and 1C

The extent to which Frank Barnwell's engaging little monoplane fighter may have been inspired by the Morane-Saulnier Type N (*see* pages 296–303) will probably never be known, but it is possible that Trenchard's disenchantment with the Morane Bullet may have been part of the reason why the Bristol was ordered belatedly and used only in small numbers in Middle Eastern theatres of war.

In its essential conception, with fully faired fuselage, closely-cowled rotary engine with large spinner on the airscrew, and shoulder-wing configuration, the Bristol monoplane that emerged in July 1916 was similar to the earlier Morane-Saulnier. The Bristol design was a larger, bulkier aircraft, however, and had ailerons for lateral control. The prototype was designated M.1A and was flying in July 1916, in which month it was tested by the CFS Test Flight. Performance was excellent (128 mph at 5,400 ft; climb to 10,000 ft in 8 min 30 sec); the appraisement was complimentary without being enthusiastic. Forward and downward view from the cockpit was criticised as poor, but the view of the entire upper hemisphere was superb.

Despite the M.1A's high speed and excellent climbing performance there was no official display of urgency to develop the type. Not until October 1916 was a contract for four further prototypes given; serial numbers were allocated on 9 October, giving the M.1A the identity A5138 and the other four A5139—5142. These four were designated M.1B and differed from the M.1A in having a pyramidal cabane, pinched-in wing roots at leading and trailing edges, and a large cut-out, covered with transparent material, between the spars at the root of the

Bristol M.1B A5139, the only Bristol M.1B to go to France, photographed at No.2 A.D., Candas. The photograph is dated 14 February, 1917, the date of A5139's arrival at that Depot.

starboard mainplane. A single Vickers gun was mounted on the port upper longeron ahead of the cockpit. The first two M.1Bs had, like the M.1A, the 110 hp Clerget 9Z engine; A5141 had a 130 hp Clerget 9B, and A5142 one of the first of the 150 hp A.R.1 engines.

On 30 December, 1916, A5139 was at Hounslow, under allocation to, of all things, the Training Brigade. On that day it was allotted to the Expeditionary Force, and went to France in January 1917. On 23 January it was tested at No.1 Aircraft Depot, St-Omer, by Lt D. V. Armstrong, Capt A. M. Lowery, and Capt R. M. Hill (later Air Chief Marshal Sir Roderic Hill, KCB, MC, AFC). Lowery and Hill submitted reports on the aircraft, Hill's being a particularly thoughtful, objective, and constructive document. Trenchard may have read Lowery's brief report but he did not wait to see Hill's (dated 26 January, 1917) before writing to the Director of Air Organisation on 23 January, stating that he did not want any more Bristol monoplanes sent out to France. Although by 31 January Maj-Gen Sefton Brancker wrote that the other prototypes were to go to France none ever did.

Curiously, however, A5139 was by no means precipitately sent back to England despite Trenchard's antipathy (and it is relevant to note that it was also at this time

The Bristol M.1Bs A5141 and A5140 with No.111 Squadron in Palestine in 1917.

that the RFC had just received two examples of the very similar Morane-Saulnier Type AC, only to send them engineless to England: (*see* pages 312–314). The Bristol M.1B went to No.2 A.D. at Candas on 14 February, 1917, and rather vague entries in official records indicate that it also went to an unspecified squadron or squadrons during that month. It was not flown back to England until 4 March, and had therefore been in France for some six weeks, quite long enough for word of its qualities to filter through to many RFC scout pilots. They were never to understand why the Bristol monoplane was not adopted for use in France.

In fact A5140 was also officially allotted to the RFC in France as late as 13 February, 1917. It never went there, however. On 6 March A5139 was at Lympne and was re-allocated to the Home Defence Wing. Two days later A5140 was similarly re-allocated, by which date it had got as far as the Southern Aircraft Depot at Farnborough. A5139 was with No.50 (Home Defence) Squadron at Bekesbourne for most of March 1917, and was flown to Farnborough on 5 April by Capt A. J. Capel. On touching down at Farnborough one of the monoplane's wheels caught in a rut, the tyre burst, and the aircraft turned on to its nose, damaging the propeller and one wing.

Bristol M.1B A5142 also went to Palestine and is seen here with its Vickers gun mounted centrally. (*Ministry of Defence*)

Parenthetically it might be noted that one or two Bristol M.1Cs did fly to France, but only as transports for senior officers, or possibly as despatch-carriers. One such was C4965, which was seen at Baizieux during No.56 Squadron's occupation of that aerodrome, either late in February or early in March 1918. It is also known that Col Lee flew C4963 from Hounslow to France on 24 May, 1918, returning via Calais, where the M.1C's tailskid was broken on landing.

Possibly because Brig-Gen W. G. H. Salmond had for some time been pressing the Director-General of Military Aeronautics to provide his Palestine Brigade with better fighting aircraft, and had been informed on 11 July, 1917, that a fighter squadron, No.111, was to be raised in Egypt, the three M.1Bs A5140—5142 were allocated to No.111 Squadron in September 1917. They had been sent to the Middle East in June. On these aircraft the clear-view cut-outs in the mainplane had no transparent covering, and on A5142 the gun was eventually mounted centrally.

M.1C of No.72 Squadron in Mesopotamia.

Typical Bristol M.1C with unusually well finished spinner and engine cowling.

A regrettably rare sight on an RFC aerodrome in France was this Bristol M.1C, C4965, photographed at Baizieux in February/March 1918 with S.E.5as of No.56 Squadron in the background. It may have been used on this occasion as a personal transport by a senior officer, and is believed to have been on the strength of a training squadron at Hounslow that February. It was also reported at Marske, home of No.4 (Auxiliary) School of Aerial Gunnery, in March 1918. (*Alex Revell*)

A contract for 125 Bristol monoplanes was placed on 3 August, 1917. The production aircraft, designated Bristol M.1C, had the 110 hp Le Rhône engine, cut-outs in both wing roots and a centrally mounted Vickers gun. Deliveries began on 19 September, and by the end of 1917 sixty-two M.1Cs had been delivered and fourteen had gone to the Middle East Brigade; a further eighteen followed in 1918, production ending early in March. These aircraft were issued in small numbers to No.72 Squadron in Mesopotamia and to Squadrons No.17 and 47 in Macedonia. Some of the latter Bristols were transferred to No.150 Squadron, RAF, on 1 April, 1918.

The numbers of M.1Cs in operational use were too small to make any significant impression on the campaign in either theatre of war. Their flight endurance was regarded as limited, and they do not seem to have been held in particularly high regard. Maj-Gen W. Sefton Brancker, who had assumed command of the Middle East Brigade of the RFC in November 1917, wrote in a personal letter to Maj-Gen J. M. Salmond: 'The Bristol Monoplanes and Vickers Bullets are not very much good except to frighten the Hun; they always seem to lose the enemy as soon as he starts manoeuvring.'

The Bristol monoplanes served on until and after the Armistice, but the majority of them were used by the RFC and RAF at training stations. Six were sent to Chile before 1917 was out, in part payment for two warships being built for Chile in British yards but commandeered by the Royal Navy; these M.1Cs included C4986 and C4987 which, respectively, were later used by Godoy and Cortinez for trans-Andean flights. That the British authorities were thus ready to part with the M.1C before the type had given any kind of operational account of itself seems to typify the official attitude of indifference towards the little Bristol.

Bristol M.1A, 1B and 1C
M.1A—110 hp Clerget 9Z
M.1B—110 hp Clerget 9Z, 130 hp Clerget 9B, 150 hp A.R.1
M.1C—110 hp Le Rhône 9J

Single-seat fighter.
M.1C: Span 30 ft 9 in; length 20 ft 5½ in; height 7 ft 9½ in; wing area 145 sq ft.
M.1C: Empty weight 896 lb; loaded weight 1,348 lb.
Maximum speed at 10,000 ft—111·5 mph, at 15,000 ft—104 mph; climb to 10,000 ft—10 min 10 sec, to 15,000 ft—19 min 50 sec; service ceiling 20,000 ft; endurance 1¾ hr.
Armament: One 0·303-in Vickers machine-gun with Sopwith-Kauper interrupter gear or Constantinesco C.C. Type B synchronising mechanism.

Manufacturer: British and Colonial.

Service use: *Western Front*—One M.1B briefly attached to various units. *Home Defence*: No.50 Sqn. *Macedonia*—Sqns 17 and 47. *Mesopotamia*—No.72 Sqn. *Palestine*—Three M.1Bs with No.111 Sqn. *Training duties*: No.1 School of Aerial Fighting, Ayr; No.5 Fighting School, Heliopolis.

Serials: M.1A, A5138; M.1B, A5139—5142; M.1C, C4901—5025.

Caproni Ca.1

In Italy Gianni Caproni began designing and building aircraft in the spring of 1910. From the experience gained by Italian aviators in the Libyan war of 1911–12 and the Balkan war of 1912–13 he concluded that in any subsequent war there would be a need for multi-engined bombers, and he at once set about designing such an aircraft. His first thoughts envisaged a twin-fuselage biplane with three rotary engines installed in tandem in a central nacelle. This design was completed in the spring of 1913 and was submitted to the military authorities. These took time to consider such a revolutionary idea, and the completed aircraft did not emerge until October 1914.

As built, the Caproni bomber was powered by three Gnome rotary engines, one of 100 hp at the stern of the central nacelle driving a pusher propeller, and two of 80 hp installed as tractors at the forward ends of the fuselage booms. Official Italian interest was aroused late in 1914, and Italian attempts to interest Britain and France in the Caproni began remarkably early: Italy was not then herself at war. On 26 December, 1914, Signor Arturo Mercanti visited the British Embassy in Rome, where he saw Mr (later Sir) Edward Capel Cure, then the Assistant Commercial Attaché. He asked for introductions to officials in London, so that he might show them drawings of the Caproni, and proposed that the British Government should acquire the rights to build the aircraft, in exchange for

The RFC's solitary Caproni Ca.1, seen at some unidentified landing field in the course of its eventful but abortive delivery flight. This rare photograph was probably taken at one of the places in France at which the aircraft landed. *(Via Jean Devaux)*

raw materials and, in particular, stationary aero-engines. Mercanti further suggested that Caproni aircraft for the British services could be built in Italy, subject to Italian needs for the type being met.

On 28 December the British Military Attaché, Col Granet, sent a telegram to advise the War Office that Signor Mercanti intended to visit London; the Admiralty was also informed. The War Office telegraphed the Military Attaché on 2 January, 1915, that they would be glad to see 'the representative' in London, and Mercanti left Italy on 10 January, 1915, armed with letters of introduction from both Col Granet and his successor, Col Lamb.

Precisely what happened in London is not clear, but things went wrong. Mercanti should have seen Col W. Sefton Brancker, but Brancker was away and his staff were unable to arrange an alternative interview before Mercanti had to leave to keep an appointment with Général Hirschauer in Paris. The Admiralty were apparently antipathetic, for a quotation from an Admiralty document states 'No encouragement was given to Mr Mercanti, who is believed to have returned to Italy.' By contrast, Mercanti's business with the French officers was completed smoothly, expeditiously and punctually, and he was understandably huffy over his reception in London and made his displeasure known to the British Ambassador in Rome. Probably as an immediate consequence of Mercanti's discussions with Général Hirschauer and Cols Bottieau, Fleury and Stammler, a French government representative, Rodolphe Soreau, went to Vizzola to see the Caproni. As a result of Soreau's visit French production of the Caproni trimotor was eventually undertaken in France under licence by Robert Esnault-Pelterie and by the Société anonyme d'Applications industrielles du Bois.

The first Italian order for the Caproni trimotor was for twelve aircraft each powered by three 100 hp Fiat A.10 six-cylinder water-cooled engines, and in March 1915 the Società per lo Sviluppo dell'Aviazione in Italia was formed for the production of the big Capronis, taking over the works at Vizzola Ticino and building a new factory at Taliedo for the purpose. On 24 May, 1915, Italy entered the war on the side of the Allies: the first of the twelve Caproni trimotors was officially accepted on 14 and 15 July, the last on 9 October. These aircraft had the official designation Ca.1; in the postwar numbering system for Caproni types they were the Ca.32. The initial dozen were followed by 150 more, numbered in various batches from Ca.703 onwards.

Despite the mismanagement of Mercanti's visit to London, the RFC returned to the question of the Caproni aircraft, and by 14 August, 1915, arrangements had been made to acquire one specimen aircraft for evaluation; it was to be powered by engines supplied by the RFC, and it was proposed to use three 120 hp Beardmores, which were of the same general configuration as the Fiat A.10. An RFC memorandum of 15 August, 1915, noted: 'M. Caproni expects to get good results from the use of Beardmore engines.'

It was hoped that the RFC's Caproni would be ready to fly by 20 November, 1915, but on 8 December Capt Lord Robert Innes-Ker of the Aviation militaire anglaise in Paris wrote to RFC Headquarters:

'I received yesterday a letter from Turin, informing me that the Société [sic] Caproni had not been able to go on with the trials of the machine, owing to the illness of the engineer and also because it has been impossible to proceed with the construction of the machine owing to the wood not being dry enough. I am, therefore, unable to give you any definite date when the machine will be ready.'

Evidently it was later expected that the aircraft would be ready to fly by 20 December. On that date Lt-Col Brooke-Popham wrote to Capt Lord Innes-Ker, advising him that Capt the Hon Maurice Baring and Capt R. A. Cooper would be going to Turin via Paris in the expectation of seeing the Caproni fly on 29 December; if all went well the aircraft was to be packed in transit cases and sent to Innes-Ker by rail. Orders to this effect were given to Cooper on 25 December; next day Trenchard personally instructed Baring that he and Cooper were to leave for Turin on 27 December but modified the orders by requiring the Caproni, if found satisfactory, to be consigned to the RFC at Farnborough and sent there via Paris.

Baring described his and Cooper's complex journeyings in his classic book *Flying Corps Headquarters 1914–18*. They finally inspected the Caproni at Gallarate, were flown in another at Malpensa by an Italian pilot (named by Baring as Pellegrini but probably Tenente Ernesto Pellegrino), and returned to RFC Headquarters on 31 December, 1915. At that date total deliveries of Ca.1 aircraft amounted to 26.

Trenchard immediately wrote, on 31 December, to the General Officer Commanding the RFC, Lt-Gen Sir David Henderson:

'. . . . Capt. Baring reports that on arrival at Milan the machine was not ready to fly.

Secondly, that nobody in the Italian Flying Corps knows anything about the Austro-Daimler engines.

Thirdly, that the Italian Flying Corps had received no instructions from the Italian Government to let this machine be exported from Italy, and they want the War Office to make an official application for this authority to the Italian Government.

He states the machine will be ready to fly in a few days, but as no one knows how to run the engines it is improbable that it will be able to start.

* * * * * * * *

I would suggest that Capt. Valentine with some Austro-Daimler mechanics be sent to Turin from home, all arrangements being made by the War Office.'

Perhaps because it had proved impossible to obtain the services of any pilot, Italian or French, to fly the aircraft out of Italy, Trenchard's suggestion was accepted: by 7 January, 1916, it had been officially decided to accept the Caproni as it stood and have it taken to England. Such was the design of the aircraft that, if it were to be dismantled, re-erection would take six weeks and could only be accomplished by Caproni personnel, who could not in any case be spared. It was therefore essential to fly the Caproni to its destination, and as Capt James Valentine was an experienced pilot it was natural that he should fly the aircraft. Evidently someone succeeded in starting the Beardmore engines, for his marathon flight began on 12 February, 1916, when, accompanied by a mechanic, he flew the Caproni from Milan to Gremaude, near Antibes, in $3\frac{1}{2}$ hr. After refuelling he left for Marseilles, possibly in the hope of reaching Lyon, but troubles began to assail the flight and general progress became slow and spasmodic.

Valentine got no farther than St-Raphaël, and the Caproni was still there four weeks later, for on 8 March Capt Lord Robert Innes-Ker reported to RFC Headquarters:

'. . . . the Caproni machine is at present at St-Raphaël. Capt. Valentine has been unable to start owing to bad weather and owing to the aerodrome being flooded. In his last attempt he tried to start from the beach and broke two of the propellers. He arrived in Paris Monday morning and leaves today for St-Raphaël with two new propellers.'

Towards the end of March the Caproni was one of several matters discussed by Brooke-Popham and Innes-Ker, and Brooke-Popham noted that

'Innes-Ker is anxious for orders to be given to Valentine to go straight on with his Caproni machine and not to stop at Paris on his way as he probably will.'

Hence, on 1 April, Brooke-Popham instructed Valentine:

'The G.O.C. directs that as soon as your Caproni machine is ready you are to take it direct to St-Omer without stopping at Paris except for unavoidable reasons connected with the machine.'

This close-up of the RFC's Caproni Ca.1 clearly shows two of its three Beardmore engines and the primitive nature of their installations. The figure in the central nacelle is said to be Captain James Valentine. *(Via Jean Devaux)*

Valentine did his best. By 29 April he had got his reluctant charge as far as Dijon, where he had 'strained two of the main struts', but he hoped to be able to set out for Paris on 2 May. Certainly he went—again—to Paris, but not in the Caproni, for he was subsequently reported to have returned to Dijon on Saturday 6 May, taking with him a new magneto obtained from Chalais Meudon.

He managed to get airborne from Dijon on 12 May, but for unrecorded reasons came down in a field some 50 miles north of the town, 'having buckled two struts on his machine'. Yet again Valentine journeyed to Paris, arriving there on 16 May, by which time it had evidently been officially decided to abandon the Caproni, for he received instructions from Lt-Col Beatty about the disposal of the aircraft, and returned to Dijon on 17 May. The three Beardmore engines were taken out and sent to Boulogne, where they arrived on 1 June, 1916. The airframe was handed over to the French authorities at Dijon, its ultimate fate being unknown. All that is certain is that the RFC never quite received its only Caproni trimotor.

Ca.1
Three 120 hp Beardmore

Four-seat bomber.
Span 22·74 m; length 11·05 m; height 3·84 m; wing area 95·64 sq m.
Empty weight 2,000 kg; loaded weight 3,000 kg. Weights for version with 100 hp Fiat A.10s.
Maximum speed 121 km/h; climb to 1,000 m—10 min, to 2,000 m—22 min; ceiling 4,000 m. Performance with Fiat engines. It appears that during its abortive delivery flight the RFC's Ca.1 reached 130 km/h with an 825 kg load.
Armament: The standard Ca.1 could carry 650 lb of bombs and two machine-guns.

Manufacturer: Società per lo Sviluppo dell'Aviazione in Italia.

Serial: As far as is known no RFC number was allotted.

Caudron 45 hp biplanes

A Government scandal of considerable magnitude occurred in 1913. The Government had promised that the RFC would have seven squadrons of twelve aeroplanes, each with six further aircraft in reserve; yet the Army Estimates for 1913 provided only £501,000 to meet all the needs of the Military Wing, Central Flying School and the Royal Aircraft Factory. The palpable inadequacy of this sum led to a public and Parliamentary outcry, and the enquiries of Members of Parliament led to the discovery that only three squadrons, none of them effectively equipped, then existed.

When Col J. E. B. Seely introduced the Estimates on 19 March, 1913, he stated that the RFC had 101 aeroplanes; by 4 June he claimed that the Service had '126 aeroplanes, of which 31 were in various stages of repair'. Searching investigations showed that some elasticity of the imagination was required to arrive at such totals, and in the spring of 1913 the War Office purchased some ill-assorted and highly unmilitary aeroplanes, presumably in an effort to give some substance to Col Seely's somewhat fanciful claims.

The 45 hp Caudron that became No.311 in the RFC, photographed at Hendon before delivery to Farnborough. (*Flight International*)

A 45 hp Caudron, almost certainly the same aircraft that became 311, photographed at Farnborough.

Contemporary journals report that the War Office bought a Caudron biplane immediately after the 1913 Olympia Aero Show ended. The two Caudron aeroplanes exhibited there, a monoplane and a 35 hp biplane, were on the stand of the W. H. Ewen Aviation Co Ltd, which had secured the British agency for Caudron aircraft in April 1912. It was reported that both of the Aero Show Caudrons were sold to private owners, hence the War Office aircraft must have been another. It may have been the two-seater with 45 hp Anzani engine that became No.308 in the RFC, for that Caudron was allotted to the Military Wing on 18 April, 1913. Its origins are obscure, but it was apparently bought from the Ewen company and may well have been far from new when acquired by the RFC, for it is believed to have done only 20 minutes' flying while with the Military Wing. Although it was officially allotted to No.4 Squadron it seems that it never left the Flying Depot at Farnborough, and efforts to make it serviceable were finally abandoned in February 1914.

Caudron No.311 photographed at Farnborough after acceptance by the RFC (*RAE*)

In the process of building up Col Seely's June total of 126 aeroplanes the War Office bought another Caudron biplane fitted with a 45 hp Anzani engine, and there is no doubt that this was delivered new. This Caudron, which approximated to the Caudron Type C single-seater then being offered for sale by the Ewen company, had been built for Ewen by Hewlett & Blondeau at their Clapham works, and incorporated a few modifications from the more-or-less standard design. An upswept top decking gave the pilot a measure of comfort and protection, and the engine was fitted with exhaust manifolds. The aircraft was seen at Hendon, its port rudder bearing the Ewen name and the number 51.

This Caudron of mixed parentage was sold to the War Office about the middle of May 1913 and acquired the official identity of 311. It was used by No.4 Squadron and was reported to be flying with that unit by mid-October. Its pilots included Maj G. H. Raleigh, Capt H. R. P. Reynolds and Lt A. H. L. Soames. By December its total flying time was 17 hr 27 min. Apparently No.311 survived until 1 April, 1914, when, following an adverse report by No.4 Squadron, it was condemned.

Caudron 45 hp biplane
45 hp Anzani

One or two-seat tractor biplane.
Type C: Span 10·365 m (upper), 6·86 m (lower); length 6·78 m; wing area 25·25 sq m.
Type C: Empty weight 260 kg.

Manufacturers: Caudron in France; No.311 built by Hewlett & Blondeau for W. H. Ewen Aviation Co Ltd, Hendon.

Service use: No.4 Sqn.

Serials: 308, 311.

Caudron G.3

The Caudron G.3 was the ultimate expression of the formula established by the earlier Caudron Types C, D and E, being the quaintly characteristic combination of the fundamentally pusher configuration with a tractor engine installation. This classic type first appeared in France in May 1914 but saw no service in the RFC before the war began.

As the war progressed, the RFC in France had to seek fresh aircraft wherever they could be obtained and naturally turned to the native French aircraft industry in its endeavours to augment the meagre supplies coming from England. Among the French aircraft that the RFC succeeded in purchasing was a modest number of Caudron G.3s. The first of these were officially numbered 1884 and 1885, which were delivered direct to No.1 Squadron on 27 March, 1915. Nos. 1886 and 1887 (C.583) were delivered to the 1st Aircraft Park on 1 April, No.1887 being issued to No.5 Squadron three days later. It stayed with that unit until 8 May and was sent to England from the 1st A.P. on 26 May.

Perhaps there was some special significance in the career of No.1886. It was struck off RFC strength on 18 May without going to a squadron, but was revivified from salvage on 1 August for use at the RFC School at Le Crotoy. It was

Caudron G.3 No.1887 photographed at Farnborough. This aircraft was received at the 1st Aircraft Park from Paris on 1 April, 1915, and was issued to No.5 Squadron three days later. It was returned to the Aircraft Park on 8 May, and was subsequently sent to England on 26 May, 1915.

despatched to England in October and was allotted to the RFC's Administrative Wing at Farnborough on 28 October; it was subsequently allotted to No.1 Reserve Aeroplane Squadron, Farnborough, on 2 November. Its service at Le Crotoy and with No.1 R.A.S. qualifies it as one of the first RFC Caudrons to serve as a trainer: not quite the first, however, for it had been preceded at Le Crotoy by 5016 and 5032, which had been delivered there on 15 May and 24 June, 1915, respectively.

In the Field the RFC employed the G.3 as a reconnaissance aircraft, but it was never numerous with operational units. By the end of June 1915 No.1 Squadron had 1884, 1885 and 1891, while 1895 had been struck off on 31 May. No.4 Squadron had had 1900 and 5003 for a few weeks, but they had gone back to the

A Caudron G.3 of No.5 Squadron at Hazebrouck Farm, early 1915. This was probably No.1887, for it was the only Caudron to be recorded on the strength of No.5 Squadron.

This Caudron G.3 had an inelegantly installed Anzani radial engine, presumably of 80 hp. It has been reported that this photograph was taken at Kinson, an airfield near Bournemouth said to have been used by the RFC

Aircraft Park on 7 June and 22 May respectively. As noted above, No.5 Squadron had had 1887 for a short time.

That was about the total operational use of the Caudron G.3 with the RFC. Last in the Field was 1885, which was finally struck off No.1 Squadron's strength on 2 October, 1915. Numbers of others were delivered to the 1st Aircraft Park in the summer of 1915, the last being 5063, which arrived on 5 September. Of these, 5020, 5031, 5037, 5038 and 5040 were sent to England; seven others were struck off without being issued to units.

That RFC Headquarters did not have a high opinion of the Caudron G.3 as an operational aircraft is reflected in a communication sent to Capt Lord Robert Innes-Ker of the British Aviation Supplies Depot on 30 November, 1915: 'The General [Trenchard] has never attached very much importance to the delivery of

At least some of the Caudron G.3s of the batch A1892—A1901, built by the British Caudron company, had the 70 hp Renault engine. One such was A1893, here seen at Hendon. (*Flight International*)

the single-engine Caudrons ..... of course the more Caudron's deliver the better. We value the 80 hp Le Rhône engines much more than the machines which contain them.'

The RFC continued to acquire and use the G.3 as a trainer, though never in large quantities, as the list of serial numbers indicates. Production in France was substantial, the Caudron company alone building 1,423; and several modest production batches were made in England by The British Caudron Company, which was the successor to the W. H. Ewen Aviation Co Ltd. Of these, A1892—1901 are believed to have been built with the 70 hp Renault engine; some aircraft had the 80 hp or 100 hp Anzani radial, but the majority of the RFC's Caudron G.3s had the 80 hp Gnome rotary. The final RFC batch, C5276—5300, were ordered from the parent French Caudron company in September 1917, the specified engine being the 100 hp Anzani.

Two Caudron G.3s were used briefly in 1915 in Mesopotamia. They arrived at Basra in July and were at once sent to Nasani to make reconnaissances of the Nasiriya positions. They made a few such flights and spotted for the British artillery, but one was lost on 30 July. The hot and dusty conditions were inimical to the 80 hp Gnome engines of these G.3s, but the second aircraft survived until 16 September, when it was brought down by rifle fire from the ground.

G.3

80 hp Gnome, 80 hp Le Rhône, 70 hp Renault, 80 hp or 100 hp Anzani
Two-seat reconnaissance and trainer.
Span 13·26 m; length 6·89 m; height 2·59 m; wing area 28·27 sq m.
80 hp Gnome: Empty weight 435 kg; loaded weight 710 kg.
Speed at sea level 105 km/h; climb to 2,000 m, 27 min; service ceiling 3,050 m; endurance 3½ hr.
Armament: Various combinations of rifles and pistols.

Manufacturers: Aéroplanes Caudron (France), British Caudron (United Kingdom).

Service use: *Western Front*—Sqns 1, 4 and 5; *Mesopotamia*—No.30 Sqn. *Training duties:* B.E.F. School, Le Crotoy, France; Reserve Aeroplane Squadrons 1, 3, 4, 6, 9 and 41. *Other units* (training purposes) Sqns 13, 14, 23 and 29.

Serials: 1884—1887, 1891, 1895, 1900, 2863, 4226, 4254, 4293, 4299, 4733, 4734, 4837, 5003, 5016, 5020, 5024, 5031, 5032, 5035, 5037, 5038, 5040, 5042, 5043, 5049, 5050, 5053, 5062, 5063, 5251—5270, 5613, 5615, 5913—5915, 7312—7320, 7741—7743,
A1892—1901, 2123—2124, 2993—3005, 3024—3048,
C5276—5300.

Caudron G.4

The Caudron G.4 was an enlarged development of the G.3 that was powered by two engines, normally 80 hp Le Rhône rotaries or 100 hp Anzani radials. It first appeared in March 1915 and saw extensive use on the Western, Italian, Aegean, and Russian fronts; a total of 1,358 G.4s were made in France, and the type was also produced in Italy and England. Those made in England by The British Caudron Company were supplied to the RNAS under Admiralty contract; they each had two 100 hp Anzani radial engines.

It seems that the RFC was introduced to the G.4 in July 1915, when Lt-Col J. W. Higgins, Commanding III Wing, and Maj Pope-Hennessy visited Capitaine St-Quentin, commanding the Aviation Group of the French Second Army. They were shown a Caudron G.4, and on 21 July Lt-Col Higgins sent a descriptive report to RFC Headquarters. He had been told that the aircraft could carry 500 kg in addition to the pilot, could achieve 120 km/h, climb to 2,000 m in 10 min with four hours' fuel, and fly on one engine without appreciable loss of height. He went on:

> 'I think it would be worth while trying this machine as it would be useful for protective work, for artillery observation and for bomb dropping.'

Evidently RFC Headquarters agreed, despite a lukewarm opinion of the single-engine G.3. On 30 November, 1915, Headquarters wrote to Capt Lord Robert Innes-Ker of the B.A.S.D.:

> 'The General has never attached very much importance to the delivery of the single-engine Caudrons. As long as we get the twin-engine machine we shall not be very dissatisfied . . .

A typical Caudron G.4 with two 80 hp Le Rhône engines. (*K. M. Molson*)

That quotation suggests that an RFC order for a specimen Caudron G.4 had been placed before the date indicated. The aircraft that arrived at No.1 Aircraft Depot, St-Omer, from Paris on 14 January, 1916, had been built by the parent Caudron company with the maker's number G.500; it was allotted the RFC serial number 7761. This was possibly the twin-engine Caudron that 2nd Lt F. G. Dunn (who was later to lose his life in the crash of the ill-fated Tarrant Tabor) was expected to fly from France to England in January 1916, and it is known that G.500 was flown to England on 23 January.

A second Twin Caudron (presumably a G.4) arrived at No.1 A.D. on 29 February, 1916. This one had the French identity G.92 and was fitted with Lewis gun No.6112. It flew on to England that same day, but no British serial is known to have been allotted to it. When, where and by whom either or both of these Caudrons were evaluated remains unknown: all that is certain is that the RFC did not adopt the Caudron G.4.

G.4

Two 80 hp Le Rhône or 100 hp Anzani

Two-seat reconnaissance-bomber.

Span 16·885 m; length 7·19 m; height 2·55 m; wing area 36·828 sq m.

Le Rhônes: Empty weight 733 kg; loaded weight 1,232 kg.

Maximum speed at sea level 130 km/h, at 2,000 m—125 km/h, at 3,000 m—124 km/h; climb to 1,000 m—6 min 30 sec, to 2,000 m—15 min; ceiling 4,300 m; endurance 5 hr.

Armament: Normally one machine-gun on nose of nacelle.

Manufacturer: Caudron.

Service use: No operational use by RFC.

Serial: 7761 (G.500).

Caudron R.11 No.4964 photographed at Martlesham Heath, where it arrived on 15 December, 1917. This aircraft later received the RFC serial number B8823, and on 23 April, 1918, it went to Orfordness. It was written off on 13 August, 1918.
(*RAF Museum P6545*)

Caudron R.11

It was a far cry from the alarmingly frail-looking and slightly bizarre Caudron G.3 and G.4 to the potent and elegant R.11 that was designed by Paul Deville early in 1917. Although conceived as a Corps d'Armée type for the French Aviation militaire the Cau.11.A 3 found its most telling employment as an escort fighter in 1918, and probably did more than any other type to consolidate French faith in the multiplace de combat concept. Technically, its design embodied several sound ideas, including an ingenious fuel system with jettisonable tanks.

Unfortunately, production was retarded by a number of difficulties, not the least of which were the various ailments of the geared 200 hp Hispano-Suiza engines that formed the standard power installation. It was stated officially in France on 27 January, 1918, that only two Cau.11.A 3s were then available to the French flying service. Nevertheless, two had been delivered to the RFC several weeks earlier.

This is believed to be one of the two Caudron R.11s acquired by the RFC in November 1917.

It appears that the RFC saw in the Caudron R.11 a potential bomber. On 2 October, 1917, the British Aviation Commission in Paris submitted a remarkably detailed report on the aircraft to RFC Headquarters; René Caudron was asked whether the R.11 could be used as a bomber, and how long it would take to effect the necessary modifications. Apparently it was hoped that the type might be able to bomb Berlin, but in the opinion of Capt Vernon Brown, RFC (later Air Cdre Sir Vernon Brown, CB, OBE, MA) who had flown the R.11 from the rear seat, it '. . . might have got there but certainly would not have got back.' Despite this RFC interest in the R.11 the Caudron design office was, in November 1917, working on the design of the Caudron CRB, a twin-Hispano-Suiza bomber specifically for the British Air Board.

The RFC ordered two R.11s. The first, No.4962, was delivered in mid-November 1917; the second, No.4964, was ready for its acceptance tests on 24 November. On that day Pierre Chanteloup flew 4962 from Issy-les-Moulineaux to Hounslow, and five days later Etienne Poulet flew 4964 over the same delivery route. Towards the end of February 1918 these R.11s were given the RFC serial numbers B8822 and B8823.

In its report dated 15 December, 1917, the Aeroplane Experimental Station at Martlesham Heath recorded its intention of collecting 4962 from Hounslow 'at the earliest opportunity'. On that day 4964 arrived at Martlesham, but engine troubles delayed its testing, and by 5 January both of its Hispano-Suizas were out for overhaul. When restored to health they were promptly put into Dolphins to speed the work being done on that type, which probably commanded higher priority in Martlesham's eyes. New engines were awaited for weeks, and by 2 March, 1918, it had been decided to re-allot 4964; by 6 April allocation to the experimental armament station at Orfordness had been decided, and the Caudron went there on 23 April.

Apparently 4962 (B8822) had gone direct to Orfordness earlier, having been flown there from Hendon by Lt F. D. Holder, and the Royal Air Force used the two for experimental purposes. In June 1918 one was fitted with a three-position installation of an 'inter-communication telephone' system, and on 13 August B8823 (4964) was written off. B8822 survived the war and was intended for preservation. Needless to say, it did not survive for long after the Armistice.

R.11

Two 215 hp Hispano-Suiza 8Bda

Three-seat general-purpose biplane.
Span 17·92 m (upper), 16·97 m (lower); length 11·22 m; height 2·8 m; wing area 54·25 sq m.
Empty weight 1,422 kg; loaded weight 2,167 kg.
Maximum speed at 2,000 m—183 km/h, at 3,000 m—178 km/h, at 5,000 m—164 km/h; climb to 2,000 m—8 min 10 sec, to 3,000 m—14 min 30 sec, to 5,000 m—39 min; ceiling 5,950 m; endurance 3 hr.
Armament: It is doubtful whether the two Caudron R.11s supplied to the RFC had any standard weapon installation, but the French Cau.11.A 3 could have five machine-guns.

Manufacturer: Caudron.

Service use: Flown experimentally at Martlesham Heath and Orfordness.

Serials: B8822—8823.

Cody V biplane

That highly individualistic and much-loved pioneer of British aviation Samuel Franklin Cody entered two aircraft in the Military Aeroplane Competition of 1912. One was to be the Cody IV monoplane, the other the Cody III biplane that had been his entry in the 1911 Circuit of Britain contest. Unfortunately the biplane crashed on 3 July, 1912, while being flown by Lt H. D. Harvey-Kelly, and five days later the monoplane was wrecked when Cody was compelled by engine failure to make a forced landing and collided with a cow that ran into his path.

The monoplane had been powered by the excellent 120 hp Austro-Daimler engine that had been fitted to the Etrich monoplane flown in the 1911 Circuit of Britain by Leutnant H. Bier; the Etrich crashed at Hatfield on 24 July, 1911, and

The Cody biplane that won the Military Trials and later had the RFC number 301. (*RAF Museum*)

The RFC's second Cody was numbered 304. Since November 1913 this aircraft has been in the care of the Science Museum at South Kensington in London.

Cody later bought its engine. It was a sound investment. This Austro-Daimler subsequently survived the crash of the Cody monoplane with little damage, so with characteristic energy and determination Cody set about the construction of a biplane that embodied re-usable components from the Cody III biplane, and in it he installed the remarkably durable Austro-Daimler. All this he accomplished before the end of July 1912.

This new biplane was subsequently designated Cody V, and was of typical Cody appearance and construction. It had warping wings, dual-action divided forward elevator, and two kite-shaped rudders generally similar to those of the monoplane, each carrying a minute horizontal surface on its central line. Seats, of contemporary agricultural-machine pattern, were provided for four persons.

A more unmilitary looking aeroplane could scarcely be imagined, yet such was the nature of the tests and the system of judging in the Military Trials that the Cody V emerged as the inappropriate winner. Its positions in the various tests were as follows:

Quick assembly	13th (1 hr 35 min)
Dismantling and re-assembling	4th (51 min)
Climbing	3rd (288 ft/min)
Gliding	4th equal (1 in 6·2)
Consumption	
(a) Petrol	8th (9 gal/hr)
(b) Oil	1st (0·42 gal/hr)
Range	5th (336 miles)
Speed	
(a) Maximum	4th (72·4 mph)
(b) Minimum	3rd (48·5 mph)
Speed-range	1st (23·9 mph)
Quick take-off	6th (320 yds)

Additionally, the Cody passed the three-hour and ploughed-field-landing tests.

These results provided no sensible basis for declaring the Cody biplane the winner of the competition, and one can only conclude that the judges must have been overly impressed by the unquestionably excellent field of view enjoyed by the Cody's crew and, perhaps, by the hand starter and the braking device, this last being an example of Neanderthal technology. It consisted of a chain wrapped round the landing skid to provide extra ground drag; this chain could be lifted by the pilot to permit take-off.

In the circumstances the RFC was compelled to accept the Cody biplane, but it did not compound the judges' misguided decision by ordering a fleet of the type: only one further example was ordered by the War Office. The competition winner itself was not formally delivered to the RFC until 30 November, 1912. In the interim Cody replaced its Austro-Daimler engine by a 100 hp Green in order to make the biplane all-British, and made a flight of 186 miles to win the British Empire Michelin Cup No.2 for 1912. The Austro-Daimler was again installed when the aircraft was handed over to the RFC. For some time before the delivery of the Cody biplane to the RFC its constructor had been instructing Lt L. C. Rogers-Harrison on it.

In December 1912 the Cody was allocated to the newly formed No.4 Squadron and received the official serial number 301, the numbers 301—350 having been allotted for aircraft used by the new squadron. The second Cody V to be delivered to the RFC was reported to be flying on 18 January, 1913. It was handed over in early February 1913 and, destined for No.4 Squadron, was numbered 304, what time the original was on display in the Olympia Aero Show, tended by RFC mechanics. Its descriptive placard announced that it had by then flown more than 7,000 miles.

Contemporary reports on No.304 indicate that on this second Cody V the elevators were placed slightly higher, the mainplanes had positive dihedral, and the rudders were closer together. Early in April 1913, No.301 was modified by re-rigging its wings with dihedral instead of their original anhedral, all the wiring of the wings being renewed in the process. At about 6.20 a.m. on Monday 28 April, 1913, while No.301 was being flown by Lt L. C. Rogers-Harrison, the elevators and mainplanes broke up at a height of 500 ft and the pilot was killed. At the enquiry it was established that parts of the aircraft were from the original Cody III and had been built in June 1911; the covering of the elevators had not been changed since July 1911. The committee of enquiry concluded that structural deterioration had taken place, and that the Cody's 'condition at the time of the flight was precarious'.

No.304 was at that time out of commission, having been wrecked on 31 March, 1913, and a decision on its repair was awaited. It had flown barely 2½ hr with the RFC. Probably the destruction of No.301 sealed the fate of No.304, for it never flew again with No.4 Squadron and in November 1913 was offered to the Science Museum, in which institution it is preserved.

Cody V
120 hp Austro-Daimler

Two/four-seat biplane.
Span 43 ft; length 37 ft 9 in; wing area 430 sq ft.
Empty weight 1,900 lb; loaded weight 2,500 lb.
Maximum speed 72·4 mph; climb to 1,200 ft, 3 min 30 sec.

Manufacturer: S. F. Cody.

Service use: No.4 Sqn.

Serials: 301, 304.

This camera-equipped Curtiss JN-3, A3277, was used by the School of Photography. One of many transferred from the RNAS to the RFC, it had the Curtiss c/n 162.

Curtiss JN-3

The first Curtiss landplanes to be supplied to either of Britain's flying Services were six two-seat tractor biplanes delivered to the RNAS in March 1915. They received the official serial numbers 1362—1367 and were apparently modified Curtiss N aircraft; they may conceivably have been the otherwise unaccounted-for Curtiss JN-1. At least five of these early Curtiss aeroplanes were allocated to the Special Flight of No.1 Squadron, RNAS, but their activities remain obscure. By 1 July, 1915, several of them were at Eastchurch, ominously recorded as 'not to be flown'.

A much larger Admiralty order was placed for 200 Curtiss landplanes and 50 flying-boats. Of the landplanes the first one hundred (3345—3444) were Curtiss JN-3s, the remainder R-2s. Deliveries began in mid-1915, and the RFC promptly cast covetous eyes on the JN-3s. On 15 July Capt F. Conway Jenkins, RFC, discussed with Capt W. L. Elder, RN, the possibility of diverting a proportion of RNAS Curtiss JN-3s to the RFC. This was formalised by a letter sent from the Army Council to the Admiralty on 20 July, asking for 100 Curtiss aeroplanes 'of the type now being delivered in England under an order placed by Their Lordships'. The Admiralty agreed on 30 July, and on that day the RFC Administrative Wing at Farnborough was instructed to 'commence the rigging of Curtiss machines, drawing them as required from the Admiralty Store, White City'. The agreement was for ten to be handed over at once, followed by one-third of each week's deliveries from the USA. Also on 30 July, 1915, Flt Lt Sidney Pickles flew JN-3 No.3348 from Hendon to Farnborough on what was apparently a performance evaluation rather than a delivery flight, for the aircraft was flown to RNAS Chingford on 31 July.

It seems that the first ex-Admiralty Curtiss to be delivered to the RFC went to Northolt. It was flown to Farnborough on 20 August, 1915, and was examined by Lt-Col J. D. B. Fulton, Chief Inspector of the AID. Fulton was not impressed and reported to the Secretary, War Office:

> 'I think it necessary to place on record that constructionally this machine leaves a great deal to be desired. The workmanship and material throughout are of a cheap and typically American kind . . . It is impossible to effect any improvement in the machines without practically scrapping all the existing fittings and making new ones . . . The use of these machines is only justified by war necessity.'

The Admiralty also had reservations about the Curtiss and on 31 August advised the War Office that they were introducing certain modifications, of which the most significant provided for the strengthening of the interplane bracing. On 17 September a lengthy list of modifications to be made to JN-3s was sent to the O.C. Administrative Wing, RFC.

Lt-Col Fulton's misgivings had been confirmed on 30 August by Capt E. N. Fuller, the officer commanding No.17 Squadron at Hounslow. Reporting to the O.C. Fifth Wing he wrote:

> 'I have carefully examined the construction of the Curtiss 5404 . . . I recommend that an effort be made to prevent any more of these machines from being allotted to the Fifth Wing.'

During the first two weeks of November 1915 structural loading tests were performed at the Royal Aircraft Factory on the JN-3 that had the Curtiss works number 72. Fully loaded, the wings proved to have a factor of only 5·3, but this was improved to 6·4 by fitting stronger lift cables in the inner bracing bay, a modification that the RNAS had been making on its JN-3s for some months.

The airscrews supplied with the JN-3s were found to be inefficient; moreover it was known that four airscrews on RNAS aircraft had disintegrated when run. The help of the Lang Propeller company was sought, and three experimental airscrews were ordered from them in November 1915. By the 12th of that month there were four Curtiss JN-3s at Hounslow and a fifth was expected.

In fact deliveries to the RFC were slow; forty-one JN-3s had been handed over by the end of November 1915, and thereafter no deliveries were made for several months. More were handed over in the spring of 1916, but these may have been aircraft that had already seen service with the RNAS. At that time the RFC made 23 modifications to each Curtiss, many of them designed to strengthen the structure, and on 8 June the RNAS advised the RFC of further reinforcing modifications that they were making. These included the fitting of enlarged tail surfaces, both vertical and horizontal, but it seems unlikely that the RFC adopted these for its JN-3s.

Although the JN-3s were intended to be trainers they were delivered without dual control, the flying controls fitted in the rear cockpit being of the so-called Deperdussin type, with lateral control from a wheel mounted centrally on a rocking bridge, the fore-and-aft movement of which actuated the elevators. It appears that dual control was not in fact installed in RFC JN-3s until June 1916. Even the first JN-4s did not at that time have dual control . It was then rather late in the day, for on 24 May, 1916, Brig-Gen J. M. Salmond, then commanding the 6th Brigade, had written to the Director of Aeronautical Equipment pointing out that many Curtiss aircraft and engines were unserviceable for want of spares, and seeking permission to cannibalise existing aircraft for spares because he understood that the type was to be allowed to die out of service. This was agreed on 27

May, when it was confirmed that the type was to die out. Nevertheless, transfers from the Admiralty continued in small numbers, the last JN-3s being issued to Training Brigade from the Ordnance Aircraft Depot at Milton on 20 October, 1916; these were A5492—5496.

Some of the JN-3s delivered to the Admiralty were made at the Canadian factory of the Curtiss company; by 10 November, 1915, the Admiralty had received a total of sixteen Canadian-built JN-3s. Because the workmanship of these was believed to be inferior to that of the American-built aircraft none was handed over to the RFC, but so much difficulty was experienced in getting any aircraft at all that the Assistant Director of Military Aeronautics indicated that the RFC would accept the Canadian JN-3s. Authority was given for the RFC to draw four Canadian-built JN-3s on 11 November, 1915, and there are grounds for believing that these were numbered 7308—7311.

No.7310, one of the four Canadian-built JN-3s to go to Egypt late in 1915, is here seen at Abu Qir, where it was used by No.22 Reserve Squadron.

It was decided late in October 1915 to send six of the RFC Curtiss aircraft to the Middle East. On 29 October it was ruled that only four were to go to Egypt, and the choice evidently fell on the Canadian-built quartet 7308—7311. These aircraft had modified centre-section bracing with an additional diagonal strut on either side, and were used by No.22 Reserve Squadron at Abu Qir.

Rather more than the originally promised one hundred JN-3s saw service with RFC training units. An attempt was made in October 1916 to obtain a further 100 from the RNAS but this was refused by the Admiralty. On 21 October the Training Brigade was informed that it would not be possible to replace the Curtiss aircraft then in use, and it is therefore doubtful whether the JN-3 survived into 1917 in any numbers with the RFC.

JN-3
90 hp Curtiss OX

Two-seat trainer.
Span 43 ft 10 in; length 27 ft 2½ in; height 9 ft 11 in; wing area 360 sq ft.
Empty weight 1,300 lb; loaded weight 1,918 lb.
Maximum speed 72·89 mph; climb to 1,000 ft—2 min 25 sec, to 2,000 ft—5 min 50 sec, to 3,000 ft—9 min 10 sec.

Manufacturer: Curtiss Aeroplane Co (Buffalo, N.Y. and Toronto, Canada).

Service use: *Training duties*: Reserve/Training Sqns 1, 4, 6, 7, 10, 11, 16, 18, 22 and 36. Other units—Sqns 17, 23, 25 and 56; No.1 Training Depot Station; School of Photography.

Serials: 2339, 5404—5412, 5606—5608, 5624—5641, 5722—5727, 5910—5912, 6116—6135, 7308—7311, A614—625, A898—903, A1254—1260, A3022, A3276—3280, A3831, A4056—4060, A5205—5206, A5492—5496.

The first Curtiss C at Farnborough. At the time of this early photograph it had cable-braced extensions and three-blade airscrews. The six-blade fan that drove the gyroscope of the Sperry Stabilizer can be seen on the prow of the fuselage. (*RAF Museum*)

Curtiss C

Although the Canadian Government, early in the 1914–18 war, showed no real inclination to become involved in aviation, the American Curtiss company saw some portents sufficiently encouraging to induce it to set up a Canadian subsidiary. On 12 April, 1915, the company started its Canadian activities in a rented factory at 20 Strachan Avenue, Toronto, where its first undertaking was the construction of eighteen Curtiss JN-3s.

The parent Curtiss company had prepared the design of a large twin-engined landplane that owed much of its geometry to the Curtiss H-1 flying-boat that had been designed to attempt the Atlantic crossing in 1914. Associated with Glenn Curtiss in the trials of the H-1 and designated as its pilot in the projected Atlantic flight was Lt John Cyril Porte, RN. The prototype of the flying-boat and a sister aircraft were acquired by the British Admiralty in the autumn of 1914 and, numbered 950 and 951, were the precursors of a sizeable number of Curtiss H-4 flying-boats for the RNAS. These small 'boats were not conspicuously successful, but they had the historical merit of providing the basis of the brilliant development work conducted at RNAS Felixstowe by Sqn Cdr Porte.

There can be little doubt that Porte played some part in the Admiralty's acquisition of 950 and 951, and his close connection with Glenn Curtiss probably ensured that he knew something about the Curtiss landplane and a contemporary flying-boat design of more advanced conception than the H-1. At some stage the RNAS ordered an example of the landplane, apparently under the name of

Curtiss Columbia, which designation is associated with the serial number 3700 in Admiralty listings of December 1915 and March 1916. By the latter month much had happened where the twin-engined landplane was concerned.

As the Curtiss factories at Hammondsport and Buffalo were fully committed to the production of JN-3s and H-4s, responsibility for the twin-engined landplane was transferred to the Canadian factory. Construction of the aircraft began about the end of June 1915 and proceeded with remarkable speed, for the big Curtiss was completed by the beginning of September. It was a bizarre structure that did not obviously owe much to the H-1 flying-boat, for the system of supports for the tail unit included two large-diameter tail-booms that grew out of the rear ends of the two engine nacelles. This arrangement in fact anticipated the design of the Curtiss H-7 flying-boat. A third and lower boom of similar aspect was built on to the stern of the crew's nacelle, which was attached to the lower wing by its upper longerons, much as was the hull of the earlier flying-boat. A wide cockpit for two gunners side by side was located well ahead of the wings. Behind the mainplanes was another cockpit for the pilot, whose forward view must have been seriously restricted; indeed, he had little outlook in any direction. The twin-engined Curtiss was intended to be a bomber and was equipped with a Sperry stabiliser, an early form of automatic pilot. When it was designed the need for effective defensive armament had not been adequately recognised, and the extraordinary configuration of the aircraft made it impossible to convert the rear cockpit into an effective gun position.

The long extensions of the upper wings were braced entirely by wires, the landing wires being anchored to rectangular kingpost structures. Ailerons were fitted to the upper wing only and had pronounced inverse taper, while the original rudder had a small balance area like that of the Curtiss H-4. The undercarriage was one of the earliest multi-wheel installations, there being a tandem pair of substantial wheels on each side. Power was to be provided by two 170 hp Curtiss VX engines but none was available in time and the aircraft was initially flown on 3 September, 1915, with two 90 hp Curtiss engines, presumably OX-5s; the pilot was Tony Jannus. Despite having little more than half of its designed power the Curtiss twin was reported to have handled well and to have reached a speed of 70 mph. It became known as the Curtiss Canada.

The first Curtiss C, with strut-braced extensions and four-blade airscrews, at Central Flying School. The Sperry Stabilizer and its fan had been removed at Farnborough.
(*National Aeronautical Collection, Ottawa*)

On 7 September, 1915, the aircraft underwent some form of acceptance test, an exercise that was witnessed by Sqn Cdr John Porte, who had apparently gone secretly to North America for this purpose. How much Admiralty interest this connoted must remain conjectural, but it appears that the question of ordering a batch of Canada biplanes was not under consideration by the Air Department of the Admiralty until January 1916. By March 1916 official serial numbers had been allotted for 101 Curtiss Canada twin-engined aircraft for the RNAS; these were for 3700 and 9501—9600.

To mention these naval serial numbers is in a way to anticipate subsequent history, for it was to the RFC that first delivery was made. A twin-engined Curtiss biplane, apparently the prototype, had reached Farnborough by 21 October, 1915, at latest, after much difficulty in transporting the enormous packing cases in which it had been shipped over England's narrow roads between Liverpool and Farnborough. At the outset it was understood that the Curtiss VX engines had developed weaknesses in the crankcase and crankshaft, and as early as 13 October it had been decided to replace them with two 150 hp Sunbeam Crusaders; these were borrowed or acquired from the Admiralty, and were sent from the RNAS storage depot at White City to the AID at Farnborough on 14 October. They were returned to White City on 27 October after a change of official mind. By the time of its arrival in England the aircraft had acquired a plain rudder without balance area.

It seems that the first Canada was not ready to fly at Farnborough until January 1916, when the War Office specified that they wanted 2nd Lt F. G. Dunn to put the aircraft through its tests. However, while it was being flown by Sgt-Maj W. B. Power on speed tests on 29 January the port airscrew disintegrated at a height of about 300 ft, breaking the port front outer interplane strut. Power had to land hurriedly and was obliged to put down on rough ground; the Curtiss was quite badly damaged. Repairs were put in hand, and in mid-February two new 170 hp Curtiss engines were delivered to the RFC Administrative Wing at Farnborough. At a relatively early stage of the Canada's RFC career the wire bracing of the upper-wing extensions was replaced by a pair of lift struts on each side, and the wire-cable interplane bracing was replaced by streamline-section Rafwires. It seems likely that the failure of the lift bracing that necessitated this modification occurred before Sgt-Maj Power's mishap.

The prototype had been delivered with the installation of the Sperry Stabilizer. As noted above, this was an early attempt at an automatic pilot, but Farnborough evidently disdained it, for the Sperry equipment was never evaluated on the Canada; on the contrary, it was removed from the aircraft and put into store at the Reserve Aircraft Park. Nevertheless, one cannot help wondering how much inspiration it may have provided for the Royal Aircraft Factory's gyro-stabilised bomb sight. The official serial number 5728 was allotted to the Canada about the beginning of April 1916, but it is doubtful whether the aircraft ever wore it.

Various airscrews of Lang and Vickers designs were tried on the rebuilt Canada, and it was to be tested at CFS Upavon with those that proved most efficient. It flew a climbing test at Farnborough on 27 April, 1916, returning figures that were abysmal despite having only fuel for two hours aboard. Eventually it left Farnborough for CFS at 4.40 p.m. on 10 May, having by then acquired two bracket mountings for Lewis guns in the front cockpit. Unfortunately the results of its trials there are not known.

By that date the ten production Canadas had been delivered to Farnborough and were awaiting erection. These aircraft were designated Curtiss C and had the maker's numbers C.2 and C.4—C.12. The Curtiss company had revised the design to save structure weight, especially in the tail unit, and instructions were given that the first of the new aircraft was to be rigged with its upper wing moved backwards through six inches in an attempt to relieve tail-heaviness. However, erection was suspended because no self-starters had been supplied with the engines.

Surviving documents indicate that the Canada had been rejected before the end of July 1916, but it seems that the Southern Aircraft Depot had not been officially informed and obviously felt a compulsion to carry out an earlier instruction to erect one of the aircraft. Erection began in mid-August but was officially halted by an instruction dated 25 August 'pending necessary modifications being made in the engines', for the 170 hp Curtiss engines were officially considered to be failures. It is very doubtful whether any of the production Canadas was ever fully assembled at Farnborough.

Production Curtiss C in the Strachan Avenue works, with F. G. Ericson, Chief Engineer of the Curtiss Aeroplanes and Motors Co. These aircraft had constant-chord ailerons. (*K. M. Molson*)

The production aircraft had lift struts to support the long extensions on the upper wings, were fitted with constant-chord ailerons, and had skids under the lower wingtips. The Sperry Stabilizer was not fitted. The serial numbers A5215—5224 were belatedly allotted to the production Curtiss Cs as late as the end of November 1916, apparently as an action of administrative tidying up, probably to facilitate write-off action.

An earlier idea that came to naught was a proposal to mount a one-pdr Vickers Automatic Aeroplane Gun Mark III in the forward cockpit. This was put forward on 4 February, 1916, and was a possibility then because the prototype was being rebuilt following Sgt-Maj Power's crash. The gun in question weighed, together with its mounting and 50 rounds, no less than 324 lb; it had a recoil force of 400 lb and a muzzle velocity of 1,200 ft/sec. Coincidentally, a similar proposal was put forward on 8 April, 1916, by Brig-Gen Brooke-Popham of RFC Headquarters, who wanted a one-pdr gun installed in such a way as to be able to fire at a downward angle of 45 deg. He optimistically wanted the modifications made and the Canada sent to France by 30 April. The Vickers company had been asked to design a suitable mounting for the gun as early as 14 February, and a layout was submitted on 15 March. The special mounting would have provided 80 deg elevation, 60 deg depression and all-round training. The design was approved on 30 March, and Vickers were asked to make a trial installation in the Curtiss nacelle that had been supplied to them. Evidently six of the one-pdr guns had been ordered.

With the abandonment of the Canada the Vickers gun installation was not wanted, and Vickers were asked on 28 August to deliver the gun to the Ordnance Aircraft Depot at Greenwich. The mounting was 'practically complete' on 8 September, 1916, and on 26 January, 1917, Vickers wrote a reminder that the mounting was still in their Bexleyheath works and asked for disposal instructions. The firm were advised on 31 January, 1917, that the mounting would be wanted by the RNAS for experimental purposes, but this is unlikely to have indicated a link with the naval service's solitary Canada (No.3700) which had been at Hendon from 6 June, 1916. This Canada was reported to be undergoing alterations to its wings in mid-July, and from 21 August was awaiting new crankshafts for its engines. It is doubtful whether these ever arrived, for the aircraft was dismantled on 12 January, 1917, and was officially deleted by a Board of Survey on 19 January. With the greatly superior Handley Page O/100 coming into service at that time the RNAS could not possibly have made worthwhile operational use of the Canada, and the cancellation of the 100 production aircraft for the RNAS was inevitable.

Curtiss C
Two 160/170 hp Curtiss VX

Three-seat bomber.

Span 75 ft 6 in (upper), 48 ft 6 in (lower); length 36 ft 3 in; height 15 ft 6 in.

Empty weight 4,712 lb; loaded weight 7,000 lb approx.

One official document attributes to the aircraft a speed of 91·5 mph and a climb to 3,150 ft in 7 min 15 sec, apparently with 220 gal of petrol, sufficient for 8½ hr at cruising speed. These are highly optimistic figures. In fact, on its climbing test at Farnborough when it carried only 60 gal of petrol, it climbed to 2,000 ft in 5 min 42 sec, to 3,000 ft in 9 min 30 sec, to 5,000 ft in 19 min, and to 6,000 ft in 24 min 7 sec.

Armament: Two 0·303-in Lewis machine-guns and bomb load.

Manufacturer: Curtiss Aeroplanes & Motors (Canada).

Serials: 5728, A5215—5224.

Curtiss JN-4A, B1926 of No.16 Reserve Squadron, Beaulieu. This aircraft was also used by No.17 Training Squadron and No.50 Training Squadron. While being flown by Lt I. Cooke of No.50 T.S. on 28 July, 1917, it crashed and he lost his life. (*Peter R. Little*)

Curtiss JN-4A

The existence of an improved Curtiss two-seater designated JN-4 was known to the War Office as early as 14 June, 1916. American sources state that the prototype JN-4 was externally indistinguishable from the JN-3, having a similar tail unit in which the lower end of the rudder projected below the fuselage, and the tailskid was mounted on a downward extension of the stern post. Ailerons, as on the JN-3, were on the upper wings only. On 24 June, 1916, the A.D.A.E. wrote to the D.A.E. that

'The JN-4 machines include all the important modifications required in the JN-3 type—except dual control. The JN-4 with dual control should therefore be the type to be obtained for the Canadian school.'

It is uncertain whether a production form of this early JN-4 was delivered to Britain. Official RNAS records indicate that, whereas the Curtiss aeroplanes numbered 3345—3423 were JN-3s, 3424—3444 and 8802—8901 were listed as JN-4s, yet photographs of aircraft in these two latter groups appear to show JN-3s. However that may be, the RNAS records state that eleven of that Service's JN-4s went to the RFC.

What the RFC in Britain did receive, in 1917, was a batch of fifty Curtiss JN-4As. These were part of the total of two hundred JN-4As that had been ordered by British authorities under contract No.N.1021—832W; the other 150 were delivered to the RFC in Canada. In the RFC the JN-4As were apparently numbered in sub-batches as they were delivered: B1901—1919, B1920—1928, B1929—1940, B1941 and B1942—1950; these numbers were allotted from 20 March, 1917. The aircraft themselves had maker's numbers running from 1 to 50*, suggesting that they

Curtiss JN-4A of the School of Aerial Gunnery, Hicks Field, Texas, with flag target, winter 1917-18. This JN-4A was made at the Buffalo works of the parent Curtiss company and assembled by Canadian Aeroplanes Ltd.
(*National Aeronautical Collection, Ottawa, Neg 2908*)

might have been the first production JN-4As.

The JN-4A differed externally from the JN-3 in having completely new tail surfaces that included a tall, enlarged rudder with its lower end flush with the base of the stern post, an enlarged tailplane with straight leading edge and parallel-chord elevators, and a simpler tailskid. The mainplanes had a large dihedral angle, and there were ailerons on the lower wings; the Curtiss OX engine was installed with a marked degree of downthrust and had stack-type exhaust pipes. Structurally the airframe embodied a number of strengthening modifications.

These JN-4As of the RFC were used by various training units in England and by No.3 Squadron, Australian Flying Corps, while working up to operational status before going to France. They were still in use late in 1917, survived into 1918, and continued in use with the RAF. The fourth official 'Statement of Aeroplanes', dated 23 February, 1918, said of the Curtiss: 'Will die out in this country when the Avro is available in sufficient quantities to take its place.' This paragraph was repeated in the next 'Statement of Aeroplanes' on 9 May; but the sixth in the series, dated 16 August, stated: 'This machine has been declared obsolete by the Air Council under A.M.O. 778, and all machines and spares should be dealt with in accordance with the procedure laid down in that Order.'

Canadian Aeroplanes Ltd of Toronto built large numbers of a JN-3 derivative under the designation JN-4(Can) for use in the RFC's Canadian flying training programme that was initiated by the decision of 12 December, 1916. This programme was expanded to incorporate the training in Canada of American pilots and to provide for the transfer of flying training activities to Texas in the winter of 1917-18. These developments increased the demand for training aircraft, and to meet it Canadian Aeroplanes Ltd undertook the assembly of a number of Curtiss-made JN-4As purchased by the RFC, presumably the 150 JN-4As ordered under Contract No. N.1021—832W, which was regarded as being completed by 8 September, 1917. These JN-4As were given serial numbers in the RFC Canada C series (which should not be confused with the basic British C-prefix series), their numbers running between and about C565—616. These

*B1901—1941 included Maker's Nos.1—9, 11, 13—21, 26—28, 23—25, 29, 30, 32, 33, 35, 37, 31, 36 and 38—43, apparently in that order; B1942—1950 were Nos.10, 12 and 44—50.

210

aircraft retained the Deperdussin form of control and apparently did relatively little pilot training in the Canadian scheme, being more frequently employed on the training of observers, target towing and associated duties. At least some of the Canadian-assembled JN-4As went south to Texas in the winter of 1917–18 with Nos.42 and 43 Wings.

JN-4A
90 hp Curtiss OX-5

Two-seat trainer.
Span 43 ft 7 in; length 27 ft 3 in; wing area 317·36 sq ft.
Loaded weight 1,870 lb.
Speed 75 mph.

Manufacturers: Curtiss Aeroplane Co (Buffalo). Some assembled in Canada by Canadian Aeroplanes.

Service use: In United Kingdom—Reserve/Training Sqns 11, 16, 17, 36 and 42; No.1 Training Depot Station, Stamford. *Other units*: No.3 Sqn, Australian Flying Corps. *In Canada*—possibly at Camp Borden, Camp Rathbun, Camp Mohawk, Leaside, Armour Heights and Beamsville. *In Texas*—Hicks, Everman and Benbrook Fields; School of Aerial Gunnery, Hicks Field.

Serials: RFC—B1901—1950; RFC Canada—between and about C565 and 616.

Curtiss JN-4 (Can)

The potential of Canada as a source of pilots was first recognised by Lt Col C. J. Burke, who had visited the Dominion in the autumn of 1915. He submitted a cogent report, urging that a Wing of the RFC should be created in Canada to train Canadian pilots. The War Office rejected the idea at that time, but in April 1916 the Imperial Munitions Board in Canada revived the subject by reporting 'that there was no reason why an aviation industry should not be organised in Canada'. Complex and protracted negotiations followed, and it was not until October 1916 that the Canadian Government in an Order in Council agreed to the proposal that an aeroplane factory and flying school should be established.

In Britain the Director of Aircraft Equipment was evidently determined to be prepared for such a development, for as early as 9 June, 1916, he minuted ADAE 1:

'As I have put forward a proposal to have 200 Curtiss machines built in Canada for training purposes, I would like you to go into the question of the design with a view to determining what alterations we shall definitely require, and to get out a design for dual control.'

Urgency was injected by Maj-Gen Trenchard's statement, made late in September, that it would be necessary to double the number of fighting squadrons in France. The Director of Air Organisation thereupon proposed a programme for the creation of 35 new training squadrons, and it was subsequently decided that 20 of them should be raised and stationed in Canada. These were Nos.78 to 97 (Canadian) Reserve Squadrons.

Typical Curtiss JN-4(Can), showing the distinctive rudder shape and inter-aileron link struts. (*K. M. Molson*)

To provide the aircraft factory, the existing Curtiss works in Strachan Avenue, Toronto, were taken over, together with the staff, tools, equipment and the rights to manufacture the Curtiss JN-3 and its OX-5 engine. Initially 100 aircraft were to be built, and on 1 January, 1917, work began on the redesign of the JN-3 to meet RFC requirements. Stick control replaced the Deperdussin-type controls of the American JN-3; a revised and characteristically rounded rudder replaced the original, together with an improved tailskid; ailerons were fitted to the lower wings, and upper and lower ailerons were linked by a pair of struts, originally parallel but on most aircraft brought together at their lower ends to form an oblique V. This aileron arrangement had in fact appeared earlier on some of the Canadian-built JN-3s. Other less obvious modifications were also incorporated.

National markings were not carried on the Curtiss JN-4s (Canadian) used by the RFC, but this one had coloured wingtips, presumably a Flight or Squadron identification. (*K. M. Molson*)

In winter the Curtiss JN-4(Can) could be flown on skis, here fitted to C247. (*K. M. Molson*)

Canadian Aeroplanes Ltd designated their creation JN-4, without regard to the fact that the American company had meanwhile evolved a revised design with the Curtiss type number JN-4. Eventually (presumably after the entry of the USA into the war) the Canadian type was redesignated JN-4 (Canadian), which came to be abbreviated as JN-4(Can). The type was popularly known as the Canuck.

Early in January 1917 Lt-Col C. G. Hoare left England for Toronto, intent on setting up the organisation of the RFC Canada. He expected that twenty aeroplanes would be delivered by the Toronto factory by the end of February, and that by the end of March a total of 80 training aircraft would be available to him from Canadian Aeroplanes Ltd and other unspecified sources. The hangars of the Curtiss company at Long Branch, Toronto, were taken over, and the first aircraft

Skis of more substantial appearance were fitted to JN-4(Can) C318. (*K. M. Molson*)

213

built by Canadian Aeroplanes was delivered there on 22 February, 1917. Presumably this was the first JN-4(Can), numbered C101. Flying instruction began at Long Branch on 27 February in a nucleus Flight designated 'X' Squadron, using three aeroplanes that probably included Canadian-built JN-3s. The nucleus Flights of Nos.78, 79 and 81 (Canadian) Reserve Squadrons reached Canada at the beginning of March; Nos.80 and 82 had arrived by 19 March, and by the middle of April all five Squadrons were settled in at Camp Borden. By mid-July nine more RFC (Canadian) Reserve Squadrons were in Canada, and 'Y' Squadron established itself north of Toronto (presumably at Armour Heights) on 6 July.

Curtiss JN-4A/JN-4(Can) hybrid of No.85 Canadian Training Squadron, Deseronto, Ontario. (*K. M. Molson*)

There can be little doubt that the JN-4(Can) was used by all of these units, for it undertook virtually all the pilot training done in Canada. By the end of 1917 Canadian Aeroplanes had built 667 complete aircraft and the equivalent in spares of another 391. Eventually, production of the JN-4(Can) exceeded 1,250 and spares production amounted to a further 1,611 equivalent aircraft. Some of the surplus JN-4(Can) wings were mated with true JN-4A fuselages to produce hybrid aircraft.

In March 1918 the War Office required the Canadian units to extend their activities to include gunnery, photography and aerial observation. This was agreed, and instructors in gunnery, photography and wireless were sent to Canada and began work late in April. These new demands led to various installations of fixed synchronised Vickers guns, Scarff ring mountings and cameras on the JN-4(Can) for the Royal Air Force (Canada).

The winter of 1917–18 produced schemes for fitting skis to the JN-4(Can), much work being done by Lt R. H. Cronyn, but the design eventually adopted has been attributed to F. G. Ericson, Chief Engineer of Canadian Aeroplanes Ltd. That same winter saw the transfer of the 42 Wing (Sqns No.78, 79, 81 and 82) from Camp Borden to Everman Field, Texas, and of the 43rd Wing (Sqns No.80, 83, 84, 85, 86 and 87) from Deseronto to Benbrook Field, Texas. These moves were made in mid-November 1917, and the Texas contingent moved back to Canada in April 1918. The 44th Wing (Sqns No.88, 89, 90, 91 and 92) continued flying training at North Toronto. This presented immense difficulty, but the aircraft had ski undercarriages and flying went on even when the temperature was as low as −22 deg F (−30 deg C).

When the US Government undertook to provide winter training accommodation for ten RFC squadrons they also undertook to buy 180 aeroplanes, with spares, from Canadian Aeroplanes to maintain that number of serviceable aircraft on the winter aerodromes. Possibly this was the earliest instance of a form of lease-lend for aircraft. The RFC (Canada) was to be responsible for all maintenance and repair, was to leave all aircraft and equipment in good order on departure, and was to train personnel for ten US squadrons in Canada during the summer of 1918. These 180 JN-4(Can) aircraft probably came from the batch of 280 ordered by the United States on 21 July, 1917. Subsequently 400 more were ordered by the USA in January 1918 and were given US serial numbers.

After the war was over the British Government presented to the Canadian Air Force the fifty-three JN-4(Can) aircraft then at Camp Borden. In the event the CAF retained only eleven and sold off the others which, with other surplus Canucks, did much pioneering and barnstormer flying in the Dominion.

JN-4(Can)
90 hp Curtiss OX-5

Two-seat trainer.
Span 43 ft 7⅜ in (upper), 34 ft 8⅝ in (lower); length 27 ft 2½ in; wing area 360·63 sq ft.
Empty weight 1,392 lb; loaded weight 1,920 lb.
Speed 80 mph.
Armament: For training purposes various installations of one fixed 0·303-in Vickers machine-gun were made; some aircraft were fitted with a Scarff ring-mounting on the rear cockpit for one 0·303-in Lewis gun.

Manufacturer: Canadian Aeroplanes.

Service use: *Training duties*: 42nd Wing, Camp Borden (No.78, 79, 81 and 82 (Canadian) Reserve Sqns, and the School of Aerial Gunnery); 43rd Wing, Deseronto (No.80, 83, 84, 85, 86 and 87 (Canadian) Reserve Sqns); 44th Wing, Armour Heights and Leaside (No.88, 89, 90, 91 and 92 (Canadian) Reserve Sqns). During the winter of 1917–18 the School of Aerial Gunnery moved to Hicks Field, Texas; the 42nd Wing to Everman Field; and the 43rd Wing to Benbrook Field, Texas.

Serials: From C101 onwards, excluding the batch of 150 JN-4A and JN-4A/JN-4(Can) hybrids, to at least C1457.

The Deperdussin monoplanes

In the pioneering years Deperdussin monoplanes were prolific and remarkably varied. They were also, by the standards of their time, successful, and enjoyed considerable popularity. The first to be used officially by the Aeroplane Section of the Air Battalion had the early number B5; it was purchased by the War Office in January 1912.

This Deperdussin had a 60 hp Anzani radial engine and was of a relatively early type, having a long and slender fuselage and very exposed accommodation for the pilot and his passenger. It was used by No.3 Squadron after the formation of the Royal Flying Corps, and had as companion another Deperdussin of very similar

This early Deperdussin monoplane with 60 hp Anzani radial engine is thought to have been the aircraft that was originally numbered B5 and later became 252.

Captain Hamilton's Deperdussin after being taken over by the Military Wing in August 1912 and numbered 257. *(RAF Museum)*

Capt Patrick W. Hamilton (wearing flying helmet) with his Deperdussin monoplane at Farnborough. This aircraft became No.257 after being acquired by the Military Wing.

appearance, likewise powered by a 60 hp Anzani, that originally belonged to Capt Patrick Hamilton and was fitted with a very external form of instrument panel between the cockpits. This latter monoplane was purchased for the use of the Military Wing in June 1912 and, when the official numbering system was introduced, B5 became 252, Hamilton's aircraft 257. Both aircraft remained with No.3 Squadron.

In mid-October 1912, No.252 was at Farnborough, where it was to be dismantled, and No.257 was at Larkhill. It is doubtful whether either ever flew again after the imposition of the War Office ban on monoplanes, though both were carried on the strength of No.3 Squadron, and, ban notwithstanding, it was reported on 25 January, 1913, that 257 was unserviceable because its rear spar required strengthening, and a week later it was again reported serviceable. By 14 March, 1913, No.257 had been packed for transport to Farnborough, where 252 still languished; and a month later both were lying in the Flying Depot there. Both were finally struck off charge on 26 November.

This photograph emphasises how well ventilated were the cockpits and instruments of Capt Hamilton's Deperdussin.

In April 1912 The British Deperdussin Aeroplane Co Ltd was registered as a successor to The British Deperdussin Aeroplane Syndicate Ltd. The new company proceeded to design and build monoplanes on general Deperdussin lines; its designer was Frederick Koolhoven. Two British Deperdussin monoplanes were entered for the 1912 Military Trials, one powered by a 100 hp ten-cylinder Anzani radial engine, the other by a 100 hp Gnome rotary. As recounted on page 19 these aircraft had the Military Trials numbers 20 and 21 respectively. After the Military Trials were over the RFC took on charge the French-built Deperdussin with the 100 hp Gnome (No.26 of the Trials) with the new identity of 258, and the Gnome-powered British Deperdussin (No.21) as 259.

No.258's career with the Military Wing was tragically short, for on Friday 6 September, 1912, while being flown by Capt Patrick Hamilton with Lt Wyness-Stuart as passenger, it broke up in the air and crashed at Graveley, near Welwyn. Hamilton and Wyness-Stuart were killed. This was the first of the fatal monoplane crashes that were to lead to the imposition of the so-called monoplane ban by the War Office.

No such fate befell No.259, for it is doubtful whether it did any serious flying with the RFC. Surviving references suggest that this British-built Deperdussin reached No.3 Squadron on 27 October, 1912. Of it a contemporary note records 'All wires and iron work rusted. Must be overhauled.' A week later it was noted that a new rudder post was to be fitted and the drift wires adjusted, and a terse record of early November states 'Not considered safe to fly, design bad.' Despite subsequent notes stating that No.259's design was dangerous it was, surprisingly, recorded as serviceable at Larkhill on 28 December, 1912. That may well have been an erroneous note, for on 4 January, 1913, it was again stated to be 'Unsafe to fly'. By 25 January it was at Farnborough and probably never left there before being struck off charge on 5 August, 1913.

A Deperdussin of uncertain identity but obviously in RFC ownership. The engine is a single-row Gnome, and the monoplane may have been 260. (*RAF Museum*)

Deperdussin No.419 (ex 260) at Farnborough in May 1913. (*RAE 47; Crown copyright*)

The Military Wing had two Deperdussins powered by the 70 hp Gnome rotary engine. These had the serial numbers 260 and 279 but, affected by the monoplane ban in common with all contemporary monoplanes in the Military Wing, they did very little. Indeed, 279 seems to have done nothing at all, for it stayed at Farnborough and was never issued to a squadron. No.260 was briefly on charge of No.3 Squadron and was evidently with that unit in October 1912, having been delivered some time between 15 October, when it was still in the British Deperdussin company's Highgate works, and 26 October. Being under the monoplane ban it could not be flown, but on 12 November, 1912, it was

218

Deperdussin No.280 had a 100 hp Gnome engine. *(RAE)*

transferred to CFS, where it apparently acquired the new identity of 419. Its original number was re-allocated to a Blériot XI in 1913. What, if anything, 419 did at CFS is not known.

At least one more Deperdussin with the 100 hp two-row Gnome rotary was delivered for use by the Military Wing. This monoplane was allotted the serial number 280. The absence of forward-projecting skids on its undercarriage suggests that the aircraft may have been built by the British Deperdussin company, but it reverted to the rectangular rudder seen on earlier Deperdussins. The tailplane, of lifting section, was braced from kingposts above the fuselage.

No.280, in the early spring of 1913, was at the Royal Aircraft Factory at Farnborough with an allocation to No.5 Squadron, in company with a Nieuport, Martin and Handasyde, and two Flanders monoplanes. None of these aircraft was ever issued, however, and No.280 was eventually struck off charge on 19 August, 1913.

Deperdussin No.421 had a 60 hp Anzani radial engine.

On Deperdussin No.437, here seen engineless at Farnborough, the fin was slightly enlarged.

Apart from No.419 (ex 260) three further Deperdussin monoplanes were purchased for use at CFS. These were numbered 421, 436 and 437, but none, it seems, ever went to CFS. No.421 was virtually identical with the British-built Anzani-powered Military Trials entrant but had only a 60 hp Anzani radial. Nos.436 and 437 were apparently struck off charge without leaving the Royal Aircraft Factory.

Deperdussin monoplanes
60 hp Anzani, 100 hp Anzani, 70 hp or 100 hp Gnome according to type
Two-seat tractor monoplane.
Span 39 ft 6 in; length 24 ft 6 in; wing area 236 sq ft. Figures for British-built aircraft with 100 hp Gnome.
Empty weight 537 kg (French-built with 100 hp Gnome), 1,226 lb (British-built with 100 hp Gnome), 1,200 lb (British-built with 100 hp Anzani); loaded weight 847 kg (French-built with 100 hp Gnome), 2,037 lb (British-built with 100 hp Gnome), 2,000 lb (British-built with 100 hp Anzani).
Maximum speed 111 km/h (French-built with 100 hp Gnome), 68 mph (British-built with 100 hp Gnome), 70 mph (British-built with 100 hp Anzani); climb to 1,000 ft (305 m)—3 min, 3 min 45 sec and 4 min 45 sec respectively.

Manufacturers: Armand Deperdussin (France), British Deperdussin (United Kingdom).

Service use: No.2 (Aeroplane) Section, Air Battalion, Royal Engineers; early 60 hp Deperdussin, Nos.252 and 257, and 100 hp type No.258 used by No.3 Sqn.

Serials: 252 and 257 (60 hp Anzani); 258 and 259 (100 hp Anzani); 260 (70 hp Gnome), became 419; 279 (70 hp Gnome); 280 (100 hp Gnome); 419 (ex 260); 421, 436 and 437.

Dunne D.8

Lieutenant J. W. Dunne was one of the earliest British experimenters with aircraft, and he shared with Cody the distinction of having an aeronautical connection with the War Office long before the Royal Flying Corps was formed. He worked at Farnborough from 1906 to 1909 and gave expression to his ideas on stability by designing a series of tailless biplanes with sharply sweptback wings. This series continued after he left Farnborough, and in 1912 he rebuilt the crashed Dunne D.5 as the D.8, replacing the twin-pusher propeller arrangement of the D.5 with a single central propeller. It was probably this D.8 that was flying at Larkhill at the time of the Military Trials in August 1912, and by early November it was being flown by N. S. Percival. Later that month it was at Eastchurch, fitted with a 60 hp Green engine. By August 1913 an 80 hp Gnome had been fitted to the aircraft, and two more D.8s were reported to be under construction at Hendon.

These may have been the two Dunne D.8s that the War Office had ordered on 19 March, 1913, for the Military Wing of the RFC. The terms of the contract (No. 87/1152) included provision for instruction in flying the Dunne to be given by The Blair Atholl Aeroplane Syndicate to two RFC officers.

The due date for delivery of the aircraft was 14 May, 1913, but construction was slow and it was not until early October 1913 that a D.8 for the RFC was reported to be nearly ready for its trials. On 3 October RFC Headquarters wrote to The Blair Atholl Syndicate, announcing that 2/Lt T. O'Brien Hubbard and Air Mechanics Nash and Ware would arrive at Hendon on 6 October 'for instruction in the details of construction and method of trueing up these machines.' As there was no Dunne instructional aircraft available the flying instruction of two officers had to be postponed. On 6 October, 2/Lt Hubbard reported his arrival at Hendon, together with the facts that the aircraft would probably be ready next day and that Commandant Felix, a French officer, had been retained by the manufacturers to put the D.8 through its tests.

The Dunne D.8 at Farnborough, 11 March, 1914. (*RAE 365; Crown copyright*)

Of the flying at Hendon on Saturday 18 October, 1913, *The Aeroplane* of 23 October reported that 'Great interest was aroused by the first public London appearance of the Dunne biplane', and printed two photographs of the aircraft. These showed that it differed in some respects from the earlier D.8, having a longer nacelle, and wingtip interplane surfaces of modified shape. The aircraft was flown briefly and, owing to a refractory engine, with difficulty by Commandant Felix.

It is uncertain whether this was one of the D.8s ordered for the RFC, and it is not clear whether any delivery was made at that time. Only two weeks later, on 3 November, 2/Lt Hubbard reported, not from Hendon but from Eastchurch:

'I beg to report that the Dunne machine is ready with the exception of the engine which has been taken down owing to a blued cylinder. It is expected to be out on Wednesday.'

The Dunne D.8 airborne at Farnborough, 11 March, 1914. (*RAE 336; Crown copyright*)

And indeed a 'new Dunne biplane' was reported to be flying at Eastchurch on Wednesday and Saturday, 19 and 22 November, piloted by Commandant Felix and N. S. Percival. However, it seems that it was not delivered to the RFC then, for in early February 1914 'two experimental Dunnes' were evidently considered to be still on order for the War Office. It appears that at some time in 1914 one of the D.8s was cancelled, for the only aircraft in respect of which evidence of delivery has been found was eventually delivered to Farnborough on 3 March, 1914, nearly ten months overdue; and an official list of aeroplanes on order for the RFC before the outbreak of war attributes only one D.8 to Contract No. 87/1152. This D.8 was flown at Farnborough on Wednesday, 11 March, making several flights, mostly with N. S. Percival at the controls. On one flight he took up a passenger and climbed to 3,000 ft. According to *The Aeroplane* of 19 March, 1914:

'The machine behaved very satisfactorily also when flown by an officer of the Royal Flying Corps, who expressed his appreciation of its stability and ease of handling. The machine's speed variation was 36 to 56 mph, and its climbing speed 500 feet per minute.'

That one brief flight with an RFC officer as pilot seems to have been the Dunne's total known flying with the Military Wing. It was allotted the official serial number 366, which suggests that it was intended for No.5 Squadron. Apparently the D.8 was still in existence in August 1914 but was at that time in a dismantled state, probably at Farnborough.

Dunne D.8
80 hp Gnome

Two-seat tailless pusher biplane.
Span 46 ft; wing area 545 sq ft.
Empty weight 1,400 lb; loaded weight 1,900 lb.
Speed 56 mph.

Manufacturer: The Blair Atholl Aeroplane Syndicate.

Serial: 366.

Henry Farman Type militaire, 1910

The basic, primitive Henry Farman biplane was well-known to Britain's aviation world in 1910, for one of its most notable exponents was the dashing Claude Grahame-White, who flew the Farman in spectacular fashion. Late in October 1910 the War Office let it be known that a Farman Type militaire and a Paulhan biplane had been purchased for use by British Army pilots. The so-called Type militaire Farman was simply a standard boxkite biplane of the classic Farman III type on which strut-braced extensions had been added to the upper wings, and a third, central, rudder fitted between the tailplanes.

After acceptance trials the War Office's first Farman was officially handed over to the British authorities at Châlons on 26 November, 1910, and it eventually arrived at Farnborough in December 1910; the price paid for the airframe, engine and propeller was £1,008, to which had to be added £5 for a Steward aneroid and £5 12s 0d for an Elliott tachometer. Assembly was undertaken in the Balloon Factory at Farnborough and the aircraft was complete early in January 1911. Its designated pilot was Capt C. J. Burke (Royal Irish Regiment), who had secured his pilot's certificate (No.260) in France on 4 October, 1910, flying a Henry Farman biplane.

On Saturday, 7 January, Capt Burke made a successful first flight on the Farman, covering some two miles at heights varying from 50 to 80 feet and landing beside the Balloon Factory. Ten minutes later he took off on a second flight but had covered only 50 yards when the Farman stalled and fell to the ground on its starboard wings. It was completely wrecked, and Capt Burke's right foot was badly crushed.

Examination of the wreckage revealed that only the tail could be used again. The rest of the airframe was entirely rebuilt in the Balloon Factory at an identified

F1, the Air Battalion's Farman III in its basic form without extensions.
(*RAE 0257; Crown copyright*)

cost of £160, using Factory labour and materials. The Farman's first flight after reconstruction was made by Geoffrey de Havilland on 6 March, 1911; apparently the extensions were not fitted and the aircraft seemed to be out of trim. An undercarriage skid was broken on landing, but this was repaired and next day the Farman flew again, this time with the extensions in place. The repaired skid broke again on landing, but the biplane was ready on 8 March for Capt Burke to make a brief flight and de Havilland some further tests. The Farman went back into the Factory to be re-rigged, and was flown again by de Havilland on 13 March, with and without a passenger. After two more successful flights on 16 March, Mervyn O'Gorman, the Superintendent of HM Balloon Factory, wrote and signed the following certificate:

'CERTIFICATE NO.1

This is to certify that Aeroplane No.F1 has been tested by me in flights of 5 to 7 miles without wing extension, both with a 14-stone passenger and without a passenger, and in two flights of 5 to 7 miles with wing extension, with a 14-stone and a 12-stone passenger respectively.

Weather calm with slight gusts.

It was subsequently examined and showed no defect.

(signed) Mervyn O'Gorman
Superintendent Balloon Factory
16 March 1911'

Although hardly a certificate of airworthiness, that document has its own claim to being historic as the first form of certification ever applied to an aircraft in Britain.

A noteworthy detail in the document is its reference to the aircraft as No.F1, signalling the introduction of a system of numbering British military aeroplanes. This may have been a concomitant of the formation of the Air Battalion announced on 28 February, 1911, and to be effective from 1 April. Pusher aircraft were numbered in the F (presumably for Farman) series, tractors in the B (presumably for Blériot) series, while new designs from the Balloon Factory were serially numbered in the F.E. and B.E. series.

Farman III F1 with its extensions fitted. Capt C. J. Burke, its pilot, is the central marked figure. (*RAE 0595; Crown copyright*)

The rebuilt F1 reappeared on Monday, 6 March, 1911; after several preliminary flights from Farnborough Common it was taken to Laffan's Plain, where it made several further flights of from 50 to 300 yards at heights of 20 to 35 ft, but a broken skid put a stop to flying for the day. It was flying again with Burke piloting on 22 March, 1911, but sustained further damage on 30 March. In June, F1 was with the Air Battalion on Salisbury Plain, its regular pilot being Capt Burke, who flew it to Farnborough on 8 June, returning two days later.

Capt Burke flew the Farman, occasionally with a passenger, until 14 July, when it was flown by Capt Massy. Massy damaged it on 17 July, and it next flew on 3 August. A crash on 28 August necessitated further repairs, but it was itself again by 14 September. On 22 September F1 was used in experiments in message-dropping and in signalling by whistle, only the latter unlikely method of communication being regarded as successful. By January 1912 the aircraft's total

Capt Burke airborne on Farman F1 at Farnborough. (*RAE 0208; Crown copyright*)

flying time was 43½ hr, but on 8 January it was again wrecked. Its map case was then transferred to B.E.1, and on 26 January it was tested after re-erection, the wing extensions having been removed and a Bristol carbine fitted in unspecified fashion. The extensions were replaced by 29 March, when Lt A. G. Fox tested the aircraft. By April 1912, F1's total flying time amounted to 58¾ hr, and it last flew on 10 June, 1912, on which date an official decision was taken that it should be dismantled and sent to Salisbury Plain. It ended its days as an instructional airframe at CFS.

Remarkably, its original Gnome engine, No.205, remained unchanged throughout 1911. Although other Gnomes were fitted briefly in January, March and April 1912, No.205 was reinstated on 3 May, 1912. This was a noteworthy record at that time.

Henry Farman Type militaire
50 hp Gnome

Two-seat pusher biplane.
Span 14 m (upper), 10·5 m (lower); length 13 m; wing area 50 sq m.
Loaded weight 500 kg.
Armament: Experimental installation of one Bristol carbine.

Manufacturer: Aéroplanes Farman.

Service use: No.2 (Aeroplane) Company, Air Battalion, Royal Engineers, later No.2 Sqn, RFC.

Serial: F1.

Henry Farman biplane

In July 1912 the RFC Military Wing acquired its first two Henry Farman biplanes. These differed from the later classic Henry Farman F.20, although there was such a resemblance between the two types that the derivation of the F.20 was obvious. The two biplanes of July 1912 had two-bay wings with long extensions; ailerons were fitted to the upper wing only and did not have a spanwise balance cable. The undercarriage was low set, and in the tail unit the rudder was of trapezium form

The early Henry Farman No.420, photographed at Central Flying School with Maj E. L. Gerrard, R.M.L.I., in the pilot's seat.

with parallel leading and trailing edges. The nacelle was of the form that became well known on the F.20 and could seat three.

The two Farmans of the RFC were taken on charge by the Military Wing on 9 July, 1912, and were numbered 208 and 209 when serial numbers were introduced. They were first allocated to No.2 Squadron, but No.208 was damaged in a crash on 26 July, 1912, and was eventually struck off the strength of the Military Wing on 4 October, and transferred to CFS. There it received the new identity of 412 and flew very occasionally in the spring of 1913, its last known flight occurring on 30 June.

Likewise No.209 was transferred to CFS, apparently on 1 December, 1912, being renumbered 420 in the process. It did some flying there but was eventually crashed and presumably written off.

Henry Farman biplane
70 hp Gnome

Two/three-seat pusher biplane.
Manufacturer: Aéroplanes Henry et Maurice Farman.
Service use: No.2 Sqn and CFS.
Serials: 208 (renumbered 412), 209 (renumbered 420).

Henry Farman (*Wake up, England*) biplane

Early in 1912, Claude Grahame-White determined to alert his country to its backwardness in aviation, choosing as his slogan 'Wake up, England!' He expanded his activities at Hendon, and organised a series of flying meetings to arouse public interest. On returning from his honeymoon in July 1912 he set about making a tour of major cities and towns with a team of pilots and aircraft, one of which was a new Henry Farman biplane that had interchangeable wheel and float undercarriages.

Claude Grahame-White disports himself on the *Wake up, England* Farman that became 434 on acquisition by the War Office in 1913. It was almost certainly a Farman F.22, a designation frequently applied in error to the much more numerous F.20.

Grahame-White flew the long-span Farman as a seaplane at coastal resorts in the summer of 1912. (*RAF Museum*)

This aircraft was generally typical of Henry Farman design, but had three-bay bracing and long extensions on the upper wing; the tall rudder had parallel leading and trailing edges; the engine was a 70 hp Gnome. Grahame-White had the Farman painted bright blue all over, and the words 'Wake up, England' were applied to the nacelle sides and the undersurfaces of the lower wing. Much of England saw this distinctive Farman during the summer of 1912 and when, late in 1912, Grahame-White went to Switzerland for a winter-sports holiday he took the aircraft with him and flew it from the frozen lake at St Moritz.

As recorded on page 252, the sudden need to give the RFC enough aeroplanes to make up Col Seely's claimed total of 101 led to the purchase, in March 1913, of a number of aircraft owned by the Grahame-White company. One of these was the *Wake up, England* Farman, which had by then done a considerable amount of flying and must have been well worn. Numbered 434, it was officially taken on charge by the Military Wing on 24 April, 1913, and was allocated to CFS. On that day it was flown from Farnborough to CFS by Capt A. G. Fox, who arrived at Upavon 'in strong wind and thick mist about 7.15 p.m.'. There it did a little flying in May 1913 but does not appear to have seen much use and presumably was abandoned after a crash in which it was quite badly damaged.

Wake up, England biplane
70 hp Gnome
Two-seat pusher biplane, interchangeable wheel and float undercarriages.
Span 15·525 m; length 8·5 m; height 3·1 m; wing area 35 sq m.
Loaded weight 600 kg.
Speed 80 km/h.

Manufacturer: Aéroplanes Henry et Maurice Farman.

Service use: Flown briefly at CFS, Upavon.

Serial: 434.

Henry Farman F.20

Best known of the Henry Farman series of designs, the F.20 appeared in the summer of 1912. First to bring an example to England was Claude Grahame-White, who sent R. T. Gates and Louis Noel to France to fly it back. They took delivery of it on 27 July, 1912, but were held up for several days by unfavourable weather. Floats replaced the wheels for the cross-Channel flight, and the Farman reached Eastchurch on 5 August. It made its public début at Hendon on 17 August and was a regular performer thereafter. In March 1913 it was one of the batch of Grahame-White's own aircraft that were hurriedly purchased by the War Office to make up Col Seely's 101 aeroplanes, and was probably the ex-Grahame-White Henry Farman that received the official serial number 435. Taken on strength by the RFC on 24 April, 1913, it was allocated to CFS, but did very little flying there.

This was the classic Henry Farman; very similar to the two biplanes originally numbered 208 and 209, yet lighter looking and better proportioned, distinguished by its characteristic ear-shaped rudder. The upper wings had long extensions and carried ailerons; power was provided by an 80 hp Gnome.

The RFC had already made the acquaintance of the F.20 when the former Grahame-White example was assimilated, for No.268 was taken on charge by the RFC Military Wing on 18 March, 1913; Nos.274, 275 and 277 followed, the last of these being taken on charge on 10 April, 1913. No.274 was built at Hendon by The

Henry Farman F.20 No.352 was first taken on charge by the Military Wing on 24 April, 1913. It was the vehicle for the early experiments with machine-guns in 1913, and saw much service with No.3 Squadron. It was serviceable on the strength of 'C' Flight of No.3 Squadron on 31 July, 1914, and went to France with the squadron on 13 August. Its war career was brief, for it was reported wrecked on 12 September, 1914, and was apparently written off then.

A Farman F.20, said to be 284, with a Vickers gun on the prow of the nacelle and its pilot in the rear seat.

No.461 was an F.20 built by The Aircraft Manufacturing Co and was delivered to Farnborough on 14 February, 1914. It was taken on the strength of the Military Wing on 3 March and went to CFS. It went to France for No.5 Squadron and was with that unit until it was wrecked on 12 September, 1914. In this photograph it has an airspeed-indicator pressure head mounted on the port kingpost.

A Henry Farman F.20, unnumbered and therefore probably newly delivered, at CFS, date unknown. This F.20 had dual control. (*RAF Museum*)

This Henry Farman armed with a Vickers belt-fed machine-gun is believed to be No.352, the vehicle for the RFC's early experiments in arming aircraft. The occupants' seats were transposed to place the gunner in front; all the modifications to seating and controls, and the gun mounting, were made at the Royal Aircraft Factory, Farnborough, where this photograph was taken.

A Henry Farman F.20, possibly 352, with a Rexer gun wielded by the observer. (*RAF Museum*)

Henry Farman No.1817 of No.3 Squadron at St-Omer in 1914. The officers are (left to right) Lt E. Conran, Capt G. Pretyman and Maj J. M. Salmond. This Farman was delivered to No.3 Squadron from Paris on 23 September, 1914. Its markings consisted of Union Flags under the mainplanes and a roundel either side of the rudder. (*RAF Museum*)

Aircraft Manufacturing Co, the first of many such British-built Farmans supplied to the RFC. Initially at least, the RFC regarded the F.20s as three-seaters, but the type was normally employed as a two-seater.

The Farman's most significant contribution to military aviation history was perhaps the employment of No.352, in June 1913, as the vehicle for experiments with machine-guns in aeroplanes. As the pilot normally occupied the front seat in the F.20, the flying-controls installation had to be modified to place the pilot in the rear seat and give the gunner a clear field of fire. The guns tested were the Hotchkiss, Rexer and Maxim (Vickers) weapons, but the trials seemed somewhat inconclusive, although they emphasised the practicability of mounting a machine-gun on a pusher aeroplane.

The F.20 No.2838 was built as a dual-control trainer by The Aircraft Manufacturing Co and was at Farnborough on 20 January, 1916. It was allotted to No.24 Reserve Squadron at Hounslow. (*RAE*)

By the standards of its time the F.20 was quite successful, and when war began in August 1914 several were on the strength of Squadrons 3 and 5, while others were serving as trainers at CFS. When the squadrons went to France on 14 August, No.3 took four Henry Farman F.20s, Nos.274, 295, 351 and 352; No.5 Squadron's complement of the type consisted of 341, 346, 364 and 393; with the Aircraft Park went 455, 456 and 462.

It was on No.341, one of the F.20s of No.5 Squadron, that Lt Louis A. Strange fitted a Lewis gun on a mounting of his own devising in August 1914. On 22 August he set off in pursuit of a German aeroplane with Lt L. da C. Penn Gaskell as his gunner, but the enemy aircraft eluded their intended attack by simply

That at least one Henry Farman went to the Middle East is suggested by this remarkable hybrid aircraft of uncertain parentage at Ismailia. Its nacelle, engine, lower wings and undercarriage were Farman components; the upper wing was of Farman origin modified with some Short seaplane parts; the tail unit was locally made and was of original design. It would seem that the RNAS contributed to the aircraft, which may have been a joint RFC RNAS creation. (*RAF Museum*)

outclimbing the Farman. Strange's commanding officer concluded that the F.20 was overloaded and required him to remove the Lewis and its mounting, adding that Strange's observer would have to make do with a rifle in future. Louis Strange was of a warlike disposition, for on 28 August, 1914, he spent the morning 'fixing up a new type of petrol bomb to my Henry Farman, and in the afternoon Penn Gaskell and I went to try it out'. In fact the Farman was carrying three bombs, with which Strange bombed enemy transport columns with modest success. He was greatly attached to No.341 and did much remarkable work on her before taking over an Avro 504; the Farman survived until it crashed on 30 October, having outlived all of the other aircraft that had gone to France with No.5 Squadron.

Although Strange made something of a warplane out of the delicate Farman F.20, the type was not really robust enough for the rigours of war. No.3 Squadron had discarded all of its F.20s before the end of 1914, as did No.6 Squadron, while No.5 Squadron's last Henry Farman, No.1824, went back to the Aircraft Park on 14 March, 1915. Thereafter the type gave long and useful service as a trainer with a considerable number of training units in England; some are known to have been in active use as late as October 1917. All of the later production batches were intended to be used as trainers; all were ordered from The Aircraft Manufacturing Co, latterly (from 7396) under a running contract that covered Maurice Farmans also, but the batches A1154—1253 and A2276—2375 were sublet to The Grahame-White Aviation Co. The final production batch, B1401—1500, was ordered about March 1917, but only B1401—1481 were delivered.

F.20
80 hp Gnome, 80 hp Le Rhône

Two- or three-seat military biplane.

Span 13·25 m (43 ft 5¾ in); length 8·06 m (26 ft 5¼ in); height 3·15 m (10 ft 4 in); wing area 35 sq m (377 sq ft).
Empty weight 360 kg (794 lb); loaded weight 660 kg (1,455 lb).
Speed 105 km/h (65 mph); climb to 500 m (1,640 ft) 8 min; endurance 3 hr.
Armament: Experimental installations of Lewis, Hotchkiss, Rexer and Vickers machine-guns; various combinations of rifles and revolvers; occasional improvised bomb loads.

Manufacturers: Aéroplanes Henry et Maurice Farman, Airco, Grahame-White.

Service use: Squadrons 3, 5 and 6. *Training duties*: Central Flying School, Upavon; Reserve Training Sqns 2, 3, 4, 6, 8, 9, 10, 19, 24 and 27; Depot Sqns 199 and 200. *Other units*: No.9 Sqn, Dover; No.15 Sqn, Dover; No.40 Sqn, Gosport.

Serials: 244, 268, 274, 275, 276, 277, 284, 286, 294, 295, 330, 339—341, 346, 350—353, 363, 364, 367, 393, 435, 440, 444, 445, 455, 456, 461, 462, 467, 502—513, 558—569, 669, 680, 685, 689, 699, 708, 719, 720, 728, 738, 1801—1805,* 1813, 1814, 1817, 1818, 1821—1824, 1826, 1835, 2832—2851, 7396—7445.
A1154—1253, 1712—1741, 2276—2375, 3023.
B1401—1481.

Henry Farman F.27

Aircraft with all-metal airframes were still relatively rare in 1914, and at that time one of the most successful steel-framed types was the Voisin pusher biplane. In 1914 Henry Farman produced his Type 27, a two-seat pusher biplane powered by a 135 hp Salmson water-cooled radial engine. It was painfully obvious that the F.27 had been inspired by the contemporary Voisin, both geometrically and structurally, but its three-bay wings were of equal span and chord and had no stagger.

Although the Farman F.27 has become moderately well known for its use by the RNAS in the Aegean, it was also used by the RFC and indeed was first introduced to the British flying Services by the RFC. When the four squadrons of the RFC took the field it was essential to find sources of supply of aircraft, for the Service had neither reserves nor an adequate native aircraft industry behind it. Such aircraft as could be obtained in France were quickly bought and numbered from 801 onwards until it was realised that this would duplicate the existing range

*At least some of this group, 1801—1805, may have been Farman F.27s.

of numbers already allotted to the RNAS. The RFC's purchases were then very simply renumbered by making them run from 1801. The first six aircraft 1801—1806 are all officially listed as Henry Farmans. It is known that 1806 was an F.27, and at least some of the other five may well have been so also.

On 1 September, 1914, No.1806 (as 806) was flown direct to No.5 Squadron, RFC, from Paris, following 1801—1804, which had gone to the squadron on 28 and 30 August; 1805 also was delivered to No.3 Squadron on 1 September. In No.5 Squadron 1806 lasted exactly a week: it was struck off squadron charge on 8 September without reason assigned. However, it must still have been in existence after 10 October, 1914, for that was the date of the instruction sent by the A.D.M.A. to the Aircraft Commander, Aircraft HQ, Expeditionary Force, requiring that 'All land machines with numbers 800 and upwards should therefore be altered to 1800 and upwards to 2000'. As 1806 is known to have borne that number at some time it must have survived its striking off by No.5 Squadron.

Henry Farman F.27 No.1806 in RFC service. It was delivered to No.5 Squadron from Paris on 1 September, 1914, was damaged on 2 September, and was struck off charge on 8 September. As indicated in the text, however, it must have been reconstructed in order to bear the number 1806.

The F.27's all-steel structure made it more suitable for use in hot countries than the contemporary lightly built wooden aircraft such as the B.E.2c. Military operations against German South-West Africa by the Union Expeditionary Force began early in September 1914. At that time no air support could be provided: although a small group of South African officers were in training at Farnborough in August they had all gone to France with the RFC squadrons. In November 1914 Capt G. P. Wallace returned to England to form a flying unit for service in German South-West Africa; early in 1915 the remaining South African officers returned from France, and two of them were sent to South Africa to recruit personnel and prepare an aerodrome.

The Admiralty had ordered a number of F.27s, and delivery was promised for the end of February 1915. Deliveries were in fact slow, and only three had been delivered by the end of March. These were handed over to the South African detachments by the Admiralty with two B.E.2cs; all reached Walvis Bay on 30 April but two of the F.27s were damaged on the voyage. The first reconnaissance over the German lines was made on 28 May by Lt K. R. van der Spuy, and early in June two more Farman F.27s arrived from England, thus enabling more reconnaissances to be made. On 29 June the three serviceable F.27s bombed Otavi, each carrying eight 16 lb bombs. A local armistice was proclaimed on 6 July, and the F.27s were considered to have performed well in the difficult conditions.

Admiralty records indicate that six F.27s were transferred from the RNAS order to the War Office, and it seems likely that these included the aircraft that went to South-West Africa. The flying element of the Union Expeditionary Force went to Cape Town to be demobilised; most of its officers then went to England; and the South African Air Corps was then mobilised with a strength of 119 NCOs and men. This body went to England early in October 1915, joined its former officers, and became No.26 (South African) Squadron of the RFC.

In November 1915 the RFC ordered eight Farman F.27s direct from the makers for use in Africa. On 26 November, RFC Headquarters wrote to Capt Lord Robert Innes-Ker of the B.A.S.D.:

> 'The eight Henry Farman machines are required for Africa and it is important that they should be not merely delivered but actually in England within six weeks. If this is not done the British Government will have to divert some of the machines intended for us and use them in Africa. Please therefore employ your utmost exertions to procure the early delivery of these machines. As stated in the telegram they should be all steel and without engines.'

These eight F.27s were probably the aircraft that were allotted the serial numbers 7746—7749 and 7752—7755 on 15 February, 1916. They were sent from France to The Aircraft Manufacturing Co to be fitted with engines, which were made in England by the Dudbridge Ironworks, who were the British licensees for the Salmson engines.

At that time No.26 (South African) Squadron had been in East Africa for just two weeks, having arrived at Mombasa on 31 January with eight B.E.2cs. The eight Farman F.27s arrived from England on 4 May, whereafter their assembly

Henry Farman F.27 of No.31 Squadron in India.

Farman F.27s and B.E.2cs of No.31 Squadron lined up at Risalpur for inspection by the Viceroy of India, Lord Chelmsford. (*RAF Museum*)

occupied several weeks, partly because faulty material used in their components made some reconstruction essential. They were not used at first, being considered too slow, but on 6 July three of them went to Mbagui to reinforce the air detachment there. These were probably the sole survivors of the original eight, for one had crashed and there was only enough reliable material to enable three sound aircraft to be assembled. Nevertheless, thanks to the reliability of their Salmson engines, they did much good work on reconnaissance. By 31 August, 1916, the squadron was at Morogoro with only the three F.27s serviceable.

Evidently the sturdy reliability of the all-steel Henry Farman was appreciated by the RFC, for more were ordered in the spring of 1916. Apparently twelve were ordered at first, the last four of this dozen leaving Paris on 6 April; and on 6 May the British Aviation Commission was instructed to order eight more. By 17 June this had increased to twelve, following a refusal by Farman to deliver any more aircraft unless the War Office paid outstanding bills. This was quickly done, and eight F.27s left for England in the last week of June. The ordered total of 24 aircraft must have been those to which the serial numbers A387—410 were allotted on 13 May, 1916; yet it seems that only 20 were delivered, for official papers state that a total of 28 all-steel Farmans were obtained during the twelve months ending on 30 September, 1916.

Nevertheless, the RFC acquired a further 27 aircraft of this type from the RNAS in May 1917. These were sent from Paris to Ascot, and at least the major group of 25 aircraft had the 160 hp Salmson engine. In the RFC these transferred Farmans were numbered A8968—8969 and A8974—8998.

In the hottest climes in which war was waged the F.27 lingered on into 1918, when some were with the Aden Flight of No.31 Squadron at Khormaksar. As late as August 1918 these were crated and sent to India, presumably to join the main body of No.31 Squadron at Risalpur, where the unit is known to have operated F.27s.

F.27
140 hp or 160 hp Salmson (Canton-Unné)

Two-seat reconnaissance bomber.
Span 16·15 m; length 9·22 m; height 3·65 m.
Total load 350 kg.
Speed at sea level 147 km/h, at 2,000 m—145 km/h, at 3,000 m—142 km/h; climb to 2,000 m—16 min, to 3,000 m—29 min; ceiling 4,800 m; endurance 2 hr 40 min.
Armament: One 0·303-in Lewis machine-gun and a varying bomb load.

Manufacturer: Henry et Maurice Farman.

Service use: *South-West Africa*: Air Detachment of Union Expeditionary Force. *East Africa*: No.26 (South African) Squadron, RFC. *Aden*: Aden Flight of No.31 Squadron. *India*: No.31 Squadron. *Mesopotamia*: one RNAS F.27 transferred to No.30 Squadron.

Serials: 1806, 7746—7749, 7752—7755.
A387—410, 8968—8969 (ex N3035 and N3029), 8974—8998 (ex 9099, 9153, N3025—3028, N3030—3034 and N3036—3049).
B3957—3983: cancelled; duplicated A8968—8969 and A8974—8998.
B3969—3970: cancelled; again duplicated A8968—8969.

No.266 was a Maurice Farman S.7 of the early type with curved tail-booms and kidney-shaped lower tailplane. It came to the Military Wing on 21 December, 1912, saw use with No.2 Squadron, and crashed on 5 May, 1913. After protracted repairs in the Royal Aircraft Factory the Farman apparently went to CFS with the new identity of 472. On the outbreak of war 472 went to Farnborough on 5 August, 1914, and was later reported to be with No.4 Squadron at Dover. It never went to France, however. The photograph shows 266 at York on 21 February, 1913, in the course of No.2 Squadron's transit from Farnborough to Montrose.

Maurice Farman Série 7 (Longhorn)

The earliest Maurice Farman biplane of 1909 was obviously derived from contemporary Voisin designs. In developing the design through 1910 and 1911 Maurice Farman evolved a remarkable configuration in which the forward elevator was carried on characteristically upswept forward extensions of the undercarriage skids. By 1911 the design developed into an unequal-span biplane with strut-braced extensions on the upper wings and a biplane tail unit in which the upper plane was rectangular and carried the single rear elevator; the lower tailplane had considerable undercamber and a liberally curved trailing edge, and the front elevator had semi-circular tips. A curious structural feature was the tail-booms' pronounced curvature in plan.

A number of Maurice Farmans of this type were acquired by the Military Wing in 1912, starting with the biplane that was numbered 403 and taken on charge on 27 July, 1912, closely followed by 207, taken on charge in August. A single example was entered for the Military Trials by The Aircraft Manufacturing Co, who held the British manufacturing rights for all Farman aircraft.

No.498 was a Farman Longhorn of the later and more numerous form, with straight tailbooms and rectilineal lower tailplane. This Farman was in use at CFS in December 1914 and was still flying at Netheravon in May 1916.

In 1913 a slightly modified development of the front-elevator pusher appeared. This variant had a rectangular lower tailplane of increased span, and the tailbooms were straight and parallel in plan. Both tailplanes had appreciably less undercamber than those of the earlier biplane. The front of the nacelle was modified to be more rounded, the forward elevator had square tips with rounded-off leading corners, and the skids just behind the landing wheels were given a shallow step, carrying at their rear extremities steel braking skids. This became the standard production form of the aircraft, eventually identified as the Maurice Farman Série 7.

The improved aircraft was supplied to the RFC, apparently before the Série 7 (or S.7) designation came into use. In the Military Wing the earlier form of the Farman was known as the Maurice Farman biplane, 1912 type; the later version was the 1913 type. Precisely when the appellation Longhorn was first applied to

No.2976 in a typical wartime setting, with a B.E.2c for company. This Longhorn was built by The Aircraft Manufacturing Co at Hendon and was apparently flown to No.9 Reserve Aeroplane Squadron at Norwich on 7 August, 1915, three days before it was formally allotted to that unit. (*Flight International 19314*)

the aircraft is not clear; obviously it followed the introduction of the slightly later Maurice Farman S.11, early known as the Shorthorn; certainly the war-time RFC thus distinguished les vaches méchaniques (the mechanical cows) from one another.

Although the earliest of the RFC's Farman S.7s were apparently built by the parent French company, early deliveries were made by The Aircraft Manufacturing Co. The first Maurice Farman to be built by the Hendon-based company went to CFS to become No.415; it was delivered to Farnborough on 7 November, 1912, and was officially allocated ten days later. All the batch-built groups of S.7s delivered to the RFC were built by Airco, with the possible exception of C4279 and C9311—9335, which were reported to be assembled by The Brush Electrical Engineering Co Ltd from spares held by Robey & Co Ltd and The Phoenix Dynamo Manufacturing Co, and to be fitted with 90 hp Curtiss engines. These contractors and engines were all normally associated with supplies to the RNAS, and those 26 Farmans may have been destined for that Service.

In the prewar period all four of the Military Wing's aeroplane squadrons had a few Farman S.7s at one time or another, and the type served as a trainer at CFS. One of these, No.426, had an unusually distinguished ferry pilot on its delivery flight: at the controls when it flew from Farnborough to CFS on Monday, 17 February, 1913, was Maj Hugh Trenchard.

On 31 July, 1914, Squadrons No.2, 3, 4 and 5 had between them 33 serviceable aircraft, none of which was a Maurice Farman S.7. The type should have formed the equipment of 'C' Flight of No.4 Squadron, which had only the promise of No.610 from the AID and three more to come from Farnborough. By 5 August, Nos.450, 472, 476 and 478 had left CFS and had been reserved for No.4 Squadron, but in the event No.4 flew only eleven B.E.2as to France on 14 August and no Maurice Farman S.7 went with the four squadrons. The RFC subsequently acquired one or two Longhorns by local purchase in France; one such was No.1830, which went to the Headquarters Wireless Section (later named No.9 Squadron) on 27 October, 1914, and survived until wrecked on 28 December.

Although the operational career of the Longhorn in France apparently ended at an early date, a few of these venerable biplanes were flying against the enemy in Mesopotamia up to at least 7 September, 1915. The formation of an RFC Aeroplane Flight for service at Basra was begun by Capt P. W. L. Broke-Smith in the second half of April 1915. Two Longhorns were sent out from England for this unit, and by the end of May these were joined by two more from Egypt; the latter pair 'had seen much service as training aeroplanes in India' but were without their engines, which did not arrive until August. The two airworthy Longhorns made useful reconnaissance flights in May and June, and one was wrecked at Ali Gharbi as late as 7 September, 1915.

If the Longhorn performed no resounding feats of arms, it gave yeoman service as a primary trainer with many Reserve and Training Squadrons, remaining in use virtually throughout the war. Some people did kill themselves while flying Longhorns, but in its day it was a safe and forgiving aircraft, and its complex structure obligingly absorbed much of the shock of the crash landings that were inseparable from its role.

Série 7 (Longhorn)
80 hp Renault

Two-seat general-purpose aircraft and trainer.

Span 15·52 m (50 ft 11 in) upper, 11·516 m (37 ft 9½ in) lower; length 11·52 m (37 ft 9½ in); height 3·5 m (11 ft 5¾ in); wing area 49 sq m (528 sq ft).
Empty weight 599 kg (1,320 lb); loaded weight 871 kg (1,925 lb).
Speed at sea level 105 km/h (65·2 mph); climb to 1,000 m (3,280 ft) 15 min, to 2,000 m (6,560 ft) 35 min; endurance 5 hr

Manufacturers: Henry et Maurice Farman, Airco, Brush.

Service use: Prewar and pre-operational service—Sqns 2, 3, 4, 5, 6 and 7. *Western Front*: Headquarters Wireless Section, later No.9 Sqn. *Mesopotamia*: RFC Aeroplane Flight at Basra, later No.30 Sqn. *Training duties*: CFS; Reserve/Training Sqns 1, 2, 3, 4, 5, 6, 7, 8, 9, 10, 11, 12, 14, 15, 19, 22, 24, 25, 26, 36, 39, 41, 49, 57, 58; No.8 Training Depot Station. *Squadrons working up*: Nos.15 and 41.

Serials: 207, 214—216, 223, 224, 246, 266 (to CFS as 472), 269, 270, 301, 302, 305—307, 322, 337, 338, 343 (to CFS as 465), 355—360, 376, 403, 410, 411, 415, 418, 425—429, 431, 450, 451, 458, 459, 463, 464, 472 (ex 266), 476, 477, 478, 490, 494—496, 498—501, 519—545 (mixture of Longhorns and Shorthorns), 546—557, 610, 661, 679, 712, 713, 735, 739, 1830, 2960—3000, 4001—4019, 5617, 6678—6727.
  A4061—4160, A7001—A7100, A9978 (ex N5010), A9979 (ex N5011).
  B393 (ex N5012), B397 (ex N5016), B398 (ex N5014), B399 (ex N5013), B400 (ex N5015), B3982.
  C4279, C9311–9335.

Maurice Farman Série 11 (Shorthorn)

On Saturday, 25 October, 1913, there appeared at Hendon a Maurice Farman biplane of a new type. It was virtually an S.7 without the forward elevator, for it retained the biplane tail unit and rectangular rudders of the earlier type. The skid members of the undercarriage ran only a relatively short way forward of the wheels, and the engine was an 80 hp de Dion. This Farman was owned by the

No.379 was one of the earliest Farman Shorthorns to be delivered to the RFC, being taken on the strength of the Military Wing on 5 April, 1914. There were no fins in its tail unit when it was photographed. It was at CFS for a time in 1914, and by June 1915 was with No.1 Reserve Aeroplane Squadron at Farnborough. On 18 May, 1916, it was at the Reserve Aircraft Park, Farnborough, allotted to the 6th Brigade.

Marquis de Lareinty Tholozan, and when it participated in some of the racing events at Hendon it was flown by Pierre Verrier.

In the issue of *The Aeroplane* dated 30 October, 1913, C. G. Grey observed:

'The new type is likely to meet with the approval of the Royal Flying Corps, who have a high regard for the Maurice Farman, despite their having nicknamed it the 'Mechanical Cow''. Perhaps, owing to the absence of the long front skids, the new type will become known as the "Shorthorn"'.

Time was to prove Grey right both as regards adoption by the RFC and as to a distinctive nickname, but the aircraft thus adopted and identified, and that first appeared in November 1913, was of a slightly different type.

Dated 16 September, 1914, this photograph was taken within the Royal Aircraft Factory at Farnborough and shows a mounting for a Lewis gun applied to a Farman Shorthorn, almost certainly No.344, which went to No.4 Squadron in France on 25 November, 1914, only to be struck off charge two days later. It is not known whether its Lewis gun ever fired a shot in anger. *(RAE 556: Crown copyright)*

The modified design had a monoplane tail unit, and its tail booms converged in side elevation to meet on the rear spar of the tailplane. Twin parallelogram-form rudders were braced to their respective upper tail-booms by a single strut on each side. The wings and undercarriage resembled those of the de Dion powered aircraft, but the nacelle was raised well above the lower wing, and the engine was an 80 hp Renault. A floatplane version of the modified design was exhibited at the 1913 Paris Aero Salon.

This quite workmanlike aircraft was, in its landplane form, standardised for production as the Maurice Farman Série 11. It was quickly adopted by the RFC, which received its first deliveries of the type in March 1914. Apparently five were acquired at that time and were numbered 342, 344, 345, 464 and 465; the two last-mentioned went to CFS, but 464 was transferred to the Military Wing in May 1914, exchanging identities with 343, an S.7. Three more, 369—371, followed in May. A civil-operated Maurice Farman S.11 was flying at Hendon late in April, and was at that time referred to as the Shorthorn, consequently that nickname must have had contemporary currency in the RFC. Officially it was the Maurice Farman 1914 type.

Photographs of operational Shorthorns of the RFC are very rare. This one is of 5019, which arrived at the 1st Aircraft Park on 26 May, 1915, and went to No.16 Squadron on 26 June. It served with that unit until it was struck off charge on 24 September, 1915, which suggests that this photograph may have been taken at La Gorgue. The aircraft has a single central roundel on the tailplane and elevator; other roundels can be distinguished on the upper wings and nacelle sides. (*RAF Museum*)

Thought to be of No.16 Squadron, this Shorthorn had what appear to be distinguishing markings round the nose of the nacelle. (*RAF Museum*)

An unusual Farman Shorthorn, photographed at Hendon. The flying controls appear to be in the rear cockpit only, and a windscreen is provided for the pilot only. (*RAF Museum*)

The three Maurice Farman Shorthorns that were flown from Basra to join No.30 Squadron at Ora on 26 April, 1916, were wrecked by a storm during the night of 2/3 May, 1916. (*W. Evans*)

In the prewar period the Military Wing's Shorthorns were used by No.6 Squadron, which was reported to have five serviceable on 31 July, 1914. No Shorthorn went to France with the four squadrons of the RFC that took the Field on 14 August, 1914, but, on 26 September, 465 and 633 joined No.4 Squadron, followed two days later by 342. The Royal Aircraft Factory had fitted a Lewis gun to No.344 on 16 September, and this Shorthorn was allocated to No.4 Squadron on 25 November. It is not known how many Farmans were similarly armed. A few more went out from England, others were bought in France, and the Shorthorn was used in small numbers by Squadrons No.2, 4, 5, 9 and 16. Apparently the last to be on the strength of an operational unit in France were those of No.16 Squadron, which soldiered on long after all other RFC units had relinquished the

This Shorthorn was used in Mesopotamia by the Australian Half Flight, Indian Flying Corps, and was photographed on the airfield at Ma'gil, near Basra. The Australian Half Flight was also known as the Mesopotamian Flight, and in August 1915 it became 'A' Flight of No.30 Squadron, RFC. (*A. E. Shorland*)

The earlier RFC Shorthorns had strut bracing to the extensions of the upper wings. This is one of the aircraft of the batch A904—A953, built by The Aircraft Manufacturing Co under their running contract, No.87/A/109, and delivered from July 1916.

type: last to leave No.16 was 5071, which was returned to the Aircraft Park on 24 November, 1915. Most of the surviving Shorthorns were sent back to Britain for use as trainers, but some remained in France and were used in that capacity at the BEF Flying School at le Crotoy.

On the great majority of production Shorthorns the area under each sloping strut supporting the rudders was covered in to provide a fin on each side. Various primitive attempts were made to carry armament, but otherwise the design changed little in operational use.

The Shorthorn remained operational longer in Mesopotamia, where No.30 Squadron had a few on its strength late in 1915. As existing maps of the Tigris area proved seriously inaccurate, aerial survey flights had to be made in order to create usable maps. Lack of equipment and personnel made photographic survey impossible, and simple instruments were devised to enable a visual survey by

Later Shorthorns had kingposts above the upper wings and wire bracing to the extensions. One such was B1957, also built by The Aircraft Manufacturing Co under Contract No.87/A/109, the batch of serial numbers B1951—B2050 having been allotted on 3 April, 1917. B1957 was later used by No.5 Squadron, Australian Flying Corps. (*K. M. Molson*)

triangulation to be made. In this work the Farmans, being slow and affording a wide field of view, were of great use, and eventually the RFC produced an excellent map of the Ctesiphon defences and the approaches to them. This map proved of great value to the ground forces before and during the Battle of Ctesiphon that began on 22 November, 1915. Four Shorthorns were sent to Basra in the spring of 1916, and such were the demands imposed by the siege of Kut-al-Imara that three of them were flown to join No.30 Squadron up country on 26 April, 1916; they included 5909 and 7346. Unfortunately all three were wrecked by a storm on the night of 2 May, a disaster that seemed to mark the end of the Shorthorn's operational career.

It is as a primary trainer that the Shorthorn is best remembered, and it served the RFC in that capacity in Britain, France and Egypt. More Shorthorns than Longhorns were built, those for the RFC being made by The Aircraft Manufacturing Co under the seemingly inexhaustible running Contract No. 87/A/109. This called for 75 per cent of deliveries to be Shorthorns, 25 per cent Henry Farmans. Virtually every training unit had some and the type survived well into 1918. Precisely how, where, and by whom the trainer Shorthorn was nicknamed the Rumpety is a fact lost in the mists of aviation's history, but as such it was widely, and not without affection, known.

Série 11 (Shorthorn)
80 hp Renault

Two-seat reconnaissance biplane and trainer.

Span 15·776 m (51 ft 9 in) upper, 11·764 m (38 ft 7⅛ in) lower; length 9·3 m (30 ft 6 in); height 3·15 m (10 ft 4 in); wing area 52 sq m (561 sq ft).

Empty weight 654 kg (1,441 lb); loaded weight 928 kg (2,046 lb).

Speed 116 km/h (72 mph); climb to 1,000 m (3,280 ft)—8 min, to 2,000 m (6,560 ft)—20 min; endurance 3¾ hr.

Armament: Some Shorthorns were armed with a 0·303-in Lewis machine-gun; various combinations of rifles and revolvers were also employed.

Manufacturers: Henry et Maurice Farman, Airco, Whitehead.

Service use: Prewar: No.6 Sqn; CFS. *Western Front*: Sqns 2, 4, 5, 9 and 16. *Mesopotamia*: No.30 Sqn. *Training duties*: BEF Flying School, le Crotoy, France; Reserve/Training Sqns 1, 2, 3, 4, 5, 6, 7, 8, 9, 10, 11, 12, 14, 15, 19, 21, 22, 24, 25, 26, 27, 29, 36, 38, 39, 41, 47, 48, 49, 57 and 68; Training Depot Stations 8 and 204; CFS. *Other units*: Wireless School, Brooklands; School of Instruction, Reading; No.2 (Auxiliary) School of Aerial Gunnery, Turnberry; No.1 School of Navigation and Bomb Dropping, Stonehenge. *Squadrons working up*: Nos.13, 14, 15, 23, 41, 83 and 97; No.5 Sqn Australian Flying Corps.

Serials: 342—345, 369—371, 379, 465, 480, 481, 514—518, 519—545 (mixture of S.7s and S.11s), 633, 707, 742, 744, 1827, 1839, 1840, 1841, 1844, 1846, 1851—1854, 1857, 1869, 1893, 2465—2468, 2856, 2940—2959, 4294, 4297, 4298, 4726—4731, 5004, 5008, 5009, 5015, 5018, 5019, 5027, 5030, 5036, 5054, 5059, 5071, 5880—5909, 7346—7395.

A324—373, 728, 904—953, 2176—2275, 2433—2532, 6801—6900.

B1951—2050, 4651—4850.

Rebuilds: 5718, A1711, A2175, A9162, A9921.

B1483, B3981, B3988, B8788, B8793, B8515, B9433, B9441—9443, B9466, B9475, B9956, B9991, B9996, B9999.

C3499, C4278.

E9986—9988, E9998.

F2183—2187, F2230.

Maurice Farman S.11 seaplane

In January 1917 two new schools of aerial gunnery were opened in Ayrshire, one at Turnberry on the coast, the other at Loch Doon, some 20 miles inland. It was intended to move the RFC's existing School of Aerial Gunnery from Hythe to Loch Doon. This loch is about six miles long and in 1917 was considered to be well suited to the operation of float seaplanes.

That the RFC intended from the outset to use floatplanes at Loch Doon is suggested by the fact that a batch of eight seaplanes was ordered as early as August 1916. These were Maurice Farman Shorthorns ordered with float undercarriages, and possibly other modifications, from The Aircraft Manufacturing Co Ltd under Contract No. 87/A/708. The serial numbers A3295—3302 were allotted on 26 August, 1916.

A floatplane version of the Shorthorn was exhibited at the 1913 Paris Aero Salon, but it is not known how much Loch Doon's aquatic Farmans owed to that early prototype. Power unit and performance details for these Farman seaplanes have yet to be found, but if only the 80 hp Renault were fitted the aircraft must have been agonisingly slow. In fact they proved to be unsatisfactory, and by early June 1917 the RFC was seeking to replace them. The chosen replacement was the F.B.A. flying-boat, and presumably the Farmans were withdrawn later in the summer of 1917 when the three F.B.A 'boats were supplied.

S.11 seaplane

Two-seat gunnery-trainer.
Manufacturer: Airco.
Service use: School of Aerial Gunnery, Loch Doon.
Serials: A3295—3302.

F.B.A. Flying-boat

When the Progress and Allocation Committee held its 88th meeting on 6 June, 1917, Gen Brancker stated that the RFC wanted to acquire a few seaplanes for use at the School of Aerial Gunnery at Loch Doon. These were required to replace the Maurice Farman seaplanes that had proved to be unsatisfactory, and it had been suggested that three or four F.B.A flying-boats might prove useful. In response to this expressed wish, Capt Vyvyan, RN, said that he thought the RNAS could probably supply three F.B.A. 'boats to start with, but in his view it would be necessary to have specially trained pilots for them. Brancker confidently said that this could be arranged (but apparently did not say how).

Capt Vyvyan pointed out that adopting aircraft of this type implied a continuing need, and that in his view the RFC should place contracts. Nevertheless, the Committee agreed that the RNAS should supply three F.B.A. 'boats for use by the RFC, and shortly after the Committee's meeting Nos.9615, 9622 and 9623 were chosen from the batch 9612—9635. As this group of F.B.A.'boats was apparently delivered as two separate dozens, 9612—9623 and 9624—9635, it is impossible to be certain that the aircraft of the first dozen were identical with those of the second, for the little F.B.A. trainer was built in two slightly different forms. If all 24 were identical, there is a reasonable probability that the three transferred to the RFC resembled 9630 illustrated here, which had rigid trailing edges on the mainplanes, small inversely tapered ailerons, and rounded leading edge on the rudder.

On transfer to the RFC the three F.B.A.s were allotted new serial numbers in July 1917; they became B3984—B3986, apparently in direct sequence. Whether, and, if so, how, they were modified for use at Loch Doon remains undiscovered.

It seems that in mid-1917 the RFC intended to use considerable numbers of F.B.A. flying-boats for, apparently agreeing with Capt Vyvyan's view, the War Office placed a production order for thirty F.B.A.s with The Gosport Aviation Co Ltd during the week ending 16 July, 1917. Presumably this production took some time to get under way, for on 11 September Capt Vyvyan asked whether the RFC wanted any of the F.B.A.s that were then about to be delivered. Apparently four were then expected from France; the RFC asked for two, and this was agreed. The identities of these two F.B.A.s remain obscure, and confirmation of their delivery to the RFC has yet to be found: they may have been wanted merely as a source of spares.

The precise configuration of the F.B.A. flying-boats used at Loch Doon is not known, but their original serial numbers were so close to that of 9630 that this photograph may be generally representative of the RFC's three F.B.As. (No.9630 itself was used at RNAS Lee-on-Solent, later No.209 Training Depot Station, RAF) (*W. Evans*)

Production and allocation problems arose with the Gosport-built F.B.A.s, for it appears that some oversight in ordering might have occurred: the RNAS had asked for ten F.B.A.s per month, yet only the thirty for the RFC had been ordered. On 10 October, 1917, it was agreed that the thirty would be shared equally by the RFC and RNAS. As a further twelve (possibly 9624—9635) had been ordered from the F.B.A. company in France, the RNAS could be allocated twenty-seven F.B.A.s in the months of October, November and December, during which period the RFC was to receive fifteen. A further order was placed on 10 October, from which the RNAS was to receive a further six in December.

It is not known how many, if any, of these later F.B.A.s were delivered to the RFC, for no known record suggests that any but the original three acquired RFC serial numbers. It is possible that all were eventually allocated to the RNAS, for the Gosport company built two consecutively numbered batches of F.B.A.s, each of thirty aircraft (N2680—2709 and N2710—2739), and as far as is known all sixty were apparently delivered to the RNAS. Similarly, it is not known how much service any of the F.B.A.s gave at Loch Doon, for the School of Aerial Gunnery was relatively short-lived, having to be closed down early in 1918, primarily because the constructional difficulties of building the station at such an inaccessible location proved to be much greater and considerably more costly than had been expected; but also because the local weather was so unfavourable that flying days were too few to be worthwhile.

F.B.A. Flying-boat
100 hp Gnome Monosoupape

Two-seat trainer.
Span 13·715 m (upper), 10·34 m (lower); length 8·61 m; height 2·97 m.
Empty weight 574 kg; loaded weight 888 kg.
Maximum speed at sea level 109 km/h.
Armament: Possibly some form of weapon installation was made in the RFC's F.B.A.s, but its nature has yet to be discovered.

Manufacturers: Louis Schreck and Gosport Aviation.

Service use: School of Aerial Gunnery, Loch Doon.

Serial numbers: B3984—3986 (ex 9615, 9622 and 9623).

Flanders F.4

L. Howard Flanders's first successful aircraft was a monoplane designated Flanders F.2 and powered by a 60 hp Green engine. It first flew on 8 August, 1911, as a single-seater, but in October 1911 it was rebuilt as a two-seater with larger wings and was redesignated F.3. This flew frequently and successfully at Brooklands in the spring of 1912, and the War Office ordered four Flanders monoplanes for the RFC's Military Wing.

The aircraft for the RFC had the 70 hp Renault engine and full dual control; they were designated Flanders F.4. Howard Flanders opened new works at

The first Flanders F.4 in its original form, at the Naval and Military Aviation Day at Hendon, 28 September, 1912. At that time the aircraft had a single cabane strut and its undercarriage was similar to that of the Flanders F.3.

A Flanders F.4 at Brooklands with the cabane and undercarriage that were standardised for the four aircraft supplied to the RFC. (*Flight International*)

Richmond in March 1912, the erecting shop being reported to be large enough for four or more aircraft to be laid down at the same time. A monoplane that was probably the first F.4 was reported to be nearing completion early in April. It arrived at Brooklands late in June and was flying by 6 July, its pilot being F. P. Raynham. What was described as 'a new Flanders monoplane, a Government order', was flown by Raynham on 11 August. At that time it had a single central cabane strut, as on the F.3, and its undercarriage was structurally similar to that of the F.3.

Raynham evidently took this Flanders monoplane to Farnborough in September 1912, for he was reported to have flown it back to Brooklands on 18 September, and on 28 September it was at Hendon for the Naval and Military Aviation Day. On 8 November Raynham was reported to have flown 'for over an hour on F4 No.1 machine with new wheel base, for official test, at Farnborough. This machine has now passed all its tests.' The undercarriage had been modified by removing the original Blériot-like arrangement of struts and shock absorbers and replacing it with individually sprung legs attached directly to the lower longerons. At the same time a new cabane structure with an inverted-V strut was fitted.

Flanders F.4 immediately after delivery to Farnborough and before being numbered. (*RAF Museum*)

The second F.4 for the RFC made its first flight on 30 November, 1912, piloted by Raynham; it was delivered to Farnborough on 3 December. It is likely that the third aircraft was the 'new military monoplane' that Raynham tested at Brooklands on 15 December and flew to Farnborough on 18 December. At Farnborough the monoplane's one-hour and speed tests were flown that day, and it performed its ground-handling and climbing tests on 23 December. The fourth and last F.4 made its first flight on Wednesday, 1 January, 1913, and was flown to Farnborough by Raynham next day, when it completed all of the official tests.

None of the Flanders F.4s left Farnborough, where this photograph of 422 was taken. (*RAE*)

The elegant and practical Flanders monoplanes had the misfortune to be delivered to the Military Wing while the official ban on monoplanes was in force. They were numbered 265, 281, 422 and 439, the first two being allotted for use by No.5 Squadron, the others for CFS. The monoplanes never reached their theoretical destinations. On 12 June, 1913, Nos.422 and 439, although allocated to CFS, were still at Farnborough, and the Officer Commanding the Military Wing asked CFS to remove or dispose of them. On 1 July the Director of Artillery instructed CFS to strike them off charge (in fact they had never been on charge of CFS) and directed the O.C. Military Wing to hand over all four Flanders to the Royal Aircraft Factory. Subsequent instructions to the Superintendent of the Royal Aircraft Factory required him to take out the Renault engines and send them to contractors building B.E.2as. The monoplanes were handed over to the Royal Aircraft Factory on 7 and 17 July, 1913, apparently never having been flown by any RFC pilot, and were never heard of again.

F.4
70 hp Renault

Two-seat military monoplane.

Span 40 ft 6 in; length 31 ft 6 in; wing area 240 sq ft.

Empty weight 1,350 lb; loaded weight 1,850 lb.

Maximum speed 67·2 mph; climb to 1,000 ft—3 min 30 sec, to 1,500 ft—5 min 30 sec, to 2,000 ft—8 min.

Manufacturer: Howard Flanders.

Service use: None.

Serials: 265, 281, 422, 439.

The Grahame-White transaction

In order to understand how some rather improbable Grahame-White aeroplanes came on to the strength of the RFC in 1913 it is necessary to look again at the remarkable episode of Col J. E. B. Seely's claims as to the number of aircraft available to the Military Wing. He was then Secretary of State for War and when he introduced the Army Estimates to the House of Commons on 19 March, 1913, he stated that the Military Wing had 101 aeroplanes. This was immediately contested by Mr Joynson-Hicks, MP, but on 4 June Col Seely stated that there were then 126 aeroplanes, of which 31 were under repair, with the Military Wing.

The political ins and outs of the situation would take too much space to examine here, but eventually Col Seely's figures were effectively proved to be grossly overoptimistic. Even the early total of 101 aircraft had only been achieved by including the fruits of the purchase by the War Office of seven aircraft from The Grahame-White Aviation Co Ltd. This deal was reported in *The Aeroplane* of 27 March and in *Flight* of 29 March, 1913, wherein it was stated that the following aircraft had been purchased:

(i) the Henry Farman biplane (70 hp Gnome) used by Grahame-White as his *Wake up, England* aircraft;
(ii) the standard Henry Farman F.20 that Richard Gates had brought to England in August 1912;
(iii) the 70 hp Grahame-White Type VIIc 'Popular' biplane;
(iv) two Grahame-White Type VII 'Popular' biplanes with 35 hp Anzani engines (yet it seems doubtful whether more than one example of this type ever existed, and some confusion with (iii) above may have arisen);
(v) the decidedly second-hand Grahame-White School biplane, a simple boxkite with 50 hp Gnome;
(vi) the 70 hp Nieuport monoplane that Grahame-White had flown with conspicuous success in the USA in 1912; it became No.282 in the RFC.

According to *The Aeroplane*, an order for the purchase of the three 'Popular' biplanes had been placed some appreciable time previously, 'but the other four were purchased at a moment's notice'.

Later, in its issue of 1 May, 1913, *The Aeroplane* reported that the RFC had also bought the landplane conversion of the Grahame-White Type VIII floatplane exhibited at the 1913 Olympia Show, together with another pusher biplane built in the Grahame-White company's workshop. The Grahame-White types will now be discussed.

Grahame-White School Biplane

The precise origins of this aircraft are obscure but it seems to have been the Grahame-White biplane that was in use at Hendon early in 1912. It was later flown, sometimes spectacularly, by Marcus Manton and Richard Gates, in between stints of routine instructional flying.

Here seen at Hendon before it was hastily acquired by the RFC in the spring of 1913, this Grahame-White box-kite biplane had first flown on 9 March, 1912, and thereafter saw much use with the Grahame-White flying school.

The aircraft was a conventional boxkite biplane with a single rudder mounted at the point of convergence of the tail-booms; this rudder was of trapezium shape, with the trailing edge shorter than the leading edge. (On a later and very similar Grahame-White biplane the rudder was in shape a parallelogram, with its leading and trailing edges of equal length.) The engine was a 50 hp Gnome rotary. Throughout 1912 it put in a good many hours of flying and was a familiar sight to Hendon's habitués. By the time when it was taken over by the Military Wing it must have given yeoman service to the Grahame-White School.

With the Military Wing it acquired the official serial number 309. It was at the Royal Aircraft Factory on 14 April, 1913, waiting to be taken over by the Military Wing, and was formally taken on charge on 24 April. Its serial number suggests that it was intended for No.4 Squadron, but it never joined that unit. Not surprisingly, it proved difficult to make serviceable and it apparently did only ten minutes' flying with the Military Wing. It was reconstructed in September 1913, but to no avail, and it vanishes from the record.

School Biplane
50 hp Gnome

Two-seat trainer.

Manufacturer: Grahame-White.

Service use: Apparently confined to RFC Flying Depot at Farnborough.

Serial: 309.

Grahame-White Popular Biplane Type VII

In the first week of January 1913 the Grahame-White company completed a remarkably small single-seat pusher biplane powered by a 35 hp Y-form Anzani three-cylinder radial engine. In its essential geometry it was a scaled-down Henry Farman; indeed the original design drawing depicts an aircraft with a nacelle and tail-booms of V form in plan, converging to the rudder post of a tail unit that showed pronounced Farman influence. This design was designated Grahame-White Popular Biplane Type VII, and the company offered it for sale at the very modest price of £400 with the Anzani engine; an alternative version with the 50 hp Gnome was priced at £660.

The aircraft that was built differed somewhat from the drawings, for it had a shallow, angular little nacelle, and its tail unit was supported on tail-booms that converged in elevation to meet on the rear spar of the tailplane. In this form it was still described by the Grahame-White company as Type VII. Of it *The Aeroplane* commented in its issue of 9 January, 1913: 'It is very light and should make a capital little mount at a low price for the private owner, or for military school work.'

Grahame-White Type VII Popular single-seat biplane, to which the RFC serial number 283 was allotted.

In fact, it seems that the War Office were interested in the little biplane at a relatively early date and began negotiations for its purchase, eventually acquiring it in March 1913. At the time it was reported that the War Office had purchased two such 35 hp biplanes. This is not impossible, because it was announced in January 1913 that six aircraft were to be built and some of them set aside for the use of pupils at the Grahame-White flying school. Confirmation that the RFC ever received two of the small Popular Biplanes has yet to be found and, as far as can now be determined, it seems that only one serial number, namely 283, was ever allocated to a Grahame-White biplane of this type. Some confusion may have arisen because the completely different Grahame-White Type VIIc was also known as the 'Popular' in the company's loose naming system.

As No.283, the aircraft was at the Royal Aircraft Factory in mid-April 1913 waiting to be taken over by the Military Wing and under allocation to No.3 Squadron along with two B.E.s (Nos.271 and 273) and two Henry Farmans (276 and 284). Although these other four aircraft all saw squadron service, it is doubtful whether No.283 ever did. This is not a matter for surprise: even in the primitive state of the squadrons in the spring and summer of 1913 it is doubtful whether anything sensible could have been done with the little Grahame-White.

Popular Biplane Type VII

35 hp Anzani

Single-seat sporting biplane.
Span 29 ft 2 in; length 20 ft 10 in; height 9 ft 2 in; wing area 230 sq ft.
Maximum speed 50 mph; endurance 4 hr.

Manufacturer: Grahame-White.

Service use: Allocated to, but probably never used by, No.3 Sqn.

Serial: 283.

Grahame-White Popular Passenger Biplane Type VIIc

In attempting an historical note on this aircraft one is obliged to employ a good deal of conjecture. In its report in the issue dated 27 March, 1913, *The Aeroplane* included in the descriptions of the aircraft bought by the War Office from the Grahame-White company 'the new 70 hp Grahame-White biplane on Henry Farman lines, designed and constructed by Mr Bill Law, chief of the G.W. aeroplane shop'. Read in conjunction with an illustration in the January 1913

The Grahame-White Popular Passenger Biplane Type VIIc, which was sold to the RFC in March 1913 and apparently received the official number 354.

issue of *Aeronautics* it appears that this aircraft may well have been the two-seat pusher that the firm called the Popular Passenger Biplane and to which it gave the puzzling designation Type VIIc. It bore no relationship or resemblance to the Type VII single-seater, but it had more than a passing connection with the original Grahame-White boxkite, being little more than an adapted boxkite in which the forward elevator was removed and a nacelle installed on the lower mainplane. Finally, the aircraft that was allotted the RFC serial number 354 is listed in official records as a Grahame-White Popular biplane with 70 hp Gnome engine.

A detail that slightly upsets this neat relationship is the fact that a contemporary Grahame-White brochure states that the Popular Passenger Biplane Type VII (illustrated as Type VIIc) was fitted with a 50 hp Gnome. However, the aircraft would have been seriously underpowered with such an engine, and it is possible that a 70 hp engine may have been necessarily substituted.

The RFC's No.354 was officially allocated to the Military Wing on 23 May, 1913. It had been purchased for a mere £465 (the Grahame-White Catalogue price was £750) and apparently never left Farnborough. It seems that it did singularly little flying, even at the Flying Depot, and was never on the strength of any squadron. Having spent most of its time unserviceable in the Depot, it was finally struck off charge on 30 December, 1913.

Popular Passenger Biplane Type VIIc
70 hp Gnome

Two-seat pusher biplane.
Span 38 ft; length 26 ft 10 in; height 9 ft 10 in; wing area 475 sq ft.
Speed 50 mph; endurance 4 hr.

Manufacturer: Grahame-White.

Service use: Flown briefly at RFC Flying Depot, Farnborough.

Serial: 354.

Grahame-White Type VIII

At the 1913 Olympia Aero Show the Grahame-White company exhibited a single-engine two-seat tractor biplane of somewhat angular appearance, powered by a 60 hp Anzani radial engine. As exhibited, it had a twin-float undercarriage, the floats being of original design. They each had a single step formed by making the underside of the afterbody concave, and air was ducted to the concavity via tubes from two air intakes on the upper surface of the float.

This aircraft had the Grahame-White designation Type VIII and was offered with an alternative wheel undercarriage, or with an 80 hp Gnome engine in place of the Anzani. No record of a Gnome installation can be found, and it is doubtful whether such a variant ever existed. With the Anzani engine the price was £900 with float undercarriage, £800 with wheels, or £950 with both. All versions were to cost a further £350 with the Gnome engine.

The construction of the Type VIII was something of a tour de force on the part of the Grahame-White company. According to a contemporary report, Richard Gates, Claude Grahame-White's manager, only decided to build the aircraft on

The Grahame-White Type VIII was exhibited in the 1913 Olympia Aero Show as a twin-float seaplane. (*RAF Museum*)

6 January, 1913, yet it was completed in time to be on the Grahame-White stand at the Olympia show, which opened on 14 February, 1913. By April the Type VIII had been fitted with a cumbersome wheel undercarriage, in which twin main-wheels on each side carried a single bumper wheel in front.

On the Type VIII the passenger occupied the front cockpit, which was placed in line with the leading edges of the mainplanes. Thus the Grahame-White company was able in its brochure to claim: 'The passenger is so placed that he has an uninterrupted view embracing 180 degrees in every direction, a point of particular value when considered from a Naval or Military aspect.' This may have been enough to commend it to a War Office perhaps driven to desperation by Col Seely's ill-advised claims in the House of Commons. Certainly the aircraft was purchased by the War Office for use by the RFC.

The Type VIII as a landplane at Farnborough after a landing mishap that apparently cut short the career it might have had in the Military Wing, possibly as No.287. (*RAF Museum*)

Its identity in the RFC remains conjectural, but it may have become No.287, an indeterminate Grahame-White biplane that was flying at Farnborough in October 1913, was tested by Lt T. O'Brien Hubbard next month, and was struck off charge on 26 November as being dangerous. What is known about the Type VIII is that it was damaged while being flown at Farnborough; in all the circumstances it seems unlikely that it would be repaired. It bore no serial number at the time of the accident.

Type VIII
60 hp Anzani

Two-seat tractor biplane.
Span 42 ft 6 in; length 25 ft; height 12 ft; wing area 380 sq ft.
Speed 55 mph; endurance 3½ hr.

Manufacturer: Grahame-White.

Service use: Flown at Farnborough.

Serial: May have been 287.

Grahame-White two-seater at Hendon. Contemporary reports state that this Pusher Biplane was also sold to the War Office in 1913.

Grahame-White Pusher Biplane

According to *The Aeroplane* of 1 May, 1913, this plain and uncompromising two-seat pusher biplane was bought by the RFC in April 1913. In general appearance it resembled the Type VIIc, but the angle of the tail-booms in side elevation was more nearly symmetrical about the thrust line, the nacelle was of a simplified form, and the ailerons were connected spanwise by balance cables. The mainplanes, tailplane, elevators and undercarriage strongly resembled those of the Type VIIc, but the rudder was taller, parallelogram-shaped, and of higher aspect ratio. As in the case of the Type VIIc, the later pusher was virtually a boxkite of the second form with the forward elevator removed and a nacelle fitted.

Unfortunately it is not now possible to isolate the history of this Grahame-White pusher biplane, for surviving official records do not identify it specifically. It might just as easily have been No.287 as the Type VIII described in the preceding pages.

Pusher Biplane
70 hp Gnome

Two-seat pusher biplane.

Manufacturer: Grahame-White.

Grahame-White Type XV

From the second of the early Grahame-White boxkite biplanes was developed in 1913 a new pusher biplane, equally skeletal, but fitted with twin rudders at the rear ends of tail-booms that were parallel in plan. The wings had conventional under-camber but straight terminal compression members at the wingtips gave a false impression of there being no aerofoil section at all: on the contrary the wings were double-surfaced. The engine was a Gnome rotary mounted behind the propeller.

This Grahame-White Type XV, A1685, is known to have been used by No.31 Training Squadron, Wyton, but is here seen at Yatesbury, which suggests that it may also have been used by Training Squadrons 13, 16, 17, 32, 55, 59, 62 or 66.

The first aircraft of the new type earned a place in history as the vehicle used in air-to-ground firing tests of the Lewis machine-gun conducted at Bisley on 27 November, 1913. The biplane was piloted by Marcus D. Manton, and the gunner was Lt Stellingwerf of the Belgian army, who was carried in a terrifyingly perilous makeshift seat underslung below the pilot's position.

With the advent of war the twin-rudder boxkite was built in numbers and supplied to the RNAS for use as a trainer. The production aircraft could be fitted with extensions on the upper wing, and all were built with protruding end fittings for these extensions, whether fitted or not. Engines varied, and the 80 hp Le Rhône and 60 hp Green were sometimes fitted. The type was also used in some numbers by the Grahame-White flying school at Hendon.

A Grahame-White Type XV of the Grahame-White School wearing its local number and with a group of typical trainees; third from the left is Mavreky Osipenko, a Russian pilot who was an instructor at the Grahame-White School. The date of the photograph is unknown, but all the military personnel are in Army or RFC uniform. (*RAF Museum*)

From the extension variant was developed a slightly refined pusher trainer with the designation Grahame-White Type XV. The first of this new development appeared at Hendon on 1 November, 1914, powered by a 60 hp Le Rhône. Its airframe was derived in the same way as those of the Type VIIc and the other pusher hastily purchased by the War Office in 1913: the forward elevator and its booms were removed and a simple nacelle was provided for instructor and pupil. Full dual control was installed. The production version had double-acting ailerons and was usually powered by an 80 hp Gnome.

Initially, production of the Type XV was for the RNAS, but in mid-1916 a batch of fifty was ordered for the RFC under Contract No. 87/A/572. These were apparently standard in all respects and were numbered A1661—1710. Deliveries were slow: twenty-three were delivered in 1916, the remaining twenty-seven in 1917. They were used as primary trainers as alternatives to the Maurice Farman Longhorns and Shorthorns then in use, but apparently were not so highly regarded for no further orders were placed for the RFC. Some of those used by the Grahame-White school continued in the service of the RFC 18th Wing School of Instruction at Hendon into 1918.

Type XV
80 hp Gnome

Two-seat trainer.

Manufacturer: Grahame-White.

Service use: Reserve/Training Squadrons 6, 31, 35 and 48; 18th Wing School of Instruction, Hendon.

Serials: A1661—1710. Those used at the 18th Wing School of Instruction had purely local numbers, *e.g.* 308.

Handley Page O/400

The Handley Page O/400 was a development of the O/100, which was originally built and produced for the RNAS in 1915–17. The Admiralty specification to which the O/100 had been built was drawn up in December 1914 and was, in its time, ambitious. Design and construction of the first O/100, No.1455, occupied most of 1915, and the aircraft made its first flight on 17 December, 1915.

A very early description of the O/100 was sent to Brig-Gen Trenchard in France on 1 January, 1916, by F. C. Jenkins of the War Office (most probably Capt F. Conway Jenkins). This was in somewhat general, and in places curiously guarded, terms; but it conveyed the essential facts that the aircraft had a span of 100 ft and two 250 hp Rolls-Royce engines, was expected to carry a crew of three, four Lewis guns, 1,200 lb of bombs, and wireless; and that the loaded weight was to be between 9,000 and 10,000 lb. On the sketchy trial flights made the level

One of the six modified Handley Page O/100s, B9446—B9451, that had Sunbeam Cossack engines.

A front view of the O/100 No.3142 at Martlesham illustrates the much-modified undercarriage and the characteristic shape of the radiators used with the Fiat engines.
(A & AEE No.7; Crown copyright)

speed was estimated to be 63 mph—not a performance that was likely to commend the aircraft to Trenchard.

All development work on the O/100 was thereafter conducted under RNAS auspices, and the troubles associated with this aircraft of unprecedented size were considerable, not the least of them being the severe tail flutter experienced on the earliest aircraft. Not until the night of 16/17 March, 1917, did the first operational O/100 flight take place. By that time the demand for the excellent Rolls-Royce engines, which were wanted for the D.H.4, F.E.2d and Curtiss H-12, as well as the O/100, far outstripped the supply. Alternative engines were therefore tested, the first apparently being the 320 hp Sunbeam Cossack; the two engines concerned were fitted with new skirtless pistons and 'improved type exhaust manifolds', and the installation was planned for a 20-hr flight-test programme. The Cossacks were replaced by two specially boosted R.A.F.3a engines that were optimistically expected to develop 260 hp each; and in November 1917 four 200 hp Hispano-Suizas were installed in two tandem pairs. The testbed for these experimental installations was the O/100 No.3117.

O/100 No.3142 with wings folded and its Fiat engines clearly seen.
(A & AEE No.3; Crown copyright)

In June 1917 the Air Board set out its Specification Type A.3.b for a night bomber that could have either two or three engines. Bomb load was to be at least 3,000 lb, and with that on board, plus fuel for 300 miles, a crew of three, and three Lewis guns, the aircraft was to have a speed of 80–85 mph at 6,000 ft, and a climb to 10,000 ft in one hour. Wing folding was specified.

It is not known how quickly details of this specification were imparted to Handley Page Ltd, but on 10 July, 1917, Frederick Handley Page wrote to the Controller of the Technical Department submitting three different design proposals 'in regard to Slow Speed Bomber designs'. These were:

(i) Modified O/100 to carry 4,000 lb of bombs, with two 300 hp engines and a revised undercarriage;
(ii) Modified O/100 to carry 3,000 lb of bombs, with two 200 hp engines;
(iii) New design for 3,000 lb of bombs, two 200 hp engines.

The first of these specifically promised a revised fuel system with the main petrol tanks inside the fuselage and consequentially shortened engine nacelles; a new undercarriage was also proposed, and the total loaded weight was estimated to be 12,250 lb.

At precisely that time Handley Page Ltd were under contract to build twenty-two Felixstowe F.3 flying-boats* but were unable to start work on them because the necessary drawings were not available. On 17 July, 1917 Frederick Handley Page wrote to Sir William Weir of the Air Board:

> 'As originally arranged, the Contract for the supply of these Flying Boats was planned to fill the gap between the end of the present contract for our type of large twin engine Bombers and the completion (1) of an improved design of flying boat, and (2) of a modified design of the large bomber.
>
> 'The delay in the supply of drawings has so delayed production of the F.3 flying boats that there will now be a gap in some of the shops between the completion of work on the large twin engine Bombers and the commencement of work on the F.3 Flying Boats . . . We would ask that an order for a further (say 6 or 12) number of our twin engine Bombers might be placed with us to fill this gap, these machines being of the same general type as at present supplied by us, but with the modifications as to different engines and greatly increased carrying capacity which have already been submitted to the Technical Department in response to their request.'

On 21 July, 1917, the Controller of Aeronautical Supplies replied, stating that instructions had been given for an order to be placed with Handley Page Ltd for six of their twin-engine bombers. Thus it was that the contract (No. A.S.20629/1/17) for B9446—B9451 was placed on 29 July, 1917.

When delivered, these six aircraft proved to be basically O/100s powered by two 320 hp Sunbeam Cossack engines housed in cylindrical nacelles with frontal radiators; the nacelle-mounted petrol tanks and interplane-strut configuration were similar to those of the O/100; and Handley Page referred to them as O/100s.

The O/400 proper differed from the O/100 chiefly in having its engines in shortened nacelles, the main fuel storage having been transferred to the fuselage. This in turn permitted the use of a single straight interplane strut directly behind the engine. The structure was improved and strengthened in various ways, but the basic geometry of the aircraft remained unchanged.

*N2160—N2179, ordered under Contract No.A.S.11426, to be powered by two 320 hp Sunbeam Cossack Engines; and N62—N63, which were to have folding wings.

The evolution of the O/400 from the O/100 now appears to have been much less clear-cut than has been believed. A false trail was started by the description of No.3138 given in the AID Aeroplane Data Book, in which the aircraft was designated as a Handley Page O/400. It has been stated that this aircraft was the subject of development at Martlesham Heath by Sqn Cdr J. T. Babington and Lt Cdr E. W. Stedman, and that by September 1917 it had been modified to become the prototype O/400. In fact there is no evidence to support this. In his memoirs,* and in his evidence to the Commission on Awards to Inventors, Stedman made it clear that from May or June 1917 onwards he was with No.7 Squadron, RNAS, at Coudekerque, commanded by Babington, and said nothing of either being at any time at Martlesham or concerned with Handley Page modification or development.

Some confusion may have arisen because Stedman himself submitted, on 28 June, 1917, his own designs for a heavy bomber. He made two proposals, the first for a triplane of 84 ft span, the second a biplane with a span of 90 ft; both were to have folding wings and two Rolls-Royce engines. He hoped for substantially better performance than that of the O/100. These proposals came to naught, partly because they were made just as manufacturers' responses to Specification A.3.b were coming in.

As for No.3138, this appears to have been a perfectly ordinary O/100, delivered under Contract No.C.P.69522. It arrived at Martlesham on 4 September, 1917, apparently fresh from acceptance trials at Hendon; it was at Martlesham for seven days only, returning to Hendon on 11 September; its performance as measured at Martlesham is recorded in Reports M.143 and M.143A dated September 1917. Although all known official reports wrongly attribute Rolls-Royce Eagle II engines to No.3138 at this time, it was in fact fitted with two Eagle VIIs, Nos.3/275/55 and 3/275/32. There seems to be no reason for believing that it ever had Sunbeam Maori engines. It was with Naval 'A' Squadron in October 1917, was later used by No. 16 (Naval) Squadron, and was finally struck off charge of the Independent Force, RAF, in November 1918 after a flying life of 155 hours.

The next Handley Page to go to Martlesham was No.3142, last of the second batch of twenty-eight O/100s. In 1917 the Russian and Japanese Governments examined the possibility of building O/100s in their respective countries, and the Russians initially considered fitting two 220 hp Renault engines to their Handley Pages. This was abandoned in favour of using the Fiat A.12bis engine. On 17 July, 1917, the Department of Aeronautical Supplies had written to Handley Page instructing the firm that one of the last three O/100s was to be allocated to the Russian Government and fitted with two Fiat engines, which were delivered to the works that same day. This may have been in some way an anticipation of the agreement for the production of Handley Pages in Russia that had been drafted and agreed by October 1917. Indeed as late as 17 October the Russian Government Committee in London intimated that sets of metal parts for fifteen aircraft were required that year. The choice for the Fiat installation fell on 3142, but Martlesham did not begin to expect this aircraft, listing it as a Naval machine, until early November 1917; and indeed it did not arrive there until 20 November, some weeks after the revolution in Russia. It was powered by two 260 hp Fiat A.12bis engines, of which Britain had ordered 1,000 on 1 September, 1917, but an

*From Boxkite to Jet, the Memoirs of an Aeronautical Engineer by Air Vice-Marshal E. W. Stedman: Canadian War Museum, August 1972.

The unpainted nacelles and the form of the national markings strongly suggest that this aircraft is the O/100 3142 after its Fiat engines had been replaced by Sunbeam Maoris. The place is Martlesham Heath. (*Gordon Kinsey*)

intention to replace these by two Sunbeam Maoris had been determined before 8 December. This O/100 also had a revised and simplified undercarriage, but this was a late modification officially requested by the Department of Aeronautical Supplies on 2 November. Work on the task of changing engines had begun by 15 December, and it is evident that 3142 did very little flying with the Fiats. The Sunbeams did not in fact arrive until mid-January 1918, and their installation in the airframe was undertaken by Handley Page personnel. This work occupied several months, and the aircraft was not reported to be flying until the Martlesham weekly report dated 4 May, 1918. After various trials and experiments it went to Netheravon on 4 July.

This means that 3142 could in no way be regarded as an O/400 prototype, for deliveries of production O/400s had been made weeks before it flew with the Maori engines; indeed, it was not even the first Maori-powered O/400. The type designation O/400 existed at least as early as the beginning of August 1917, and the initial order for the improved design had been given on 14 August, 1917, when, under Contract No.A.S.17573, one hundred Handley Page bombers

A later Maori-powered production O/400 with the rudder flash that was standardised for the type.

265

(C3381—C3480) were ordered from Handley Page Ltd. This was a direct consequence of the meeting of the Air Board on 30 July, 1917, at which, following the determined intervention of Sir William Weir, the Board renounced its previous disbelief in night bombing and decided to order one hundred Handley Pages. Amid much argument and indecision (see *The War in the Air*, Vol. VI, pp 166–8) the order for the big bombers was cancelled on 20 August, but on that same day Contract No.A.S.22434/1 for one hundred Handley Pages, C3281—C3380, was given to Boulton & Paul of Norwich. This in turn was cancelled on 20 September, but the same serial numbers were re-allocated to the Norwich firm as late as 13 April, 1918, for Camels. Although Trenchard preferred D.H.4 day bombers he had to accept the evidence of the statistics produced by Capt V. Vyvyan, RN, relating to the greater effectiveness of RNAS Coudekerque's O/100s than the same unit's D.H.4s. Trenchard's qualified acceptance came on 7 September, 1917, and on 10 September Sir Douglas Haig wrote to the War Office asking that 25 per cent of additional bombing squadrons for France should be night bomber units. Significant and influential though these two communications were, neither had the brutal conviction of the effectiveness of night bombing that had been provided by the German night raid of 2 September. Four days later the Air Board decided to order 200 more Handley Pages, a number that was further increased by another hundred.

C9636, the first Handley Page built production O/400, had Sunbeam Maori engines, as did several other aircraft of the same batch. Markings were of O/100 proportions.
(*RAF Museum*)

Decisions were one thing, contracts another. Merely finding manufacturing firms physically capable of producing such large aeroplanes in quantities must have presented many difficulties. On 21 September, 1917, Contract No.A.S.27644 called for 150 Handley Pages (C9636—9785) from the parent firm, while D4561—4660 were ordered from the Metropolitan Carriage, Wagon and Finance Co on 2 October, and D5401—5450 from The Birmingham Carriage Co on the same date. D8301—8350 were ordered from The British Caudron Co late in 1917 but the serial numbers were later allocated to Handley Page, who assembled the 50 aircraft from components made by British Caudron and by Harris Lebus. Fifty more (D9681—9730) were ordered from Clayton & Shuttleworth in January 1918, and the last RFC contract was for twenty (F301—320) ordered from The Birmingham Carriage Co on 29 March, 1918.

A word needs to be said about the O/400s that were built, or at least assembled, at the Royal Aircraft Establishment, Farnborough. It has been said that a batch of

266

An Eagle-powered O/400 with the early cup-anemometer form of wind-driven fuel pumps, one of which can be seen immediately under the inboard blade of the port airscrew.
(*W. Evans*)

twelve Handley Pages numbered B8802—8813 were ordered by the War Office in February 1916, these to be built by what was then the Royal Aircraft Factory. In fact, such an order was not possible. The batches of serial numbers known to have been allotted in February 1916 ran between 7744 and 7760; even the A-prefix series did not begin until early April 1916.

The block of serial numbers B8581—8830 was first allotted on 31 July, 1917, for Avro 504Js ordered from Parnall under Contract No.A.S.20353. When it was realised that the contract called for only 200 Avros, the last 50 numbers, B8781—8830, were cancelled; their re-allotment did not begin until 24 January, 1918, when B8781—8783 were allotted for the three A.E.3s.

Similarly, C3451—3506 had apparently first been allotted about the end of July 1917 for fifty-six Sopwith 1½ Strutters to be built by the Sopwith company, but it seems that these numbers were cancelled almost as soon as they had been allotted. Their re-allocation began on 11 January, 1918, two weeks earlier than the re-allocation of B8781—8830. Thus it was that the first dozen Handley Pages ordered from the Royal Aircraft Factory under Contract A.S.1198 were allotted C3487—3498 on 12 January, 1918; about five weeks later a further twelve aircraft were ordered from the R.A.F and only then, on 16 February, 1918, were the numbers B8802—8813 allotted for Handley Pages. This allocation was annotated 'Increase on previous contract. Now 24 machines'; and was apparently accompanied by the introduction of the new contract number 35a/88/C.43.

It appears that C3487 had been completed late in March 1918, and it is known that C3489 was complete by 26 April, C3490 by 3 May. They were probably preceded by the first Handley Page O/400s to be built by Handley Page, for, as noted later, C9637 (which had Maori engines) was officially allotted to the Expeditionary Force as early as 13 March, 1918.

It had been expected that deliveries of O/400s would begin in March 1918 and attain a delivery rate of 80 per month by August. On 12 December, 1917, Handley Page were instructed that the first twenty O/400s built by them were to be fitted with Sunbeam Maori engines, but it was initially intended that all other O/400s were to have the Rolls-Royce Eagle VIII. By August 1918, however, plans had changed, and it was intended to fit the 400 hp Liberty 12* to most production O/400s; while some of the aircraft intended for training puposes had Eagle IVs. Just how many did in fact have Sunbeam Maoris is uncertain, but it is known that on 13 March, 1918, the D.A.E. wrote to the G.O.C., RFC:

'I am to inform you that it is hoped to allot to you this month for delivery to Dunkirk four Handley Page machines, of which two will be fitted with Sunbeam Maori engines and will be for training purposes, and two will be fitted with Eagle VIII Rolls-Royce engines and will be for No.16 and No.7 Squadron respectively. All the above machines will be sent to Dunkirk by air.

The first of the Handley Pages fitted with Sunbeam Maoris, namely C9637, had been allotted to you today. It is not certain that the two machines with Rolls-Royce engines will be ready in time to reach you this month, but they will be despatched as soon as possible, as also will all other Handley Page machines coming forward with Rolls-Royce engines until No.7 and No.16 Squadrons have been brought up to strength.'

Thus it was that the RFC never used the O/400 operationally: that was the subsequent prerogative of the first heavy bomber squadrons of the Royal Air Force once deliveries made against RFC orders came forward in quantity. The RAF placed later contracts for the type and it survived somewhat obscurely into the postwar period, but from the summer of 1918 it was the intention to replace the O/400 by the Vickers Vimy once production of the latter was in being.

O/400

The six modified O/100s each had two 320 hp Sunbeam Cossacks. The O/400 had two 375 hp Rolls-Royce Eagle VIIIs, two 284 hp Rolls-Royce Eagle IVs, or two 275 hp Sunbeam Maoris; later, in RAF service, some O/400s had two 350 hp Liberty 12-N engines.

Four-seat heavy bomber.

Span 100 ft; length 62 ft 10¼ in; height 22 ft; wing area 1,648 sq ft.

Empty weight 8,502 lb; loaded weight 13,360 lb. Figures with Eagle VIIIs.

Maximum speed at sea level 97·5 mph, at 6,500 ft—84·5 mph, at 10,000 ft—80 mph; climb to 6,500 ft 27 min 10 sec; service ceiling 8,500 ft. Performance with Eagle VIIIs.

Armament: Up to 2,000 lb of bombs and up to five 0·303-in Lewis machine-guns.

Manufacturers: Handley Page; Birmingham Carriage Co; Clayton & Shuttleworth; Metropolitan Carriage, Wagon and Finance Co; National Aircraft Factory No.1; Royal Aircraft Establishment. Components made by the British Caudron Co, and by Harris Lebus.

Service use: Production aircraft used by the following Royal Air Force units until the end of the war: *Western Front*: IX Brigade, Squadrons 58, 207 and 214. VIII Brigade (the Independent Force), Squadrons 97, 100, 115, 215 and 216. *Palestine*: One O/400 with No.1 Squadron, Australian Flying Corps. *Training duties*: Schools of Navigation and Bomb-dropping, Nos.1 and 2; No.8 Training Depot Station.

Serials: Modified O/100s—B9446—9451.
  O/400s—B8802—8813, C3487—3498, C9636—9785, D4561—4660, D5401—5450, D8301—8350, D9681—9730, F301—320, F3748—3767, F5349—5448, H4370—4419 (cancelled; duplicated D8301—8350), J2242—2291, J3542—3616, J6578 (cancelled 15 October, 1919).

*O/400s known to have the Liberty were C9739, F3751 and F5349.

Capt Maitland's Howard Wright biplane with extensions fitted, being flown by Lt Watkins. (*RAF Museum*)

Howard Wright biplane

In the summer of 1910 Capt E. M. Maitland, Essex Regiment, purchased one of the first of several Howard Wright biplanes that were built at that time. It was an equal-span boxkite powered by a 60 hp E.N.V. engine installed as a pusher. Its upper and lower tail-booms did not meet either in plan or elevation; from the rear of the lower tail-booms the upright spacers leaned backwards slightly to meet the rear spar of the tailplane, which was mounted above the upper tail-booms. The single central rudder was made in two similar rectangular halves that moved above and below the tailplane; an unusual structural feature of the mainplanes was the use of small kingposts along the leading edges, two between each pair of compression members. The aircraft had a sensible, workmanlike appearance.

Unfortunately Capt Maitland was at that time recovering from the injuries he had sustained in a crash on 1 August, 1910. His aircraft on that occasion was described as 'a new biplane built by Mr Howard Wright', but whether this was the same aircraft that appeared at Brooklands as his property late in September is not known. It might quite well have been the biplane of 1 August repaired.

On 30 September Maitland's Howard Wright biplane was flying at Brooklands, piloted by a brother officer of his regiment, Lt H. E. Watkins, who 'made some good straight flights' on it. Watkins flew the aircraft regularly at Brooklands, occasionally taking a passenger with him, and made the test flights for his Royal Aero Club aviator's certificate (No.25) on 4 and 5 November. A modest improvement in the Howard Wright's performance was achieved a few days later by fitting a White and Poppe carburettor to the aircraft's E.N.V.; and by mid-March 1911 extensions had been added to the upper mainplane, these were braced from kingposts above the outermost interplane struts.

When the Air Battalion of the Royal Engineers was formed on 1 April, 1911, one of its original members was Capt E. M. Maitland. Evidently he took his

Howard Wright biplane to Larkhill, for it was recorded that on 20 June, 1911, Lt D. G. Conner, R.A., 'was testing a Howard Wright biplane, fitted with an E.N.V. engine'. It had gone there via Farnborough, for Geoffrey de Havilland noted in his log-book entry for 16 May, 1911: 'Howard Wright out in evening; flew on Common and to Jersey Brow. Requires adjusting.' In its issue of 15 June, 1911, *The Aeroplane* reported the arrival at Larkhill of 'Capt. Maitland's old Howard Wright . . . that has been reposing at the aircraft factory at Farnborough for a considerable time.'

Capt Maitland's Howard Wright biplane after its acquisition by the Air Battalion, bearing its number F3 on the upper rudder. (*RAF Museum*)

On 23 June, 1911, Capt Maitland sold his Howard Wright biplane to the War Office for £625, plus £5 10s 6d for an Elliot air-speed indicator. It is not clear whether the aircraft had an engine at the time of purchase, for an official document records that a 60 hp E.N.V. was fitted on 26 June, 1911. The Howard Wright was allotted the official number F3, but its service with the Air Battalion was to be short. On Thursday, 6 July, it was recorded that Lt Conner was practising on the Howard Wright, and Capt J. D. B. Fulton made a 20-minute flight on it; next day the aircraft was flown by Lt H. R. P. Reynolds, R.E. While Reynolds was flying the Howard Wright on Saturday, 8 July, 1911, one of the E.N.V.'s cylinders parted company from the crankcase. This incident apparently marked the end of F3's flying career, for the engine was sent to the Army Aircraft Factory at Farnborough and the aircraft was dismantled for storage on 10 July. It languished at Larkhill until January 1912, when it, too, was taken to the Army Aircraft Factory. Its total flying time with the Air Battalion was 1 hr 56 min. Eventually it was used as the pretext for the building of a B.E.2-type biplane that was originally numbered B.E.6 and later 206.

Howard Wright biplane
60 hp E.N.V.

Two-seat pusher biplane.

Span 36 ft (without extensions), 48 ft (with extensions); length 36 ft 6 in; wing area 435 sq ft

Empty weight 800 lb; loaded weight 1,200 lb.

Speed 45 mph.

Manufacturer: Howard T. Wright.

Service use: No.2 (Aeroplane) Company, Air Battalion, R.E., Larkhill.

Serial: F3.

Martin-Handasyde monoplane

In the pre-1914 era the Martin and Handasyde firm produced a series of monoplanes that became progressively more handsome as the design developed. Their basic inspiration was the supremely beautiful French Antoinette monoplane, but the Martin and Handasyde aircraft managed to suggest greater solidity while yet retaining impressive elegance of form. The Martin-Handasyde monoplanes embodied the mid-span kingposts that typified Antoinette structure, and George Handasyde long remained faithful to the Antoinette aero-engine.

Although the Martin-Handasyde monoplanes were never numerous and were not extensively flown, an example was ordered for the Royal Flying Corps Military Wing in the spring of 1912. Its precise appearance seems not to have been recorded in word or picture, but it is known to have been powered by a 65 hp Antoinette engine, and to have had a new type of spar construction in its mainplane. This was one of the earliest box spars ever used, having ash longitudinals connected by plywood webs.

This Martin-Handasyde destined for the RFC made its first flight on Thursday, 27 June, 1912, with Gordon Bell at the controls and, according to witnesses, only five cylinders of its engine firing. The testing of this aircraft continued for some time at Brooklands, and its military destination did not prevent its participation in sporting contests there, for it was flown in a handicap race on 13 July, 1912, but had to retire when a water connection parted. An improvement in performance was achieved on 25 July by fitting a new airscrew.

During the Military Trials Gordon Bell was the nominated pilot of the Chenu-powered Martin-Handasyde monoplane that was the firm's entry in the Trials, but he did relatively little flying at Larkhill because the Chenu engine gave endless trouble. On Friday, 9 August, 1912, he flew the Antoinette-powered monoplane from Brooklands to Larkhill but not, it seems, with any intention of delivering the aircraft to the Military Wing. Bell flew the monoplane again at Larkhill in a strong wind on Monday, 12 August.

The RFC's Martin-Handasyde monoplane, officially 278 in the Military Wing.

On Thursday, 24 October, Edward Petre set out to fly to Farnborough on a Martin-Handasyde monoplane, but damaged the aircraft extensively in a forced landing. Whether this was the RFC monoplane is not known, but Petre finally got to Farnborough on a Martin-Handasyde on Tuesday, 19 November. He was reported to be flying the aircraft at Farnborough on 27 November and again on 3, 4 and 5 December. It is uncertain whether this was the RFC monoplane of mid-1912 or a second, later, aircraft.

It is, however, likely that this Martin-Handasyde was the aircraft that was actually delivered to the RFC and received the official serial number 278. It arrived while the monoplane ban was still in force and seems to have done no flying with the Military Wing. No.278 was in the Royal Aircraft Factory in mid-April 1913, waiting to be taken over by the Military Wing. It never was, however, and was eventually struck off charge on 29 August, 1913.

Two-seat tractor monoplane
65 hp Antoinette
Serial: 278

Martinsyde S.1

In the period preceding the war the Bristol Scout and Sopwith Tabloid made profound impressions on the aviation world. The Martin and Handasyde partnership at Brooklands, well-known and deservedly admired for its series of elegant monoplanes, began work on a small single-seat biplane in the summer of 1914. A brief report in *Flight* of 24 July, 1914, recorded that '. . . a tractor biplane of the small fast scouting type is now in the course of building at their Brooklands works'. There can be little doubt that the aircraft was intended to be a sporting single-seater, but by the time when, four weeks later, *Flight* reported that the little

No.710 was one of the earliest Martinsyde S.1 scouts to be delivered to the RFC. It is here seen at Farnborough on 28 October, 1914, and saw limited service with Nos.1, 12, 14 and 17 Squadrons at various times in 1915. Its life ended ignominiously as a ground target at Orfordness. (*RAE 622; Crown copyright*)

biplane would be completed in about a week's time Europe was at war and there were sterner things for aeroplanes to do.

Designated S.1, the little Martinsyde single-seater was a neat single-bay biplane that, owing to its engine installation, bore some resemblance to the Sopwith Tabloid, but in its original form was distinguished by a remarkably functional four-wheel undercarriage of sturdy construction but clumsy appearance.

Inevitably the Martinsyde S.1 was taken over for military purposes, and it seems likely that the first acquired the serial number 696,* for that aircraft was impressed at Brooklands at a price of £1,050. After undergoing tests on 25 September it went to No.1 Squadron on 12 October, and was still in use for training purposes with No.1 Reserve Aeroplane Squadron at Farnborough in September 1915. More S.1s were ordered quickly, the initial deliveries being of a dozen aircraft, most of them having reached Farnborough before the end of 1914. The rapidity of production of these S.1s was attributable to the foresight of H. P. Martin, who was so convinced that the RFC would need fighting scouts that, on the outbreak of war, he planned for the production of 50 without waiting for official orders.

*The number 599, also of a Martinsyde S.1, was almost certainly a later allocation than 696.

No.2449 was the Martinsyde S.1 on which Louis Strange had his adventure of 10 May, 1915. This S.1 was first reported in the AID shed at Farnborough on 20 January, 1915, and went to the Aircraft Park at St-Omer on 29 January. Its first allocation was to No.4 Squadron on 9 February; it returned to the Aircraft Park on 6 April, was issued to No.6 Squadron on 15 April, and again went back to the A.P. on 29 June. It remained there as a reserve aircraft until it was sent back to Farnborough on 1 August; it was subsequently allocated to No.9 Reserve Aeroplane Squadron at Norwich. This photograph was taken at St-Omer while 2449 was with No.4 Squadron. The aircraft had been crudely camouflaged and its upper surfaces bore no national markings; only the rudder carried a small Union Flag.
(*Peter Liddle's 1914–18 Personal Experience Archives, presently housed within Sunderland Polytechnic*)

The first S.1 to join the squadrons in France was 1601, which went to No.5 Squadron on 5 January, 1915; two days later 749 joined the same unit, and on 10 January 748 went to No.6 Squadron. Further small batches of the type were ordered, and a few more examples (743, 2449, 2822 and 2823) followed in the early spring of 1915. The S.1s were therefore never numerous in France and did nothing to influence the course of the war in the air. A distinguished former member of No.6 Squadron, Louis Strange, had adventures in the S.1s flown by that unit, and wrote of the Martinsyde in his book *Recollections of an Airman*: '. . . . experience showed it to be a very instable machine both fore and aft, with not much aileron control. . . . Although a single-seater, it was hardly superior in speed and climbing power to the Avro which carried two men, but in my eyes all these defects were outweighed by the fact that it had a Lewis gun mounted on its top plane, which could be fired forwards and upwards'.

That Lewis gun nearly cost Strange his life on 10 May, 1915, when, having engaged an Aviatik over Menin, he had to change ammunition drums. The drum on the gun jammed, and in his efforts to dislodge it Strange half stood in the cockpit, stick between his knees. The Martinsyde, in a climbing attitude at its ceiling, stalled and went into an inverted spin, whereupon Strange found himself dangling from the Lewis-gun drum, which fortunately remained immovable. Eventually he succeeded in getting back into the cockpit and regained control just in time to avoid disaster. The aircraft concerned could only have been 2449, which was No.6 Squadron's only Martinsyde Scout from 15 April until 29 June, 1915.

Martinsyde S.1 No.5452 was used for training purposes at Netheravon, possibly with No.7 Reserve Aeroplane Squadron. It had the later V-strut undercarriage. In this photograph the occupant is Lt C. F. A. Portal, later Marshal of the Royal Air Force Lord Portal of Hungerford.

The Martinsydes were withdrawn from Squadrons No.1, 5, and 6, in July and August 1915, but 2823 was belatedly issued to No.12 Squadron on 14 September. It was struck off charge on 1 October, whereafter no S.1 was operational in France. On 16 October, 4236 arrived at No.1 Aircraft Depot from England but returned on 22 November without going to any unit.

In Mesopotamia an RFC Aeroplane Flight began to form at Basra late in April 1915; it was intended to integrate this unit into No.30 Squadron, which had begun to form at Ismailia in March. On 26 August, 1915, four Martinsyde S.1s, including

274

As noted in the narrative, the four Martinsyde S.1s that formed the second Flight of No.30 Squadron were given secondary identities. Here is 4244 (MH 6) about to be unloaded from a barge at Basra. The officer may be Capt H. A. Petre.
(*A. E. Shorland, via Jim Clarke & Col Brian P. Flanagan*)

4243, 4244 and 4250, arrived at Basra to form the nucleus of a second Flight of No.30 Squadron. In the hot and dusty conditions the Martinsydes' rotary engines suffered from many troubles, but a few useful flights were made. The aircraft were, for some unexplained reason, given the auxiliary identities of MH5, MH6, MH8 and MH9 by the squadron.

When the decision to occupy Kut-al-Imara was taken, No.30 Squadron sent to Ali Gharbi on 7 September, 1915, an advanced Flight consisting of a Maurice Farman Shorthorn, a Caudron G.3 and two Martinsyde S.1s, MH5 and MH6 (4243 and 4244 respectively). MH5 crashed on 13 September, leaving 4244 as the only serviceable aircraft with Maj-Gen C. V. F. Townshend's division. On 16 September Maj H. L. Reilly flew 4244 to make a brilliant reconnaissance that gave Townshend all the information he needed to enable him to issue Battle

Four of the aircraft of No.30 Squadron picketed in the open and, in the case of the two Martinsydes, awaiting engines. These were the entire equipment of the squadron's advanced Flight, which arrived at Ali Gharbi from Amara on 7 September, 1915. The aircraft were the two Martinsyde S.1s 4243 (nearest) and 4244 (locally numbered MH5 and MH6), the Caudron G.3 numbered C3 and the Maurice Farman MF1. The Farman crashed on 11 September, the Martinsyde 4243 on 13 September, and the Caudron forced-landed behind the Turkish lines three days later. The other Martinsyde did some useful work before being shot down on 21 November, 1915.

Instructions. No.4244 again gave valuable service on 6 October, when Capt H. A. Petre made the first reconnaissance of Baghdad; on 22 October it bombed an Arab camp successfully. This most warlike of all the Martinsyde S.1s was at last brought down by anti-aircraft fire on 21 November; its pilot, Maj Reilly, being made prisoner. Next day MH8 was similarly shot down and captured with its pilot, Capt E. J. Fulton. Last of the four to survive, MH9, apparently fell into Turkish hands in besieged Kut.

The majority of the S.1s were used by various training units and ended their days in obscurity. The later aircraft all had a simple V-strut undercarriage in place of the original clumsy four-wheel affair, and some of the earlier S.1s were converted to have the later undercarriage. A few of these home-based Martinsydes were regarded as anti-airship aircraft, and were theoretically supposed to be armed with bombs, Hales grenades and incendiary darts in such quantities that they could scarcely have left the ground, still less bring a Zeppelin to action.

S.1
80 hp Gnome

Single-seat scout.
Span 27 ft 8 in; length 21 ft; wing area 280 sq ft.
Maximum speed at sea level 87 mph.
Armament: One 0·303-in Lewis machine-gun above the centre-section. Home Defence S.1s had a specified weapons load of six bombs, twelve Hales grenades, 150 incendiary darts and five powder bombs.

Manufacturer: Martin and Handasyde.

Service use: Western Front—Sqns 4, 5, 6, 12 and 16. *Mesopotamia*: RFC Aeroplane Flight, later No.30 Sqn, Basra. *Home Defence*: aircraft stationed at Brooklands, Dover, Farnborough, Hounslow, Joyce Green, Northolt and Shoreham; No.18 Sqn, Northolt and Mousehold Heath. *Training duties*: CFS; Reserve Aeroplane Sqns 1, 2, 3, 4, 5, 6, 7, 9, 11, 14 and 18. *Other units*: Sqns 1, 15 and 29.

Serials: 599, 696, 702, 710, 717, 724, 730, 734, 741, 743, 748, 749, 1601, 2448—2455, 2820—2831, 4229—4252, 5442—5453.

Martinsyde G.100 and G.102

In the summer of 1915 the Martinsyde works built the prototype of a new single-seat biplane that was apparently intended to be a long-range escort fighter or reconnaissance aircraft. Its large wing area betokened an ability to lift a considerable load of fuel, and it was unusually powerful for a single-seater, having a Beardmore-built Austro-Daimler engine of 120 hp. On the prototype this power unit drove a three-blade airscrew.

The aircraft was designated Martinsyde G.100 by its makers and clearly inherited some of the elegance of the pre-war Martinsyde monoplanes, the slender fuselage terminating in a shapely tail unit. On the prototype, which was allotted the serial number 4735 under Contract No.94/A/298, the external form of the engine cowling was ugly and angular. A curious feature of the engine installation was the placing of the radiator behind the engine.

After undergoing official performance trials at CFS, Upavon, on 8 September, 1915, No.4735 went to Farnborough. It was in the AID shop there on 25 September, when it was officially allotted to the Expeditionary Force. Not until 29 October did Lt Dunn fly the aircraft to No.1 Aircraft Depot at St-Omer, whence it was allocated to No.6 Squadron on 5 November. It went to No.20 Squadron on 1 February, 1916, and returned to the Depot at St-Omer on 16 March. No.6 Squadron must have reported favourably on it, for one hundred production aircraft were ordered in November 1915 under Contract No.87/A/192; for these the serial numbers 7258—7307 were allotted on 12 November and 7459—7508 on 22 November.

Martinsyde G.100 No.7281 is seen here with camera mounted on the starboard side of the cockpit and dark-painted upper wingtips. This Elephant was at Farnborough on 14 March, 1916, subsequently went to France and was issued by No.1 A.D. to No.23 Squadron on 25 March. It was involved in two crashes in May 1916 and went to No.2 A.D. for repair on 29 May.

Production G.100s of the first batch began to come forward early in February 1916. They had a much improved engine cowling, the engine driving a conventional two-blade airscrew, and were armed with a single Lewis gun on a simple overwing mounting; an extension handle was fitted to the gun to facilitate lowering and reloading, and a re-cocking lanyard was provided. This installation was officially known as the Martinsyde No.5 Mark I mounting.

Most of the earliest production G.100s were delivered to No.27 Squadron, a new unit that was at Dover preparing to go to France. Ten Martinsydes of No.27 Squadron flew to St-Omer on 1 March, 1916, and arrived at the unit's aerodrome at Treizennes next day. Their initial operational employment was as fighting scouts and photographic reconnaissance aircraft, the latter being one of the earliest instances of high-speed single-seaters being used for this task. Less than a fortnight after No.27's arrival in France a need was felt to augment the Martinsyde's armament, and in a primitive expedient a second Lewis gun was fitted on a form of spigot mounting behind the cockpit on the port side.

In service the Martinsyde acquired the nickname 'Elephant', but precisely when or how this happened remains uncertain: the late Maj Oliver Stewart associated the name with the later 160 hp version of the aircraft. Whatever its origin, the name remained unofficial.

On the Western Front only No.27 Squadron was completely equipped with Elephants, but the early practice of allotting one or two single-seat scouts to two-seater squadrons had not been abandoned in early 1916, and single Martinsydes went briefly to such units. Thus 7279 went to No.18 Squadron, 7260 to No.20, 7263 to No.21 and 7281 to No.23.

Early operational experience of the Martinsyde showed that it was not ideal for the kind of aerial combat that by then had become commonplace. Not only did its large size make it less tractable than the contemporary Bristol and Morane scouts, but the view from the cockpit was somewhat restricted by the long, high nose and the broad chord of the mainplanes. By the middle of April 1916 Maj-Gen Trenchard had decided to use the Martinsyde as a bomber and reconnaissance aircraft, but this change of role was not formalised until 9 July. As a bomber the Elephant normally carried one 230-lb bomb, or two 112-pounders or an equivalent load of smaller bombs.

Second Lt E. R. Pennell with the two Lewis guns of his Martinsyde Elephant, No.27 Squadron. An Aldis optical sight was fitted for aiming the overwing gun, which was on a Martinsyde No.5 Mark I mounting, but accurate shooting with the rearward-firing gun was clearly impossible. (*R. C. Bowyer*)

More Martinsydes were ordered in June 1916, when a further batch of fifty (A1561—1610) was ordered. As the 160 hp version of the Beardmore engine had begun to become available in February 1916 it was adopted as the standard power unit for the aircraft, and the intention was to fit it to all Martinsydes from A1561. The majority were so equipped, but some of the later aircraft had to make do with the 120 hp engine. For the 160 hp engine some structural modifications had to be made, and conversion kits were provided to enable existing aircraft to be modified for the more powerful engine. The 160 hp Martinsydes had the type number G.102. Later aircraft also had an improved overwing gun mounting, the Martinsyde No.5 Mark II.

Deliveries from A1561 began in mid-July 1916, and in September seventy more (A3935—4004) were ordered, followed by a final batch of fifty (A6250—6299) ordered in October. All 170 Martinsyde G.102s were ordered under Contract No.87/A/487, and production continued into the summer of 1917. The majority of the production Elephants went to No.27 Squadron, which unit went on flying the

The Martinsyde Elephants of 'A' Flight, No.27 Squadron, photographed in July 1917. Aircraft 2 was 7507, which was initially allotted to the RFC in France on 9 October, 1916, arrived at No.1 A.D. on 14 October, was transferred to No.2 A.D. on 20 October, and survived its operational career to fly back to England on 12 December, 1917. It had the 120 hp engine. (*Canadian Forces RE.19608*)

type until November 1917. Much gallantry, determination and skilled airmanship underlay that remarkable operational career, and on a surprising number of occasions the Elephants shot down enemy aircraft.

The Elephant also saw service in the Middle East in smaller numbers. In Palestine Nos.14, 67 and 142 Squadrons each had a few Martinsydes, and in Mesopotamia some were used by Nos.30, 63 and 72. In these remote and unsympathetic theatres of war the Martinsydes performed well but without distinction, and it was there that the type remained in service with the Royal Air Force until and indeed beyond the Armistice of November 1918.

The aircraft's armament remained unchanged to the end, but official references

Martinsyde G.100 A4001, which had the 120 hp Beardmore, was at Farnborough on 7 April, 1917. Later that month it joined No.49 Squadron at Dover as a training aircraft. (*RAF Museum*)

exist to two installations of fixed forward-firing guns. In one (No.8 Mark I) a fixed Lewis gun was mounted on the fuselage side within a fairing and the airscrew was protected by deflecting pieces of armour, as on the Morane Bullet. The No.7 Mark I installation consisted of a fixed Vickers gun with the Vickers-Challenger interrupter gear; apparently it was similar to the weapon installation of the B.E.12. Evidence of these having been made has yet to be found, but it is known that A6299 was fitted with an experimental installation of the Eeman triple Lewis-gun mounting. In this the guns were arranged to fire upwards and forwards at an angle of 45 deg. A Home Defence application was evidently intended, but only one Elephant was officially allotted to such duties.

G.100 and G.102
G.100—120 hp Beardmore
G.102—160 hp Beardmore

Single-seat scout and bomber.
Span 38 ft; length 26 ft 6½ in; height 9 ft 8 in; wing area 430 sq ft.
Empty weight 1,759 lb (G.100), 1,793 lb (G.102); loaded weight 2,424 lb (G.100), 2,458 lb (G.102).
Maximum speed at 6,500 ft—95 mph; at 10,000 ft—87 mph (G.100).
Maximum speed at 6,500 ft—102 mph; at 10,000 ft—99.5 mph (G.102).
Armament: One or two 0.303-in Lewis machine-guns; bomb load of up to 336 lb.

Manufacturer: Martinsyde.

Service use: Western Front: Sqns 6, 18, 20, 21, 23 and 27. *Home Defence*: No.51 Sqn. *Mesopotamia*: Sqns 30, 63 and 72. *Palestine*: Sqns 14, 67 (Australian) and 142. *Training duties*: Central Flying School; Reserve/Training Sqns 11, 19, 22, 23, 44, 49, 51 and 61. *Squadrons working up*: Nos.39, 49, 57, 58, 61 and 110.

Serials: 4735, 7258—7307, 7459—7508, A1561—1610, A3935—4004, A6250—6299.
Rebuilds: A5204, B851, B852, B860, B864, B865, B866, B872, B873.

Martinsyde F.3 and F.4

Prompted no doubt by the Elephant's lack of agility in combat, the Martinsyde company built in 1916 a modified variant that had single-bay wings while retaining the Beardmore engine. It was followed early in 1917 by a somewhat similar single-seater powered by the 190 hp Rolls-Royce Mark I engine; this new type had good flying qualities but the pilot's view was very limited and the armament makeshift. Further development and a complete redesign produced the Martinsyde R.G. with a 275 hp Rolls-Royce Falcon III and twin synchronised Vickers guns. The R.G. was officially tested in June 1917 and won the enthusiastic approval of the Martlesham Heath pilots who flew it.

At that time a new single-seat fighter design was in hand at the Martinsyde works. First thoughts of May 1917 tentatively envisaged a B.H.P. six-in-line engine as the power unit, but the type that appeared with the designation Martinsyde F.3 had a version of the Falcon III in an installation that resembled that of the R.G. At first the guns were enclosed in an unsightly and view-obscuring hump ahead of the cockpit, but this was quickly removed. After

The prototype Martinsyde F.3 at Brooklands in its original form, with four-blade airscrew, multiple exhaust manifolds, and humped fairing over the guns. (*RAF Museum*)

The prototype F.3 with two-blade airscrew and plain exhaust stubs. The humped fairing over the guns has been removed. (*RAF Museum*)

First prototype F.3 at Martlesham Heath in November 1917.

manufacturer's trials in September 1917 armament was installed and the first F.3 went to Martlesham Heath on 9 November for official trials.

The Martlesham pilots recognised in the F.3 a fighter of exceptional quality and reported on it in glowing terms. Within days of the completion of Report No.M.158, which was dated 26 November, 1917, an official decision was taken to build the F.3 in quantities sufficient to equip and maintain two operational squadrons. Initially 150 were ordered, to be numbered D4211—4360, together with a development batch of six F.3s, B1490—1495. The prototype was given the number X2 early in January 1918.

The first prototype F.3 was allotted the experimental-type serial number X2 early in January 1918.
(*Peter Liddle's 1914–18 Personal Experience Archives, presently housed within Sunderland Polytechnic*)

In mid-December 1917 the 300 hp Hispano-Suiza engine was regarded as an alternative power unit for the Martinsyde F.3, but in that month Maj Gen Trenchard unexpectedly turned against the Martinsyde and, in effect, recommended that it be abandoned. Why he did so remains a mystery, but his opposition to the F.3 was disregarded: it is clear that it had been decided early in January 1918 to adopt the 300 hp Hispano-Suiza engine for the production aircraft, a decision that must have been influenced by the knowledge that Falcon production was inadequate even to provide power units for the Bristol Fighters then on order. A proposal to use the 240 hp Lorraine 8Bb seems not to have gone beyond an experimental installation in one F.3 airframe, but it is not known whether this F.3 ever flew with that engine. By the last week of January 1918 modifications to the lower wings to improve downward view had evidently been agreed, thus effectively establishing the design that was in April designated Martinsyde F.4.

Official documents make it clear that the aircraft of the initial major production batch were considered to be F.3s, and that in mid-August 1918 forty-four production F.3s numbered D4211—4213 and D4215—4255 had been completed but were lying engineless in store for want of Falcon engines. The first 300 hp Hispano-Suiza engine to go to the Martinsyde works had been delivered there

282

First of the six development F.3s, B1490, photographed at Martlesham.

The second F.3 of the development batch, B1491.

B1492 photographed at Biggin Hill.

during the first half of March 1918, thus enabling D4256 to be completed and flown as the first Martinsyde F.4. The last major modification of the basic design was the moving of the cockpit aft by one bay, further to improve the pilot's view, and it is known that this alteration appeared on D4214, the apparently 'missing' aircraft from the batch of production F.3s that were consigned to store. D4256 was officially delivered on 21 June, 1918.

It had been hoped that the first Martinsyde F.4 squadron would go to France in April 1918, the second in May. Lack of engines made the realisation of these dates impossible, and it was late in June when the first F.4, D4256, arrived at Martlesham Heath to undergo official trials. It had been preceded by a matter of weeks by the first development F.3, B1490, which arrived at Martlesham on 26 April, 1918. This F.3 consolidated the excellent reputation that X2 had won, and proved itself superior in manoeuvre and performance to all of the contemporary types of single-seat fighters against which it was matched in mock combat.

Martinsyde F.4 D4256 at Martlesham, with a Fairey IIIA in the background.

Four of the development-batch F.3s were allocated to Home Defence squadrons in 1918. Two were with No.39 Squadron, RAF, on 8 June, and No.141 Squadron had B1492 for a time. These were the only Martinsydes of this basic type to be on an operational footing of any kind, for production of the F.4 did not proceed fast enough to enable any squadron of the RAF to be equipped with it.

Initial deliveries of the 300 hp Hispano-Suiza were slow, and expectations that six F.4s would be delivered in August, 25 in September and 40 in October were not realised. Such engines as were received lacked vital parts of the Constantinesco gear; the AID rejected much of the timber intended for the F.4s; final approval of the production drawings was long in forthcoming; and the completion of final equipment schedules was delayed. Although the September estimate of deliveries was revised to only ten it was not achieved.

Had the Martinsyde F.4 been able to see operational use it must have made a great impact on the war in the air, for it was incomparably the best British single-seat fighter in existence at the time of the Armistice. Some lingered on in the RAF for a few years after the war, but the F.4 was unjustly passed over in favour of the inferior Snipe.

F.3 and F.4
285 hp Rolls-Royce Falcon experimental
275 hp Rolls-Royce Falcon III
240 hp Lorraine 8Bb
300 hp Hispano-Suiza 8Fb

Single-seat fighter.

F.3: Span 32 ft 10 in (upper), 31 ft 6 in (lower); length 25 ft 6 in; height 8 ft 8 in; wing area 337 sq ft.

F.3: Empty weight 1,859 lb; loaded weight 2,446 lb.

F.3: Maximum speed at 10,000 ft—129·5 mph, at 15,000 ft—123·5 mph; climb to 6,500 ft—4 min 40 sec, to 10,000 ft—8 min 5 sec, to 15,000 ft—15 min; service ceiling 21,500 ft; endurance 2¼ hr.

Armament: Two fixed synchronised 0·303-in Vickers machine-guns. Provision for one 0·303-in Lewis gun on the upper wing envisaged, and apparently B1492 had such a mounting.

Manufacturers: Martinsyde. Later contractors for the F.4 were Boulton & Paul, Hooper, and Standard Motor Co.

Service use: Four F.3s were allocated to Home Defence units in 1918, possibly not until after the formation of the RAF. These included Sqns 39 and 141.

Serials: X2, B1490—1495, D4211—4360, H6540—6542 (F.4 Buzzard Ia), H7613—8112, H8413—8512, H8763—9112, J1992—2141, J3342—3541, J5592—5891.

Morane-Saulnier Types G and H

La Société anonyme des Aéroplanes Morane-Saulnier rose to almost instant prominence within weeks of its formation on 10 October, 1911, by displaying at that year's Salon de l'Aéronautique in Paris four monoplanes of clean lines and advanced design. From these progenitors descended a series of monoplanes, all powered by rotary engines, that were flown with considerable success, notably by Roland Garros, who joined Morane and Saulnier as their pilot in 1912.

Like B C. Hucks's looping Blériot, Gustav Hamel's looping Morane-Saulnier Type G had roundels on the upper surface of its wing. This aircraft was impressed at the outbreak of war and received the official number 482. (*RAF Museum*)

Believed to be the only Morane-Saulnier Type H to see service with an operational unit of the RFC, No.587 arrived at the 1st Aircraft Park on 23 April, 1915, and was issued to No.4 Squadron on 6 June. The aircraft is seen in this photograph at Bailleul with a bomb rack under the cockpit. It went to the 3rd Aircraft Park on 26 September, was transferred to the 1st A.P. next day, and on 23 October joined No.12 Squadron. No.587 was sent back to the 1st A.P. on 17 November, 1915, where it was written off that day.
(*Peter Liddle's 1914–18 Personal Experience Archives, presently housed within Sunderland Polytechnic*)

In the spring of 1913 there appeared the Morane-Saulnier Types G and H, two typical shoulder-wing monoplanes of such similarity that they were all but indistinguishable from one another. The Type G was a two-seater with a span of 9·63 m; the Type H was normally a single-seater and had a wing span of 9·12 m; their basic geometry was virtually identical. In the type G the pilot and his passenger sat close together in tandem on a single lengthwise seat in a communal cockpit.

Both types were taken into service by the French Aviation militaire, the Type H becoming the first Morane-Saulnier aircraft to be given an official designation under the system introduced by the Service des Fabrications de l'Aviation (SFA): this was MoS.1. In French service, possibly the greatest contribution made to military aviation by a Morane-Saulnier type was provided by the Type G that acted as a testbed for the Garros-Hue gun installation and deflector airscrew.

W. L. Brock taxies in at Hendon after winning the great London-Paris-London race of 11 July, 1914. His Morane-Saulnier Type H, No.6 in the race, was impressed for the RFC and is believed to have been numbered 623. It saw no service, for it was rejected by the AID before 22 August and had been smashed before 17 September, 1914. (*RAF Museum*)

Morane-Saulnier No.629 was the aircraft in which R. H. Carr participated in the London–Paris–London race of 11 July, 1914, and retained its racing number 8 under the wings after impressment into the RFC. Here it is seen at Farnborough with Lt James Valentine in its cockpit. (*RAE 500; Crown copyright*)

Claude Grahame-White acquired the British licence for production of the Morane-Saulnier monoplanes, and his company had built several before the outbreak of war in August 1914. Three of these were among the heterogeneous collection of privately-owned aeroplanes that were impressed by the RFC when the war began; they were of Type H, were purchased for £750 each, and received the official serial numbers 623, 627 and 629. Their careers were brief and totally unproductive. Nos.623 and 627 were issued to 2nd Lt P. B. Joubert de la Ferté, and Trenchard recorded that they were in his possession on 26 August, 1914, having gone to him from the AID at Farnborough. No.623 had been struck off by 24 September, and a similar fate had befallen 627 before 13 October. The former is believed to have been the Morane flown by W. L. Brock, and 629 was the aircraft flown with the racing number 8 by R. H. Carr, in the London–Paris–London race of 11 July, 1914. It seems that No.629 had the shortest career of any of the ex-Grahame-White Moranes, having been rejected by the AID on or before 22 August.

With these three Type H monoplanes a Type G was also impressed. Its owner in August 1914 was F. E. Etches, but it had earlier been the looping Morane of

Hamel's Morane at CFS as No.482, seen here with Avro Type E biplanes and B.E.2as.

Gustav Hamel and retained the tall cabane specially fitted for inverted flying. The RFC bestowed on it the number 482, and it went to CFS on 8 August, 1914. Apparently it went to Farnborough after only a short time at Upavon, for it was recorded as being at Jersey Brow on 17 September. This historic Type G was later used by No.1 Reserve Aeroplane Squadron at Gosport and as late as May 1916 was lent to No.60 Squadron, shortly before that celebrated unit went to France.

The Grahame-White Aviation Co had a batch of Morane-Saulnier monoplanes in production early in 1915. They had been ordered by the RFC under Contract No.A2610 and were apparently of Type H. Deliveries began in April. No.587 is known to have been at Farnborough on 14 April, 1915, and seems to have been the only Morane-Saulnier of its type to see operational service with the RFC in the Field. It went to the First Aircraft Park on 23 April, 1915, and was allocated to No.4 Squadron on 6 June, 1915, with which unit it remained until its return to the Third Aircraft Park on 26 September. Next day 587 went to the First Aircraft Park, whence it was re-issued to No.12 Squadron on 23 October. Its career ended on 17 November, when it was returned to No.1 Aircraft Depot and struck off charge.

Morane-Saulnier Type H of the batch 5693—5716 built for the RFC by The Grahame-White Aviation Co under Contract No.87/A/26. Like most of its fellows this aircraft was used for training purposes. (*RAF Museum*)

As far as is known, all of the other Grahame-White-built Morane-Saulniers were used by various Reserve Aeroplane Squadrons for training purposes. A second batch of twenty-four Type H monoplanes was ordered from the Grahame-White company under Contract No.87/A/26, apparently about August or September 1915. These were numbered 5693—5716, and likewise found employment with the training squadrons. Possibly the last to survive was 5706, which was still in use as a non-flying instructional airframe at the School of Military Aeronautics, Reading, in February 1918.

Types G and H
80 hp Gnome

Type G—two-seat monoplane; Type H—single-seat monoplane.
Type G: Span 9·63 m (31 ft 7 in); length 6·38 m (20 ft 11 in); height 3 m (9 ft 10 in).
Type H: Span 9·12 m (29 ft 11 in); length 6·28 m (20 ft 7 in); height 2·3 m (7 ft 6½ in); wing area 14 sq m (150·7 sq ft).
Empty weight 314 kg (693 lb) Type G; loaded weight 544 kg (1,199 lb) Type G, 470 kg (1,034 lb) Type H.
Type G: Climb to 1,000 m (3,280 ft) 7 min.
Type H: Maximum speed 135 km/h (85 mph).
Armament: Small load of bombs or grenades.

Manufacturers: Morane-Saulnier, Grahame-White.

Service use: Western Front: Sqns 4 and 12. *Training duties:* Reserve Aeroplane Sqns 1, 2, 4, 10 and 11. *Other units:* Sqns 7, 15 and 60.

Serials: 482, 587—598, 623, 627, 629, 5693—5716.

Morane-Saulnier Type L

The Morane-Saulnier company made a marked departure from their successful shoulder-wing monoplane configuration in August 1913 when they built the Type G-19. This consisted of a slightly modified Type G fuselage to which was fitted a parasol wing. Landing loads were borne by cables anchored to a tall central cabane structure, and lateral control was by wing warping. It has been said that the aircraft was designed at the request of Alberto Santos Dumont, but this remains unconfirmed.

From the Type G-19 was evolved the Type L, geometrically similar but having its wing span increased by one metre and providing separate seats for the two occupants instead of the communal lengthwise seat that the G-19 inherited from

Sgt Frank Courtney with a Morane-Saulnier Type L of No.3 Squadron, Auchel, in June 1915. The aircraft has a fin.

289

the basic Type G. The Aviation militaire did not order the Type L in the prewar period but Morane-Saulnier were officially permitted to build 50 of the type for Turkey. These were to have only 50 hp Gnomes, all available 80 hp engines being reserved for French aircraft. When war began in August 1914 the Turkish Type L parasols were promptly commandeered by the Aviation militaire, fitted with 80 hp Gnome or Le Rhône engines, and used to form two new escadrilles, MS.23 and MS.26, which began their operational careers on 15 August. In French service the Type L's official designation was MoS.3.

Just as the RFC had set about acquiring Blériots and Farmans shortly after its arrival in France, so did the Service seek to buy Morane-Saulnier parasol monoplanes. The first to join an operational unit of the RFC was 1829, which was delivered from Paris direct to No.3 Squadron on 2 December, 1914. It lasted only a week for it was wrecked on 9 December. Two more Moranes joined No.3 Squadron before the year ended: 1843 arrived from Paris on 24 December, 1845 two days later. Thereafter most deliveries of the early Parasols were to No.3 Squadron, which by 1 April, 1915, had on charge fourteen Moranes and one Blériot XI, the latter for instructional use only. Of the Moranes, 1849 was transferred to No.1 Squadron on 10 April, being followed by three more delivered from the First Aircraft Park, two in April and one in May. By the end of 1915 No.1 Squadron was effectively an all-Morane squadron, having on charge thirteen Parasols and one Type N single-seat scout. At that time No.3 Squadron was slightly under strength, having struck off four Parasols between 26 and 29 December, and was equipped with eleven Parasols, one Morane-Saulnier Type N and one Morane-Saulnier Type BB biplane. The later Type L Parasols had the addition of a triangular fin in the tail unit.

The Morane Parasols then with Nos.1 and 3 Squadrons at the end of 1915, and No.12 Squadron's singleton, 5064, were all of the later Type LA, but the last of the Type L monoplanes left No.1 Squadron only on 27 and 28 December, when 5052 and 5051 respectively went back to No.1 Aircraft Depot at St-Omer. The Type L had given good if somewhat unspectacular service throughout 1915 as a reconnaissance aircraft with the RFC: its more memorable exploits had been performed with the Aviation militaire, in which it served as Garros's pioneering fighting vehicle, carrying into combat the first fixed forward-firing machine-gun to be used operationally on a tractor aircraft; and in the RNAS as Flt Sub-Lt R. A. J. Warneford's mount that destroyed the LZ.37 on 7 June, 1915.

A Morane-Saulnier Type L, possibly No.1881, in RFC service in 1915. No.1881 was delivered from Paris direct to No.3 Squadron on 22 March, 1915, and was returned to the Aircraft Park on 25 September, 1915. (*RAF Museum*)

Morane-Saulnier Type L No.5051 was delivered to the 1st Aircraft Park on 12 August, 1915, and went to No.1 Squadron on 20 August. It was returned to No.1 A.D. on 28 December and was sent to England on 30 December, 1915. When photographed it had a four-blade airscrew.

Nevertheless, it should be remembered that one of the RFC's earliest combat successes was won by 2nd Lt V. H. N. Wadham and his observer Lt A. E. Borton on a Morane Type L of No.3 Squadron on 5 February, 1915. Although Borton was armed with only a rifle, his sustained fire drove down an Aviatik encountered over Merville, the conclusive shots being fired at a range of only 50 ft.

As far as can now be determined, just over fifty Type L Parasols were delivered to the RFC. By the standards of its time it was regarded as being tricky to fly, probably because it had its own characteristic imbalance of sluggish lateral control combined with sensitive elevator reaction. On 26 December, 1915, Maj-Gen Trenchard wrote to the Assistant Director of Military Aeronautics:

> 'I am sending home as soon as possible the remainder of the old type (warped wings) Morane parasols for instructional purposes.'

By then, however, there must have been few Type L monoplanes left in the RFC in France. One of the last to go was 5056, which was sent to England as late as 3 February, 1916, and was allotted to No.25 Squadron for training duties. For such work the Type L could only have been of limited use, needing considerable skill to handle yet lacking the dual control necessary for the imparting of such skill. Few variations of the type existed, but in the RFC 5051 differed from standard in having a four-blade airscrew.

Type L
80 hp Le Rhône 9C

Two-seat reconnaissance monoplane.
Span 11·2 m; length 6·88 m; height 3·93 m; wing area 18·3 sq m.
Empty weight 393 kg; loaded weight 677·5 kg.
Speed at sea level 125 km/h; climb to 1,000 m—8 min, to 2,000 m—18 min 30 sec; endurance 4 hr.
Armament: Normally one 0·303-in Lewis machine-gun.

Manufacturer: Morane-Saulnier.

Service use: *Western Front*: Sqns 1, 3 and 12. *Training duties*: Sqns 15 and 25.

Serials: 1829, 1843, 1845, 1848, 1849, 1855, 1859, 1861, 1862, 1863, 1866, 1870—1875, 1878, 1880—1882, 1888, 1892, 1894, 1896, 1897, 5002, 5005—5007, 5012, 5021—5023, 5029, 5033, 5034, 5039, 5041, 5044—5048, 5051, 5052, 5055—5058, 5060, 5061.

Morane-Saulnier Type LA

In the summer of 1915 the Morane-Saulnier company produced a refined development of the Type L Parasol monoplane. The new type retained the 80 hp Le Rhône as its standard engine, but the fuselage was fully faired over its entire length, having full-length side fairings and both dorsal and ventral deckings, the whole being of approximately circular cross-section. A sizeable spinner was fitted to the airscrew, and the engine cowling was of a revised design. Internally, the tank capacity was reduced from the 120 litres of the Type L to 105 litres, oil tankage from 34 litres to 20 litres. The mainplane was entirely new, being rigidly braced by cables, and having tapered ailerons in place of the warping wings of the Type L. In the tail unit the plain rudder was hinged to triangular fin surfaces above and below the fuselage, but the balanced 'all-flying' elevator was retained as the only horizontal surface.

These modifications altered the appearance of the aircraft radically, and its makers designated it Type LA, the suffix A signifying ailerons. It was put into production without delay and was built in substantial numbers.

It is now difficult to determine with precision the point at which deliveries of the Type LA to the RFC began, but it is possible that the RFC's first Moranes of the

A Morane-Saulnier Type LA, photographed before being delivered to the RFC. This type had a faired fuselage, ailerons and revised fin and rudder.

Morane-Saulnier LA No.5120 of No.3 Squadron.
(Peter Liddle's 1914–18 Personal Experience Archives, presently housed within Sunderland Polytechnic)

new type were 5064 and 5065, which arrived at No.1 Aircraft Park on 17 September, 1915. Both were allocated to No.3 Squadron, 5064 on 20 September, 5065 six days later. On 25 November, 5064 went to No.12 Squadron via No.1 Aircraft Depot, while 5065 stayed with No.3 until it was struck off charge on 7 December. Many more Type LA Parasols followed these two, and rather more than 100 were delivered to the RFC in 1915 and 1916.

The Type LA consolidated the Parasol's reputation for being tricky to handle, as Cecil Lewis recorded in his classic *Sagittarius Rising*:

'There was a machine standing quietly in a corner of the hangars which had been pointed out to me casually, as one points to a rattlesnake at the Zoo and passes on to more congenial creatures, as a Morane. I had heard of it, of course. It was one of the recognised death-traps which pilots in training prayed they might never have to fly.'

Of its flying qualities Lewis wrote:

'. . . . the elevator was as sensitive as a gold balance; the least movement stood you on your head or on your tail. You couldn't leave the machine to its own devices for a moment; you had to fly it every second you were in the air. The other controls, just to make it more difficult, were practically non-existent. There was a rudder, too small to get you round quickly, and ailerons which were so inefficient that sometimes, if you got a bump under one wing taking off, it was literally seconds before you could get the machine on an even keel again.'

Early in January 1916, Maj-Gen Trenchard instructed Capt Lord Robert Innes-Ker of the B.A.S.D. that fifty Morane Parasols would have to be obtained during the first quarter of the year. At that time thirteen of the previous quarter's allocation remained undelivered, and only twenty (without engines) had been allocated to the RFC for the first three months of 1916. Evidently some modification to the mainplane of the Morane was introduced at about this time, for a note dated 7 January required the fifty Parasols to have '. . . . improved type wings to keep squadrons going.' Official thinking then envisaged that the Morane squadrons would be maintained until the end of September.

Just how greatly improved the wings were remains conjectural, for on 14 January the Morane-Saulnier company asked to be allowed to substitute pine for ash in Parasol wings, because stocks of ash were so depleted that lengths of dried wood were not available.

Morane-Saulnier LA No.5120 of No.3 Squadron, January 1916. (*D. R. Neate*)

In the RFC the Type LA Parasol progressively replaced the Type L in Nos.1 and 3 Squadrons, and one or two were used by No.7 Squadron. The Type LA continued the reconnaissance, photographic and artillery-spotting activities of the earlier type, and occasionally undertook bombing missions. One of the most noteworthy of these was the attack on the airship sheds at Brussels, made by five B.E.2cs and three Moranes of No.1 Squadron on 2 August, 1916. Bombs were dropped from only 1,000 ft but no direct hits were registered.

When No.60 Squadron went to France it was equipped shortly after its arrival there with three different types of Morane-Saulnier aircraft: 'A' Flight had the Type N Bullet, 'B' Flight the Type BB biplane, and 'C' Flight received three Type LA Parasols. Nos.5131, 5140, 5141 and 5146 all served with No.60 in June 1916; on 8 June, 5141 crashed at Boulogne, and the other three were returned to the Depot on 16 June, being replaced by Type N Bullets on that date. The Type LA Parasol of No.60 Squadron lost on 3 August on a spy-dropping mission, flown by 2nd Lt C. A. Ridley, was A143. Ridley's subsequent evasion and escape was one of the great personal adventures of the war.

Despite their frail appearance the LA Parasols could be, by the standards of the time, remarkably durable. No.3 Squadron recorded that No.5114 had put in 260 hours, having been with the unit since 28 December, 1915. It was returned to No.2 A.D. for repair on 24 August, 1916, but was struck off two days later.

At Filton Morane-Saulnier Type LA No.5121 (MS509) was used for training purposes and had had its spinner removed when photographed there. It was first delivered to No.1 A.D. at St-Omer on 15 December, 1915, and was flown to England on 9 February, 1916. This Parasol also saw service at Gosport, possibly with No.1 Reserve/Training Squadron.

In their turn, the Type LA Parasols were progressively replaced by the Type P as the new and improved type became available. Yet the LA was slow to disappear and lingered on in dwindling numbers into 1917: perhaps the last operational survivors of the type were A181, which was withdrawn from No.3 Squadron as unfit for further service on 18 January, 1917, and struck off charge at No.2 A.D. on 4 February, and the veteran 5174, returned from No.1 Squadron to No.1 A.D. on 23 January to be struck off on 15 February. A handful remained in use for a few more weeks at the pilots' school of No.1 A.D.; of these, 5189 was struck off on 15 February, 1917.

Morane-Saulnier Type LA A168 (MS700) was at Avonmouth, allotted to the 6th Brigade, on 5 August, 1916, and was destined to go to Gosport. (*The late Peter W. Moss*)

Thereafter the remaining LAs were used by training units in England, the last of them being sent from France in packing cases; these included 5133, 5188 and A124, which were despatched in this fashion from No.1 A.D. in March 1917. It had been decided on 8 March that Morane Parasols and Biplanes were no longer to be used by the School at No.1 A.D. but were to be replaced by Bristol Scouts. Of those that were sent to England A192 was allocated for experimental purposes. It was shipped from Rouen on 8 November, 1916, with Avonmouth as its primary destination.

Type LA
80 hp Le Rhône 9C

Two-seat reconnaissance monoplane.
Span 10·9 m; length 7·078 m; height 3·85 m.
Empty weight 400 kg; loaded weight 650 kg.
Maximum speed at sea level 138 km/h, at 2,000 m—135 km/h; climb to 1,000 m—6 min 10 sec, to 2,000 m—15 min 25 sec, to 3,000 m—29 min 25 sec; endurance 2½ hr.
Armament: One 0·303-in Lewis machine-gun; small load of bombs.

Manufacturer: Morane-Saulnier.

Service use: Western Front: Sqns 1, 3, 7, 12 and 60. *Training duties*: France: Pilots' School at No.1 Aircraft Depot, St-Omer. Britain: Reserve Aeroplane Squadron No.1.

Serials: 5064, 5065, 5070, 5072, 5073, 5076, 5077, 5080—5082, 5085—5094, 5096, 5098—5103, 5105—5121, 5123—5125, 5128, 5129, 5131—5136, 5138—5141, 5143—5148, 5150—5155, 5174, 5175, 5178, 5179, 5186—5190, 5198, 5199.
A123, A124, A140—142, A144—146, A152, A153, A157—159, A168—170, A180—182, A192, A194.

Morane-Saulnier Types N, I and V

The first operational use of a fixed machine-gun firing through the plane of rotation of a tractor airscrew was made in the early spring of 1915 by that great French pioneer pilot Roland Garros. His aircraft was a Morane-Saulnier Type L parasol monoplane fitted with a special airscrew that was protected by the steel bullet deflectors developed by Garros's mechanic Jules Hue from the basic primitive device made by Raymond Saulnier when ammunition inequalities thwarted his prewar experiments with an interrupter mechanism. Garros was flying a Type L Parasol with the Hue deflectors when he was shot down and captured on 18 April, 1915.

In the summer of 1914, a few weeks before war started, Garros had been flying a new and highly refined Morane-Saulnier monoplane at the flying meeting held at Aspern, Vienna. This was the original Morane-Saulnier Type N. Shortly after Garros had been shot down in 1915 an armed type N was flown operationally by Garros's friend Eugène Gilbert, who named the aircraft *Le Vengeur* (*The Avenger*). The gun was a Hotchkiss and the airscrew was protected in the same way as that of Garros's Type L. Production on a modest scale followed, and the military Type N monoplanes, which differed in certain respects from the original, formed part of the equipment of a few French escadrilles.

Morane-Saulnier Type N No.5191 was originally MS643. It was delivered to No.2 A.D. at Candas by Guillaux on 10 April, 1916, saw some use with the Scout School at No.1 A.D. from 10 May, and subsequently went to No.60 Squadron. It is here seen while with that famous unit, without a gun and being used as a practice aircraft. It was eventually returned to No.2 A.D. on 5 September and was sent to England crated on 21 September, 1916.
(*Lord Balfour of Inchrye*)

The Morane-Saulnier Type N 5069 was the third of the first three monoplanes of this type to be delivered to the RFC. It was delivered from Paris to the 3rd Aircraft Park on 19 September, 1915. On 8 October it was issued to No.1 Squadron, returning to No.1 Aircraft Depot on 26 December and being again issued to No.1 Squadron on 24 January, 1916. Its career ended on 9 March, when it was shot down by an aircraft of Fliegerabteilung (A) 213 manned by Leutnant Patheiger and Unteroffizier Gröschler. The Morane's pilot, 2/Lt R. P. Turner, lost his life. *(A. E. Ferko)*

To assist the cooling of the 80 hp Le Rhône engines in the summer of 1916 No.60 Squadron removed the spinners of their Morane Bullets. The aircraft in this photograph was 5191, its pilot Lt A. D. Bell-Irving (*Lord Balfour of Inchrye*); and (right) in operational use, providing windscreens on the RFC's Lewis-armed Moranes presented problems. This Bullet of No.3 Squadron had no windscreen. (*Peter Liddle's 1914–18 Personal Experience Archives, presently housed within Sunderland Polytechnic*)

In that summer of 1915 there were nine squadrons of the RFC in the Field with the BEF in France. Between them on 30 June they mustered 105 aircraft of fourteen different types, of which only six Bristol Scouts and two Martinsyde S.1s could be described as true single-seat scouts, and neither could be effectively armed. In such circumstances the new Morane-Saulnier with its forward-firing machine-gun must have seemed an aircraft of lethal potential. The RFC ordered three; deliveries were made on 18 and 19 September, 1915, and the monoplanes were numbered 5067—5069. Single examples went to Nos.1 and 3 Squadrons.

On 7 January, 1916, the British Aviation Supplies Department (B.A.S.D.) in Paris was instructed by RFC Headquarters to order twenty-four Morane-Saulnier Scouts. A week later, further instructions were given to the B.A.S.D. to order one similar aircraft that was to be powered by the 110 hp Le Rhône engine and to have a fuel capacity sufficient for an endurance of three hours.

Deliveries of the 80 hp monoplanes to the RFC began when MS639 was received by the B.A.S.D. at Villacoublay on 18 March, 1916; the aircraft was flown to No.1 Aircraft Depot at St-Omer on 30 March and, now with the RFC identity 5180, was allocated to No.3 Squadron on 23 April. It was preceded by 5194 (MS641) and 5195 (MS640), which had been sent to No.3 Squadron on 4 April. Last to be delivered was A196 (MS668), which was flown to No.2 A.D. at Candas during the third week of June 1916. Thus the RFC had received all of its Morane-Saulnier Type N monoplanes before the Battle of the Somme began, but the type had already been withdrawn from operational use by the French

An experimental windscreen on another of No.3 Squadron's Moranes. This elaborate fitting was apparently intended to house the gun's magazine and thus facilitate the changing of drums.
(*Peter Liddle's 1914–18 Personal Experience Archives, presently housed within Sunderland Polytechnic*)

Aviation militaire, which at no time made much use of the little Morane.

In the RFC the Type N quickly became known as the Morane Bullet, but in official documents it was more usually the Morane Scout. A few references record the misleading (but popular) name of Morane Monocoque. With the RFC the

Morane-Saulnier Type N A173 of No.60 Squadron photographed in German hands after its capture on 28 August, 1916; its pilot, 2/Lt B. M. Wainwright, becoming a prisoner of war. Originally MS662, this Type N was delivered to No.2 A.D., Candas, on 30 May, 1916, and was issued to No.60 Squadron on 16 June. (*A. E. Ferko*)

All four Type I monoplanes acquired by the RFC were used by No.60 Squadron. A199 was returned to No.2 A.D. on 18 October, 1916, and was sent to England in a packing case on 2 November. On 27 January, 1917, when A199 was at the Southern Aircraft Depot, the Central Aircraft Depot was asked to arrange storage for it. (*RAF Museum*)

type retained the bullet-deflector airscrew, but the gun was a Lewis in place of the Hotchkiss that had armed the French air service's Type N. Some of the RFC aircraft from A166 (MS650) had wings with a revised aerofoil section; these added some 5 mph to the aircraft's speed but reduced the climbing performance somewhat.

The Type N never formed the complete equipment of any RFC squadron. It entered British service at a time when two-seater squadrons still had a number of single-seat fighting scouts to provide protection for the reconnaissance aircraft. With Squadrons 1 and 3 the Morane Bullets did little to establish much of a reputation, but the aircraft later achieved some prominence, if not notoriety, in its service with No.60 Squadron. This unit went to France without aircraft on 28 May, 1916, receiving Morane-Saulnier Bullets for 'A' Flight, Morane-Saulnier Type BB biplanes for 'B' Flight, and LA Parasols for 'C' Flight three days later. The Parasols were soon replaced by more Bullets.

An unnumbered Morane-Saulnier Type I, possibly the prototype, showing the scalloped trailing edge of the mainplane. The Vickers gun is mounted centrally.

In their relatively brief operational career during the savagely contested summer of 1916 the RFC's Bullets had many combats and registered a number of successes. The type never became popular, however, for it combined the heavy lateral response of warping wings with the disproportionately sensitive control provided by its 'all-flying' elevators, and was considered to be difficult to fly.

If the Type N was disliked, the 110 hp developments designated Types I and V were actively detested by those who had to fly them. This dubious privilege fell to the pilots of No.60 Squadron, the only British unit to use the 110 hp monoplanes. The Morane-Saulnier Type I was virtually a Type N re-engined with the more powerful Le Rhône and had a flight endurance of only 1½ hours. On the other hand, the Type V was a complete redesign, being a larger aircraft with an endurance of three hours. Although it appears that these aircraft were intended to have Lewis guns fitted with the French Alkan synchronising mechanism, in service they fought with a single synchronised Vickers gun mounted centrally ahead of the cockpit.

Four Type I and twelve Type V monoplanes were supplied to the RFC. The first 110 hp aircraft known to have been delivered was MS747, a Type V that was

A207 was a Morane-Saulnier Type V that was issued to No.60 Squadron on 13 September, 1916. It was returned to No.2 A.D. for repair on 23 September, possibly as a result of the mishap illustrated, and was struck off charge on 26 September, 1916. (*RAF Museum*)

officially accepted by the B.A.S.D. on 13 May, 1916. It was flown to No.1 A.D. three days later and became A160. The four Type I monoplanes were accepted in mid-July, and subsequent deliveries of Type V went on until the week ending 26 August, 1916. All of the Type I were used by No.60 Squadron at some time, but some of the later Type V did not leave the Depot. The operational career of the 110 hp Morane Bullets lasted only until 19 October, 1916; they were used by only one Flight of No.60 Squadron, for the other two Flights had been equipped with Nieuports when the unit returned to operational flying on 23 August. Thereafter thirteen Type N, two Type I and nine Type V were sent to England without engines, and it seems that a few aircraft saw brief and limited use with training units, but the Bullet was too unpopular to last long.

Types N, I and V
Type N—80 hp Le Rhône 9C
Types I and V—110 hp Le Rhône 9J
Single-seat fighting scout.

	Type N	Type I	Type V
Span	8·146 m	8·242 m	8·75 m
Length	5·83 m	5·815 m	5·815 m
Height	2·25 m	2·5 m	—
Wing area	11 sq m	11 sq m	—
Empty weight	—	334 kg	—
Loaded weight	444 kg	510 kg	—
Maximum speed—			
at sea level	144 km/h	168 km/h	165 km/h
at 3,000 m	—	156 km/h	—
Ceiling	—	4,700 m	—
Endurance	1½ hr	1 hr 20 min	—

Armament: One fixed 0·303-in Lewis (Type N) or Vickers (Types I and V) machine-gun.

Manufacturer: Morane-Saulnier.

Service use: Western Front: Sqns 1, 3 and 60; some of No.3 Sqn's Type Ns attached to No.24 Sqn during opening stages of the Battle of the Somme.

Serials: Type N: 5067—5069, 5180, 5191, 5194—5197, A122, A127, A128, A148, A166, A167, A171—179, A186, A196.
Type I: A198, A199, A202, A206.
Type V: A160, A204, A207, A209, A219, A236—238, A245, A246, A252, A254.

Morane-Saulnier Type BB

In the summer of 1915 the Morane-Saulnier company, so long faithful to the monoplane configuration, produced a two-seat military biplane. Its relationship to the Type N monoplane was evident in its large spinner and fully-faired fuselage, and the characteristic Morane form of undercarriage was fitted. The equal-span wings were braced as a single-bay structure, and lateral control on the prototype was by wing-warping.

The French authorities displayed no interest in the type, but both the RFC and the RNAS turned their attention to it. Maj D. S. Lewis, DSO, at that time the Commanding Officer of No.3 Squadron, RFC, then equipped with Morane Parasols, flew as passenger in a Morane biplane, possibly the prototype, and gave it a good report. His report is dated 2 August, 1915, and includes the statements: 'It appears to be a most excellent machine, the best "general purposes" machine I have yet seen . . . It appears to be very strong . . . should be fairly easy to fly and can be landed very slowly.' At an earlier point in his report he had said: 'It should be much easier to fly than a Parasol, when the ailerons are fitted.' The 110 hp Le Rhône engine naturally gave it quite a good performance.

A single example of the biplane was supplied to the RNAS as 3683. This

This Morane-Saulnier BB, No.5167 (MS583), was delivered to No.1 A.D. from Paris on 16 March, 1916. It was sent to England on 21 April and was the subject of the official CFS performance report on the type. On 4 May it was with the AID at Farnborough and returned to France, being allotted to No.60 Squadron on 19 May, 1916. It crashed on 24 June, 1916, and was returned to Depot.

biplane was at Dunkerque as early as 1 October, 1915; it had a modified fuselage but retained wing warping. Some weeks were to elapse before any delivery was made to the RFC, but on 4 November Capt Lord Robert Innes-Ker of the B.A.S.D. wrote to Maj-Gen Trenchard:

'Yesterday afternoon I saw our first biplane flying with a new 110 hp Le Rhône motor belonging to us. Mr Morane himself told me that he considered this to be the best machine that he has ever flown . . . weather permitting, it will be flown to St-Omer tomorrow.'

The biplane was allotted the Morane-Saulnier designation Type BB and the French official identity of MoS.7.

The RFC's initial request was for 92 Morane biplanes but the French authorities would allocate only 26 for the final quarter of 1915. Two had been delivered by 9 November and the remaining 24 were expected by the end of 1915. With the threatened cessation of deliveries of the 110 hp Le Rhône (see page 306) Trenchard instructed Lord Innes-Ker on 6 January, 1916, to reduce the RFC's demand for Morane biplanes to 36, but some official notes prepared on the following day suggest that 46 were required for delivery by 31 March, 1916. Deliveries for that quarter fell short of the allocation by 18; for the second quarter 36 (apparently later reduced to 24) were demanded; and for the third quarter of 1916 a further 24 were demanded but only ten allotted.

Deliveries continued in small weekly quantities throughout the spring and summer of 1916, terminating during the week ending 14 October with the delivery of MS816, MS818, MS849, MS852, MS853 and MS854. In the RFC these became respectively A301, A299, A302, A304, A300 and A303. Despite French resumption of production of the 110 hp Le Rhône engine these aircraft were dismantled and their engines returned to the Morane works for installation in Type P Parasols; the BB airframes were sent to No.1 Aircraft Depot in packing cases and were despatched, still packed, to England. The basic fuselage frame of A301 survives in the Royal Air Force Museum at Hendon, the only known relic of the Morane-Saulnier Type BB.

At an early stage the RFC asked for the Type BB to be modified in various ways to meet operational requirements. The major modification required was the enlargement of the side fairing of the fuselage on the starboard side to house a camera and a wireless transmitter mounted on the outer side of the basic fuselage frame. This produced an asymmetrical fuselage but apparently had no adverse effect on performance or handling qualities. A vertical tube for a gun mounting was to be provided behind the cockpit, and other minor modifications were to be made. It was intended to fly the RFC's first Morane biplane, incorporating the widened fuselage fairing and the rear gun mounting, back to Paris to serve as a sample, and Morane and Saulnier personally visited St-Omer on 25 November, 1915, to see the modified aircraft for themselves. By 20 February, 1916, it had been decided to fit the wireless and its accumulator behind the observer's seat.

Lateral control was regarded as inadequate, and wider ailerons were suggested. The manufacturers replied by stating that the ailerons had originally been larger but the biplane was then found to be tiring to fly in rough weather, and the tapered surfaces were then adopted. RFC Headquarters accepted this view and decided on 16 February, 1916, not to press for broad-chord ailerons. Nevertheless the problem did not go away and on 19 April Brig-Gen Brooke-Popham asked 'what has happened regarding the experiments in connection with the widening of the ailerons on the Morane Biplane 587?'.

The firm had, by 2 May, tried four different types of aileron, all of which were unsatisfactory. A fifth type was tried, but was still regarded as worse than the standard design. Morane-Saulnier then proposed to increase the span of the ailerons, but at Brooke-Popham's instruction an RFC aircraft was fitted with fairing strips over the aileron hinge gap, with beneficial results. On 24 August, 1916, he instructed both Aircraft Depots that all Morane biplanes and parasols were to be fitted with these fairing strips.

In July 1916 MS578 was delivered with what was described as 'a special type of aileron to give it greater lateral control'. As A195 it went to No.3 Squadron and, on 23 July, 2nd Lt E. M. Pollard reported that it had 'slightly more lateral control than the old type with smaller ailerons . . . The new ailerons give more lateral control but the machine is heavier to handle'. It was decided that the modified ailerons did not bestow a great enough improvement on the Type BB to be worth while.

No.5137 was a Morane-Saulnier BB of No.3 Squadron. It was received at No.1 A.D. on 27 January, 1916, and issued to the squadron on 1 February. On 23 February it was shot down at Souain by Leutnant Max Immelmann as his 9th combat victory; the pilot, Lt C. W. Palmer, was wounded and made prisoner; his observer, Lt H. F. Birdwood, was killed. The aircraft was officially struck off charge on 2 March, 1916. (*D. R. Neate*)

The Morane Biplane was issued to Squadrons 1 and 3, a policy statement of 29 January, 1916, announcing that 'When Morane Parasols are struck off they will be replaced by biplanes as these become available, until there are four biplanes in each of the two Morane Squadrons . . . These four biplanes will be formed into one flight in each squadron.' Despite slow deliveries of the type this evidently came to pass, for on 1 July, 1916, No.1 Squadron had five Morane Biplanes and Nos.3 and 60 had four each.

In service the Morane BB was usually armed with two Lewis guns. One was fixed on the upper wing firing forward; it was fired by the pilot but loaded by the observer, who himself had a second Lewis gun on a rear mounting very similar to that of the Martinsyde Elephant. The Morane biplane performed the same duties as the contemporary Parasols, apparently without attracting praise or blame. More than 80 were delivered to the RFC, and some were issued to No.60

Squadron on its arrival in France. These outlived the few Morane Parasols used by No.60 but sustained a fair number of losses in action. One or two biplanes went to No.12 Squadron for a time. Six were still with No.1 Squadron in January 1917, perhaps the last in the Field with the RFC, but were soon replaced by Nieuport Scouts. Thereafter the type was used for a time by training units.

Type BB
110 hp Le Rhône 9J

Two-seat reconnaissance biplane.
Span 8·585 m; length 6·935 m; height 2·615 m; wing area 22·32 sq m.
Empty weight 491 kg; loaded weight 761 kg.
Maximum speed at 3,050 m—134 km/h; climb to 1,980 m—13 min, to 3,050 m—26 min 48 sec; service ceiling 3,660 m.
Armament: Two 0·303-in Lewis machine-guns.

Manufacturer: Morane-Saulnier.

Service use: *Western Front*: Sqns 1, 3, 12 and 60; Pilots' School at No.1 Aircraft Depot, St-Omer. *Training duties*: CFS.

Serials: 5104, 5126, 5130, 5137, 5142, 5149, 5156—5170, 5176, 5177, 5181—5185, 5192, 5193, 5200, A119, A132, A137—139, A147, A149—151, A155, A161, A163, A183, A189—191, A195, A217, A218, A220, A222, A226, A227, A230—233, A242—244, A251, A256, A257, A282—284, A286—290, A293—296, A299—304.

Morane-Saulnier Type P

Something of a supply crisis arose at the end of 1915 when the French authorities informed the RFC that production of the 110 hp Le Rhône 9J engine would cease early in 1916, and that only a further 66 engines would be delivered to the RFC. On 6 January, 1916, Maj-Gen Trenchard instructed Capt Lord Robert Innes-Ker to reduce the RFC's demand for Morane biplanes to 36 in order to conserve engines. This was in reply to a letter sent by Innes-Ker on 5 January in which he advised Trenchard: 'I think it would be as well to concentrate on the Parasols. Morane is making alterations in the Parasol which will make it as fast as, if not faster than, the Morane biplane.'

Indeed this was essentially true, but the alterations amounted to a complete redesign, and the new type's improved performance was produced by employing the about-to-be-discontinued 110 hp Le Rhône. Fortunately it proved possible to avert the cessation of this engine's production, and the British forces eventually received no fewer than 2,448 engines of this type from French contractors and 953 from W. H. Allen of Bedford between 1916 and 1918.

On 18 March, 1916, the Morane-Saulnier company wrote to Capt Lord Innes-Ker of the B.A.S.D.:

'Order for 36 Morane-Saulnier biplanes'

'We understand that the principal advantage of this type of machine consists of the speed. With regard to this, we would inform you that we are going to experiment from Sunday, March 26th, with a 110 hp Parasol, from which we are expecting a higher speed than from the biplanes. We therefore consider it more to your advantage to pass the order in the following form

36 110 hp machines. These machines will be either biplanes or parasols according to the results obtained.'

This was communicated to RFC Headquarters. On 25 March it was decided to let the order for 36 biplanes stand as such, but three days later this was reduced to 24 and Capt Lord Innes-Ker was instructed to order four 110 hp Parasols and to place a provisional order for 24 more if the first four proved satisfactory.

The new 110 hp Morane Parasol was designated Type P by its makers and was adopted by the Aviation militaire as the MoS.21. It was a wholly new design in which the more corpulent lines of the fuselage suggested the influence of the Types BB and N, for it was designed from the outset to be a fully-faired structure, and the large spinner on the airscrew was very similar to those fitted to the biplane and the single-seater. The tail unit was in the tradition of the Types N and LA, but the system of struts connecting the mainplane to the fuselage was more elaborate than on the LA. The landing wires ran up to a simple inverted-V cabane, and the ailerons were operated by a torsion rod linked by a crank to the vertical actuating rod from the pilot's controls.

A Morane-Saulnier Type P Parasol in France with P.C.10 finish. (*E. Harlin*)

On 21 March, 1916, Innes-Ker reported to RFC Headquarters that 'the new type aileron Morane is ready and will be tried this afternoon, weather permitting', almost certainly a reference to the Type P. The prototype underwent its French official tests on 31 March, 1916. The RFC managed to secure an example remarkably quickly, doubtless because Brig-Gen Brooke-Popham had lost no time in ordering the first four for the RFC as early as 25 March, before the prototype had been officially tested. It appears that the first of these was officially handed over to the B.A.S.D. in mid-April. Its French identity was MS746, and it was flown to No.2 A.D. at Candas on 24 April by Guillaux. In the RFC it became A120.

Without waiting for trial results, Brooke-Popham, on 25 April, increased the initial order for the Type P to 36 in addition to the initial four. At the same time, he specified that the second aircraft must be fitted with a circular rotating gun mounting for the observer's gun (the original French design provided only a fixed pillar mounting placed centrally immediately behind the cockpit), that the third must have the circular gun mounting and the starboard side fairing of the fuselage modified as on the Morane biplane to house a camera, and that the fourth machine should have the circular gun mounting, provision for a camera, and an Alkan interrupter gear to enable a fixed forward-firing gun to be installed.

Presumably these modifications caused some delay and may account for the fact that the next three Type P monoplanes were not delivered until July 1916, becoming A193, A197 and A205 in the RFC. Because the French authorities would not allow the RFC to have enough 110 hp engines Brooke-Popham instructed the B.A.S.D. on 18 May to cancel the thirty-six Type P Parasols due to be delivered in the third quarter of 1916; and on 24 May he sent a telegram to Lt-Col W. Sefton Brancker at the War Office, telling him that supplies of Nieuport aircraft and spares for 110 hp Le Rhônes had been stopped, and asking that the French authorities be told that British aviation material such as steel, Vickers and Lewis guns, ball bearings, and so on, would be withheld unless the engines and spares that had been promised were delivered. It is not known whether such action was taken, but the French authorities later agreed to give the RFC twenty-four Le Rhône 9Js before the end of August 1916. On the strength of this promise Brooke-Popham instructed the B.A.S.D. on 9 June to ask whether the RFC could be allowed to order twenty Morane Type P monoplanes, ten Nieuport single-seaters and twenty Nieuport two-seaters.

Deliveries of the Type P in appreciable numbers began in mid-August, and MS843, MS844 and MS769—MS781 were delivered for the use of the RFC. Subsequent orders and deliveries suffered so many alterations that it is now virtually impossible to follow them in any systematic fashion. The position at the end of September 1916 was that twenty Type P monoplanes had been delivered and 44 more had been authorised for delivery but were still to come forward.

The RFC asked for modifications to be made to the later production Type P Parasols, among the earliest and most significant being the provision of a fixed

The true Morane-Saulnier Type P had a two-strut cabane and a large spinner. A268 (MS781) was delivered to No.2 A.D. in a packing case on 4 October, 1916, and was issued to No.3 Squadron on 25 February, 1917. It was shot down near Ligny on 6 March, 1917, and its pilot, Lt C. W. Short, was killed; his observer Lt S. McK. Fraser was wounded. The aircraft's destruction was completed by German shell-fire. *(D. R. Neate)*

The 80 hp (MoS.24) version of the Type P Parasol does not seem to have been separately identified in the past, doubtless owing to confusion with the Type LA. Here A6629 clearly displays its 80 hp Le Rhône engine. Originally MS919, it was delivered to No.2 A.D. on 22 November, 1916, went to No.1 Squadron on 16 December, and was sent back to No.1 A.D. on 23 January, 1917. Evidently it was subsequently used for training purposes, and still existed in March 1918, when it was selected for preservation. (*W. Evans*)

Vickers gun for the pilot and a sound rotating mounting for the observer's gun. This was communicated to the Morane-Saulnier company in July 1916; not surprisingly, they pointed out that the provision of the synchronised front gun would demand so many modifications that deliveries would be delayed by a month. They proposed to deliver the first fifteen with an overwing gun for the pilot and await the RFC's judgment before proceeding with the balance of the order. The RFC accepted this suggestion, and in August A205 was modified at No.2 A.D. to have a fixed Vickers, a rotating rear gun mounting, camera and wireless installations, and was flown to Paris on 29 August to serve as a model for the various alterations to be incorporated in the last twenty Parasols of the first major production order.

During the week ending 4 November, 1916, the B.A.S.D. recorded that the first Parasol with modifications incorporated had been received. This was MS885, which went to No.2 A.D. on 19 November and became A6628; it was allocated to No.3 Squadron on 5 January, 1917. The camera and wireless installations became standard, the distended side fairing that enclosed the former making the fuselage of the RFC's Type P Parasol asymmetrical in plan. The fixed Vickers was, strangely enough, not welcomed by pilots, at least in No.3 Squadron. On 29 January, 1917, the Commanding Officer of that unit reported that the overwing Lewis gun was preferred because the Vickers obscured forward view, tended to jam or fail to fire, fired slowly with the engine off, and its tracers could not be seen. On 13 February Brooke-Popham informed No.2 A.D. that it had been decided that the overwing Lewis was preferable to the Vickers, and next day he instructed the B.A.S.D. to tell the Morane company not to fit Vickers guns. Nevertheless,

Type P Parasols continued to be delivered with the Vickers gun, and its removal was one of eight modifications that, in May 1917, No.3 Squadron was still making to its aircraft on delivery. By that time the usual rear gun mounting was a Scarff No.2 Ring.

In mid-1916 the continuing shortage of 110 hp Le Rhône engines obliged the RFC to ask whether the current order for twenty-four Morane Parasols could be met by delivering Type LA Parasols in place of the Type P then being supplied. On 23 June the Morane-Saulnier company wrote to say that the Type LA was no longer in production, but suggested that the 80 hp Le Rhône might be fitted to the Type P airframe, suitably modified, instead of the 110 hp engine. This was agreed on 27 July, and MS905—MS928 were delivered to the RFC with the 80 hp engine, starting in the latter half of September 1916. The existence of these aircraft has led to confusion with the Type LA.

The 80 hp Type P differed from the 110 hp version in more than its power unit. The visible difference lay in the engine and the smaller spinner used on the 80 hp version, but it also had structurally different wings, undercarriage, bracing struts, fuel tanks and aileron linkage. Apparently this version (which must have been peculiar to the RFC alone) was given the separate French official designation of MoS.24.

At least nine of these 80 hp Type P aircraft were issued to No.3 Squadron, four to No.1 Squadron, at different times; but on 17 January, 1917, RFC Headquarters ruled that no more were to be issued to squadrons. Thereafter most of the 80 hp Type P were packed and sent to England to serve as trainers. Belatedly and for no obvious reason a dozen of the 80 hp Type P Parasols were delivered, seven in September and five in October 1917. These were sent from France direct to Farnborough without being given British serial numbers in France. On arrival in England they were numbered B9902—B9908 and B9933—B9937, but whether they were ever used is unknown.

B1604 was one of the few MoS.26 Parasols to reach the RFC. This variant retained the 110 hp Le Rhône engine but differed from the MoS.21 in having a fully-cambered conventional engine cowling without spinner. B1604 was one of the last parasols to be used by No.3 Squadron and, in company with thirteen others, was sent to No.2 A.D. before 12 October, 1917, when the squadron became an all-Camel unit.

B1611 was another MoS.26 Parasol of No.3 Squadron, and this photograph shows the position of the pilot's Lewis gun on the wing. This aircraft was wrecked on 30 September, 1917, after a flying life of 34 hr 35 min. (*The late H. H. Russell*)

The final tally of Type P Parasols delivered to the RFC cannot now be determined with certainty, but at least 106 of the 110 hp version and 36 of the 80 hp are known to have been accepted.

The true 110 hp Type P was further developed, and the later aircraft dispensed with the large spinner, having a well-cambered, full circular engine cowling. In French service this version was designated MoS.26. The first of this variant to reach the RFC was B1594, which was issued to No.3 Squadron from No.2 A.D. on 27 May, 1917. It was found that its throttle and fine adjustment were on the port side of the cockpit (apparently the standard position was to starboard), and the throttle had to be pulled backwards to open. The mounting for the overwing Lewis gun was unsatisfactory, and the new engine cowling made access to the plugs and valves more difficult. On 1 June, RFC Headquarters ruled that Parasols similar to B1594 were not to be issued without instructions from Headquarters. However, more came forward, and this, in some ways the most handsome, version of the Parasol, was used by No.3 Squadron in small numbers.

Although No.1 Squadron had begun to re-equip with Nieuport Scouts early in January 1917, No.3 Squadron went on using the Type P Parasol until October. It was originally intended to re-equip No.3 with S.E.5as by the end of August 1917, but it is probable that the difficulties with the 200 hp Hispano-Suiza engine thwarted this intention. The substitution of the Camel for the S.E.5a was delayed by some two months, and by 5 October No.3 Squadron had received three of its new Sopwith Camels but still had fourteen Moranes on its strength. These were A6643, A6659, A6700, A6708, A6716, A6717, A6730, B1521, B1596, B1604, B1612, B1668, B3451, and B3476, but a week later all had gone back to No.2 A.D. and the squadron was an all-Camel unit. The day of the Morane Parasol in the RFC on the Western Front was over.

Some had maintained the standards of durability set by a few of the Type LA Parasols. A234 (MS770) had been delivered to No.2 A.D. at Candas on 23 August, 1916, and was allotted to No.3 Squadron on 26 October. It was given new wings on 4 June, 1917, after flying 179 hr 20 min, and went on to build up 286 hr 10 min of operational flying with that unit until it was reported missing on 19 September, 1917. This constituted a new record for No.3 Squadron, surpassing the 260 hr total of the Type LA 5114.

With the general replacement of the final variant of the Morane Parasol two-seaters imminent, the Director of Equipment on 29 August, 1917, had instructed RFC Headquarters to return all serviceable Parasols to England, because it was thought that some might be of use to the Russians. Late in September 1917, No.2 Aircraft Depot was instructed to send all MoS.26-type Parasols to England in cases and without engines. The revolution in Russia frustrated any ideas of sending any Parasols there, and on 11 October Brig-Gen Brooke-Popham instructed No.2 A.D.: 'All Morane Parasols can now be sent to England.' It is doubtful whether many of them saw much use on training duties. In March 1918 the 80 hp Type P A6629 was selected for preservation, but it was to perish along with all the other types similarly chosen.

Type P
110 hp Le Rhône 9J, 80 hp Le Rhône 9C

Two-seat reconnaissance.
Span 11·2 m; length 7·2 m; height 3·47 m; wing area 18 sq m.
Empty weight 433 kg; loaded weight 733 kg.
Maximum speed at sea level 162 km/h, at 2,000 m—155·8 km/h; climb to 2,000 m—8 min 45 sec, to 3,000 m—15 min 50 sec; ceiling 4,800 m; endurance 2½ hr.
Armament: One 0·303-in Vickers machine-gun and one 0·303-in Lewis machine-gun, or two Lewis guns, and a small load of bombs as necesary.

Manufacturer: Morane-Saulnier.

Service use: *Western Front*: Sqns 1 and 3. *Training duties*: No.1 Reserve Sqn, Gosport. Pilots' School, No.1 Aircraft Depot, St-Omer, France.

Serials: *110 hp version*: A120, 193, 197, 205, 221, 234, 235, 239, 240, 241, 247—250, 255, 261, 266, 267, 268, 6628, 6635, 6636, 6637, 6638, 6643, 6648, 6650, 6651, 6652, 6653, 6655, 6659, 6660, 6698, 6699, 6700, 6702, 6708, 6715, 6716, 6717, 6719, 6722—6725, 6727—6730, 6750, 6757, 6758, 6760, 6799, 6800.
B1521, 1594, 1596, 1599, 1604, 1611, 1612, 1614, 1615, 1668, 1673, 3451, 3476, 3477, 3480, 3517, 3521, 3525, 3527, 3545, 3547, 3548, 3549, 6757, 6759, 6760, 6763, 6764, 6771, 6781, 6782, 6783, 6801, 6806, 6811.
80 hp version: A260, 264, 265, 269, 270, 277, 280, 297, 298, 308, 315, 6601, 6606, 6607, 6608, 6612, 6626, 6629, 6630, 6631, 6632, 6639, 6656, 6666.
B9902—9908, 9933—9937.

Morane-Saulnier Type AC

In mid-1916 the Morane-Saulnier company built two single-seat fighter versions of the Type P Parasol monoplane. The first had one Vickers gun and the pilot sat in the normal position under the wing; the second had twin Vickers guns, the pilot sat farther aft, and the wing was brought down close to the fuselage to be more or less level with his eyes. Details of both versions were sent to the RFC in July 1916 but this did not lead to any orders for the parasol fighter.

Although the Morane company thus looked ahead to the design of their brilliant Type AI of 1917, itself months ahead of the much-vaunted German parasol-wing fighters, they had not at that time abandoned the shoulder-wing configuration employed in the Types N, I and V. The design of a new shoulder-wing monoplane designated Type U had been completed by the middle of April 1916. This was an advanced design, having a fully-faired fuselage that terminated in a narrow horizontal knife-edge some way behind the tail unit. The rudder was divided by the fuselage, there were triangular fins above and below the fuselage, and the usual 'all-flying' horizontal tail was fitted, the elevators being shaped to match the contours of the rear fuselage. The wing structure was highly original and unusual. It was rigidly braced by an underlying truss of steel-tubing struts that, in effect, represented the bracing structure of a biplane. This departure enabled ailerons to be fitted instead of wing warping.

One of the RFC's two Morane-Saulnier AC monoplanes after assembly at Gosport.

Proof that the Type U was ever built has yet to be found, but its design provided the basis for the very similar Type AC single-seat fighter that appeared in the autumn of 1916. The fighter was somewhat larger than the Type U but had the same wing-bracing truss, its ailerons being built on to spanwise torque tubes that ran behind the rear spars to provide internal control runs. Similar careful design ensured concealed control runs to the tail surfaces. The fully-faired fuselage tapered to a sharp point behind the tail unit. Power was provided by a 110 hp Le Rhône; the armament was a single fixed Vickers gun synchronised to fire forward through the airscrew disc.

The type AC's performance proved to be good and it gave its pilot a superlative field of view in all upward directions. Possibly as an insurance against failure of the Spad 7, the French authorities ordered a small batch (believed to be of about 30 aircraft) of the new Morane, giving it the official designation MoS.23.C 1, and deliveries began late in 1916. Two Morane-Saulnier AC monoplanes were delivered to the British Aviation Supplies Department in Paris during the first week of January 1917. Their French official identities were MS878 and MS879, and they were evidently intended for evaluation by the RFC.

It seems that RFC Headquarters valued the two 110 hp Le Rhône engines more highly than the AC airframes, for on 3 January, 1917, the B.A.S.D. was instructed to take out the engines, send them to No.2 Aircraft Depot at Candas, and despatch the monoplanes to England. The aircraft were dismantled, crated, and sent to No.1 Reserve Squadron at Gosport on 30 January, 1917. It was during

Morane-Saulnier Type AC MS878 became B1395 in the RFC but was incorrectly marked as C1395, as seen here.

this same period in that January that the Bristol M.1B monoplane, of essentially the same configuration as the Type AC, was evaluated in France and incontinently condemned by Trenchard before he had read all of the reports on it. As Trenchard had earlier come to dislike the AC's precursors, the Types N, I and V, it seems there was no chance that the Type AC could hope for a dispassionate assessment from the RFC in France.

The serial numbers B1395 and B1396 were allotted to the two ACs in England, and the incidence of such allotments suggests that this one was not made until early April 1917. However, MS878 at least was given the erroneous identity C1395 and apparently was flown as such at Gosport, an engine having been found for it. The career of MS879 (B1396) is not known, but the likelihood is that it would be held as a source of spares for MS878.

The Morane AC was used operationally in small numbers by some French Escadrilles de Chasse in the spring of 1917. Indeed, it took until 2 May, 1917, for the RFC's Fourth Brigade to send to RFC Headquarters silhouettes of the type, described as 'the new French Morane'.

Type AC
110 hp Le Rhône 9J or 120 hp Le Rhône 9Jb

Single-seat fighter.

Span 9·8 m; length 7·05 m; height 2·73 m; wing area 15 sq m.

Empty weight 435 kg; loaded weight 658 kg.

Maximum speed at sea level 178 km/h, at 2,000 m—174 km/h, at 3,000 m—171 km/h; climb to 2,000 m—5 min 55 sec, to 3,000 m—10 min 15 sec; ceiling 5,600 m; endurance 2½ hr.

Armament: One fixed, synchronised 0·303-in Vickers machine-gun.

Manufacturer: Morane-Saulnier.

Service use: *Training purposes*: No.1 Reserve Sqn, Gosport.

Serials: B1395 (MS878), B1396 (MS879).

Nieuport monoplanes

One of France's greatest aviation pioneers, Edouard de Niéport,* entered the field of aircraft design as early as 1905. His first aircraft was a monoplane and all of his designs were of that configuration. The Nieuport monoplanes were distinguished by a remarkable and sensible simplicity and were, by the standards of their time, successful.

On 8 September, 1911, Capt J. D. B. Fulton, the Commandant of the Air Battalion, R.E., sent a memorandum to the Director of Fortifications and Works, setting out the considerations relating to the provision of aircraft for 1912. He recommended the purchase of one each of the contemporary Breguet biplane, and the Nieuport, Deperdussin and Sommer monoplane types. A Nieuport monoplane was, remarkably, purchased that month and was given the official number B4; it had a 50 hp Gnome engine, and was used by the Aeroplane Company of the Air Battalion. When the aircraft of the Royal Flying Corps received serial numbers in 1912, B4 became 253. By 23 November, 1912, it was reported that its '. . . wings and fuselage want renewing', a reference to the fabric covering of these components, not their structure. By then it was under the War Office ban on the flying of monoplanes in the Military Wing, but by 28 December it was again serviceable at Larkhill. Wing re-covering was again considered necessary by 4 January, 1913, and by 14 March the Nieuport had been packed for transport to the Royal Aircraft Factory.

Two more Nieuport monoplanes were taken on charge by the RFC on 19 June, 1912, one having a 70 hp Gnome engine, the other a 100 hp Gnome. Some official

*For his aviation activities de Niéport adopted the name Nieuport.

A Nieuport monoplane with Gnome engine, almost certainly B4 of the Air Battalion, photographed at Larkhill. (*RAF Museum*)

Possibly the same Nieuport as in the previous illustration, again at Larkhill, with crew aboard and undercarriage wheels at or near full deflection. (*RAF Museum*)

This aircraft can be positively identified by its rudder markings as B4, the Air Battalion's 50 hp Nieuport monoplane. (*RAF Museum*)

On the introduction of the RFC numbering system the Nieuport B4 became 253. It was used by No.3 Squadron but did not fly after the imposition of the monoplane ban. It lay for months in the Royal Aircraft Factory and was finally struck off charge on 13 August, 1913. (*RAF Museum*)

No.254 was the RFC's 70 hp Gnome-powered Nieuport monoplane, having been B7 with the Air Battalion. Here it is seen near Baldock during Army manoeuvres in 1912, while it was in use with No.3 Squadron. Like the Military Wing's other monoplanes it was grounded when the ban was imposed and was finally struck off charge on 1 September, 1913. (*RAF Museum*)

documents suggest that the former was at one time numbered B7, but that identity has also been connected with the B.E.1 and its correctness is open to question. What is certain is that the two Nieuport monoplanes, as aircraft of No.3 Squadron, were given the official serial numbers 254 and 255.

No.254 flew from Farnborough to Larkhill on 2 August, 1912, piloted by Lt B. H. Barrington-Kennett. It did a little flying at Larkhill in the hands of Barrington-Kennett and Capt D. G. Conner, but was damaged on 14 August, 1912, in a landing mishap. It was sent to the Royal Aircraft Factory for repair but was recorded as being again at Larkhill on 15 October. By 21 December, 1912, it was reported to be at Farnborough, having long since ceased to fly in consequence of the ban on monoplanes in the Military Wing, imposed in September 1912.

Little seems to have been recorded about the 100 hp monoplane No.255; indeed it is doubtful whether it ever flew with No.3 Squadron. In mid-November 1912 it still lacked flying wires but was regarded as serviceable by the 23rd of that month and in December was at Larkhill. The monoplane ban ensured that it did no flying there, and by 15 February, 1913, it was at Farnborough.

The only Nieuport monoplane in the RFC to have the 100 hp Gnome engine was No.255, which was first taken on charge on 19 June, 1912. It was at Larkhill in mid-October 1912 and was on the strength of No.3 Squadron; by the end of March 1913 it was languishing in the Royal Aircraft Factory, and it was struck off charge on 5 August, 1913. (*RAE*)

The Nieuport monoplane sold to the War Office by Claude Grahame-White in 1913, photographed in the USA in 1912. In the cockpit is Reginald H. Carr, then Grahame-White's mechanic but later to be a pilot of great talent, notably in the RFC.

Grahame-White's Nieuport monoplane after acquisition by the War Office as 282. It was at the Royal Aircraft Factory on 14 April, 1913, but never joined any squadron of the Military Wing.

Most obscure of all the RFC's Nieuport monoplanes was a single-seater with 28 hp Nieuport engine. This aircraft, despite its lack of military potential, was taken on charge by the Military Wing on 22 October, 1912, and was numbered 264. Despite the monoplane ban then in force it was promptly allotted to No.3 Squadron and was on that unit's strength on 26 October. It may have been intended as a practice aircraft but it seems never to have flown, although its engine was being overhauled early in November. By 1 February, 1913, it was at Farnborough. In mid-April it was reported to have passed the structural-safety tests imposed by the Monoplane Committee (though it was simultaneously reported to have a twisted fuselage) and it was in fact the only one of all the monoplanes then in the Royal Aircraft Factory to have done so. Despite this fact No.264 seems never to have flown.

Its companions on the strength of the Flying Depot at Farnborough on 14 April, 1913, included Nos.253, 254 and 255, but it is highly doubtful whether they ever left the Royal Aircraft Factory or flew again. On that same date another Nieuport monoplane was reported to be at the Royal Aircraft Factory waiting to be taken over by the Military Wing. This was another two-seater with the 70 hp Gnome and had the serial number 282. It was in fact quite an historic aircraft, having originally belonged to Claude Grahame-White, and was the aircraft of which he had taken delivery at Cherbourg in August 1912 on his departure for the USA. Flying the Nieuport, he won many prizes at the Boston and Nassau Boulevard flying meetings. Thus it had much distinguished flying to its credit long before it became No.282 with the Military Wing. It was one of several aircraft sold by Grahame-White to the War Office in March 1913 when there was an undignified official scramble to acquire aircraft in order to support Col Seely's claims as to the Military Wing's strength. The Nieuport went to Farnborough, where it was apparently flown at least once but never went to a squadron.

None of the Military Wing's Nieuport monoplanes survived beyond the autumn of 1913. No.253 was written off charge on 13 August, having done a total of 40 hr 29 min flying; No.254 on 1 September; No.255 on 5 August; No.264 on 15 October; and No.282 on 20 August.

Nieuport monoplanes

50 hp, 70 hp or 100 hp Gnome; one aircraft (No.264) with 28 hp Nieuport
Two-seat tractor monoplane (No.264 single-seat).
Span 11 m; length 8·4 m; height 2·6 m; wing area 21 sq m.
Dimensions for 50 hp aircraft.
Loaded weight 325 kg (50 hp).
Speed 88 km/h (50 hp).
Armament: None.

Manufacturer: Nieuport.

Service use: No.2 (Aeroplane) Company, Air Battalion, R.E.; No.3 Sqn, RFC.

Serials: 253 (50 hp Gnome), formerly B4; 254 (70 hp Gnome), formerly B7; 255 (100 hp Gnome); 264 (28 hp Nieuport); 282 (70 hp Gnome); 409.

Nieuport 11 and 21

Contrary to popular belief, the RFC never used the Nieuport 11. It is necessary, however, to refer to that type and to the somewhat similar Nieuport 21 because the RFC serial numbers A8738—8743 were allotted late in 1916 for six RNAS Nieuport single-seaters (respectively 3956, 8751, 8750, 3986, 3957 and 3958) that were to be taken over from the Admiralty at Dunkerque. Although this transaction never took place, at least one RFC document solemnly but erroneously lists these six Nieuports as being on charge of the British Expeditionary Force on 15 March, 1917.

Of these six RNAS aircraft, 3986 was a Nieuport 11 that was actually with No.9 Squadron, RNAS, when it was supposed to be on the BEF's strength. The other five were of a type known to the RNAS as 'Type 17B', a confusing designation in view of the existence of Nieuport Types 17 (110 hp Le Rhône) and 17bis (130 hp Clerget), of which the latter was used by the RNAS. Various references to the RNAS Type 17B indicate that these aircraft had the 80 hp Le Rhône engine and wings of the same area as the true Nieuport 17, thus identifying them as, almost certainly, Nieuport 21s.

Of these five Nieuports intended for transfer to the RFC, 3957 had been lost as early as 8 December, 1916, when Flt Sub-Lt The Hon A. C. Corbett of No.8 Squadron, RNAS, was shot down and killed. Nos.8750 and 8751 had an operational link with the RFC, for by 2 November, 1916, they were on the strength of the RNAS Detached Squadron that was formed at the end of October 1916 to join the 22 Wing, RFC. This RNAS unit was subsequently named No.8 (Naval) Squadron and won considerable renown for its gallant work with the RFC. The Nieuport 8750 returned to Dunkerque on 15 January, 1917, and both it and 8751 later served with No.9 (Naval) Squadron; they transferred to No.11 (Naval) Squadron together on 27 March, 1917, and the deletion of both Nieuports was approved on 30 June, 1917. They thus outlived 3956 and 3958, the deletion of which had been approved on 16 May, 1917.

Beardmore-built Nieuport 12, originally 9214 and renumbered A3270 on transfer from the RNAS. It was one of nine Nieuport 12s accommodated at No.1 A.D. for No.46 Squadron on 21 October, 1916, and it was still with No.46 when the unit was re-equipped with Pups in April 1917.

Nieuport 12 and 20

The RFC made no use of the basic Nieuport military biplane, the Type 10, but a number saw quite extensive service with the RNAS. In 1915 there appeared an enlarged development, powered by a 110 hp Clerget 9Z and intended for use as a fighter-reconnaissance two-seater. This was the Nieuport 12, which in its turn was ordered by the RNAS, and production aircraft were in service early in 1916. By 23 March deliveries of the Nieuport 12 to the RNAS totalled thirty.

In addition to its orders placed in France, the RNAS ordered fifty Nieuport 12s (9201 — 9250) from William Beardmore & Co, and the first of these was test-flown by A. Dukinfield-Jones on 10 May, 1916. Deliveries were slow and occupied a full year, the later aircraft having modified engine cowlings, fuselage flank fairings, interplane bracing and undercarriage; some had a new vertical tail assembly embodying a fixed fin and plain rudder.

When the Admiralty responded to Trenchard's plea for help in making up the deficiency in the strength of the RFC during the period preceding the Battle of the Somme, the RNAS aircraft transferred to the RFC included a number of Beardmore-built Nieuport 12s. An official document dated 28 June, 1916, stated that the transfers were to include:

'Starting from 1 July, if possible, three Beardmore Nieuport two-seater 110 Clerget machines per week, but not less than two per week up to a total of 20 machines. These machines will be supplied without guns. Army to make arrangements as to their transference from Messrs Beardmore, Dalmuir.'

The aircraft concerned were 9213—9232, and all were delivered to No.46 Squadron, RFC, which was at Wyton preparing to go to France as a Corps squadron. The first transfers were of 9214, 9215, 9216, and 9218, which were handed over on 4 September, 1916, and deliveries continued until 30 September. These twenty Nieuports were renumbered A3270—3275 and A3281—3294,* and in the RFC they were armed with a fixed Vickers gun and the Scarff-Dibovsky interrupter gear.

The RFC subjected A3288 (ex 9226) to climbing and speed trials, conducted by CFS Testing Flight at Upavon on 3 and 7 October, 1916. The trial report was highly critical, as the following extracts indicate:

'Pilot's view not good on account of the high fairing round his cockpit.
When carrying full reconnaissance load it is so heavily loaded as to be not only useless in the matter of performance, but dangerous to fly, while in any case it is much too bad in climb and speed to be any use as a fighter.
Vibrates badly.
Tiring to fly owing to cramped position.
Landing: very difficult in a confined space as there is not sufficient elevator control to get the tail properly down.
Owing to the very bad performance of this machine, it appears that no alterations practicable would render it fit for service overseas.'

*This renumbering was as follows: 9213 became A3281; 9214—9219 became A3270—3275; and 9220—9232 became A3282—3294.

The Nieuport 12 A3291 (ex 9229) was used by No.46 Squadron while working up at Wyton before going to France.

Beardmore-built Nieuport 12 with 100 hp Gnome Monosoupape engine. (*RAF Museum*)

Few other aircraft of the period were so roundly condemned as this Nieuport 12, but the CFS report did nothing to save No.46 Squadron from having the type. In fact, A3282 had arrived at No.1 A.D., St-Omer, on 5 October, 1916, destined for that squadron, while A3288 was still under test at Upavon. By 21 October ten Nieuport 12s were at St-Omer, allocated to No.46 Squadron, and the unit arrived at its aerodrome at Droglandt on 26 October with twelve aircraft. Its unspectacular operational work began on 4 November and continued until its re-equipment with Sopwith Pups in April 1917. By then most if not all of the squadron's Nieuports were of the Type 20 described later.

The RFC received a further twenty Beardmore-built Nieuport 12s, numbered A5183—5202. Their origin is obscure, for there is no obvious connection with an earlier intention to transfer 9239—9250 to the RFC as A5157—5168: these RFC serial numbers were re-allocated, and dates of delivery to RNAS destinations for 9242—9249 are known, the last of these going to Cranwell as late as 8 March, 1917; whereas allocations of the batch A5183—5202 began in early December 1916. It appears that these Nieuport 12s went to training units and squadrons

A5200, a Nieuport 12, had also been built by Beardmore, originally for the RNAS, and still had its characteristic Nieuport-type mounting for the observer's Lewis gun. This aircraft was used for training purposes by No.84 Squadron at Lilbourne, though this photograph is believed to have been taken at Sedgeford.

working up to operational status, some of them acquiring 100 hp Gnome Monosoupape engines for their non-operational duties.

In 1916 two variants of the Nieuport 12 appeared. The first of these, which had the French official designation Nie.12bis.C 2, was a modified Nieuport 12 with the 130 hp Clerget 9B that was supplied only to the French Aviation militaire and does not concern us here. The second variant may, in terms of prototype existence, have preceded the Nie.12bis.C 2, for it appears that at least one existed by mid-June 1916. This version of the design had the 110 hp Le Rhône 9J engine, which Gustave Delage, Nieuport's chief designer, much preferred to the Clerget 9Z. The two-seater with the Le Rhône 9J was given the French S.F.A. designation Nieuport 20, and during the week ending 5 August, 1916, thirty were allotted to the RFC by the French authorities; deliveries were to be made during the July—September quarter of 1916. The Nieuport 20 was not adopted by the

Possibly the first Nieuport 20 in the RFC, A154 had only vestigial flank fairings behind the engine cowling. Originally N1169, this aircraft was delivered to No.1 A.D. on 9 May, 1916, and was issued to No.1 Squadron on 24 June. It was tested with a Vickers gun in July 1916 and on 4 January, 1917, was transferred to No.46 Squadron. By July 1917 it was at Gosport.

French and the production aircraft were specially constructed for the RFC; nevertheless they were delivered wearing French serial numbers. The first two whose delivery was recorded were N1816 and N1829, which underwent reception tests at Villacoublay in mid-September 1916 and were flown to No.2 A.D. at Candas on 15 September to become A258 and A259 respectively. In fact they had been preceded by a few single aircraft, exemplified by A154 and A156, of which the latter was on the strength of No.1 Squadron as early as 29 June, A154 in July. It is known that A154 differed in detail from the production Nieuport 20 and may have been a kind of prototype.

By September 1916 it was intended to change No.1 Squadron's equipment with the object of creating a unit with twelve Nieuport two-seaters and six Morane-Saulnier BB biplanes. Two other Nieuport 20s went to the squadron in the

The standard Nieuport 20 had full flank fairings behind the engine cowling. N1829 was flown to No.1 A.D. on 15 September, 1916, to become A259 in the RFC. It was issued to No.1 Squadron on 23 September but was lost with Lts C. C. Godwin and P. C. Ellis on 17 October, 1916. (*Imperial War Museum Q67922*)

summer of 1916. These were A185 and A188, the former having the unusual installation of the Arsiad* interrupter gear for its Vickers gun.

Deliveries of the main batch, once started, continued at a modest rate, but by mid-October Trenchard was understandably anxious to obtain as many Nieuport 17 single-seaters as possible; and on 22 October Capt Lord Robert Innes-Ker of the B.A.S.D. was requested to

'. . . arrange to exchange as far as possible the 23 Nieuport two-seaters still due to us for Nieuport scouts.'

A reduction was in fact made. By 28 October Innes-Ker was able to report:

'I have definitely arranged for the outstanding order for 23 double-seater machines to be converted to one of 10 single-seater and 13 double-seater machines.'

By 4 November seven aircraft had been delivered and the outstanding balance was recorded as only fifteen; and as far as can now be established, only twenty-one of the thirty Nieuport 20s originally ordered were in fact delivered. The last known delivery was of N2996, which was sent to No.1 A.D. by road on 13 March, 1917.

It is known that A228, A229, A258, A259 and A291 were used by No.1 Squadron. With the exception of A259, which was lost on 17 October, 1916, all later served with No.46 Squadron, as did A285, A291, A292, A309, A314, A6602 and A6625. While with No.1 Squadron A258 had an overwing gun fitted in mid-October 1916. When No.45 Squadron was in difficulties in April 1917 it was given a few Nieuport two-seaters to tide it over; one of these was the Nieuport 20 A6736.

*Designed by Major A. Vere-Bettington, the officer commanding the Aeroplane Repair Section, No.1 Aircraft Depot. The mechanism's name was an acronym derived from the title of that unit.

What might be termed the production form of the Nieuport 20 differed from the Nieuport 12 primarily in having the 110 hp Le Rhône engine. A horseshoe cowling was faired smoothly into the fuselage sides by full flank fairings. Internally, an engine back-plate of a different shape was used; the first vertical steel-strut spacers in the fuselage sides were farther aft; the front fuel tank was larger and of different form.

The Nieuport 20 was no more effective operationally than the Nieuport 12, for the necessary performance and manoeuvrability simply were not there. When No.46 Squadron was re-equipped with Sopwith Pups in April 1917 the Nieuport two-seaters were withdrawn to training units, and remained in service in that capacity for several months. One or two were briefly allocated for Home Defence duties: A292 was with No.39 Squadron at Woodford but was re-allocated to Training Brigade on 30 June, 1917, and A309, although allocated to No.39 Squadron, was found unfit and transferred to No.1 (Southern) Aircraft Repair Depot.

A6740 was a Nieuport 20 that was first recorded at No.1 A.D. on 13 March, 1917. It was used by No.45 Squadron and was flown to England on 8 May, 1917. Thereafter it was used on training duties, retaining its No.45 Squadron marking, as seen in this photograph.

Nieuport 12

110 hp Clerget 9Z or 100 hp Gnome Monosoupape
Two-seat fighter-reconnaissance.
Span 9 m; length 7 m; height 2·7 m; wing area 22 sq m.
Empty weight 550 kg; loaded weight 850 kg.
Maximum speed at 2,000 m—146 km/h; climb to 1,000 m—5 min 40 sec, to 2,000 m—14 min 15 sec; ceiling 4,000 m; endurance 3 hr.
Armament: One fixed 0·303-in Vickers machine-gun and one 0·303-in Lewis machine-gun on Etévé ring mounting.

Nieuport 20
110 hp Le Rhône 9J

Two-seat fighter-reconnaissance.
Dimensions as Nieuport 12.
Empty weight 453 kg; loaded weight 752 kg.
Maximum speed at sea level 157 km/h, at 2,000 m—152 km/h; climb to 1,000 m—5 min 12 sec, to 2,000 m—12 min 2 sec.
Armament: As Nieuport 12.

Manufacturers: Nieuport (France), William Beardmore (United Kingdom).

Service use: *Western Front*: Sqns 1, 45 and 46. *Home Defence*: No.39 Sqn. *Training duties*: Sqns 65 and 84; Reserve/Training Sqns 31, 43, 45 and 55 and at Gosport.

Serials: Nie.12: A3270—3275, 3281—3294, 5183—5202.
 Nie.20: A154, 156, 185, 188, 228, 229, 258, 259, 285, 291, 292, 309, 314, 6602, 6625, 6707, 6731, 6732, 6735—6737, 6740—6743.

Nieuport 16

Early in 1916 the Nieuport company produced a development of the Nieuport 11 in which the 110 hp Le Rhône 9J engine replaced the 80 hp Le Rhône 9C of the earlier type. In French service the 110 hp aircraft received the official designation Nie.16.C 1 and was soon introduced to operational duties in the Escadrilles de Chasse of the Aviation militaire.

The Nieuport 16 so closely resembled the Nieuport 11 that from most angles the two were indistinguishable: this fact may account for the widely-held but mistaken belief that the Royal Flying Corps used the Nieuport 11. The RFC was enabled to use the Nieuport by courtesy of the Royal Naval Air Service, which had ordered the new type as soon as it became available. Deliveries to the RNAS began surprisingly early, the aircraft being numbered from 9154. However, the Naval service did not use the Nieuport 16s but diverted them instead to the RFC. No.9154 went from the RNAS station at Dunkerque to No.1 Aircraft Depot at St-Omer on 18 March, 1916, being numbered 5171 and achieving the distinction of being the first Nieuport single-seat scout to go to the RFC. It was quickly followed by 9155 and 9158, which in the RFC became respectively 5172 and 5173.

The significance of these transfers from the RNAS was overlooked by the official historian (indeed they remained undiscovered for 60 years), but in fact they began three months before Maj-Gen Trenchard appealed for RNAS assistance to the RFC in the period preceding the Battle of the Somme. Fourteen further Nieuport 16s followed the first three, all seventeen having been originally ordered for the RNAS. The total represented a real and substantial sacrifice by the Naval service, for the Nieuport transfers began before even the prototypes of the Sopwith Pup and Triplane had gone to France, and virtually denied the RNAS effective and up-to-date single-seat fighters for a time.

These first Nieuport 16s came to the RFC at a time when most single-seat scouts were distributed in ones and twos to two-seater squadrons, and the first allocations were to Squadrons 1 and 11. One or two subsequently went to No.3 Squadron, but further deliveries in the summer of 1916 made it possible to issue substantial numbers of Nieuport 16s to No.60 Squadron, which must have been

During its brief career this Nieuport 16, A126 (N1133), was flown by Lt Albert Ball. It was received at No.2 A.D. on 25 April, 1916, and went to No.11 Squadron next day. Damaged on 27 May, it went to No.2 A.D. for repair on 30 May, 1916.

almost completely equipped with the type by mid-October 1916. In No.11 Squadron the Nieuport found one of its most aggressive and effective exponents in Lt Albert Ball, who first flew a Nieuport 16 on 15 May, 1916.

In the RFC the Nieuport 16's Lewis gun was carried on the simple but highly effective Foster mounting. This ingenious piece of equipment had been invented by Sgt R. G. Foster of No.11 Squadron and embodied a quadrantal slide along which the gun could be lowered rearwards for changing magazines. This also permitted the gun to be fired vertically upwards, a facility of which Ball made spectacular and victorious use on several occasions. The Foster mounting remained standard equipment on all the Nieuport types used by the RFC up to and including the Nieuport 27. A few of the RFC's Nieuport 16s were delivered with a fixed Lewis gun synchronised by the French Alkan mechanism; these included A125, which was tested with near-catastrophic consequences on 19 May, 1916, by Capt W. J. C. K. Cochran Patrick. His gun gear obviously malfunctioned, and the violent vibration that followed the damage to his airscrew all but shook the Nieuport to pieces.

One or two of the RFC Nieuport 16s were fitted with cameras and employed on high-speed photographic-reconnaissance missions. One such was A208 of No.60 Squadron, which failed to return from a photographic sortie on 6 March, 1917. This must have been one of the last Nieuport 16s in operational use with the RFC, the last known being A131, which was returned to No.2 Aircraft Depot from No.29 Squadron on 7 April, 1917.

Thereafter some Nieuport 16s served with the Scout School attached to No.1 Aircraft Depot at St-Omer, and it appears that the type ceased to be used in France by the autumn of 1917.

Nieuport 16
110 hp Le Rhône 9J

Single-seat fighting scout.
Span 7·52 m (upper), 7·4 m (lower); length 5·64 m; height 2·4 m; wing area 13·3 sq m. Empty weight 375 kg; loaded weight 550 kg.

Maximum speed at sea level 165 km/h, at 2,000 m—156 km/h; climb to 2,000 m—5 min 50 sec, to 3,000 m—10 min 10 sec; ceiling 4,800 m; endurance 2 hr.

Armament: One 0·303-in Lewis machine-gun, supplemented by eight Le Prieur rockets.

Manufacturer: Nieuport.

Service use: *Western Front*: Sqns 1, 3, 11, 29 and 60. *Training*: Scout School at No.1 Aircraft Depot, St-Omer.

Serials: 5171—5173.
A116, 117, 118, 121, 125, 126, 130, 131, 133—136, 164, 165, 184, 187, 208, 210—212, 214, 216, 223—225.

Nieuport 17 and 23

There is a fair amount of evidence to suggest that the combination of the 110 hp Le Rhône engine with the basic Nieuport 11 airframe was not an unqualified success, and that the Nieuport 16's handling qualities left something to be desired. Gustave Delage's next design embodied the Le Rhône 9Ja engine but provided increased wing span and area and a somewhat refined airframe. The new type, designated Nie.17.C 1 by the Service des Fabrications de l'Aviation, nevertheless retained the principal Nieuport design characteristics, being a single-bay sesquiplane with single-spar lower wing, V-form interplane struts and the typical Nieuport rudder. In the fuselage fuller flank fairings behind the full-circular engine cowling blended its cross-section smoothly into the flat sides of the rear fuselage.

Some of the early Nieuport 17s were fitted with a hemispherical fairing fixed to a forward extension of the stationary crankshaft of the rotary engine. Although this fairing resembled a spinner it was in fact stationary and was described as a

One of the first Nieuport 17s to be delivered to the RFC, this may have been A200. When photographed it was fitted with the fixed spinner-like cône de pénétration in front of the airscrew.

The Nieuport 17 A200 (N1553) was transferred from the RNAS to the RFC on 19 July, 1916. Initially it went to No.11 Squadron but was transferred to No.60 Squadron on 28 August. It was flown by Albert Ball on 15 September, when he scored a victory on it, but the Nieuport was so shot about that it had to go to No.2 A.D. for repair next day. A200 was again issued to No.60 Squadron on 16 December, 1916, finally leaving the squadron on return to No.2 A.D., again for repair, on 24 March, 1917.

cône de pénétration in the relevant French patent specification. A few of the first Nieuport 17s supplied to the RFC had this fitting, but it was evidently soon discarded in both French and British service.

The Nieuport 17 came to the RFC in mid-July 1916: as with the Nieuport 16, the RFC received their first examples as transfers from the Naval service. The first three to come from the RNAS had the French identities of N1553, N1494 and N1561; they were taken over by the RFC on, respectively, 19, 21 and 22 July, being numbered A200, A201 and A203, again respectively. A200 and A201 were

Attention being given to the Foster-mounted Lewis gun of a Nieuport 17 of No.60 Squadron.

briefly used by No.1 Squadron, and all three were at one time on the strength of No.60 Squadron.

An RFC order for ten Nieuport 17s had been placed a few days before the RNAS transfers took place, and direct deliveries apparently began late in July 1916. Lack of engines caused some delay, but by mid-October the RFC had ordered a further 50, a total that was soon increased to 60. Of these, 39 had been delivered by the end of 1916, and apparently 50 more were ordered early in 1917.

Many Nieuport 17s in French service were armed with a single fixed Vickers gun, but on the RFC's aircraft the standard armament was a Lewis gun on an overwing Foster mounting. Occasionally Le Prieur rockets were carried in the characteristic strut-mounted tubular launchers. At least one RFC Nieuport 17, A6678, was tested with a dual installation of a synchronised Vickers gun and a Foster-mounted Lewis, but this was not favourably reported on.

Nieuport 17 or 23 of No.40 Squadron. The pilot is believed to be Lt Gordon T. Pettigrew. (*K. M. Molson*)

In service several alarming incidents involving structural failure or deformation occurred. These, without exception, were failures of the wing structure, several being caused by twisting of the single spar of the lower wing in the unsatisfactory collar attachment at the base of the interplane struts. In one such incident on 19 April, 1917, the lower starboard wing of B1540 of No.40 Squadron came away completely while the Nieuport was being flown by 2nd Lt Edward Mannock. He managed to land the disabled aircraft, turned over on touching down, but was uninjured. Others were less fortunate. Engine cowlings also gave trouble, and various forms were fitted.

Nevertheless, the RFC held a high opinion of the Nieuport 17 and were specifically grateful to the French authorities for releasing a further 60 aircraft for the use of the RFC in mid-March 1917. This was a timely consignment, for losses were high and the costly fighting of April 1917 was close at hand; during the first three weeks of that month wastage of RFC Nieuport Scouts totalled 55. In May a further 100 Nieuports were handed over to the RFC at Buc, near Paris.

Whether A6680 was a Nieuport 17 or 23 cannot be determined from this photograph, which shows the aircraft taxi-ing out while with No.40 Squadron at Treizennes in March 1917, wearing the three white bars that then constituted the squadron's marking. A6680 was at No.1 A.D. on 24 February, 1917, went to No.40 Squadron on 12 March, and returned to the A.D. for repair on 13 April. It was subsequently issued to No.1 Squadron on 15 July but was lost on 28 July, when 2/Lt G. B. Buxton was shot down by Leutnant Schmidt of Jagdstaffel 3 as his 7th victim. (*I. P. R. Napier, via R. C. Bowyer*)

B1575 was a Nieuport 23 of No.60 Squadron. On 15 July, 1917, while being flown by 2/Lt G. A. H. Parkes, it was shot down near Douai by Hauptmann Adolf Ritter von Tutschek as his 19th combat victory. The aircraft is here seen in German hands: Parkes had been wounded and was made a prisoner of war. (*Egon Krueger*)

This batch included a number of Nieuport 23s. The Nieuport 23 in French service was distinguished from the Nieuport 17 by having its Vickers gun offset to starboard by a few centimetres because a different form of synchronising mechanism was employed. This distinction could not be made on the RFC's aircraft because they were all armed with the standard overwing Lewis gun, and only the closest scrutiny could reveal detail differences between the Nieuport 17 and 23. Not surprisingly the RFC's wartime records are frequently confused and confusing in their references to the two types, and it is no longer possible to differentiate accurately between them. In practice this mattered little.

In mid-July 1917 Nieuports were still regarded as priority aircraft in the RFC, preceding all types except D.H.4s and S.E.5s. Nevertheless, by the end of that month the Nieuport 17/23 design was outdated, yet the disappointment initially experienced with the Nieuports 24 and 24bis led to a directive dated 29 July that stated that only Nieuport 17s, 23s and 27s were to be issued to squadrons. These squadrons were Nos.1, 29, 40 and 60, of which the last-mentioned began to re-equip with S.E.5s in that month of July.

Thirty Nieuport 23s were allotted to the RFC in August 1917. Deliveries ended in October, including four aircraft described as Nieuport 23bis (an unidentified variant) and one Nieuport 27. As far as can now be determined the RFC received about eighty Nieuport 23s. In RFC service many of these came to be powered by the basic 110 hp Le Rhône 9Ja instead of the 120 hp Le Rhône 9Jb usually fitted to the aircraft used by the Aviation militaire. As Nieuport 27s became available to the RFC in France, the 17s and 23s were withdrawn from operational use but in January 1918 orders were given for sixteen Nieuport 17s and four Nieuport 23s to be packed to go to Egypt; additionally two Nieuport 17s and five 23s were sent as a

At Treizennes in the spring of 1917 Sgt Ditchburn of No.40 Squadron titivates the engine of A6781. This Nieuport 23 had the S.T.Aé identity N2911 and was allotted to No.40 Squadron on 21 March, 1917. Later it saw service with Nos.1 and 29 Squadrons; from the latter unit it returned to No.1 Repair Park on 24 November, 1917.
(*I. P. R. Napier via R. C. Bowyer*)

A Nieuport 23 of No.111 Squadron, Palestine. (*RAF Museum*)

source of spares. Later, at least one of the mysterious Nieuport 23bis, B6799, went to the Middle East also. In that theatre of war the RAF's assorted Nieuports remained operational until the war ended.

Nieuport 17 and 23
Nieuport 17: 110 hp Le Rhône 9Ja
Nieuport 23: 120 hp Le Rhône 9Jb

Single-seat fighting scout.
Span 8·16 m (upper), 7·8 m (lower); length 5·8 m; height 2·4 m; wing area 14·75 sq m.
Nieuport 17: Empty weight 375 kg; loaded weight 560 kg.
Maximum speed at sea level 165 km/h, at 2,000 m—160 km/h, at 3,000 m—154 km/h; climb to 2,000 m—6 min 50 sec, to 3,000 m—11 min 30 sec; service ceiling 5,300 m; endurance 1 hr 45 min.
Armament: One 0·303-in Lewis machine-gun, usually on Foster mounting. Eight Le Prieur rockets could be carried.

Manufacturers: Nieuport, Sté pour la Construction et l'Entretien d'Avions, R. Savary et H. de la Fresnaye, Sté anonyme française de Constructions aéronautiques.

Service use: *Western Front*: Squadrons 1, 11, 29, 40 and 60. *Palestine*: Sqns 14 and 113. *Training*: Heliopolis.

Serials:A200, 201, 203, 213, 215, 271—276, 278, 279, 281, 306, 307, 311, 313, 6603—6605, 6609—6611, 6613—6624, 6644—6647, 6657, 6658, 6664, 6665, 6667—6680, 6684, 6689, 6691—6694, 6701, 6718, 6720, 6721, 6726, 6733, 6734, 6738, 6739, 6744, 6745, 6751, 6752, 6754, 6755, 6756, 6761—6798.
B1501—1520, 1522, 1523, 1539—1559, 1566—1572, 1574—1579, 1582—1585, 1590, 1595, 1597, 1598, 1600—1603, 1605—1610, 1613, 1616—1619, 1621, 1624—1626, 1629—1652, 1654—1659, 1662, 1665, 1666, 1670—1672, 1674—1694, 1699, 1700, 3452—3456, 3458, 3459, 3461—3463, 3465—3470, 3473, 3474, 3478, 3481—3487, 3494—3497, 3500, 3540, 3541, 3554, 3555, 3558, 3561, 3577, 3581, 3583—3587, 3589, 3593, 3597, 3598, 3636, 3643, 3644, 6799.

Nieuport 24.C 1 N4662 photographed shortly after its arrival at No.2 A.D., Candas; the photograph was dated 28 July, 1917, and the Nieuport became B3601 in the RFC. It was issued to No.40 Squadron on 15 August, 1917, subsequently went to No.29 Squadron on 30 March, 1918, and was lost on 7 April while being flown by Lt A. G. Wingate-Grey on a special mission.

Nieuport 24 and 24bis

In about the late summer of 1916 the Nieuport company built a variant of the Nieuport 17 that had the 130 hp Clerget engine and a fuselage provided with full-length stringers and formers to create a fully-faired fuselage. The traditional Nieuport tail unit was retained. This variant was designated Nieuport 17bis but apparently found no favour with the French Aviation militaire, though one slightly modified example was flown for a time by that great French pilot Charles Nungesser.

Britain's Royal Naval Air Service used the Nieuport 17bis, and No.6 (Naval) Squadron was possibly the only unit of any air service to be fully equipped with this type. A few were also used by No.11 (Naval) Squadron. In operational service the Nieuport 17bis was a disappointment; this was the greater because the French original had had a promising performance.

The promise of the first Nieuport 17bis may have persuaded the Nieuport company that the full-length fairing of the fuselage might produce performance sufficiently improved to assure the continued service of the basic design. Certainly all their high-powered single-seaters coming thereafter retained the faired fuselage, and by February 1917 the French Section Technique de l'Aéronautique was testing a new Nieuport designated Nie.24.C 1. The precise form of this aircraft at that time is uncertain, but it seems likely that it incorporated a tail unit of completely new form and construction, having elegantly curving outlines and a built-up ply covering.

Although this Nieuport 24 had a good performance it did not immediately come into service. It was in fact preceded by a type bearing the designation Nie.24bis.

This Nieuport 24bis, B3592, of No.111 Squadron in Palestine bore the name *Demoiselle* and was flown by Lt R. J. P. Grebby. (*W. S. Lighthall via K. M. Molson*)

This appeared to be something of a stop-gap, having a fully-faired fuselage and the traditional Nieuport tail unit. It differed from the Nieuport 17bis only in having a 120 hp Le Rhône engine instead of the earlier type's Clerget.

In French service the Nie.24bis was short-lived, for Nieuport 24s were in quantity production by early May 1917. The production Nieuport 24 had the faired fuselage and Le Rhône engine but retained the characteristic centre-section bracing and faired tailskid of the earlier types. Both the Nieuport 24 and 24bis had a stiffened strip of canvas fairing over the gap between aileron and mainplane.

Lt A. Eckley in B3591, another Nieuport 24bis of No.111 Squadron. Although the aircraft here has both a fixed Vickers and an overwing Lewis gun on a Foster mounting this armament proved too heavy and the Vickers was removed.
(*W. S. Lighthall via K. M. Molson*)

Both types arrived in the RFC late in July 1917. Among the earliest arrivals were the Nieuport 24bis examples that became B3591 and B3592 (these do not appear to have passed through the B.A.S.D. acceptance system), and on 27 July the Nie.24 N4662 and the Nie.24bis N4569 were flown to No.2 Aircraft Depot at Candas to become respectively B3601 and B3602. They came with a poor reputation and their French ferry pilots' reports were highly critical of both aircraft's lateral control. This was emphatically confirmed by Capt F. G. Dunn at No.2 A.D. Immediately, RFC Headquarters issued instructions that the Nieuport 24 and 24bis were, as far as practicable, to be confined to the Scout School at No.2 A.D.. These two aircraft were followed almost immediately by a further nine Nie.24 and four Nie.24bis.

Fortunately it was found that the lateral control deficiency was attributable to the canvas fairing strips covering the aileron hinge gaps: once these were removed the Nie.24, 24bis and 27 became as light on the ailerons as the Nie.17 and 23. Eventually most of the RFC's few Nieuport 24s and 24bis saw some operational service with Squadrons 1, 29 and 40 in France in 1917. Early in 1918 the Nieuports that were sent to the Middle East included three Nie.24 and three Nie.24bis. Of these at least one Nie.24bis went to No.113 Squadron in Palestine. One Nie.24 and three Nie.24bis were used at the Scout School at No.2 Aircraft Depot, and three Nie.24 and one Nie.24bis were flown to England between 22 April and 3 May, 1918, presumably for training purposes.

Nieuport 24 and 24bis
120 hp Le Rhône 9Jb or 130 hp Le Rhône 9Jby

Single-seat fighting scout.

Span 8·21 m (upper), 7·82 m (lower); length 5·87 m; height 2·4 m; wing area 14·75 sq m.

Nie.24: Maximum speed at sea level 176 km/h, at 2,000 m—171 km/h, at 3,000 m—169 km/h; climb to 1,000 m—2 min 40 sec, to 3,000 m—9 min 25 sec, to 5,000 m—21 min 30 sec; ceiling 6,900 m; endurance 2¼ hr.

Nie.24bis: Speed at sea level 170 km/h, at 2,000 m—170 km/h, at 3,150 m—167·7 km/h; climb to 1,000 m—2 min 40 sec, to 3,000 m—9 min 40 sec, to 5,000 m—21 min 40 sec.

Armament: One 0·303-in Lewis machine-gun on Foster mounting.

Manufacturers: Nieuport, Sté pour la Construction et l'Entretien d'Avions, R. Savary et H. de la Fresnaye, Sté anonyme française de Constructions aéronautiques.

Service use: Western Front: Sqns 1, 29 and 40. *Palestine*: Nie.24bis used by Sqns 111 and 113. *Training*: Scout School at No.2 Aircraft Depot, Candas.

Serials: Nie.24: B3601, 3604, 3606, 3607, 3609, 3610, 3612, 3613, 3614, 3617. (Also reported as a Nie.24 was B3582, but this remains unconfirmed.)
Nie.24bis: B3591, 3592, 3602, 3603, 3605, 3608, 3611.

Nieuport 27

The Nieuport 24 was closely followed by the Nieuport 27, which was the final variant of the Nieuport V-strut biplanes to see operational service. It came into use in France in the early summer of 1917 and soon outnumbered the Nieuports 24 and 24bis in the French escadrilles de chasse.

In appearance the Nieuport 27 closely resembled the Nie.24, having the same fully-faired fuselage, rounded tail surfaces and curving aileron tips. The later design had the form of undercarriage that had first appeared on the aircraft believed to be the Nie.24 prototype. This incorporated two centrally pivoted half-axles for the mainwheels, while the tailskid was a simplified unit devoid of the characteristic triangular fairing. As initially delivered, the Nieuport 27 had the troublesome canvas fairing strips along the aileron hinge lines and, like the Nie.24 and 24bis, suffered from poor lateral control.

B6768 was a Nieuport 27 of No.1 Squadron that retained its French camouflage in RFC service. It was lost on 9 January, 1918, when its pilot was Lt R. C. Sotham.

The RFC began to provision for the Nieuport 27 in June 1917, but it was late in July when the French S.F.A. authorised the allocation of 100 aircraft for delivery in August and September, together with spares equivalent to 30 additional aircraft. The French subsequently sought to substitute Nie.24bis but these were not accepted by RFC Headquarters, and the French authorities then offered thirty Nieuport 23s and seventy Nieuport 27s to be delivered in August, September and October 1917. Deliveries began on 14 August when the first Nie.27, N5405, was flown from Villacoublay to No.2 Aircraft Depot at Candas. It was followed by 15 more Nieuport 27s, and by the third week of October a total of twenty-nine Nieuport 23s and seventy-one Nieuport 27s were delivered to complete the allocation.

By that time the French Aviation militaire had been forced to recognise that the Nieuport single-seaters, including the Nie.27, had become outclassed on the Western Front. They themselves had to go on using the type into 1918, largely because their early experience with the Spad 13 was disappointing, but by 1 April, 1918, the Nieuport 27 was no longer operational in French units in France. In the RFC the type survived in No.29 Squadron until after the formation of the RAF, for it was not until 20 April, 1918, that the last Nieuport 27 left the squadron.

In January 1918 four Nieuport 27s were among the fourteen assorted Nieuports packed to go to Egypt, and it was at one time intended to send all remaining Nieuport 27s to the Middle East when the re-equipment of No.29 Squadron had been completed. This was largely frustrated by the confusion that followed the German offensive of March 1918, and many Nieuport 27s were sent to England instead. No doubt some of them served with training units thereafter; one or two were used at the Royal Aircraft Establishment, Farnborough, and were still in use in early 1919.

Nieuport 27
130 hp Le Rhône 9Jby

Single-seat fighting scout.
Span 8·21 m (upper), 7·82 m (lower); length 5·87 m; height 2·4 m; wing area 14·75 sq m. Loaded weight 535 kg.
Maximum speed at sea level 172 km/h, at 2,000 m—170 km/h, at 3,000 m—167 km/h, at 4,000 m—165·7 km/h; climb to 2,000 m—5 min 40 sec, to 3,000 m—9 min 25 sec, to 4,000 m—14 min 40 sec; ceiling 6,850 m; endurance 2¼ hr.
Armament: One 0·303-in Lewis machine-gun on Foster mounting.

Manufacturers: Nieuport, Sté pour la Construction et l'Entretien d'Avions, R. Savary et H. de la Fresnaye, Sté anonyme française de Constructions aéronautiques.

Service use: Western Front: Sqns 1 and 29. *Palestine*: possibly No.111 Sqn.

Serials: B3600, 3621—3635, 3637, 3647—3650, 6751—6756, 6765—6770, 6774, 6778, 6779, 6784—6786, 6789—6793, 6797, 6798, 6800, 6803, 6804, 6807, 6809, 6810, 6812—6815, 6818—6832, 6836, 6837.

Paulhan biplane

Louis Paulhan wrote his name indelibly in the history of aviation in Britain in April 1910 when he succeeded in completing the first flight from London to Manchester ahead of Claude Grahame-White, thereby winning the *Daily Mail* prize of £10,000. He devoted part of his winnings to the construction of an extraordinary biplane, in the design of which he collaborated with Henri Fabre of Marseilles who, earlier in 1910, had built and flown the first successful floatplane.

The Paulhan biplane made its first flight at St-Cyr on 11 September, 1910, and was exhibited at the Paris Salon de l'Aéronautique in October 1910, when it created a sensation by its unconventionality. It was a bizarre creation, each mainplane being a monospar structure consisting of an open W-girder spar to

The Paulhan biplane at Farnborough. (*RAE 0373; Crown copyright*)

which curved ribs were attached. Each rib slid into a tailored pocket in the single thickness of fabric that constituted the wing surface proper, the fabric being lashed to wooden cleats on the spar. The structure of the rudder, rear tailplane and forward elevator followed that of the mainplanes. This type of construction had been evolved by Fabre, who had employed it in his floatplane of 1910. Four single interplane struts separated the wings, across which lay two parallel beams of constant depth; these were two W-girder structures covered with fabric, and carried at their forward ends the elevator, at the rear the rudder and tailplane. The tailplane was a fixed surface, but its incidence could be altered on the ground. The pilot occupied a small central nacelle, to the rear of which was attached the 50 hp Gnome engine. Possibly the most extraordinary feature of the aircraft was the extensive use of chrome leather to make many of the joints and hinges in the airframe. The Paulhan's price was quoted as Fr 30,000 (£1,200).

Paulhan's agent in Britain was George Holt Thomas, and it was probably his persuasiveness that induced the War Office to buy the Paulhan biplane. Although it was known that the aircraft had flown successfully for $1\frac{1}{2}$ hours, piloted by Caillé with Paulhan as passenger, and that it could be packed in a crate measuring 15 ft 6 in by 3 ft 3 in by 3 ft 3 in, the War Office was reported to have stipulated that it must be capable of flying for two hours with a passenger plus 440 lb of ballast, in a 25-mph wind, and of gliding down with the engine stopped from a height of 200 ft. To represent the War Office at the reception tests of the Paulhan, Capt J. D. B. Fulton was sent to St-Cyr late in December 1910, despite the fact that he had only taken his aviator's certificate (No.27) some seven weeks earlier, on 12 November, 1910. It was intended that he should learn to fly the aircraft after it had passed the tests.

Caillé flew the Paulhan in the first of its War Office tests at St-Cyr on 31 December, 1910, and successfully concluded the series of tests on 11 January, 1911. Capt Fulton accepted delivery of the biplane and himself flew it at St-Cyr on 16 and 17 January, damaging it slightly on the second occasion. The Paulhan arrived at Farnborough on 16 February, 1911, and was deposited in the aeroplane erecting shed at the Balloon Factory, where it aroused the curiosity of the Superintendent, Mervyn O'Gorman. Although Maj Sir Alexander Bannerman, Commandant of the Balloon School, was in no hurry to have the aircraft

This three-quarter rear view emphasises the bizarre structure and appearance of the Paulhan. (*RAE 0336; Crown copyright*)

unpacked, he agreed on 23 February, 1911, to allow O'Gorman to unpack and erect it, ostensibly to enable drawings of it to be made.

The Advisory Committee for Aeronautics became interested in the Paulhan and apparently asked O'Gorman to conduct certain experiments with it. As far as flying was concerned it did not do so at Farnborough until 1 May, 1911. Under that date Geoffrey de Havilland recorded in his log-book: 'Paulhan out for first time. Rolled and made short flights on Common—very gusty. Several adjustments required.' De Havilland flew the aircraft again on 4 May, found it improved, and handed it over to Capt Fulton. Unfortunately it crashed next day while being flown by Fulton and had to go into the Factory for repair. It re-emerged on Friday, 7 July, 1911, to be flown by Geoffrey de Havilland, whose log-book records that the aircraft had been extensively modified while being repaired. A completely new elevator had been made; it did not retain the original W-girder spar and was pivoted on its centre of pressure. At the rear of the aircraft a new rudder was mounted on the tailplane and was in upper and lower parts about the horizontal surface. The rudder-post carried the tailskid, an arrangement that twisted the tailplane spar sufficiently to reduce the tailplane's angle of incidence. De Havilland found the rudder control improved and that the biplane was more manageable on the ground. More damage was sustained on 12 July, when failure of a cable splice brought the aircraft down on its starboard lower wing: a main spar, the rudder and the propeller were broken.

Geoffrey de Havilland's log-book next records the Paulhan on 16 October, 1911, when it taxied on the Common and bent an axle. Later that month it was handed over to the Air Battalion and was with No.2 (Aeroplane) Company by 27 October, Apparently it did no flying thereafter and had been dismantled in, or possibly before, December 1911. Indeed, it seems there was no intention to fly it again, for on 18 January, 1912, Mervyn O'Gorman, inspired perhaps by the success of B.E.1 and his subterfuge in 'reconstructing' it from the Duke of Westminster's Voisin, as described in the following section, asked whether the Factory could be allowed to reconstruct the Paulhan. Bannerman was not in favour of a reconstruction, though he seemed to recognise the wisdom of putting a 50 hp Gnome engine to better use. Apparently no reconstruction took place, but on 1 February, 1911, the Director of Fortifications and Works asked the Commander of the Air Battalion to hand the Paulhan over to the Superintendent of the Army Aircraft Factory. This was done on 3 February and the aircraft was never heard of again.

Paulhan biplane
50 hp Gnome

Two-seat pusher biplane.
Span 12·2 m; length 8·5 m; wing area 30 sq m.
Empty weight 400 kg.
Speed approximately 80 km/h.

Manufacturer: Louis Paulhan.

Service use: No.2 (Aeroplane) Company, Air Battalion, R.E.

Serial: F2.

B.E.1 in its original form with 60 hp Wolseley engine. (*RAF Museum*)

Royal Aircraft Factory B.E.1

His Majesty's Balloon Factory at Farnborough had first built an aeroplane in the spring of 1911 by recourse to a form of subterfuge on the part of the Factory's Superintendent, Mervyn O'Gorman, and his Engineer in charge of design, Frederick M. Green. Late in 1910, and at Green's suggestion, O'Gorman engaged the young man who was one of Britain's most talented and versatile pioneers of aviation, Geoffrey de Havilland. Such a combination of men naturally had ambitions to design aircraft, and an opportunity was seized late in 1910 when the Blériot XII was sent to the Balloon Factory for repair. O'Gorman sought authority for reconstruction which, when received, was interpreted as permitting total mutation into a weird pusher canard biplane, the S.E.1.

B.E.1, still with Wolseley engine but with its identity marked on the rudder and dihedral rigged into the wings. (*RAF Museum*)

Later in 1911 a similar opportunity arose when a somewhat shadowy Voisin biplane was sent to the Factory (since 26 April the Army Aircraft Factory) for repair. This aircraft had been presented to the War Office by the Duke of Westminster and was powered by a 60 hp Wolseley water-cooled V-8 engine. Again the 'reconstruction' ploy was invoked and a wholly new aircraft arose, ostensibly from the Voisin but in fact entirely designed by Geoffrey de Havilland and designated Blériot Experimental No.1, or B.E.1. The choice of Blériot's

B.E.1 with Renault engine and serial number 201. (*H. F. Cowley*)

342

name was made because he was regarded at Farnborough as the leading pioneer of the tractor form of aeroplane, just as Farman was considered to be the pioneer of the pusher configuration and Santos Dumont of the canard.

B.E.1 was apparently completed in October 1911 and made its first flights, with Geoffrey de Havilland at the controls, on 4 December. His remuneration, as test pilot, for a number of short flights on that day seems to have been the princely sum of 2s 6d. By 27 December B.E.1's Wolseley carburettor had been replaced by a Claudel, and de Havilland made six flights on that day, carrying Capt C. J. Burke, S. Heckstall Smith and one Peters as passengers, and taking a speed test that gave an average speed of 55 mph. With its landing wheels moved 12-in aft the aircraft flew again on 1 January, once with Mervyn O'Gorman in the front seat. By 7 January the upper wing had been rigged 3-in back, and three days later the mainplanes were given 1 deg of dihedral.

B.E.1 flying with smaller tailplane. (*RAF Museum*)

The B.E.1 continued to fly with agreeable regularity, and underwent various minor modifications and refinements. On 11 March, 1912, Geoffrey de Havilland handed it over to Capt C. J. Burke of No.2 Aeroplane Company, Air Battalion, R.E., who apparently flew it for the first time three days later. The aircraft was mostly flown by Burke, occasionally by Lt A. G. Fox, on relatively brief flights. While with No.2 Company, B.E.1 was allotted the number B7.

On 18 June the Wolseley engine was replaced by a 60 hp Renault, a modification that enhanced performance and greatly improved the forward view from the cockpits, for the advent of the air-cooled Renault allowed the cumbersome vertical radiator to be dispensed with. On the formation of the Royal Flying Corps the Farnborough detachment of No.2 Company became No.2 Squadron in the Military Wing, and B.E.1 continued to fly with that unit. By the end of July 1912 the aircraft had flown 36 hr 3 min in the hands of its military pilots. In August 1912 it was given the official serial number 201, following a period during which its rudder had borne the marking B.E.1, a form of identification that it shared for a few months with several other early B.E. types.

In the Military Wing the B.E.1 saw relatively brief periods of service with Squadrons 4 and 5, its sojourns with the three squadrons being separated by returns to the Royal Aircraft Factory, presumably for repair or modification. As time went on No.201 became virtually indistinguishable from contemporary B.E.2as, acquiring deckings ahead of and behind the front cockpit and, latterly, the narrower-chord tailplane that was standardised for the B.E.2a.

By the standards of its time, B.E.1's longevity was remarkable. Writing of it in Volume I of *The War in the Air* Sir Walter Raleigh said:

> 'It had a long and adventurous career, and was often flown at Farnborough for the testing of experimental devices. When at last it was wrecked, beyond hope of repair, in January 1915, it had seen almost three years of service, and had perhaps known more crashes than any aeroplane before or since. . . . The first machine of its type, it outlived generations of its successors, and before it yielded to fate had become the revered grandfather of the whole brood of factory aeroplanes'.

In fact, the B.E.1 did better than Raleigh knew, for it was still flying at Farnborough in May 1915 and at CFS in July 1916. By that time it had acquired a very up-to-date B.E.2b-type fuselage, and indeed it is doubtful whether any of the original structure remained in the airframe that then bore the serial number 201. The date and nature of its final demise have yet to be discovered, but its place in the history of the RFC is secure.

B.E.1
60 hp Wolseley, 60 hp Renault

Two-seat military biplane.
Span 38 ft 7¼ in (upper), 34 ft 11⅝ in (lower); length 29 ft 6½ in; wing area 374 sq ft.
Speed 55 mph.
Armament: None.

Manufacturer: HM Army Aircraft Factory, Farnborough.

Service use: Prewar, Sqns 2, 4 and 5.

Serial: Successively numbered B7, B.E.1 and 201.

B.E.2, 2a and 2b

A second B.E. biplane followed B.E.1 remarkably quickly. Against the date of 1 February, 1912, in his log-book Geoffrey de Havilland noted: 'B.E.2 first time out 11 a.m. Flew straight to L.P. [Laffan's Plain] and back. Made 4 flights in all. Green passenger. Gets away and climbs much faster than B.E.1. Engine revs 1720. Speed slightly above that of B.E.1'.

The airframe of the B.E.2 was virtually identical with that of B.E.1, but its engine was a 60 hp Renault, which was not only lighter than the B.E.1's Wolseley but also did not need the large radiator of the water-cooled engine. These factors largely accounted for the B.E.2's better performance. In its original form the B.E.2 had mainplanes of unequal span, their aerofoil being the rather heavily cambered N.P.L.3a section; there was no decking ahead of or between the

B.E.2 at Larkhill during the Military Trials.

cockpits, and the downward lengths of the exhaust pipes were taken inside the longerons to pass through the floor of the fuselage.

By the third week of March 1912 a wireless installation was being made in B.E.2, and the earliest recorded flights on which this was tested were made on 26 March. Tests continued regularly thereafter, the pilot being Geoffrey de Havilland. He did most of the flying of the aircraft at that time, including the testing of various experimental devices. On 28 April he flew the aircraft for the first time after a 70 hp Renault had replaced the original 60 hp engine, and on 1 May he tried the aircraft 'fitted with floats and wheels' (apparently a kind of amphibian undercarriage) but found it 'not satisfactory for rolling' (*i.e.* taxi-ing). Nevertheless, on 11 May he flew the B.E.2 to Fleet Pond, where it was fitted with floats. Trouble was experienced because the draught of the floats was too great for the depth of the pond: he succeeded in getting the aircraft into the air that evening but damaged the floats on alighting.

Restored to its wheel undercarriage, B.E.2 continued to fly throughout the summer of 1912. On 3 June its passengers were Gen Henderson, Capt Godfrey Paine, Maj F. H. Sykes and Mervyn O'Gorman; and later that day Geoffrey de Havilland, flying solo, dropped a 112 lb weight from the aircraft. From 8 to 14 August the aircraft was at Larkhill during the Military Aeroplane Competition, where it flew in most of the tests hors concours and outperformed most of the competing aircraft. It was beyond doubt incomparably more suitable as a military aircraft than the Cody biplane that was declared the official winner.

This is almost certainly why the B.E.2 was ordered in some numbers whereas the Cody was not: despite the predictable brayings of the professional detractors of the Royal Aircraft Factory, the decision was entirely sensible. According to surviving official records only five further B.E.2s were built at Farnborough itself (but to these have to be added the two aircraft that began life designated B.E.5 and B.E.6 but were ultimately B.E.2s in all essential particulars). All the other production aircraft were ordered from commercial aircraft-manufacturing firms, most of whom had entered aircraft in the official competition. The earliest orders went to such firms as The British and Colonial Aeroplane Co, Vickers, The Coventry Ordnance Works, Handley Page, and the Armstrong Whitworth concern.

B.E.2a No.206 was extensively used for experimental purposes and had one of the earliest British oleo undercarriages. After much use at Farnborough, 206 went to France on 18 December, 1914, to join No.6 Squadron. On 31 December it went to the Aircraft Park but returned to No.6 on 17 January, 1915, and survived with the squadron until 2 September, 1915, when it was returned to the A.P. and struck off charge. It is seen here at Farnborough and has its serial number painted under the lower mainplanes. (RAE)

A clear definition of what precisely constituted a B.E.2a as distinct from a B.E.2 has yet to be found. The B.E.2a, as represented publicly in R & M No.66 dated 12 June, 1912, was depicted as having wings of unequal span and a broad-chord tailplane. It is also certain that the production B.E.2as delivered under the initial contracts had the unequal span wings, whereas the later deliveries (which constituted the majority) had wings of equal span measuring 36 ft 11⅛ in. The equal-span wings had a new aerofoil section with reduced camber, and were rigged at only 3 deg 30 min incidence, whereas the unequal-span surfaces had been rigged at 4 deg 30 min. With very few exceptions the production B.E.2as had a new tailplane of substantially reduced chord and 34·6 sq ft in area that was roughly semi-circular in plan. It is of interest to note that the official general-arrangement drawings that bear the designation B.E.2a are dated as early as 20 February, 1912; they show deckings behind the engine and between the cockpits, refinements that did not in fact appear on a few of the earliest production aircraft. The date of the drawings strongly suggests that all the contractor-built, and probably all production, aircraft were in fact B.E.2as. The fact that many are recorded as B.E.2s in some contemporary documents is not particularly signifi-cant: at the time there were only B.E.2as, and to call them B.E.2s was simpler, saved a modicum of writing or typing, and was quite enough to distinguish them from B.E.3s and 4s.

As the production B.E.2as came forward they were allocated to the Military Wing. In the pre-1914 period they saw most extensive use in Squadrons No.2 and 4, flying many hours with commendable regularity and fair dependability. With such extensive and intensive use, there were some accidents, and the Factory's

No.205 was originally B.E.5 and first appeared with a 60 hp E.N.V. engine, as seen in this photograph. Later, when powered by a Renault, this aircraft was the B.E. on which Lt Desmond Arthur was killed at Montrose on 28 May, 1913.

virulent critics lost no opportunity to make capital out of these occurrences, although they remained conspicuously silent when similar mishaps befell aircraft of other makes. One of the most notorious disasters was the crash of No.205, originally built as B.E.5 with the 60 hp E.N.V. engine that had powered the S.E.1 and had earlier come from the Army's Blériot XII. This power unit was replaced in the autumn of 1912 by a 60 hp Renault; later still, a 70 hp Renault was fitted. On 27 May, 1913, while being flown at Montrose by Lt Desmond Arthur of No.2 Squadron, a faulty repair in one wing failed, the B.E. crashed and Arthur was killed. This crash was later said to be a possible origin of the so-called 'Montrose Ghost', though it is uncertain whether that phantom truly existed outside the

Bristol-built B.E.2a No.240 at the RFC Concentration Camp, summer 1914. This B.E. was delivered to Farnborough on 25 October, 1913, and allotted to the Military Wing on 10 December, having made its first flight on 29 November. It had a Rouzet wireless set, the aerial drum for which can be seen beside the front cockpit. No.240 went to France with No.4 Squadron on 13 August, 1914, but went to the Aircraft Park on 16 August and was struck off charge on 29 August. The aircraft was scrapped at Amiens.
(*Flight International 038*)

Capt C. A. H. Longcroft in the modified B.E.2a, No.218, that he flew from Montrose to Portsmouth and back to Farnborough, nonstop, on 22 November, 1913. The B.E.2a was built by Bristol and was taken on charge by the Military Wing on 11 February, 1913. On 2 May, 1914, it crashed while on loan to No.6 Squadron and was apparently written off.

imagination of the late C. G. Grey who missed no opportunity to vilify or calumniate the Royal Aircraft Factory and all its works.

When No.2 Squadron had gone to Montrose its C.O., Maj C. J. Burke, had decided that five of its aircraft should fly there from Farnborough; this proved to be something of an endurance test for the pilots and aircraft concerned. The adventure began on 13 February, 1913, when Capts C. A. H. Longcroft and J. H. W. Becke on B.E.2as, and Capt G. W. P. Dawes and Lts F. F. Waldron

Three B.E.2as and a Farman Longhorn of No.2 Squadron, fitted with flotation bags for the flight across the Irish Sea. (*Harold Dawes, via Marvin Skelton*)

348

Like No.240, B.E.2a No.336 was equipped with wireless. This Vickers-built aircraft was taken on charge by the Military Wing on 5 May, 1914, and on 31 July was, with No.240, considered to be available to support No.4 Squadron. Indeed, No.336 went to France with No.4 Squadron, but on 5 October was transferred to the Headquarters W/T unit that was designated No.9 Squadron on 4 December. On 25 January, 336 went to the Aircraft Park and was issued to No.2 Squadron on 4 March, 1915, only to be damaged next day. Repaired, 336 returned to No.22 Squadron on 19 March and was transferred to No.4 Squadron on 1 August, moving on to No.8 Squadron on 11 August and finally retiring to the A.D. on 8 October, 1915.

and P. W. L. Herbert on Maurice Farmans set out on the 450 mile journey. All finally arrived, after many vicissitudes, on 26 February, Longcroft's B.E. No.218 having behaved perfectly throughout the long journey. At that time it still had the unequal-span wings.

Longcroft later used No.218 to perform one of the greatest of all prewar flights. His aircraft had been extensively modified by having an additional 54 gallon fuel

The deeper fuselage decking of the B.E.2b is seen in this photograph of 487, which was flown direct from England to No.4 Squadron on 13 October, 1914. It went to the Aircraft Park on 28 December and returned to No.4 Squadron on 22 January, 1915. On 11 March, in company with 321 and 703, this B.E.2b set off to bomb Lille from Bailleul, an enterprise that resulted in the loss of all three B.Es. No.487 was captured intact by the Germans and taken back to Germany for evaluation.

349

tank mounted in the front cockpit, which was then faired over. Equal-span wings were fitted. On 22 November, 1913, Longcroft flew No.218 from Montrose to Portsmouth and back to Farnborough, a nonstop flight of 650 miles that was accomplished in a flying time of 7 hr 20 min. For this outstanding achievement Longcroft was made the 1913 recipient of the Britannia Trophy.

At Farnborough, B.E.2 and 2a aircraft were used for many experimental installations, notably of oleo undercarriages, and in Edward Busk's series of experiments aimed at achieving automatic stability, then regarded as a highly desirable quality in military aircraft. It was held that, as the aeroplane was primarily, if not exclusively, a reconnaissance vehicle, it would be at its most efficient if its pilot could leave it to fly itself, thus enabling him to devote his whole attention to reconnaissance and observation.

Another B.E.2b of No.4 Squadron was 705, which was built by Hewlett & Blondeau, and was at Farnborough on 24 November, 1914. It joined No.4 Squadron on 10 February, 1915, and stayed with that unit until it was struck off on 6 July, 1915. In this photograph 705 is seen on No.4 Squadron's airfield at St-Omer, with a Martinsyde S.1 and an Avro 504 for company.
(*Peter Liddle's 1914–18 Personal Experience Archives, presently housed within Sunderland Polytechnic*)

In March 1914 structural difficulties with the virtually unsupported rudders of the B.E.2a led to the type being grounded. A conference held on 26 March decided that flying could be resumed provided that all bent rudders were replaced by new ones. A modified design was got out and was to be introduced as soon as rudders built to it were available. Additionally, new-type compression ribs were to be fitted as opportunity arose.

When No.2 Squadron flew to Ireland on 1 September, 1913, to participate in manoeuvres, its five B.E.2as (including 217, 218, 272 and 273) were fitted with a form of flotation gear for the sea crossing from Stranraer to Island Magee. This equipment, known at the time as floats, consisted of cylindrical air bags mounted

Although not mentioned in any known history of No.30 Squadron, this unusual B.E.2b was apparently used by 'C' Flight of the squadron when it was based at Ismailia and operated against the Turks during the unsuccessful Turkish attack on the Suez Canal. This can only have been the aircraft, one of two originally sent to India as B.E.2as, that was sent from the Indian Central Flying School to Ismailia in December 1914, and is here seen in company with Maurice Farman Shorthorn No.369. No numerical identity is known for this B.E. and its only discernible marking seems to have been the letters I.C.F.S. borne on its rudder.
(*RAF Museum*)

inflated under the lower wings; these may have been supplemented by an arrangement of air bags in the fuselage designed at the R.A.F. When it was proposed that a Flight of B.E.s should again go to Ireland in 1914, No.2 Squadron enquired whether floats were to be used, because 'it was universally considered by pilots who used these floats last year, that they not only made the B.E. very sluggish in getting off, but also made her very difficult to handle if the weather was at all rough'. In the event, war intervened, and none of the B.E.s that crossed the English Channel was burdened with floats of any kind.

It is noteworthy that early in January 1914 No.2 Squadron was flying No.228 equipped with dual control. It is not known how many B.E.2as were equipped in this way, but it seems likely that No.228 may have been one of the earliest operational aircraft to have dual control.

Development of the design produced, in 1914, the B.E.2b, which had an

The I.C.F.S. B.E.2b acquired a B.E.2c-type fin, possibly being the only aircraft of its type to do so. Later still its rudder was painted with red, white and blue stripes and its I.C.F.S. marking was then transferred to the fin. (*RAF Museum*)

improved arrangement of top deckings about the cockpits, revised flying controls and a new fuel system. A fin was designed for this variant but found no general use. Later production contracts called for B.E.2bs, and quite large orders were placed after the start of the war.

When the four squadrons of the RFC flew to war on 13 August, 1914, No.2 Squadron took twelve B.E.2as and No.4 Squadron eleven. It has for long been believed that No.347, alleged to have been flown by Lt H. D. Harvey-Kelly, was the first British aircraft to land on the Continent after the outbreak of war. The truth is that No.347, although on the strength of 'A' Flight of No.2 Squadron on 31 July, 1914, did not go to France on 13 August or at any other time. Harvey-Kelly's aircraft was No.471, but it seems equally possible that the first aircraft to land in France was the B.E.2a No.327, flown by Capt F. F. Waldron with Air Mechanic Skerritt as his passenger. This historic B.E. lasted just a week, being struck off charge on 21 August, and a similar fate befell 471 on 9 September, 1914.

With the squadrons in the Field the B.E.2a and 2b gave surprisingly long, varied and satisfactory service during the early stages of the war. It is significant that the squadrons clearly preferred the B.E.s to, for instance, the Sopwith 80 hp tractors. When No.6 Squadron went to France in October 1914, it had eight B.E.s and four Henry Farmans, and among B.E.s that were later allocated to the squadron was the veteran 206, which joined the unit on 18 December, went to the Aircraft Park on 31 December and returned to No.6 on 17 January, 1915. It served on until 2 September, when it returned to the A.P. and was finally struck off charge after a remarkable career.

The B.E.2b No.687 of No.2 Squadron was the aircraft on which the first Victoria Cross to be awarded for an aerial action was won. The pilot concerned was 2nd Lt W. B. Rhodes Moorhouse, who on 26 April, 1915, dropped a 100 lb bomb on the railway line near Courtrai station. Because he came down to 300 ft to drop his bomb he was severely wounded by ground fire, yet he managed to fly his B.E. back to his own aerodrome at Merville. He died of his wounds next day.

At the end of September 1915 few of the early B.E.s remained in France. The B.E.2a 368 and the B.E.2b 746 were still with No.4 Squadron; No.8 Squadron had one B.E.2a (336) and one B.E.2b (492); and the B.E.2b 484 was with No.12 Squadron. The only others still flying in France were the B.E.2a 385 and the B.E.2b 396, which were with the Pilots' School attached to the Aircraft Park at St-Omer. Apparently the last in the field was 493, which went to No.12 Squadron on 8 October from the 1st A.P. and returned to England on 4 November.

A B.E.2b of the Joucques-built batch (2770—2819) with a training unit.

This B.E.2b of a training squadron had a two-blade airscrew, presumably of type T.5572, which normally denoted an 80 hp Renault engine. The aircraft's unit markings are of interest. (*RAF Museum*)

Thereafter the type served with training squadrons in Britain, and it was for training purposes that the Joucques and Whitehead contracts for B.E.2bs were given in 1915. A belated B.E.2b was A376, built by the R.A.F. under War Office authority, Extract 113, dated 9 May, 1916. It was tested at Farnborough on 19 May, 1916, was reconstructed in July 1918, and was reported ready for flight as late as 30 January, 1919. It had been dismantled by 27 February, 1919, and next day was handed to the Southern Aircraft Repair Depot, presumably to be scrapped. The survival of this B.E.2b marked the culmination of the career of a basic design that went back to the B.E.1 of 1911.

B.E.2, 2a and 2b
60 hp Renault, 70 hp Renault

Two-seat military biplane.

Span originally 37 ft $7\frac{1}{8}$ in (upper), 34 ft $10\frac{1}{2}$ in (lower), later equal-span wings of 36 ft $11\frac{1}{8}$ in; length 29 ft $6\frac{1}{2}$ in; height 10 ft 2 in; original wing area 357 sq ft, later 376 sq ft.

Empty weight 1,100 lb; loaded weight 1,600 lb.

Maximum speed 70 mph at sea level, at 6,500 ft—65 mph; climb to 3,000 ft—9 min, to 7,000 ft—35 min; service ceiling 10,000 ft; endurance 3 hr.

Armament: One rifle, carbine or pistols, bomb load of about 100 lb. Experimental loadings of one or two Fiery Grapnel weapons, or three 20 lb bombs.

Manufacturers: Royal Aircraft Factory, Armstrong Whitworth, British and Colonial, Coventry Ordnance Works, Grahame-White, Handley Page, Hewlett & Blondeau, Joucques Aviation, S.E. Saunders, Vickers, Whitehead Aircraft.

Service use: *Prewar*: Sqns 1, 2, 4 and 6; CFS. *Western Front*: Sqns 2, 4, 6, 8, 9 (previously HQ W/T Sqn), 12 and 16; Pilots' School at 1st Aircraft Park, St-Omer. *Egypt*: Ismailia Flight, RFC; No.30 Sqn, Ismailia. *Training duties*: CFS; Reserve Aeroplane Sqns 1, 4, 5, 8, 16, 18, 19, 24, 50 and 52. *India*: Indian CFS, Sitapur. *Other units*: Reserve Aircraft Park, Farnborough, for photographic experiments; Squadrons forming—Nos.15, 29 and 58.

Serials: **B.E.2/2a**: 205, 206 (ex B.E.6), 217, 218, 220, 222, 225—242, 245, 248—250, 267, 271—273, 299, 314, 316—318, 320, 321, 327—329, 331, 332, 336, 347—349, 368, 372, 383—385, 441, 442, 447, 449, 452—454, 457, 460, 461 (renumbered 328), 466, 468—471, 474, 475, 601, 602, 612, 622, 635, 667, 668, 709.

The following are believed to have been B.E.2bs—396, 484, 487, 488, 492, 493, 646, 650, 666, 676, 687, 703, 705, 722, 733, 746, 2175—2180, 2770—2819, 2884—2889, A376.

B.E.2c, 2d, 2e, 2f and 2g

In the history of the R.E.1 mention is made of the experiments aimed at creating an automatically stable aeroplane that were conducted at the Royal Aircraft Factory by Edward Busk in 1913–14. Application of the principles thereby established to a development of the B.E.2b produced the B.E.2c, in which staggered mainplanes with ailerons were employed, together with the large rectangular tailplane that evolved as T.P.3 on experimental B.E.2as. A standard B.E. rudder was used but there was in front of it a braced triangular fin.

These modifications were initially made to No.602, a B.E. of indeterminate origin that retained the 70 hp Renault as power unit, and had a twin-skid undercarriage similar to that of the B.E.2a/2b. It had been completed in May 1914, and its first recorded flights were made on 30 May, when it flew four times. On all but one of these flights its pilot was Edward Busk, and thereafter it was flown frequently at Farnborough.

On 9 June, 1914, No.602 was flown to the RFC Concentration Camp at Netheravon by Sefton Brancker, who was an indifferent pilot. Of this flight he

The B.E.2c prototype at Farnborough. By the end of July 1914 this aircraft was, it seems, included among a number of B.Es regarded as supporting aircraft for No.4 Squadron. It arrived at the Aircraft Park in a packing case on 20 August and was issued to No.2 Squadron, apparently as 807, on 2 September. Inevitably renumbered 1807 in October, this B.E.2c was despatched back to England on 24 December, 1914. Thereafter it was used for training purposes with No.1 Reserve Aeroplane Squadron and No.15 Squadron, and was eventually struck off charge on 14 December, 1915.

354

recorded:

> 'On this occasion, which was quite a bumpy day, I climbed to about 2,000 ft over Farnborough and then flew the forty odd miles to Netheravon without touching the controls with my hands. During the journey I wrote a full report of the country flown over, somewhat punctuated with dots and dashes when I had met the more violent bumps. On arrival at Netheravon I throttled the engine down, and spiralled and glided to within twenty feet of the ground before I caught hold of the joy stick in order to land'.

Brancker flew the aircraft back to Farnborough later that afternoon, but at noon on 19 June he again left Farnborough on 602, his destination being recorded as Salisbury. It seems likely that he was in fact again visiting the Concentration Camp, for 602 did not return to Farnborough until 26 June, when its pilot was Lt Sheppard, RN. It has been recorded that a number of RFC pilots flew the B.E.2c while it was at the Concentration Camp, and these flights may have been made during its week's absence from Farnborough.

It appears that this first B.E.2c was sent to No.4 Squadron in July 1914 but had evidently left that unit before the end of the month. In the preparations that preceded the RFC's departure to France in August it was dismantled, put into a packing case and sent to France with the Aircraft Park. In its reconstruction as the B.E.2c, No.602 had apparently lost the visible evidence of its true identity, for shortly after being erected at the Aircraft Park it was given the number 807[*] in the only series of numbers of which the operational RFC in France had any control or record. It went to No.2 Squadron on 2 September, 1914, and was renumbered 1807 on 19 October; it remained with the squadron until its return to England on 24 December. Thereafter it was used on training duties by No.1 Reserve Squadron and by No.15 Squadron; its last known flight was made on 11 October, 1915, and it was finally struck off charge on 14 December, 1915.

Another early conversion was that of No.601, a B.E.2a that had seen extensive experimental use at Farnborough. It played a part in Busk's stability experiments of 1913–14, being flown with two 'floating' fins each of 6 sq ft, above the centre-section, somewhat after the fashion of the fixed overwing fins tried on the R.E.1s. The floating fins were, however, pivoted on their leading edges and could be manipulated by the pilot. While fitted with these fins 601 also had an experimental tailplane, known as T.P.2, that was braced from a central kingpost. The results obtained with these fins led to the adoption of an increased dihedral angle on the mainplanes. Later, 601 was given an oleo undercarriage and was used as a testbed for the prototype R.A.F.1 engine. When this B.E.2a was under repair shortly after the start of the war the opportunity was taken to convert it to a B.E.2c. At 4.35 p.m. on 5 November, 1914, 601 left the ground with Edward Busk at the controls; shortly afterwards over Laffan's Plain the aircraft caught fire and was destroyed, claiming the life of its pilot.

By then the B.E.2c was in large-scale production, this having been facilitated by the fact that the R.A.F.'s Drawing Office was able to provide the necessary drawings with little or no delay. Contracts were divided between British and Colonial Aeroplane Company, Vickers, and Armstrong Whitworth, from which

---

[*]The RFC in the Field had begun to number locally acquired aircraft from 801 onwards until it was realised that these numbers duplicated those of the RNAS allocation 801—1600. Following an instruction dated 10 October, 1914, the RFC aircraft in France were renumbered as 1801 *et seq.*

Properly No.1748, the first Vickers-built B.E.2c was one of the earliest named presentation aircraft and is here seen at Farnborough, where it was recorded on 19 December, 1914. It went to France to join No.6 Squadron on 29 January, 1915, with which unit it remained until 12 June, when it went to the Aircraft Park; it was struck off charge on 2 July, 1915.

companies a total of 178 B.E.2cs were initially ordered. The first production aircraft to reach Farnborough was Vickers-built; proudly named *Liverpool* and numbered 1748, it was with the AID at Farnborough by 19 December, 1914. Bristol's first B.E.2c, 1652, was at Farnborough by 4 January, 1915, Armstrong Whitworth's 1780 by 2 February, 1915. The early production aircraft resembled 602 in having 70 hp Renault engines, twin-skid undercarriages and mainplanes of R.A.F.6 aerofoil section.

The first production B.E.2c to go to France was the Bristol-built 1652, which went to No.9 Squadron on 25 January, 1915. Four days later 1748 went to No.6 Squadron, where it was joined by 1780 on 10 February. On 13 February, 1652 was transferred to No.2 Squadron, to which unit 1658 and 1660 went on 1 March. By 1 April, 1915, there were twelve B.E.2cs in France, four each with Squadrons 2, 6 and 16. By the end of 1915, Squadrons 2, 4, 5, 6, 7, 8, 9, 10, 12, 13, 15 and 16 had

B.E.2c No.1748 again, but looking somewhat war-weary after operational use with No.6 Squadron in the Field. (*A. E. Ferko*)

between them a total of 119 B.E.2cs, and six were in reserve at the 1st Aircraft Park.

In March 1915 the first production R.A.F.1a engines began to come forward from the Daimler company, and the engine was adopted as standard for the B.E.2c. It is likely that the first R.A.F.-powered aircraft to go to France was 1687, which went to No.2 Squadron on 11 May, 1915.

Other modifications were made to the B.E.2c airframe as time went on. Having no B.E.2c of its own making available, the Royal Aircraft Factory apparently took the second Vickers-built aircraft, 1749, and adopted it as a testbed for modifications. One of the earliest of these was the V-type undercarriage, which had been installed by 7 January, 1915, and it was fitted with a 90 hp R.A.F.1a engine early in that year. At one time it bore an early camouflage scheme, and by 10 June it was being used in tests of Rouzet wireless equipment. Six days later it underwent climbing trials while carrying a 350 lb bomb, presumably a missile related to the 336 lb R.A.F. bomb. Several other B.E.2cs, including 1688, 2015, 4305 and 4550, were used as versatile work-horses at the R.A.F. during the war.

No.1741 was a Bristol-built B.E.2c that, with nine others, went to France as the original operational equipment of No.12 Squadron. It was damaged in January 1916 but was restored to serviceability by 26 January. It was sent back to England and was fitted with an experimental gun mounting at Farnborough in February. This photograph, dated 6 February, 1916, was taken at Farnborough, and portrays the aircraft armed with two Lewis guns, the rear one on what came to be known as the B.E.2c No.10 Mark I Mounting. No.1741 went back to France, where it was wrecked on 1 March, 1916, and was struck off three days later. (*RAE 004507; Crown copyright*)

The V-type undercarriage tested on 1749 was adopted as standard for the B.E.2c, but a few production aircraft were fitted with the R.A.F.-pattern oleo undercarriage. There can be little doubt that the weight and drag of the oleo installation must have severely reduced the B.E.'s undistinguished performance. A further major modification was the introduction on later production B.E.2cs of wings of R.A.F.14 section. This modification was made simultaneously with the introduction of a revised design of airscrew, and on and after 1 February, 1916, all B.E.2cs despatched to the RFC in France had the new wings and airscrews.

In operational use in France the B.E.2c was used for reconnaissance, artillery spotting and bombing. When used in this last capacity it was usually flown solo to permit a worthwhile load of bombs to be carried. On such duties the aircraft

B.E.2c with oleo undercarriage.

proved to be satisfactory until fighting in the air became commonplace, whereafter its great stability made it almost unmanoeuvrable and rendered it fatally vulnerable to more agile aircraft. This drove B.E. squadrons to exercise their ingenuity in devising installations of defensive armament. At least eight different mountings for Lewis guns were in use at one time or another and are recorded in an official manual; they include the Strange, Medlicott and Albemarle mountings, but doubtless there were other local devices in use. The official manual records an installation of a fixed Vickers gun with an unspecified interrupter gear, mounted on the port upper longeron and having a belt with 250 rounds; but this found no known operational application.

The B.E.2c's greatest combat successes were won in the Home Defence

In Mesopotamia No.30 Squadron converted B.E.2c No.2702 into a single-seater, optimistically regarded as a scout and known in the Squadron as 'Oo-er'. (*RAF Museum*)

squadrons, for whose nocturnal duties the B.E.'s stability proved to be an asset. Some of the Home Defence B.E.2cs were built as or modified to be single-seaters by fairing over the forward cockpit. As early as October 1915 the defences of London included six B.E.2cs, two of which were at Northolt, two at Hainault and two at Sutton's Farm. The spring of 1916 saw increasing use of the B.E.2c by Home Defence units and, on 31 March, 2nd Lt A. de B. Brandon attacked the crippled Zeppelin L.15 near Brentwood, and doubtless contributed to its eventual demise about an hour later. On 3 September, 1916, Lt W. Leefe Robinson of No.39 Squadron, on B.E.2c No.2092, shot down the Schütte-Lanz airship S.L.11 over Cuffley, an action for which he was awarded the VC. Three weeks later, 2nd Lt F. Sowrey, also of No.39 Squadron, shot down the Zeppelin L.32 while flying B.E.2c No.4112, armed with a single Lewis gun firing upwards behind the trailing edge of the centre-section. Shortly afterwards, Brandon, again on a B.E.2c, hastened the destruction of L.33. In combat later that year the Zeppelins L.31, L.34 and L.21 all fell to B.E.2cs, in the last case aircraft of RNAS Great Yarmouth. The B.E.2c's stability greatly aided night-flying, and the type justified its existence by these actions alone

No.2028 was one of several B.E.2cs that were fitted with heavy armour. It was issued to No.6 Squadron from No.2 A.D. on 9 September, 1916, flew back to England from France on 23 March, 1917, and again on 10 July, 1917. By September 1917 it was being used, presumably without armour, by No.19 Training Squadron. (*RAE 1324; Crown copyright*)

An armoured version of the B.E.2c was created by applying slabs of armour plate about the engine and cockpits. The RFC was surprisingly slow to accept these armoured aircraft, of which at least 15 were made. These followed the armouring of the fuselage underside of No.2030 of No.8 Squadron, which was in use in July 1915 and was well reported on; nevertheless, the Battle of the Somme was well under way a year later when rather tentative use of the more extensively armoured B.E.s began. By early October 1916 only five were operational in France, and a request made on 1 November by Brig-Gen C. A. H. Longcroft, commanding the 5th Brigade, for three to be supplied for each squadron was never acted upon. The handful of Heavy Armour B.E.2cs remained in use throughout the winter of 1916–17, but by mid-March 1917 only two were still on the strength of the 5th Brigade.

The many modifications made to or experimentally tested on B.E.2cs were legion, but they included rotating interplane strut fairings that acted as air-brakes (January 1916), some valuable early work with Guardian Angel parachutes and with superchargers, and, strangest of all, a centre-float seaplane variant used at the School of Aerial Gunnery at Loch Doon in Ayrshire. This aircraft, No.4721, was officially allotted to Loch Doon on 22 November, 1916, at which time it was despatched first to the Isle of Wight to be fitted with its float undercarriage by S. E. Saunders of Cowes.

Possibly the only B.E.2c to operate as a floatplane, No.4721 was used in this remarkable form at the School of Aerial Gunnery, Loch Doon. The float installation was made and fitted by S. E. Saunders Ltd.

The directional controllability of the B.E.2c left something to be desired, and a tendency to spin in certain circumstances was noticeable. To counteract this an enlarged fin with curved leading edge, 8 sq ft in area, was designed. Early in January 1916, No.2026 went to France fitted with the new fin; it was allotted to No.12 Squadron for evaluation. By March it was agreed that the new fin was an improvement and it was adopted as standard for the B.E.2e. Later deliveries of the B.E.2c and 2d had the new fin, but not until 16 January, 1917, was an instruction given that all still in service were to be given the enlarged fin.

In view of all the obloquy that has been heaped upon the B.E.2c (largely by people who had nothing to do with it) it is pertinent to observe that in mid-1916 official statistics for the second quarter of the year indicated that the B.E.2c was the most durable of the eight main types then in service. The monthly wastage rate of B.E.2cs was, at 17 per cent, the lowest of all;* of the total of eighty-three B.E.2cs struck off, eleven had fallen to hostile action, whereas thirteen F.E.2bs

*Corresponding figures for the other seven types were F.E.2b, 22 per cent; Bristol Scout, 50 per cent; D.H.2, 30 per cent; Morane biplane, 28 per cent; Morane Parasol, 38 per cent; Morane Scout, 33 per cent; Martinsyde Elephant, 22 per cent.

A night-flying B.E.2c of No.50 (H.D.) Squadron, fitted with navigation lights and flare brackets, and carrying the squadron's skull-and-crossbones marking under each upper mainplane.

had been lost in this way. The RFC's Command was therefore entirely justified in retaining faith in the B.E.2c, for the Fokker monoplanes had at that time been operational for a full year and were now likely to be contained by the D.H.2 and F.E.2b.

Such considerations render more reasonable the RFC's first thoughts on the use of the excellent new 150 hp Hispano-Suiza engine when it became available in 1916. The earliest RFC installation of this engine was made in a B.E.2c (No.4133) at No.1 A.D. St-Omer, and its test flying began on 13 March, 1916: this was in the light of an official intention, never realised, to employ the RFC's first fifty Hispano-Suizas as B.E.2c power units.

The Royal Aircraft Factory created two developments of the B.E.2c—the B.E.2d and B.E.2e. The B.E.2d was externally very similar to the B.E.2c but was fitted with dual controls and had a revised fuel system that incorporated an

B.E.2c A8948 was assembled from spares by No.18 Reserve Squadron, Montrose, and bore the inscription *Montrose No.8*. It had the enlarged fin that became standard for all B.Es, and when photographed had a camera mounted on the starboard side at the rear cockpit.

An early and very clumsy installation of a Hispano-Suiza engine in a B.E.2c, photographed at Farnborough. (*RAE 1306; Crown copyright*)

A somewhat neater Hispano-Suiza installation in B.E.2c No.2599 at Farnborough. The photograph bears the date 13 December, 1916. (*RAE 1967; Crown copyright*)

Although No.4451 had the underwing gravity tank that normally denoted a B.E.2d it is doubtful whether it had the full B.E.2d fuel system. (*T. Heffernan*)

external gravity tank under the port upper wing and linked to a tank in the fuselage top decking immediately behind the front cockpit. In August and October 1915, four batches of B.E.2ds were ordered, two from the British and Colonial Aeroplane Co and one each from Vulcan and Ruston, Proctor. Considerable numbers of these were modified, either in production or subsequently, to have the unequal-span mainplanes that were introduced with the B.E.2e.

In fact the prototype B.E.2e was flying before any production B.E.2d became available. This prototype was created by modifying the Bristol-built B.E.2c 4111, which had acquired its unequal-span wings and new tailplane by 14 February, 1916, and was test-flown on 18 February. The initial results were regarded as promising, and on 1 March Mervyn O'Gorman, the S.R.A.F., wrote to M.A.2 to say that an average of 97 mph low down had been obtained on two speed tests, with an indicated speed of about 90 mph at 6,000 ft. Three days later he informed

Production B.E.2d, still with the original triangular fin. This Bristol-built aircraft was officially allotted to the RFC in France on 30 May, 1916, and was subsequently used by No.2 Squadron. (*E. F. Cheesman*)

B.E.2e No.7216, Bristol-built, was initially allotted to the E.F. on 11 November, 1916, but a revised allotment of 17 November allocated it to the Home Defence Wing. For these duties it was fitted with navigation lights, flare brackets, an R.L. Tube and, most conspicuously, launching tubes for Le Prieur rockets. (*R. C. Bowyer*)

the ADMA that the B.E.2e would be 2 mph faster than the B.E.2d at ground level, and probably 3 mph faster at 8,000 ft, with a 7 per cent to 10 per cent improvement in climbing performance, while retaining the same factor of safety, offering 'very much better lateral controllability', and landing 'substantially more easily'. These were estimated figures, comparing the B.E.2d and 2e each fitted with the 105 hp R.A.F.1b engine. Tests conducted at CFS in May 1916 with aircraft having the standard 90 hp R.A.F.1a seemed to confirm that the B.E.2e did indeed have an appreciably better performance than the 2d. Brancker was obviously impressed by the B.E.2e's performance. Late in March 1916, in an undated note to Trenchard, he wrote:

B.E.2e No.5858, also Bristol-built, saw service in 1916 with No.34 Squadron, and in October 1918 was in use at the Wireless and Observers School, Winchester. (*RAF Museum*)

'My dear Trenchard,
This should do you—B.E.2e at C.F.S. has done 90 over a measured course at 6,000 feet, and they hope to do better.
Yours W.S.B.'

As far as the R.A.F.1b engine was concerned, it was officially decided early in April 1916 to fit this engine only to the B.E.2e. In fact only 115 R.A.F.1b engines were produced and the R.A.F.1a remained the standard power unit.

The first Bristol-built B.E.2ds of the batch 5730—5879 were delivered in May 1916, but it is known that by 27 May, 5733, 5734, 5738 and 5739 had all been fitted with B.E.2e wings. Most were delivered as 2ds, but considerable numbers were modified in service to have the 2e mainplanes and tailplane.

Surviving Farnborough drawings indicate that the B.E.2e was, like the B.E.2d, designed to have an external gravity tank under the port upper mainplane, and dual control. In service, the B.E.2d's gravity tank was frequently dispensed with, presumably in the interests of performance: its removal added 11 mph to the

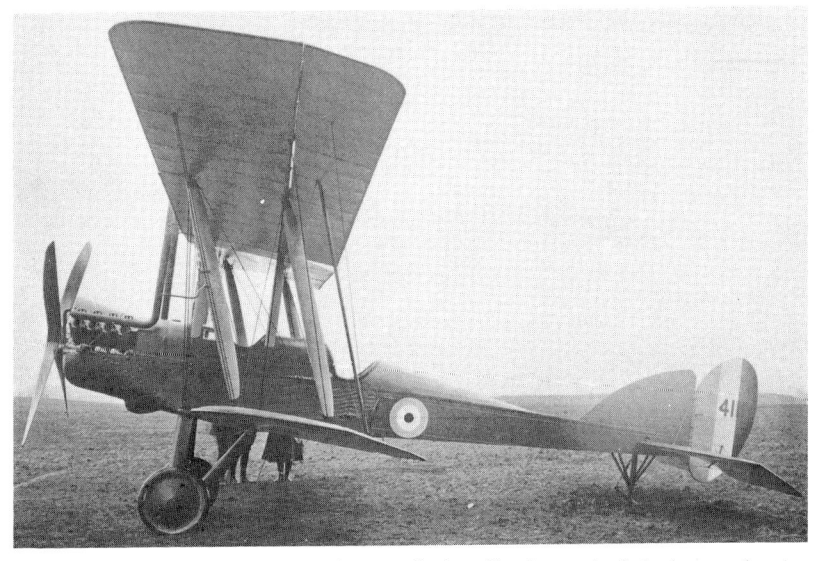

The prototype B.E.2e, No.4111, photographed at Farnborough. It had no underwing gravity tank when photographed.

speed at 6,000 ft and greatly improved climbing performance. On the other hand, the B.E.2e was, initially at least, regarded as a long-range aircraft and official preference was for the retention of the gravity tank on operational B.E.2es. It seems likely that in some circumstances the dual control was removed from the front cockpit. As already noted it was with the B.E.2e that the large curved fin that was eventually standardised for all B.E. aircraft began to appear in quantity.

The B.E.2ds were, with few exceptions, used in France and in the training squadrons at home, but the B.E.2e configuration was to be seen on every front. Evidently the 2e's slightly better performance prompted the production of the type on a larger scale than applied to any other B.E. variant. Many left the factory with the 2e wings and large numbers of 2cs and 2ds were modified to have the unequal-span cellule.

All such aircraft were in Britain described as the B.E.2e and allotted as such to the RFC in France. Soon after their arrival there it was found that all B.E.2es were not identical: those converted from B.E.2cs had two petrol tanks, a main pressure tank holding 18 gallons under the observer's seat and an auxiliary gravity tank of 14¾ gallons: the aircraft converted from B.E.2ds or built as B.E.2es with B.E.2d fuselages had a main pressure tank of 19 gallons above and behind the engine, an auxiliary gravity tank of 10 gallons in the decking between the cockpits, and a service gravity tank of 12 gallons under the port upper wing. The different sizes and dispositions of tanks and the provision of dual controls in B.E.2d fuselages led immediately to spares and maintenance problems, and it became essential to differentiate between what were, in effect, two different forms of B.E.2e. This was resolved by designating as B.E.2f those aircraft that had the B.E.2c fuselage, and as B.E.2g those that had the B.E.2d fuselage. These designations were decreed in an instruction dated 2 October, 1916, sent by RFC Headquarters to Nos.1 and 2 Aircraft Depots. Thus, aircraft that had left England

Installation of a 150 hp Hispano-Suiza engine in a B.E.2e at Farnborough.

for France as B.E.2es were usually returned under the designations of B.E.2f or 2g as appropriate.

The B.E.2e configuration was adopted as standard and production of nominal B.E.2es far exceeded that of B.E.2cs and 2ds. Deliveries of B.E.2es to France did not begin until late in June 1916, when 5739, 5814 and 5817 were allotted to the RFC with the Expeditionary Force. Others had evidently gone to No.34 Squadron, which went to France on 15 July, 1916, equipped throughout with B.E.2es. Allocations in growing quantities did not start until August, but thereafter virtually all allotments of B.E. two-seaters to France were of B.E.2es and these continued well into 1917. By that time aerial fighting on the Western Front had advanced markedly from the Fokker era, for the RFC was opposed by the redoubtable Albatros and Halberstadt single-seaters, and the B.E.2e was no match for these fast and well-armed fighters. RFC casualty reports for April 1917

B.E.2e 5836 would be regarded as a B.E.2g by the RFC in France. It was first allotted to the E.F. while at Filton (it was Bristol-built) on 26 June, 1916, and was allotted to No.34 Squadron, whose markings it bears in this photograph. It was reported missing on 29 August, 1916; its crew, 2/Lts D. S. Cairns and K. E. Tulloch, became prisoners of war.
(*Egon Krueger*)

('Bloody April') mention four B.E.2cs, eight B.E.2ds, and no fewer than forty-eight B.E.2es, fs and gs.

Losses on this scale were not the fault of RFC Headquarters. As early as July 1916, HQ had proposed to make good shortfalls in B.E. deliveries by employing instead Armstrong Whitworth F.K.8s and Vickers F.B.14s. Later that year, too, the R.E.8 was looked to as a B.E. replacement. However, F.K.8s were slow to appear, the Vickers F.B.14 never saw operational use, and the early R.E.8s were beset with difficulties. Thus the B.E.s, mostly 2es, had to battle on well into obsoleteness. The B.E.2e fared no better than the 2c and 2d for in its time it was

The Experimental Station at Orfordness did much work on the camouflaging of aircraft. One of its early experimental schemes is here seen on the B.E.2e A8636 which, with the similarly painted B4520, was allotted to the RFC in France on 18 February, 1918. Unfortunately A8636 crashed while being flown to Lympne and was replaced on 20 February by the B.E.12a A562, which had also been camouflaged in the Orfordness scheme. A8636 had been built by the British and Colonial Aeroplane Co and on 7 May, 1917, while still at Filton, was allotted to the RFC in France. This was amended on 5 June, when it was allocated to the Training Brigade. Following a subsequent re-allocation to the Controller of the Technical Department it was transferred to Orfordness later in 1917. (*C. H. Shelton, from Peter Liddle's 1914–18 Personal Experience Archives, presently housed within Sunderland Polytechnic.*)

An example of the little-known night-bombing B.E.2e was the Bristol-built No.5844. As a B.E.2d it was at Farnborough allotted to the E.F. on 4 July, 1916, and was issued to No.4 Squadron from No.2 A.D. on 4 September, 1916. At some stage it became a B.E.2g with the fitting of B.E.2e wings, and on 2 April, 1917, it was at the Southern Aircraft Repair Depot allotted to the Director of Aircraft Equipment, but this was amended and 5844 again went to France, where it was used by No.100 Squadron as a night bomber. In the photograph it is carrying a 230-lb bomb under the fuselage, with sundry flares in other racks.

more sternly opposed, was no more manoeuvrable and was no more effectively armed. Its normal armament was a single Lewis gun, fired with difficulty by the front-seat observer, for which were provided assorted mountings. Wolseley-built B.E.2es of the batch A3049—3148 were leaving the factory with a Strange mounting with toothed arc fitted as a No.4 Mark I installation between the cockpits for rearward firing, and two outrigged spigot mountings, one on each forward centre-section strut.

There were some armoured B.E.2es late in 1916 and early 1917. Known examples are 2122 (evidently a B.E.2f conversion from 2c), 2567, 2568, 5786, 5801, 6739, and A2757.

The B.E.2e saw operational use in Palestine, Africa, Aden, Macedonia, India, and on Home Defence, apart from its long agony on the Western Front. Nowhere did it register any marked success, possibly its nearest approach to spectacle being the gallant attempts made by 2nd Lt W. W. Cook of No.76 (Home Defence) Squadron to close with the Zeppelins L.55 and L.41 successively in the course of a single patrol in the early hours of 25 September, 1917. He was frustrated by his B.E.2e's inadequate performance.

Until the war ended the B.E.2e remained in service, mostly as a ubiquitous trainer, in which form it frequently discarded its underwing gravity tank. One or two of those used at Schools of Aerial Fighting had a fixed synchronised Vickers gun firing through the propeller, while some of the 2es used by Home Defence squadrons were armed with Le Prieur rockets carried in tubes attached to the interplane struts.

Latterly in France some B.E.2es were flown solo as night bombers, notably by Nos.10 and 100 Squadrons. With the latter unit the B.E.2e was still operational late in August 1917: A3059 was struck off charge on 25 August after a forced landing following a night raid on which it had carried a 230 lb bomb, presumably on its Michelin bomb rack, and it also had two 20 lb bomb racks.

B.E.2c, 2d, 2e, 2f and 2g
70 hp Renault, 90 hp R.A.F.1a, 105 hp R.A.F.1b (experimental), 150 hp Hispano-Suiza (experimental)

Two-seat reconnaissance or artillery-observation biplane; single-seat bomber or Home Defence fighter; two-seat trainer.

B.E.2c and 2d: Span 36 ft 10 in; length 27 ft 3 in; height 11 ft 4 in; wing area 396 sq ft.
B.E.2e: Span 40 ft 9 in (upper), 30 ft 6 in (lower); length 27 ft 3 in; height 12 ft; wing area 360 sq ft.

Empty weight 1,370 lb (B.E.2c with R.A.F.1a), 1,375 lb (B.E.2d), 1,431 lb (B.E.2e); loaded weight 2,142 lb (B.E.2c with R.A.F.1a), 2,120 lb (B.E.2d), 2,100 lb (B.E.2e).

B.E.2c with 70 hp Renault: Maximum speed at sea level 75 mph.

B.E.2c with 90 hp R.A.F.1a: Maximum speed at 6,500 ft—72 mph, at 10,000 ft—69 mph; climb to 6,500 ft—20 min, to 10,000 ft—45 min 15 sec; service ceiling 10,000 ft; endurance 3¼ hr.

B.E.2d: Maximum speed at sea level 88·5 mph, at 6,500 ft—75 mph, at 10,000 ft—71 mph; climb to 6,500 ft—17 min 35 sec, to 10,000 ft—33 min 40 sec; service ceiling 12,000 ft; endurance 5½ hr.

B.E.2e: Maximum speed at sea level 90 mph, at 6,500 ft—82 mph, at 10,000 ft—75 mph; climb to 6,500 ft—23 min 50 sec, to 10,000 ft—53 min; service ceiling 9,000 ft; endurance 4 hr.

Armament: On early B.E.2cs one rifle or carbine, fired by observer; later from one to four 0·303-in Lewis machine-guns were carried on a wide range of mountings. When flown solo as a bomber the B.E. with R.A.F.1a engine could carry one 230 lb bomb or two 112 lb bombs or equivalent. Some Home Defence B.E.2cs and 2es were armed with Le Prieur rockets carried on the interplane struts, and experiments with Fiery Grapnels were conducted. A few B.E.2es at Schools of Aerial Fighting had a fixed, synchronised Vickers gun.

Manufacturers: Armstrong Whitworth; British and Colonial; Daimler; Ruston, Proctor; Vickers; Vulcan; G. & J. Weir (some assembly by Barclay, Curle; British Caudron; Napier & Miller; William Denny); Wolseley Motors.

Service use: **B.E.2c**:*Western Front*: Sqns 2, 4, 5, 6, 7, 8, 9, 10, 12, 13, 15, 16, 19, 21 and 25; HQ Communications Flight. *Home Defence*: Sqns 33, 34, 39, 50, 51, 76 and 77; Flights at Chelmsford, Cramlington, Farnborough, Hounslow, Northolt, Hainault, Hendon, Sutton's Farm and Farmingham. *Macedonia*: No.17 Sqn. *Arabia*: No.14 Sqn. *Palestine*: No.67 (Australian) Sqn. *Mesopotamia*: No.30 Sqn. *India*: No.31 Sqn. *East Africa*: No.26 Sqn. *Training*: Reserve/Training Sqns 1, 2, 4, 5, 8, 9, 10, 11, 13, 15, 16, 17, 18, 19, 20, 22, 23, 24, 31, 35, 36, 44, 50, 51, 52, 57, 58, 59, 64, 66, 69, 189, 190, 194, 195, 197, 198. Squadrons working up—Nos.14, 15, 22, 23, 24, 38, 43, 49, 54, 56, 57 and 83. Central Flying School; Wireless School, Brooklands; Schools of Aeronautics at Oxford and Reading; School of Aerial Gunnery, Loch Doon; No.2 Auxiliary School of Aerial Gunnery, Turnberry; Machine-Gun School, Hythe; Artillery Observation School, Almaza.
B.E.2d: *Western Front*: Sqns 2, 4, 5, 6, 7, 8, 9, 10, 12, 13, 15, 16 and 42; HQ Communications Flight. *Home Defence*: No.77 Sqn. *Training*: Training Sqns 1, 6, 13, 17, 34, 35, 49, 51, 58 and 191; Sqns working up, Nos.43 and 49; Wireless School, Brooklands.
B.E.2e, 2f, 2g: *Western Front*: Sqns 2, 4, 5, 6, 7, 8, 9, 10, 12, 13, 15, 16, 19, 25, 34, 42, 52, 53 and 100. Aerial Musketry Range, Berck-sur-mer. *Home Defence*: Sqns 33, 36, 37, 38, 39, 44, 50, 75, 76, 77 and 78. *Macedonia*: No.17 Sqn. *Palestine*: Sqns 67 (Australian) and 111. *Arabia*: No.14 Sqn. *East Africa*: No.26 Sqn. *Training*: Training Sqns 6, 8, 9, 11, 13, 15, 16, 17, 18, 19, 20, 23, 24, 25, 26, 31, 35, 36, 38, 42, 44, 46, 49, 50, 51, 52, 53, 57, 58, 59, 60, 61, 64, 65, 66, 187, 189, 190, 191, 194, 195, 198; Training Depot Stations Nos.1, 2 and 5. Squadrons working up—Nos.47, 49, 56, 62, 82, 83, 104 and 110. Wireless and Observers Schools, Brooklands and Winchester; No.1 School of Aerial Navigation and Bomb-dropping, Stonehenge; No.2 Wireless School, Penshurst; Artillery Observation School, Almaza; Aerial Observers School, Heliopolis.

Serials: *Built under production contracts—* **B.E.2c**:* 601, 602, 1652—1697, 1698—1747, 1748—1779, 1780—1800, 1807 (ex 602), 2001—2009, 2010—2029, 2030—2129, 2470—2569, 2570—2669, 2670—2769, 4070—4219, 4300—4599, 4700—4709, 4710—4725, 5384—5403, 5413—5441 and 5616.
B.E.2d:* 5730—5879, 6228—6327, 6728—6827, 7058—7257, 7321—7345, A1792—1891.
B.E.2e: A1261—1310, A1311—1360, A1361—1410, A2733—2982, A3049—3148, A3149—3168, A8626—8725, B3651—3750, B4401—4600, B6151—6200, C1701—1750, C6901—7000, C7001—7100, C7101—7200.
Rebuilt aircraft: **B.E.2c**: A3269, A4816, A4817, A5207, A5208, A5488, A5497—5499, A8896, A8897, A8948, A8949, A9973, A9974, B386, B387, B705, B707, B708, B771, B3951, B3978, B3979, B3993, B3995, B4009, B4020, B9453, B9459, B9993, F9627.
B.E.2d: B3954, B8825, C9988, C9989, C9995, F4218, F9571, H8256.
B.E.2e: B702, B706, B709, B710, B712, B717, B719, B722—725, B728, B748, B752, B770, B772, B790, B797, B3970, B4004, B4006, B4009, B4010, B4022, B4023, B4026, B8787, B8794, B8816, B8817, B8820, B8828, B8831, B8835, B8849, B8853, B8854, B8863, B9444, B9454, B9457, B9469, B9474, B9938, B9940, B9941, B9943, B9957, B9998, C9987, C9996, E9985, F9567.

* Many conversions to B.E.2e (B.E.2f and 2g).

B.E.3 and B.E.4

The B.E.3 was first sketched out in December 1911 as a straightforward version of the B.E.1/2 in which a rotary engine was to replace the Renault. Its mainplanes were to be of unequal span and rigged without stagger; the rudder was an upright ear-shaped surface hinged to a conventional rudder post. Design work was under the leadership of John Kenworthy, and the biplane that was eventually built differed somewhat from the original rough layout.

Two aircraft were built to the design, and were numbered serially, not as types, B.E.3 and B.E.4, for they followed B.E.1 and B.E.2. At the time the Army Aircraft Factory was still without authority to build aircraft to its own designs so,

The B.E.3, No.203, photographed on a visit to Hendon, possibly on 21 September, 1912.
(RAF Museum)

as with B.E.1, B.E.2, and S.E.1, the two new biplanes were theoretically reconstructions of, respectively, the Paulhan biplane F2 and a Bristol Boxkite. The new rotary B.E.s had mainplanes that were basically similar to those of the B.E.2 but were rigged with heavy stagger. A hooded cowling covered the 70 hp Gnome engine, giving a deep forward fuselage; behind the cockpits the fuselage tapered to a point and was overhung aft by the one-piece full-span elevator. The rudder was roughly of B.E. shape but was mounted with its major axis horizontal on a single unbraced steel tube with a small proportion of its area ahead of the axis of rotation to serve as a balance.

On the morning of Friday, 3 May, 1912, Geoffrey de Havilland took off in B.E.3 for the first time, at 10.30 a.m. Evidently he was well satisfied with it, for he proceeded to give flights to a succession of passengers (F. M. Green, Lt A. G. Fox, F. T. Hearle, P. W. L. Broke-Smith and Col Cooke). He flew the aircraft again on 6 May, breaking two undercarriage struts, and again four days later with B.E.2 skids and front struts fitted.

Within a few days of its initial flights, B.E.3 was taken over by No.3 Squadron and made its first subsequent flight on 13 May, 1912, piloted by Lt A. G. Fox. He

B.E.4 photographed at Farnborough.

felt that the B.E. did not lift well with a passenger, but the aircraft subsequently made a fair number of flights with passengers. On one of these, made on 22 June, Sapper Lewis tested signalling flags, and in that month B.E.3 flew a total of 15 hr 36 min. This time was in fact put in only from 19 June, for earlier in the month the aircraft had been back at Farnborough, presumably for repairs. It returned thither yet again between 7 July and 26 August, and it seems that on this occasion its flying controls were modified from what may have been an unconventional arrangement. A note in a surviving Farnborough document states that the B.E. was 'Handed over to R.A.F. to have controls altered. Footwarp taken off and ordinary controls fitted.' At about this time a wireless installation was made in the aircraft. On 28 August, Maj R. Brooke-Popham flew B.E.3, now renumbered 203, to Larkhill after an abortive attempt on the previous day, but two days later it was flown back by Fox to have a new petrol tank fitted and an engine change.

During August 1912 a dual-control installation was designed for the B.E.3/4 but it is not known whether this was installed in B.E.3 at that time or later.

B.E.4, numbered 204, having its airscrew swung. (*Flight International 0163*)

B.E.3 seems to have been a hard-working aircraft, for it had to be repaired on several subsequent occasions, the repairs needed in March 1913 involving extensive rebuilding. During this time 203 was on the strength of No.3 Squadron, and in January 1913 was used in inconclusive experiments in visual signalling, using Véry lights and a small 2-c.p. electric lamp that proved to be 'invisible at 300 yards on a dull day'.

B.E.4 made its first flight on Monday, 24 June, 1912, piloted by Geoffrey de Havilland. Its engine was a somewhat erratic 50 hp Gnome that was wont to miss. Apparently this aircraft was handed over to the RFC on 8 August, but its refractory engine brought about a crash near Basingstoke, and this put B.E.4 out of action until 30 August. By now numbered 204, it came back to Farnborough on 3 September to have its engine replaced by a 70 hp Gnome. It did a considerable amount of flying with No.3 Squadron and underwent various minor modifica-

One of the later B.E.4s at Farnborough. (*RAE*)

372

B.E.4 No.416 of Central Flying School after overturning on a downwind landing. Its pilot was Lt Arthur.

After the fatal crash of 204 on 11 March, 1914, the B.E.4 416 was given a new vertical tail assembly of outstanding ugliness.

tions, including a 4-in increase in stagger. By April 1913 it had been fitted with an 80 hp Gnome engine.

At least two and possibly three further rotary-powered B.E.s with 50 hp Gnome engines were built. Two, numbered 416 and 417, were built late in 1912 and, being fitted with dual control, were delivered to CFS in December; another imperfectly identified B.E. with the 50 hp Gnome was numbered 303, and may well have been the 'new B.E. type staggered plane machine' that Geoffrey de Havilland flew at Farnborough on 14 January, 1913. As No.303 it was taken on charge by No.4 Squadron on 20 January, 1913, and was flying at Farnborough with that identity on 25 January. Its career seems to have been brief: after only 5 hr 49 min flying time it was wrecked on 10 February and seems not to have been restored. Nos. 416 and 417, however, gave extended service at CFS.

All of these B.E.s, as originally built, had the same form of rudder, pivoting on a single, unbraced length of 20-gauge steel tubing; there was neither fin nor fixed rudder-post. As a control surface this was apparently satisfactory, but the absence of bracing or fin surface had tragic consequences on 11 March, 1914, when the

To the right of this photograph stands B.E.4 No.417, at Central Flying School, with its original tail unit.

rudder axis tube of No.204 broke over Netheravon and Capt C. R. W. Allen and Lt J. E. G. Burroughs were killed in the ensuing crash. The findings of the subsequent enquiry were not conclusive, but the disaster may have been one of the first air accidents to be caused by metal fatigue.

On 14 March the Adjutant of the Military Wing instructed No.3 Squadron: 'Please note that no flying is to take place on B.E.3 type machines pending the receipt of further instructions.' A revised form of vertical tail assembly was hurriedly designed, apparently by adaptation from that of the H.R.E.2; this had been fitted to 203 by 27 March, and 416 and 417 were similarly modified. A triangular fin was mounted above the tailplane with a B.E.-form rudder hinged to it, the rudder's major axis being upright. The resulting combination was unbelievably ugly. In this form 416 was still in use at CFS in the summer of 1914, but it is doubtful whether 417 survived the extensive damage it suffered in a crash when Sgt Rigby sideslipped into the ground from 50 ft while doing straights.

No.203 was last reported to be lying within the Royal Aircraft Factory, packed in a transit case, up to 23 November, 1914. It is not known whether it ever flew with its modified tail unit, and it apparently ended its days as a ground instructional airframe.

B.E.4 No.417, still at CFS, with its revised tail unit. (*RAF Museum*)

B.E.3 and B.E.4
50, 70 or 80 hp Gnome

Two-seat military biplane.
Span 39 ft 6 in.
Manufacturer: Royal Aircraft Factory.
Service use: Sqns 3 and 4; CFS.

Serials: 203 (previously B.E.3), 204 (previously B.E.4), 303, 416, 417.

B.E.7

In July 1912 John Kenworthy drew up the design of a development of the B.E.3/4 to be powered by the 140 hp fourteen-cylinder two-row Gnome rotary engine. The new design was evidently the seventh in the B.E. series, for it was numbered B.E.7. Its structural geometry and dimensions were those of the B.E.3/4, but the forward fuselage was shortened by one bay to compensate for the heavier engine, and the undercarriage skids were correspondingly shorter.

The B.E.7 was completed early in 1913. On 28 February it was tested by Geoffrey de Havilland, who made several circuits in it and flew it over Camberley for 20 minutes at 1,000 ft. Later that day it was flown from Farnborough to CFS by Maj E. L. Gerrard, having at that time been allotted the serial number 408. That number had earlier belonged to a Bristol Boxkite (originally F7), a fact that might prompt one to speculate whether B.E.7 might have been created as yet another of the R.A.F's 'reconstructions'; however, as indicated below, the aircraft did not retain that number for long.

On 2 May, 1913, Maj Gerrard established a new British altitude record when he flew the B.E.7, with two passengers, Air Mechanics Sharp and McDonald, to a height of 8,400 ft. By that time the aircraft had been renumbered 438: perhaps the earlier application of 408 had been simply a painter's error.

In June the aircraft was returned to the Royal Aircraft Factory and was reconstructed. When its reconstruction was complete it was flown from Farnborough to Upavon on 5 August, 1913, again by Maj Gerrard; thereafter it flew regularly at CFS throughout the remainder of the summer and autumn of 1913.

There can be no doubt that the RFC's Military Wing regarded the B.E.7 as a successful and satisfactory aircraft, for on 19 May, 1913, the Officer Commanding the Military Wing wrote to the Secretary of the War Office:

'Sir,
I have the honour to request that authority be given to the Superintendent, Royal Aircraft Factory, to design and construct a B.E.3 type two-seater biplane fitted with a 160 hp Gnome engine.

My reasons for this recommendation are as follows:

One of the most important lines of development which should in my opinion be pursued, is towards machines of large radius of action for strategical reconnaissance. Such machines should, in order to enable them to evade hostile aircraft, be possessed of the highest practicable speed; both these requirements involve the use of an engine of high power.

The 140 hp Gnome B.E. machine has now been some time at the Central Flying School, and is understood to be found very satisfactory. I understand

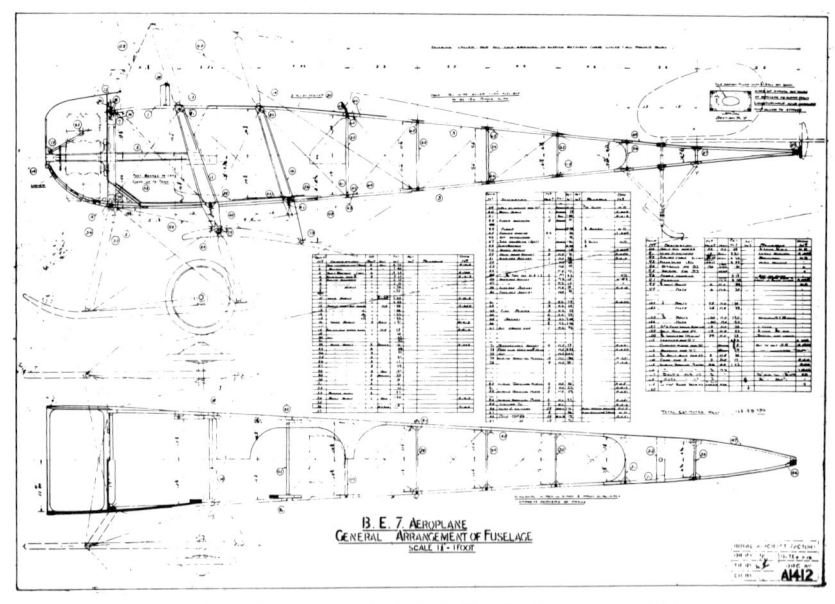

Fuselage drawings of the B.E.7 (*RAE; Crown copyright*)

that the 160 hp Gnome engine is proving a successful type and it is not very much heavier than the 140 hp engine.

A further argument in favour of the policy of providing a few fast high-power machines in the military wing is the necessity of training officers to handle such machines.

Fourteen-cylinder Gnome engines are a very suitable type for the purpose, as the machines fitted with them can be treated for landing purposes as twin-engine machines, landings being made at comparatively low speed with only 7 cylinders in use. With this point in view, it is very necessary that the controlling surfaces should be of sufficient area to enable the machine to be controlled at comparatively low speed.'

It seems that this approach proved fruitless, but the R.A.F. went on to make spectacular use of the 160 hp Gnome as the power unit of the brilliant S.E.4 single-seater of 1914.

B.E.7
140 hp Gnome

Two-seat military biplane.
Dimensions and weights not known.
Armament: None.

Manufacturer: Royal Aircraft Factory.

Service use: CFS, Upavon.

Serial: 408, later renumbered 438.

B.E.8 and B.E.8a

Last of the rotary-powered B.E.s was the B.E.8, known to the RFC as The Bloater. Two were built at the Royal Aircraft Factory in 1913, but how far they can be regarded as having been prototypes is largely a matter for conjecture, for the type was apparently ordered off the drawing board. Both had the 70 hp Gnome engine; the first was handed over to the Flying Department at the R.A.F. on 20 August, 1913, the second on 8 September. The B.E.8 bore a strong family resemblance to the B.E.3, 4 and 7, but had its fuselage attached directly to the lower wings instead of in mid-gap, and its rudder was similar in its shape and mounting to that of the B.E.2. Lateral control was by warping.

A decision to build some numbers of the type seems to have been reached before the first prototype was passed to the Flying Department at Farnborough. There seems little doubt that the first orders were for four B.E.8s under Contract No.A/2238 dated as early as 18 August, 1913: two were to be built by Vickers, against delivery dates of 17 November, 1913, and 25 May, 1914, the other two by The British and Colonial Aeroplane Co, with delivery dates of 8 and 22 January, 1914. One surviving Vickers document suggests that the company built fifteen B.E.8s against contracts up to about mid-January 1914, whereas in fact official sources indicate that the total was no more than eleven. It would seem that the compiler of the Vickers listing may have included repair contracts with construction orders.

Up to 31 March, 1914, orders were placed for B.E.8s as follows:
The British and Colonial Aeroplane Co   6
The Coventry Ordnance Works   7
Vickers Ltd   11

It seems that the Coventry Ordnance Works order was reduced to six, and it is known that the Vickers total included two B.E.8s destined for India, presumably

One of the B.E.8 prototypes in original condition with finless tail unit, undivided cockpit and 70 hp Gnome engine.

Bristol-built B.E.8 No.373 about to take off at Farnborough, on 30 July, 1914. It went to Belgium with No.6 Squadron on 7 October, 1914, but was wrecked on 17 October and was apparently struck off charge then. (*RAE 465; Crown copyright*)

for the Central Flying School that was being prepared at Sitapur for 1914.

The six B.E.8s ordered from The British and Colonial Aeroplane Co were built with the Bristol works sequence numbers 201, 202 and 208–211. The first production B.E.8s to be delivered were the two Bristol-built aircraft 201 and 202; in the RFC they became 365 and 373, the former being taken on charge on 10 March, 1914, the latter on 28 March. The first Vickers-built B.E.8, No.377, was taken on strength on 24 April, 1914.

Production B.E.8s differed from the two prototypes in having separated cockpits for the two occupants, a triangular fin in the tail unit, and four-blade airscrews. The Bristol-built and Vickers-built aircraft were preceded in service by several months by two B.E.8s numbered 423 and 424 that evidently appeared in November 1913 and went to CFS. These were built by the Royal Aircraft Factory and may well have been the two prototypes brought up to production standard and fitted with the 80 hp Gnomes from the Short tractor biplanes that had been the original but short-lived recipients of the serial numbers 423 and 424.

Further deliveries were just coming forward when the war began. These included the two aircraft intended for India, but the war led to the immediate abandonment of the planned Central Flying School at Sitapur, and the Indian B.E.8s were taken on charge by the RFC as 624 and 625. Both were at Farnborough early in August 1914, and it appears that realised production from the Vickers factories consisted of seven B.E.8s built at Erith between April and September 1914 and only four at Dartford between August 1914 and February 1915. If this is correct, it would seem that several other B.E.8s were built by other contractors. These may have included the Coventry Ordnance Works, which are known to have been the contractors for 2131—2132.

The RFC made little operational use of the B.E.8. Three went to France with the Aircraft Park on 14 August, 1914; these were 377, 391 and 625. The first two went to No.5 Sqn, and 625 was allocated to No.3 Sqn. This last B.E.8 had left Farnborough on 11 August piloted by Lt G. I. Carmichael (the late Group Capt Carmichael, DSO), who was later to describe the B.E.8 as 'a short-lived nasty contraption with insufficient fin surface'. Short-lived 625 certainly was, for it crashed at Amiens on 16 August, killing Lt E. W. Copland Perry and his mechanic H. E. Parfitt. Perry had been a test pilot at Farnborough. Two days later, on 18 August, 391 crashed at Péronne and Corporal F. Geard died; his pilot, 2nd Lt

R. R. Smith-Barry, was injured but recovered, later to command No.60 Sqn and create the School of Special Flying at Gosport.

No.3 Sqn received 479 on 1 September, 1914, as a replacement for the wrecked 625, but it was reported missing only a week later. No.5 Sqn had returned 377 to the Aircraft Park on 3 September, and this B.E.8 was struck off charge on 30 September. Thereafter the only B.E.8 in France was 636 which No.6 Sqn had taken to France on 8 October; it survived until 16 January, 1915. No.1 Sqn arrived on 7 March, 1915, bringing with it four B.E.8s: these were 399, 740, 2131 and 2132, which mounted the B.E.8's most warlike operation on 12 March, 1915. On that date Capt E. R. Ludlow-Hewitt, Lt E. O. Grenfell, Lt V. A. Barrington-Kennett, and Lt O. M. Moullin, flew the B.E.s solo to bomb a railway bridge at Douai and the junction at Don. These targets survived unscathed, but Ludlow-Hewitt dropped a 100-lb bomb on the railway at Wavrin in error for Don. Moullin failed to return and was later reported to be a prisoner of war; presumably his aircraft was 2132, which the squadron recorded as wrecked on 14 March. It was replaced a week later by 2130, but 399 and 740 were returned to the Aircraft Park on 27 March, the former to be struck off on 3 April, the latter to return briefly to No.1 Sqn on 3 June and move on to No.8 Sqn on 8 June together with 2130.

Some five weeks later these two B.E.8s, the last in the field, returned to the Aircraft Park, 740 on 13 July, 2130 four days later.

In August 1915, Nos.638, 663, 693 and 727 were at CFS, evidently as trainers. How long they survived is uncertain, but they were probably outlived by the B.E.8 No.643 that was transferred to the RNAS. In March 1915 it was on the strength of No.2 Sqn, RNAS, at Eastchurch, and later went to Westgate, remaining in use into 1916.

The basic B.E.8 design underwent few modifications and little development. In January 1915 an installation of a 100 hp Gnome Monosoupape was made at Farnborough in 658, but this power unit was much in demand at the time for

No.625 was the second of two B.E.8s built for India by Vickers Ltd and is here seen at Farnborough wearing crude camouflage. It left Farnborough on 11 August, 1914, piloted by Lt G. I. Carmichael, and went to No.3 Squadron via the Aircraft Park on 14 August. Two days later it crashed at Amiens, killing Lt E. W. Copland Perry and AM H. E. Parfitt.
(*RAE 561; Crown copyright*)

aircraft of greater lethality than the B.E.8 and the Monosoupape was not adopted for the type. No.658 was the subject of experiments with wireless at the Royal Aircraft Factory on 24 May, 1915.

A variant of the B.E.8 appeared early in 1915. Designated B.E.8a, this consisted essentially of a slightly improved B.E.8 fuselage combined with B.E.2c wings, ailerons, tailplane and elevators. The standard power unit was the 80 hp Gnome, but the 80 hp Clerget was from an early stage regarded as an alternative. Eventually at least eight of the B.E.8as had the Clerget but their conversions seem not to have been made until February and March 1916. As on the B.E.8, an experimental installation of a 100 hp Gnome Monosoupape was made in a B.E.8a in July 1915. The aircraft concerned was 2153, but no development followed. An installation of the experimental 120 hp R.A.F.2 radial engine in a B.E.8a has been reported.

The B.E.8a had B.E.2c mainplanes and tail unit. No.2134, Vickers-built, had been completed by 26 March, 1915. It was used in wireless trials in April and May, and by 27 September, 1915, had been allotted to No.9 Squadron at Dover.
(*RAE 748; Crown copyright*)

Production of the B.E.8a was undertaken by Vickers at Erith and by the Coventry Ordnance Works. Each contractor received a contract for 21 aircraft, the respective batches of serial numbers being 2133—2153 and 2154—2174. Deliveries of the Vickers-built B.E.8as began in March 1915, those from the Coventry Ordnance Works in May. Although 2135 was used in bomb-dropping trials in May 1915, no B.E.8a ever became operational and none went to France. Most of them went to training units or to squadrons preparing to go to France.

In service some of the B.E.8as were fitted with the enlarged curved fin that became standard for all B.E.s, and this was apparently standardised for the type. The B.E.8a remained in use in 1916, and in late March and early April of that year 2145—2148 were dismantled at Farnborough to provide spares. As late as 17 January, 1917, No.2167 was at Farnborough. In February 1916 a B.E.8a was experimentally fitted with snow skis mounted inboard of the wheels, which were retained.

B.E.8 and B.E.8a

80 hp Gnome, 100 hp Gnome Monosoupape. Some B.E.8as had the 80 hp Clerget 7Z, one an experimental installation of the 120 hp R.A.F.2.

B.E.8—two-seat military biplane
B.E.8a—two-seat trainer.
B.E.8: Span 37 ft 8 in; length 27 ft $4\frac{1}{2}$ in.
B.E.8a: Span 37 ft $8\frac{1}{16}$ in; length 27 ft $4\frac{1}{2}$ in; height 10 ft $3\frac{1}{2}$ in; wing area 368 sq ft.
B.E.8: Maximum speed at sea level 70 mph; climb to 3,000 ft—10 min 30 sec.
B.E.8a: Maximum speed at sea level 75 mph.
Armament: Bomb load of about 100 lb.

Manufacturers: Royal Aircraft Factory, British and Colonial, Coventry Ordnance Works, Vickers.

Service use: **B.E.8**—*Western Front*: Sqns 1, 3, 5, 6 and 8. *Training duties*: CFS, Upavon.
B.E.8a—*Training duties*: CFS; Reserve Aeroplane Sqns 7 and 8; Squadrons working up, Nos.9 and 13.

Serials: **B.E.8**—365, 373, 377, 391, 399, 423, 424, 479, 624, 625, 632, 636, 643, 645, 656, 658, 663, 670, 693, 725, 727, 729, 736, 740, 2130—2132.
B.E.8a—2133—2153, 2154—2174.

B.E.9

On 27 February, 1915, the French Société Anonyme pour l'Aviation et ses Dérivés (SPAD) applied for French patent No.498.338 to protect the design of a tractor aircraft in which a gunner's cockpit was carried ahead of the airscrew. The drawings accompanying the specification depict an aircraft having recesses in the leading edges of the wings to accommodate the tractor airscrew; spanwise spar members aligned with the leading edges carried the supports for the gunner's nacelle, which was wholly clear of and not connected to the airscrew shaft. On 7 June, 1915, Addition No.22.088 to Brevet d'Invention No.498.338 sought protection for a modified design in which the gunner's nacelle was supported on two V-struts, each of which was attached to a forward extension of the apex of each undercarriage V-strut, and the back of the nacelle was coupled to the forward end of the airscrew shaft by a ball-bearing mounting. Clearly this modification had been necessitated by the need to be able to swing down the nacelle to gain access to the engine and airscrew.

It is not recorded whether the design staff of the Royal Aircraft Factory had any knowledge of the SPAD patent or its Addition but they had given practical proof, long before war began, that they understood clearly the value of an effectively mounted forward-firing gun on an aeroplane. By the early summer of 1915 the first production F.E.2as were available, and there may have seemed to be advantages in trying, in the absence of any British machine-gun synchronising device, to combine the performance of a tractor aircraft with the unimpeded hemisphere of fire obtainable with a pusher. Whatever the reason, the B.E.9 was designed in the early summer of 1915, more or less when the first Fokker E I monoplanes went to the Western Front, and parts of the new type were undergoing inspection in the Royal Aircraft Factory by late July. The completed aircraft was sent for its pre-flight inspection on 14 August, 1915.

The B.E.9 was basically the Bristol-built B.E.2c No.1700 with the engine moved aft, wide-span centre-sections and an enlarged fin; in front, secured to the modified undercarriage and the forward end of the airscrew shaft, was the hapless gunner's nacelle. It made its first flight, of five minutes, on 14 August, piloted by Frank Goodden, and was proudly shown to King George V when he visited the Royal Aircraft Factory on 18 August. Thereafter it flew several times at Farnborough, occupants of the plywood nacelle including W. S. Farren, F. M. Green and Keith Lucas. On 23 August Frank Goodden and B. M. Jones tried bomb-dropping from the aircraft, and two days later it was twice flown to Netheravon, remaining there after its second flight until 31 August. Apparently it underwent its official trials while at Netheravon, for CFS Report No.132 bears the date 30 August, 1915. This recorded a low-level speed of 82·05 mph in a 15 mph cross-wind, a performance that, though unremarkable, was surprisingly little worse than that of the standard B.E.2c with the same R.A.F.1a engine. The B.E.9 went to CFS proper on 1 September, returning to Farnborough that evening.

The narrative portion of the CFS Report No.132 was signed by Capt Godfrey Paine, RN, the Commandant of CFS himself. It contains scarcely a hint of criticism of this bizarre aircraft, calling for an adjustable strap seat for the observer, the provision of dual control for rudder and elevators and an efficient telephonic intercommunication device between pilot and observer. Otherwise, enthused Paine: 'The design is, I consider, excellent and the vision and range of action for a gun better than any other type seen here. Stability very good indeed. Improvement is possible in the direction of fitting an adjustable tailplane on the D.F.W. lines. This would allow pilot to adjust for large variations in load carried'.

The aircraft underwent a further speed test at Farnborough on 4 September, was tested with various ASI pressure heads on 7 September, and performed compass and climbing tests on the 8th. On that day two sets of dual control gear for the aircraft were inspected and approved, and on 11 September its gun mounting was inspected. At 3.25 p.m. that afternoon Lt Breeze flew the aircraft to No.1 Aircraft Depot at St-Omer.

Some word of the B.E.9 had evidently preceded it to France, for as early as 20 August, 1915, Lt-Col Brooke-Popham circulated to squadron commanders a very

B.E.2c No.1700 as delivered to Farnborough in June 1915, with 70 hp Renault engine. (*Peter Liddle's 1914–18 Personal Experience Archives, presently housed within Sunderland Polytechnic*)

No. 1700 after conversion to B.E.9 configuration. This photograph is dated 17 August, 1915. (*RAE 872; Crown copyright*)

rough sketch of a 'B.E.2c with 12-cylinder R.A.F. engine' having a crude cockpit for the observer depicted ahead of the airscrew. He invited opinions on the idea, and the responses that were swiftly forthcoming were mostly highly unfavourable: although it was recognised that the field of view and of fire was excellent, squadron and wing commanders commented adversely on the extreme difficulty of communication between pilot and observer, on the danger of the observer's position, loss of propeller efficiency and risk of engine overheating. Lt-Col Ashmore (I Wing), who had previously seen the B.E.9 and had sat in the nacelle, thought the aircraft should at least be tried out.

Despite the almost unanimous criticism of the type nothing prevented its arrival in France, and on 13 September Brooke-Popham himself went up in the observer's seat, with Capt Norman Spratt piloting. He found the view excellent, the proximity of the airscrew 'not in the least disconcerting', observed that there was very little vibration in the forward nacelle, and generally thought it comfortable. In later years, R. R. Money was to write in *Flying and Soldiering*: 'Major [*sic*] Brooke-Popham . . . lives in my memory because he went up in a horrible contraption called the Pulpit B.E.2c and refused to condemn it'. Directly after Brooke-Popham's flight the B.E.9 went to No. 6 Sqn on 13 September, being flown to its new unit by Capt L. G. Hawker. It spent eight days with the squadron, during which time it flew one operational reconnaissance and patrol in addition to various test flights; engine trouble prevented it from doing much flying. The B.E.9 moved on to No. 16 Sqn on 21 September, 1915, on which date Lt-Col J. M. Salmond, commanding II Wing, submitted a report on the aircraft, listing its advantages and more numerous disadvantages and concluding: 'I do not consider it worth while'.

At that time No. 16 Sqn was commanded by Maj H. C. T. Dowding, who was, 25 years later, to win immortality and deserve (if not actually to receive) the undying gratitude of the British nation as Air Marshal Sir Hugh C. T. Dowding, creator of RAF Fighter Command and the victorious British commander in the

Battle of Britain. Dowding made it his business to fly the B.E.9 as soon as it arrived on his squadron, and on 22 September he submitted an objectively neutral report on it, commenting on the aircraft's very poor performance and recording his opinion that it was '. . . an extremely dangerous machine from the passenger's point of view'. More engine trouble occurred, but Dowding managed to submit a second and substantially more adverse report on 29 September.

On 2 October, 1915, the B.E.9 left No.16 Sqn and went to No.8 Sqn. One of its pilots in that unit was Lt W. Sholto Douglas, later Marshal of the Royal Air Force Lord Douglas of Kirtleside, who flew the B.E.9 for the first time on 8 October. Several other pilots of No.8 Sqn flew the aircraft, and on 13 October it was involved in a brief and inconclusive combat with a Fokker monoplane; its pilot on this unique occasion was Lt D. A. Glen. On 26 October Sholto Douglas took with him Air Mechanic Walker on a reconnaissance and offensive patrol over the front, lasting 1 hr 20 min but mercifully encountering no opposition.

At that time Lt-Col W. Sefton Brancker was in command of III Wing and he submitted a report on the B.E.9 in which he reached the astonishing conclusion that it was 'worth proceeding with in moderation *if the engine trouble theory can be disproved* and that three of these machines in a B.E.2c squadron would be very useful for patrolling and escort duty'. In this view Brancker was conspicuously alone. On 20 October Trenchard sent to the ADMA a masterly and elegant report (obviously written by Maurice Baring), setting out the well-merited criticisms levelled against the design and concluding with firm clarity ' . . . this type of aeroplane cannot be recommended'.

How much flying B.E.9 did after that report is uncertain, but on 11 December, 1915, RFC HQ wrote to ADMA:

'With reference to B.E.9 No.1700 at present on charge of one of the squadrons, this machine has never been satisfactory, the engine constantly overheating. It is therefore no longer desired to retain it in the Royal Flying Corps, Expeditionary Force. Please inform us early if you require the machine flown home or whether the engine may be taken out and retained here and the machine scrapped'.

This evoked an instruction dated 15 December, requiring the B.E.9 to be flown to Farnborough, and, on 22 December, Sholto Douglas flew the thing from Marieux to No.1 A.D. at St-Omer. The Pulpit B.E. crossed the Channel back to England on 9 January, 1916, piloted by Sgt Maj W. B. Power, and was never heard of again.

It will be recalled that Brooke-Popham had mentioned a frontal-cockpit B.E. with twelve-cylinder R.A.F. engine in inviting criticisms of the configuration as early as 20 August, 1915. It may have been intended all along to build the aircraft with a 140 hp R.A.F.4 as power unit, for the R.A.F. design office drew up a development with that power unit under the designation B.E.9a. The rejection of the B.E.9 ensured that work on the B.E.9a did not proceed.

B.E.9
90 hp R.A.F.1a

Two-seat reconnaissance biplane.
Span 40 ft 10½ in; length 29 ft; height 11 ft 5 in.
In CFS trials: Maximum speed at sea level 82·05 mph; climb to 4,000 ft—12 min 40 sec.
With No.16 Sqn in France: Maximum speed 70 mph at 6,000 ft; climb to 3,500 ft—16 min.

Armament: One 0·303-in Lewis machine-gun.

Manufacturers: Basic airframe by British and Colonial; modifications by Royal Aircraft Factory.

Service use: Western Front: Sqns 6, 8 and 16.

Serial: 1700.

B.E.12, 12a and 12b

To understand the original reason for the creation of the B.E.12 it is necessary to recall that in the middle of 1915 the B.E.2c was still giving a reasonably good account of itself and was standing up to the rigours of operational flying rather better than most of its contemporaries. Its 90 hp R.A.F.1a engine was, by the standards of the time, not unsuccessful, and in December 1914 the Royal Aircraft Factory had designed a larger twelve-cylinder air-cooled engine that used R.A.F.1a components and gave about 140 hp. This was designated R.A.F.4 and was subsequently put into production as the R.A.F.4a, with stroke increased from 120 to 140 mm.

In mid-1915, with the depredations of the Fokker monoplane still to be experienced, it must have seemed sensible to combine the new and powerful engine with the sturdy B.E.2c airframe. Such a decision had been reached before the end of June, for by that time No.1697, a Bristol-built B.E.2c with Renault engine and twin-skid undercarriage, had begun to undergo modifications at Farnborough. On 28 July, with the R.A.F.4 engine installed and a V-type undercarriage fitted, the aircraft was submitted for its pre-flight inspection and subsequently made its first flight. Its performance inspired Mervyn O'Gorman, the Superintendent of the R.A.F., to write to Brancker that same day:

'I have had a preliminary run on B.E.12 today . . . she is as stable as B.E.2c, alights as slowly—climbs like mad—and flies in the neighbourhood of 100 mph.

I do not see why all the pilots who can fly a B.E.2c should not use it as soon as they get it. They can throttle down at first so that it *is* a B.E.2c and after a few alightings open up throttle'.

At that time the B.E.12 prototype had no armament and there is no evidence that there were any thoughts of providing a gun installation, though doubtless bomb-carrying was envisaged. In September 1915 No.1697 was in fact being used for experiments in bomb-dropping and dart-dropping at Farnborough, and there is not a shred of evidence to suggest that anyone anywhere at that time contemplated using the B.E.12 as a single-seat fighter.

Later that month No.1697 went to France and had to be recovered from Boulogne on 30 September, having been obliged to land there after blowing out No.9 cylinder of its engine. It was noted that at this time the aircraft had an enlarged fin, apparently with curved leading edge.

The first contract for the production of the B.E.12 was given to the Standard Motor Co in September 1915. It was for a modest 50 aircraft (6136—6185) and was soon followed by a further order for 200 (6478—6677) from The Daimler Co.

No.1697 was another Bristol-built B.E.2c that was converted at Farnborough; it became the prototype B.E.12. This photograph is dated 30 July, 1915. *(RAE 852; Crown copyright)*

In due course No.1697 received the enlarged fin and was painted with P.C.10.
(*A&AEE; Crown copyright*)

Both firms had been contractors for the B.E.2c. Production of the R.A.F.4a engine was similarly concentrated in Coventry, the only production contracts having been given to the Daimler and Siddeley-Deasy companies. The latter manufacturer began delivery of engines in December 1915, and Daimler production became available in February 1916.

Few modifications were made to the production B.E.12s, the earliest of which retained the original triangular fin and rectangular tailplane of the B.E.2c. Fuel capacity was increased by providing a larger main tank that was accommodated by sloping down the upper longerons ahead of the cockpit; additionally, a 12-gallon gravity tank was attached under the port upper mainplane. Deliveries had begun

Production B.E.12 No.6478, showing revised upper longeron arrangement ahead of the cockpit. This B.E.12 was at Farnborough on 26 March, 1916, and later was used by No.76 (H.D.) Squadron.

by March 1916 and evidently continued in some quantities in ensuing weeks.

The earliest potentially operational allocations of B.E.12s were to Home Defence squadrons. For such duties 6484 was allotted to No.52 Squadron on 18 May, 1916, while on 26 May 6489 was allotted to No.51 and 6490 to No.53 Squadron (in the event, No.52 went to France as an R.E.8 squadron on 21 November, 1916). The first B.E.12 to fly operationally in France was 6483 which, with 6479, had first been allotted to the Expeditionary Force on 31 May, 1916, and went to No.10 Squadron on 19 June. It was tested that day by Lt R. G. Gould, who found that at 300 ft it would climb at 100 mph, and he reached 6,000 ft in 10 minutes. With No.10 Squadron the B.E.12 did relatively little flying. Its first operational sortie was made on 25 June, when Maj W. Mitchell, the squadron's C.O., flew it as an escort to the unit's B.E. two-seaters on a bombing mission. On a few occasions 6483 itself dropped bombs, and on 1 July Capt P. E. L. Gethin reported that the B.E.12, armed with Vickers gun, Lewis gun and three ammunition drums, and four 20-lb bombs, took 38 minutes to reach 10,000 ft.

B.E.12 No.6562 was initially allotted to No.48 Squadron mobilising at Rendcombe on 31 July, 1916, but was re-allotted to the E.F. on 5 August. It was issued to No.19 Squadron, but its brief career ended on 26 August, 1916, when 2/Lt S. P. Briggs was obliged to land in German-held territory. His B.E.12 was taken to Adlershof for evaluation.

On that fateful date, which saw the beginning of the Battle of the Somme, 6483 was the only B.E.12 in operational use in France, despite the fact that ten others* had gone there during June 1916. On 5 July, while 6483 was being tested by 2/Lt C. E. W. Foster, its interrupter gear malfunctioned and the gun shot through the propeller. The B.E.12 was apparently grounded until 16 July, and on 18 July it was flown to No.1 A.D. by Capt J. T. Rodwell, subsequently seeing service with No.21 Squadron, to which it was issued on 6 October, 1916.

It is of interest to observe that 6483 had been fully armed as early as 1 July, 1916. Various attempts had been made to fit armament to the B.E.12 from an early stage; these included a synchronised Lewis gun and an installation of an

*6144, 6508, 6512, 6517, 6529 and 6531—6535.

unsynchronised gun with deflector-protected airscrew. However, the standard operational armament consisted basically of a single Vickers gun with 250 rounds, synchronised by the Vickers-Challenger interrupter gear. The precise nature of 6483's Lewis-gun mounting is not known, but it may well have been the rearward-firing installation that, as mentioned later, was proposed by Brooke-Popham. The shortcomings of the B.E.12 as a military aircraft were recognised early by the RFC's command in France. On 30 June, 1916, a month before the first B.E.12 squadron went to France, Brooke-Popham wrote to the D.A.E.:

'As the result of trials with the B.E.12 it is found that they are at present unsuitable for use as fighting machines for the following reasons:

The machine cannot dive with the engine full on, since owing to the stable tail the pilot cannot exert enough force on the elevators to keep the nose of the machine down at a steep angle.

If on the other hand the engine is throttled down the interrupter will probably fail to operate and the bullets will strike the propeller.

Will you please say if this matter has been considered in England and whether you can suggest any means of rectifying it'.

In practice the gun worked well enough as long as the engine rpm exceeded 800; below that figure the gun's rate of fire was seriously retarded. On 7 July, 1916, Brooke-Popham again wrote to D.A.E., this time to suggest that a double cam be fitted in place of the single cam of the interrupter gear in order to double the rate of fire. Eventually this useful modification was introduced on the B.E.12.

The cumbersome engine installation with its big air scoop and twin exhaust stacks made it very difficult to fit an effective sight for the Vickers gun. Early in July 1916, No.10 Sqn were experimenting with a Bellieni sight on their B.E.12 but apparently without consequence, and experiments with a periscope sight were requested by Brig-Gen Brooke-Popham of No.1 A.D. Ultimately, an arrangement of a ring-and-bead sight on a bar outrigged from the port centre-section struts was more or less standardised. The pilot was obliged to lean out into the slipstream when he wanted to sight his gun.

Before No.19 Sqn went to France, Brooke-Popham realised that some rear defence would be needed and proposed the fitting of a Lewis gun mounted on the port side abaft the cockpit, as on the Martinsyde Elephant. This, too, was adopted, and its fitting to the B.E.12s of No.19 Sqn, then still in England, was requested.

In July 1916 it was found that the tractability of the B.E.12 was greatly improved by fitting the B.E.2e tailplane and elevators in place of the rectangular B.E.2c surfaces. Both Aircraft Depots were immediately instructed to indent for twelve of the small tailplanes. On 18 July, 6533 was tested at No.1 A.D., fitted with B.E.2e tail surfaces and a periscopic gunsight. Diving presented no problems, but the pilot (2nd Lt W. Pendavis) did not care for the periscopic sight.

No.19 Sqn, equipped with B.E.12s, finally arrived in France on 30 July, 1916, at which time it had been decided to re-equip No.21 Sqn with B.E.12s. As far as can be determined, No.19's aircraft had R.A.F.4a engines fitted with the new double oil pump and extra air intake that improved the engine's performance. No.21 Sqn had suffered from the inadequacies of the earlier unmodified R.A.F.4a engines in its R.E.7s, and Brooke-Popham had advised D.A.E. on 28 July that the squadron would remain out of action until B.E.12s with reliable engines could be provided.

As observed in the text, the configuration of the R.A.F.4a engine and its tall air scoop made it difficult to fit gun sights on the B.E.12. This photograph shows the bead sight mounted on the front centre-section strut, the ring sight on the rear. The difficulty and discomfort of using these sights are obvious.

At that time No.48 Sqn was working up at Rendcombe, having a complement of ten B.E.12s. These were forthwith sent to France for the re-equipment of No.21 Sqn, and No.48's operational début was thereby postponed for many months until it went to France as the first Bristol Fighter squadron. The B.E.12s of No.21 Sqn had the double oil pump and ammunition boxes holding 350 rounds for the Vickers gun; the squadron soon devised their own rear gun mounting for the Lewis gun, and it was adopted as standard from 25 August, 1916, only to be superseded by a 19 Sqn mounting from 5 September.

No.6579, here very obviously in France, had its ring-and-bead sights mounted on a rod that was in turn attached to the centre-section struts. This B.E.12 was allotted to the E.F. on 4 August, 1916, and was issued to No.19 Squadron from No.1 A.D. on 27 August. It was reported missing on 24 September, 1916, being one of two B.E.12s of No.19 Squadron shot down in combat over Havrincourt Wood. (*Jean Noël*)

Operational experience with the B.E.12 in France was unhappy and losses were heavy. On 24 September, 1916, Maj-Gen Trenchard wrote to the Director of Air Organisation:

> 'I have come to the conclusion that the B.E.12 aeroplane is not a fighting machine in any way. There are only one or two pilots who can do any good with it, and even those could do much better with some other type of fighting machine.
>
> I have lost a very large number of them, and I am afraid that we are losing more of these machines than we can afford in pilots, and certainly more, I am afraid, than they bring down.
>
> I realise fully that I shall lose two squadrons if I stop using the B.E.12 and delay, I suppose for some considerable period, two other squadrons. Although I am short of machines to do the work that is now necessary with

B.E.12 No.6592 was originally allotted to No.48 Squadron at Rendcombe but was re-allotted to the E.F. on 26 August, 1916. It went to No.21 Squadron from No.1 A.D. on 2 September and apparently remained with the squadron until flown back to England on 8 February, 1917.

the large number of Germans against us, I cannot do anything else but to recommend that no more be sent out to this country'.

Replacements were not forthcoming and the B.E.12 remained in use as a bomber. In November 1916 efforts were being made to adapt 336-lb bomb carriers from Martinsydes to fit the B.E.12, but these had not been delivered by the end of the year. By then the original contracts for the B.E.12 were running out and on 27 November, 1916, RFC HQ were informed that aircraft coming forward from the end of that month would be B.E.12as.

B.E.12 of No.50 (H.D.) Squadron at Bekesbourne, fitted with navigation lights, flare brackets and launching tubes for Le Prieur rockets. (*Peter Liddle's 1914–18 Personal Experience Archives, presently housed within Sunderland Polytechnic.*)

B.E.12a, A597, seen here without armament. (*Imperial War Museum Q67526*)

B.E.12a A597 at Martlesham Heath with a 112-lb bomb under the fuselage. This aircraft was written off on 21 July, 1918.

An operational B.E.12a, most probably in Palestine, with an upward-firing Lewis gun mounted high up on the port rear centre-section strut.

B.E.12b of No.77 (H.D.) Squadron, Penston, East Lothian, with two 100-lb bombs on underwing racks. (*E. F. Cheesman*)

This view of the cockpit area of the same B.E.12b (as in the previous illustration) of No.77 (H.D.) Squadron shows the nature of the overwing mounting for the Lewis gun, the reversed fuel tank, and the external release lever for the bombs. (*E. F. Cheesman*)

The B.E.12a had B.E.2e mainplanes and tail unit, which bestowed a small improvement in performance. Very few ever went to France, and the type's operational employment was principally in Palestine, where No.67 Sqn had a few. These were later handed over to No.142 Sqn in February 1918, but all had gone by the autumn of that year. A few saw service with No.17 Sqn in Macedonia, and some went to Home Defence squadrons.

On Home Defence duties the B.E.12 saw quite extensive use, for its stability enhanced its ability to fly successfully at night. As early as the summer of 1916 it had been decided that the B.E.12 would replace the B.E.2c on Home Defence duties. By 7 July, 1916, three had been allotted to No.33 Sqn, one to No.54 (future No.38) Sqn. Three months later, on 16 October, fifteen B.E.12s had been despatched to Home Defence units and three others allocated. Various arrangements of armament were fitted to Home Defence B.E.12s, some being armed with as many as ten Le Prieur rockets. On 17 June, 1917, No.6610 of No.37 Sqn, piloted by Lt L. P. Watkins, shot down the Zeppelin L.48.

By then, a few more B.E.12s had been asked for by the RFC in France, apparently for special missions: the document asking for them, dated 7 April, 1917, was classified as secret at the time, and asked for B.E.12s in preference to B.E.12as. To meet this request a number of B.E.12s had to be sent from training units and proved to be unfit for operational flying on arrival in France. The belated B.E.12s were used by No.101 Squadron and the Special Duty Flight, as were a small number of B.E.12as. These were in service at least until March 1918, and 6511was not flown back to England until July 1918.Before its operational use with No.101 Squadron and the Special Duty Flight, this B.E.12 had seen much experimental work at Farnborough, where it tested an installation of a 6-pdr Davis gun and was at one time fitted with B.E.2e wings that had horn-balanced ailerons on the upper mainplanes.

It is a remarkable fact that production of the B.E.12 was revived in the summer of 1917 to provide Home Defence aircraft. Contract No.A.S.11453/1 was placed with the Daimler Co on 18 August, 1917, calling for two hundred B.E.12s, for which the serial numbers C3081—3280 were allotted on 21 August. A considerable number of the aircraft of this batch were completed as B.E.12bs. This variant had a 200 hp Hispano-Suiza engine in an installation very similar to that of the S.E.5a, driving a two-blade airscrew, and the long exhaust pipes were in some cases fitted with flame dampers.

Early in September 1917 the Royal Aircraft Factory was arranging to fit a 200 hp Hispano-Suiza engine (1,500 gear) in a B.E.12, an initial request for the supply of one such engine having been made on 30 August. The Hispano-Suiza arrived at the Factory on 10 September, 1917, but the Southern Aircraft Repair Depot, also located at Farnborough, was likewise busily modifying a B.E.12 to take the Hispano engine. Although some official displeasure was expressed at this duplication it was agreed on 17 September that the S.A.R.D. should proceed with the task and that the R.A.F. should give all possible assistance in fitting the engine. The S.A.R.D. conversion had been completed by 22 September, 1917, and a month later the aircraft was reported to be under test with the Home Defence forces. It may have been B718, but this is unconfirmed.

Deliveries of the B.E.12s built under the 1917 Daimler contract were expected to begin in October, and by the 20th of that month Daimler had completed ten, apart from having to change one Rafwire on each aircraft. It was intended that the first 100 aircraft should have the Hispano-Suiza engine, the following 50 the R.A.F.4a, and the remaining 50 the Hispano-Suiza. The installation of the Hispano-Suiza engines in the B.E.12 airframes was done at the Northern Aircraft Repair Depot at Coal Aston, near Sheffield, but it is uncertain whether as many as 150 were completed as B.E.12bs. The 200 hp Hispano-Suiza was critically scarce at the time, and the RFC in France must have had serious misgivings about the diversion of such sorely needed engines to an untried development of the obsolete B.E.12. The decision thus to employ so many Hispano-Suizas must be interpreted as reflecting the Government's awareness of the seriousness of the Home Defence situation, for none of the aeroplanes then available for such duties was capable of climbing to the altitude that Zeppelins could attain with ease.

B.E.12bs and B.E.12s of No.77 (H.D.) Squadron at Penston. (*RAF Museum*)

Effective deliveries of completed B.E.12bs began late in 1917; and 130 had been officially accepted and twelve had been issued to Home Defence units by the end of December. Despite the official anxiety to expedite production of the B.E.12b it saw relatively little use: only 24 more were issued to squadrons in 1918, and only 17 were still with units of the VI Brigade on 31 October, 1918, on which date 67 were in store and 31 at Aircraft Repair Depots. This situation probably arose because the Zeppelin raids had virtually ceased by the time when the B.E.12b entered service, and it is doubtful whether any B.E.12b ever had the opportunity of trying to intercept a Gotha or a Giant bomber, for these concentrated their attentions on London and the South-East.

It seems that the evaluation of the B.E.12b was performed wholly within the Home Defence organisation: as far as is known no example of the type was tested at Martlesham Heath. Any kind of test report on the type has yet to be found, but it has been reported that the B.E.12b's performance was, compared with that of the B.E.12, quite spectacular.

B.E.12, 12a and 12b
B.E.12 and 12a—140 hp R.A.F.4a
B.E.12b—200 hp Hispano-Suiza

Single-seat general-purpose aircraft; B.E.12b single-seat Home Defence fighter.

Span 37 ft (B.E.12 and 12b), 40 ft (upper) and 30 ft 6 in (lower) (B.E.12a); length 27 ft 3 in (B.E.12 and 12a); height 11 ft 1½ in (B.E.12), 12 ft (B.E.12a); wing area 371 sq ft (B.E.12 and 12b), 360 sq ft (B.E.12a).

Empty weight 1,635 lb (B.E.12), 1.610 lb (B.E.12a); loaded weight 2,352 lb (B.E.12), 2,327 lb (B.E.12a).

Maximum speed at sea level 102 mph (B.E.12), 105 mph (B.E.12a), at 6,500 ft—97 mph (B.E.12), 91·5 mph (B.E.12a), at 10,000 ft—91 mph (B.E.12), 80·5 mph (B.E.12a); climb to 5,000 ft— 11 min 5 sec (B.E.12), 9 min 10 sec (B.E.12a), to 10,000 ft—33 min (B.E.12), 24 min 15 sec (B.E.12a); service ceiling 12,500 ft (B.E.12); endurance 3 hr (B.E.12).

Armament: B.E.12—one 0·303-in Vickers machine-gun and one 0·303-in Lewis machine-gun. Home Defence aircraft had up to four Lewis guns or up to ten Le Prieur rockets. Bomb load of up to 336 lb. B.E.12a—one 0·303-in Vickers gun, two 112-lb bombs. B.E.12b—one, sometimes two, 0·303-in Lewis machine-guns; bombs could be carried.

Manufacturers: Coventry Ordnance Works, Daimler, Standard Motor. B.E.12b conversions by Northern Aircraft Repair Depot, Coal Aston.

Service use: **B.E.12**: *Western Front*: Sqns 10, 19, 21 and 101; Special Duty Flight. *Home Defence*: Sqns 33, 36,37, 38, 39, 50, 51, 53, 75, 76 and 77. *Macedonia*: Sqns 17, 47 and 150; Composite Fighting Squadron at Hadzi Junas. *Palestine*: 'B' Flight formed from No.23 Training Sqn; 'X' Flight at Aqaba; No.144 Sqn. *Training*: Reserve/Training Sqns 11, 12, 13, 19, 20, 31, 36, 46, 50, 51, 66 and 69; CFS. *Squadrons mobilising*: Nos.48, 49, 54, 56 and 58.
B.E.12a: *Western Front*: No.101 Sqn; Special Duty Flight. *Home Defence*: Sqns 37, 39, 50 and 76. *Macedonia*: No.17 Sqn. *Palestine*: Sqns 14, 67 and 142. *Training*: No.15 Training Sqn. *Squadrons mobilising*: Nos. 49, 56, and 58.
B.E.12b: *Home Defence*: Sqns 37, 50, 51, 76 and 77.

Serials: **B.E.12**: 1697, 6136—6185, 6478—6677, A4006—4055.
C3081—3280 (Many conversions to B.E.12b).
B.E.12a: A562—611, A6301—6350.
Rebuilds: **B.E.12**: B701, B720, B721, B726, B749, B753, B1500, B8826, B9462, B9950.
B.E.12b: B718, C9992.

F.E.2a

The staff of the Royal Aircraft Factory had quite considerable experience of mounting machine-guns on pusher aeroplanes by the time the war began, for they had been closely associated with the RFC's early experiments in firing such weapons from aeroplanes. These followed a decision of January 1913 to try the Vickers belt-fed gun on some of the aeroplanes then in use in RFC squadrons, and on 3 April, 1913, the Superintendent of the Royal Aircraft Factory was instructed to provide an installation on a Henry Farman F.20 to enable the Vickers, Hotchkiss and Rexer machine-guns to be tested in flight. On 27 March,

F.E.2a No.1 as it first appeared, with fully cowled Green engine. This official photograph is dated 26 January, 1915. (*RAE 692; Crown copyright*)

1913, the Superintendent had been told that a Rexer gun was to be sent to Farnborough to be mounted on an aeroplane and, presumably in the absence of any specific instruction, he had expressed the opinion that the reconstructed F.E. biplane would be suitable.

It is uncertain whether the F.E. biplane in question was the original F.E.2 or whether Mervyn O'Gorman was looking ahead to the completely revised design with the 70 hp Renault engine. That the latter may have been the case is suggested by a memorandum dated 3 April, 1913, from the Director of Artillery in which he noted that the F.E. would not be ready for some time. The original F.E.2 had been designed by Geoffrey de Havilland as an improvement on his earlier F.E.1, and had attempted its first flight as early as 16 August, 1911; the aircraft's first truly successful flight was made two days later. The F.E.2 was a two-seat pusher of primitive appearance with a biplane tail unit, a 50 hp Gnome engine, and a rudimentary fabric-covered nacelle. The F.E.2 underwent many minor modifications and did much flying. Late in September 1911 the lower tailplane was removed, and the aircraft had been fitted with basic dual controls. A 70 hp

F.E.2a No.1 on the airfield at Farnborough, presumably just before or just after a flight. Squatting on the nosewheel is Frank Goodden. (*RAE 707; Crown copyright*)

Gnome replaced the original 50 hp engine by 26 April, 1912, and next day the F.E.2 was flown by Geoffrey de Havilland to Fleet Pond, where it was fitted with a centre-float undercarriage and flew in its floatplane form on 29 April.

Later still, this first F.E.2 was armed with a Maxim gun on the prow of the nacelle, much as the experimental gun installations had been made on the Henry Farman F.20. In its armed form the early F.E.2 reverted to single control, and the nose area of the nacelle was strengthened by being clad with plywood.

The second aircraft to bear the designation F.E.2 was an entirely new design specifically intended to mount a Vickers 1-pdr quick-firing gun in the forward cockpit of the deep nacelle. Its outer wing panels appeared to be similar to those of the B.E.2a, and its engine was a 70 hp Renault. This reconstructed F.E.2 (thus was it described) was completed in the summer of 1913. How much weapon-development work it did remains unknown but probably did not amount to much. It crashed at Wittering on 23 February, 1914; its pilot, Ronald Kemp, was injured, and his passenger, Ewart Haynes, was killed.

F.E.2a No.3, the first to be completed with the 120 hp Beardmore engine. This F.E. was numbered 2864 and joined No.6 Squadron in France on 20 May, 1915. It returned to the 1st Aircraft Park on 22 May and went back to No.6 Squadron on 29 June. It was struck off charge on 22 September, 1915. (*RAE 732; Crown copyright*)

It is not known precisely when the design of the F.E.2a was begun, but it must have been about the time of the start of the war. This was a completely new and larger aircraft, having nothing in common with the F.E.2 but the pusher configuration. Its three-bay wings had wide-span centre-sections and the outer panels were identical with those of the B.E.2c, while the entire trailing-edge portion of the upper centre-section was hinged along the rear spar and could be lowered to act as an air-brake; a substantial oleo undercarriage with a small nosewheel was fitted. In the deep nacelle the pilot sat at a higher level than his observer, who occupied the forward cockpit in near-total exposure but with a

F.E.2a No.3 at Central Flying School. (*RAF Museum*)

remarkably wide field of fire for his Lewis gun, which was eventually carried on the No.2 Mark I mounting. This consisted of a horizontal steel tube fitted with a runner, mounted ahead of the front cockpit and giving the Lewis gun limited traverse. On the first F.E.2a the power unit was a 100 hp Green six-in-line water-cooled engine, closely cowled with small air intakes for the radiator and driving a two-blade propeller.

F.E.2a No.1 had been completed by 22 January, 1915, and made its first flight, of 15 minutes duration, on 26 January with Frank Goodden at the controls. An engine change was made on 29 January, but with the Green the aircraft was underpowered and an alternative power unit was sought. At that time the Royal Aircraft Factory had had some experience of the 120 hp Austro-Daimler engine, then being manufactured by William Beardmore & Co, which was the standard power unit of the R.E.5, and its six-cylinder inline configuration must have made it seem to be a natural alternative to the Green. A suitable installation was designed and was first applied to the third F.E.2a, which was submitted for final inspection on 15 March and was flown by Frank Goodden next day. Modifications to the first and second aircraft took some months to make, the former being sent for final inspection on 17 April, the latter on 3 May.

The third F.E.2a went to CFS for official trials on 20 March. While there it was matched against a Voisin LA and was well reported on, being described as 'extremely easy to fly and very easy to land'. Stability and control response were regarded as good, and its performance was held to be considerably better than that of the contemporary Vickers F.B.5. However, the RFC command were at that time aware that there was no likelihood of the Austro-Daimler engine being available in sufficient quantities to permit early deliveries of F.E.s, and it was decided to go ahead with a proposed order for fifty Voisins.

All the F.E.2as had the 120 hp Austro-Daimler, but production was, as expected, greatly delayed by belated deliveries of engines, for Beardmore did not begin to deliver the 120 hp Austro-Daimler in quantities until October 1915. Some modifications were made to the aircraft as time went on. The first three were fitted with new-type ailerons in May 1915, and these may have been the horn-balanced surfaces that were designed for the type. The covering of the nacelle varied on some of the later F.E.2as and at least some of them embodied armour plate on the underside of the nacelle; No.11 was fitted with a simple V-strut undercarriage, its wings were of a different aerofoil section, and it had no air-brake on the upper centre-section. These modifications made it the last of the

One of the later F.E.2as with nacelle covering as standardised for the F.E.2b. The standard F.E.2a could be distinguished from the F.E.2b by its circular-section gravity tank and the air-brake flap on the centre-section.

twelve F.E.2as to be completed, for it was not submitted for inspection until 5 October, 1915. On the standardised F.E.2a the Austro-Daimler engine was wholly exposed and the air intakes to its radiator were larger than those provided for the Green. The fuel system incorporated a streamlined gravity tank of circular cross-section carried under the centre-section to port.

The F.E.2a had the alternative official designation Fighter Mark I and all twelve received official serial numbers. First to go to France were, appropriately, the F.E.2as Nos.1 and 2, with the new respective identities of 4227 and 4228. These went direct to No.6 Squadron at Abeele on 14 May, 1915, but 4228 lasted only two days and was struck off charge on 16 May. Its replacement was 4253 (No.4) which arrived direct from England on 19 May, and next day 2864 (No.3) was flown in from the 1st Aircraft Park by Lt Louis Strange.

F.E.2a No.11 had new wings without the air-brake, and was the first F.E. to have the plain V-strut undercarriage. This photograph illustrates the No.2 Mark I mounting for the observer's Lewis gun. (*RAE 905; Crown copyright*)

All twelve of the F.E.2as went to France and all but 5647 (No.10) were used by No.6 Squadron: the exception, which arrived at the 1st Aircraft Park on 20 October, 1915, went to No.16 Sqn on 26 October. The last to arrive in France was 5648 on 3 November, which went to No.6 Squadron on 7 November and was struck off charge on 21 November.

By 31 December, 1915, the only F.E.2as remaining in service in France were 5642, 5643 and 5645 with No.6 Squadron, while 5647 was still on the strength of No.16 Squadron. How many were lost in combat is uncertain: 4227 was shot down by anti-aircraft fire on 5 September, 2864 was struck off charge on 22 September, and 5644, 5646 and 5648 were likewise struck off in November and December.

Nos.4253 and 4295 were returned to No.1 A.D., the former to be struck off, the latter to be rebuilt. Of the few that remained operational into 1916, 5642 was transferred from No.6 to No.20 Squadron, but on 7 March, 1916, the latter unit reported that the F.E.2a was no longer satisfactory for service.

Louis Strange flew the F.E.2as of No.6 Squadron and in his book *Recollections of an Airman* he recorded that the squadron grew to like the aircraft, quoting a diary note that he made in 1915:

'The F.E. is a fine machine and would be ideal if it were only a little faster. As it is, the Hun always seems to see us in time and avoids us like the plague, so that we seldom catch him. It is, however, a splendid reconnaissance and photographic machine.'

References in surviving official documents indicate that one of the later F.E.2as was modified to become the prototype F.E.2c and evidently went to France in that form. In the F.E.2c the positions of pilot and observer were transposed; the pilot's seat was somewhat farther aft than that of the observer in the F.E.2a, and the F.E.2c's observer sat high and somewhat cramped with limited movement for his Lewis gun. For the pilot a remotely controlled Lewis gun was to be mounted in the nose of the nacelle.

Word of the F.E.2c's crew disposition reached France in mid-September 1915, and apparently indicated or implied that all future F.E.s would be built in this form. This so alarmed Lt-Col J. M. Salmond, then commanding the Second Wing, that on 20 September he wrote to RFC Headquarters pointing out that the supposed advantages claimed for the reversed crew positions were insubstantial, and concluding 'Indeed, as a fighting machine, if the Observer had to sit behind, the F.E. would be of little value'. Shortly after he wrote, however, the converted F.E.2a was sent to No.6 Squadron in his Wing, and on 7 October, 1915, he reported to Headquarters:

'The new type F.E. with Observer's seat behind has now had about five hours flying including one patrol and reconnaissance flight.'

He went on to describe disadvantages heavily outweighing the type's few barely discernible advantages, and next day Trenchard wrote to ADMA:

'It is not desired to alter the previous recommendations regarding this machine. The F.E.2b is the type required. The F.E.2cs under construction could, however, be sent out and altered locally if necessary.'

Such evidence as can now be considered seems to suggest that the first F.E.2c may have been 5644, which went to No.6 Squadron direct from England on 23 September, 5645 which was sent to the same unit from the Aircraft Park on 1 October, or 5646, which went from the A.P. to No.6 Squadron on 2 October. Of these, 5644 was struck off charge on 11 November, 5646 on 6 December, while 5645 was still with No.6 Squadron in February 1916. The F.E.2c history did not end there, however, for others were built later as converted F.E.2bs, and their story is related in following pages as part of the F.E.2b history.

F.E.2a
100 hp Green, 120 hp Austro-Daimler (Beardmore)
Two-seat fighter-reconnaissance.
Span with plain ailerons 47 ft 10 in, with horn-balanced ailerons 50 ft 1 in; length 32 ft 3½ in; height 12 ft 7½ in; wing area 494 sq ft.
Loaded weight 2,680 lb.
Maximum speed at sea level 80·3 mph; climb to 3,000 ft—8 min 10 sec, to 6,000 ft—18 min, to 9,000 ft—34 min.
Armament: One 0·303-in Lewis machine-gun.

Manufacturer: Royal Aircraft Factory.

Service use: *Western Front*: Sqns 6, 16 and 20.

Serials: 2864 (No.3), 4227 (No.1), 4228 (No.2), 4253 (No.4), 4295, 5642—5648 (5647 was No.10, 5648 was No.11).

No.6364 was a production F.E.2b built at the Royal Aircraft Factory. Named *Junagadh No.2* as a presentation aircraft, it was in the AID inspection shed at Farnborough on 15 February, 1916, and later saw operational use in France with No.23 Squadron. It crashed on 26 April, 1916, and was struck off charge.

F.E.2b, 2c and 2h

The modifications required to adapt the F.E.2a design to accept the 120 hp Beardmore engine were hastily sketched out under pressures of war never before experienced. To facilitate production by contractors these changes were properly drawn up, and the designation F.E.2b was applied to the full-production form of the design. In the F.E.2b the air-brake flap on the centre-section was abandoned and the gravity tank was of a simpler but less streamlined shape; the total fuel capacity was slightly reduced from the 36 gallons of the F.E.2a to 33 gallons.

By the end of 1915 a total of 450 F.E.2bs were on order, the first contractors being G. & J. Weir, who were to build 200 (4256—4292 and 4838—5000), and Boulton & Paul, whose initial 50 (5201—5250) were augmented by a further 150

(6928—7027 and 7666—7715); additionally 6328—6377 were to be built at the Royal Aircraft Factory.

Despite the fact that the F.E.2b was the first aeroplane of any kind to be made by Boulton & Paul, 5201 was handed over to the RFC on 2 October, 1915. After examination by the AID at Farnborough it went to France on 20 October, 1915, the first true F.E.2b to go to war. On 30 October it was allotted to No.16 Squadron, in which unit it was joined by 5202 on 20 November. No.5204 arrived at No.1 A.D. on 22 December and on Christmas Day went to No.6 Squadron. The Royal Aircraft Factory had completed 6328 by 3 December, but the first two R.A.F.-built F.E.2bs to go to France (6330 and 6331) did not arrive there until 26 December. Both went to squadrons two days later, 6330 to No.16, 6331 to No.6.

In service the F.E.2b's revised petrol system was criticised in some detail by Lt-Col Brooke-Popham on 2 January, 1916, and on 1 March Brig-Gen Trenchard wrote to A.D.M.A., complaining that the F.E.2b's fuel tankage and system allowed the aircraft to spend only $2\frac{1}{2}$ hours over its reconnaissance duties, whereas $3\frac{1}{2}$ hours' operational duration was needed. He therefore asked that the F.E.'s total endurance be increased by fitting an additional 8-gallon tank. Official reaction to this request was commendably swift and sensible. Taking the view that the impending introduction of the 160 hp Beardmore would need yet more fuel, an 18 gallon tank was designed and deliveries thereof began before the end of March 1916. This additional tank was installed under the pilot's seat.

As in the case of the B.E.2c, the earliest production F.E.2bs had wings of R.A.F.6 section, and the F.E. changed over to R.A.F.14 mainplanes in the spring of 1916. The new wings were not regarded as an unmixed blessing, for it was found that above 9,000 ft they were less efficient owing to the increase in the angle of attack needed to continue climbing. By that time there were four F.E.2b squadrons in the Field, Nos.20, 22, 23 and 25, all of which had gone to France with the F.E.2b as their main equipment, and the re-equipment of Squadrons 11 and 18 followed shortly. In these two last-named units the F.E.s replaced the sturdy old Vickers F.B.5 and F.B.9.

On the earliest F.E.2bs the observer's Lewis gun was carried on the horizontal rail of the No.2 Mark I or Mark II mounting, but later aircraft had a considerable

The V-strut undercarriage on this F.E.2b suggests that it may have been 6372, which went to France on 14 April, 1916, and was issued to No.22 Squadron on 23 April. It had the 120 hp Beardmore engine and served with No.22 Squadron until 18 December, when it returned to No.1 A.D. On 24 December it was returned to England. (*T. Heffernan*)

Night-bomber F.E.2b A5478 of No.100 Squadron, a presentation aircraft, *Gold Coast No.10*, with 230-lb bomb under the nacelle and two Michelin parachute flares. The observer's Lewis gun is on a No.4 Mark IV mounting. A5478 was first allotted to the E.F. on 22 December, 1916, was issued to No.23 Squadron from No.1 A.D. on 12 January, 1917, and was withdrawn in exchange for a Spad 7 on 12 February. By June 1917 it was with No.100 Squadron, (*Ministry of Defence H.1766*)

variety of mountings. By 10 April, 1916, it had been decided to standardise the Anderson (officially the F.E.2b No.4 Mark I or Mark II) mounting, which was a fixed pillar normally (on the F.E.2b) installed centrally on the front edge of the observer's cockpit. This was followed by the No.4 Mark III mounting, which was a pillar that could be rocked transversely in a vertical plane. In turn this was superseded by the No.4 Mark IV or Clark mounting, a swivelling pillar on a universal joint that was secured to the floor of the cockpit, where it could be secured in one of several spring clips. Various mountings, mostly adaptations of the basic Anderson type, were provided between the cockpits to enable the observer to fire rearwards over the top wing. The use of these gun installations demanded great agility from F.E. observers, who frequently had to stand to shoot, regardless of the gyrations or attitude of their aircraft, to which they were usually anchored by a safety cable.

The sturdiness and resilience of the oleo undercarriage proved valuable on the sketchy, unprepared fields that had to serve as aerodromes, but the structure was heavy and created much drag. It had been suggested early in October 1915 that a plain V-strut undercarriage should be fitted instead, but RFC opinion in the Field was opposed to this. On 7 October Trenchard advised A.D.M.A. that the V-type undercarriage should not be fitted to the F.E. Nevertheless 5648, the F.E.2a with the V-strut undercarriage, went to France on 3 November, 1915, and its undercarriage was almost immediately put to the test at No.1 Aircraft Depot. The F.E.2a was deliberately dropped in a heavy landing: the tailskid broke and the axle bent, but the nacelle longerons were not damaged, whereas in the recent past heavy landings with the oleo undercarriage had resulted in cracked or crushed longerons. On 23 January, 1916, Lt-Col Brooke-Popham wrote to A.D.M.A. agreeing to accept all F.E.s with V undercarriages. By then 5648 was no longer available to continue tests. It had gone to No.6 Squadron on 7 November, 1915, but was struck off squadron strength on 21 November after a crash; although it was returned to No.1 A.D. on 27 November the Depot did not trouble to take it on charge.

On 31 March, 1916, D.A.D.M.A. wrote to Trenchard that an F.E.2b with V-strut undercarriage was ready for test and more undercarriages were being made. This F.E. was apparently 6372, which was tested at Farnborough on 7 April, was sent to France on 14 April, and was issued to No.22 Squadron on 23 April. A week later Brig-Gen E. B. Ashmore, commanding the 4th Brigade, reported to RFC Headquarters:

'The F.E.2b with the V type undercarriage climbs much better than the type with the oleo undercarriage. With full service load it does 6,000 ft in 14

These remarkable photographs graphically portray the kind of attitudes that an F.E.2b observer could be required to adopt in combat, although the retention of the shoulder stock (in the first three photographs) was unusual. In these photographs the forward mounting is a No.4 Mark III, which was officially considered obsolete by mid-1916. The rear mounting is a No.10 Mark I Anderson arch mounting with swan-neck sliding telescopic tube. The double-yoked pair of Mark II Lewis guns were not widely used operationally.
(*All photographs RAF Museum*)

minutes, and 11,000 ft in 38 minutes. It is more comfortable to fly as it is faster at all heights, quicker on controls, and does not roll so much as the type with the oleo undercarriage. From the observer's point of view there is no difference. One great advantage is that the camera is carried inside the nacelle, thus reducing head resistance.

The oleo undercarriage, however, no doubt saves many machines, particularly with inexperienced pilots.'

The V-strut undercarriage might have been adopted for general use (indeed it was specified for the one hundred F.E.2bs, A778—A877, ordered from G. & J. Weir in June 1916) had not a report of 12 May, 1916, been submitted (on 26 May)

The cockpits of an F.E.2b, the observer's gun mounting being a No.4 Mark IV, for which the port and starboard spring forks can be seen at the sides of the cockpit. There is a telescopic pillar mounting just ahead of the pilot's windscreen and what appear to be two Anderson mountings, one each side of the windscreen. Note the specially curved elevator lever at right: this curve had to be provided on the elevator levers of F.E.2bs fitted with the V-strut undercarriage. (*Imperial War Museum Q56133*)

An F.E.2b of No.51 (H.D.) Squadron flown as a single-seater night fighter from Tydd St Mary. The lower Lewis gun was apparently fixed, and the upper was on a pillar mounting. Both guns had Hutton illuminated sights and, curiously, only the 47-round ammunition magazine. (*R. C. Bowyer*)

to RFC Headquarters by the 11th Wing. The report had been made by Maj C. J. Malcolm, the C.O. of No.20 Squadron, and advised that Lt Trafford Jones, a pilot of the squadron, had successfully modified the oleo undercarriage of 6359 by removing the nosewheel and taking up its horizontal struts to the lower longerons of the nacelle. Subsequent tests in No.23 Squadron indicated that the modification added 2 mph to the speed at 6,000 ft and added 1,500 ft to the height climbed in 60 minutes.

Elaborate single-seat conversion of F.E.2b into single-seat fighter for Home Defence duties made by No.51 (H.D.) Squadron. The cockpit was moved forward and two fixed Lewis guns were carried internally; navigation lights and flare brackets were fitted. (*'The Eye', RAF Wyton*)

B704 was a rebuilt F.E.2b from the Southern Aircraft Repair Depot. When photographed it had the undercarriage simplification devised by Lt Trafford Jones of No.20 Squadron. This F.E.2b was used by No.199 Night Training Squadron. (*RAF Museum*)

Headquarters quickly agreed to the general adoption of the modification, and on 3 June Brooke-Popham wrote to A.D.A.E. asking that all F.E.2bs with the oleo undercarriage could be thus modified before being sent to France. Sadly, neither Lt Jones nor 6359 survived to see the general use of his ingenious idea. On 16 May, 1916, while flying 6359, Jones was killed in combat; his observer, Capt E. W. Forbes, was wounded in the shoulder and lung yet managed to bring the F.E. down in the British lines. The aircraft was wrecked and had to be struck off charge.

On 8 August Brooke-Popham advised A.D.A.E. that it had been decided to adopt the converted oleo undercarriage as standard in preference to the V-strut type; three days later he instructed both Aircraft Depots that oleo undercarriages would no longer be replaced by V-type undercarriages. The effect of these decisions was that at that time only a small proportion of the F.E.2bs in operational service had the V-strut undercarriage. As a further measure to improve the F.E.'s durability all aircraft overhauled in the A.D.s after 28 August, 1916, were fitted with ash lower longerons and engine bearers instead of the spruce originals, and from September all contractors for the type were instructed to substitute ash for spruce in the lower longerons.

That the F.E.'s performance was less than adequate had been apparent from the advent of the F.E.2a. A natural alternative to the 120 hp Beardmore was its 160 hp development, one of the earliest products of the great talents of Frank Halford, in which the bore of the Beardmore was increased from 130 mm to 142 mm and the stroke from 175 mm to 176 mm; the dry weight went up from 538 lb to 615 lb but the normal bhp at ground level rose from 133 to 180, an increase of 35 per cent. Precisely when it was decided to adopt the 160 hp Beardmore for the F.E.2b is not known, but as early as 22 January, 1916, D.D.M.A. wrote to Trenchard, sending design data of an F.E.2b fitted with 250 hp Rolls-Royce, an advance intimation of the F.E.2d which, as indicated on page 419, was intended to be merely a stop-gap until the 160 hp F.E.2b became available.

On 11 May, 1916, F.E.2b No.6357 was tested at Farnborough with Capt B. C.

Hucks as its pilot. This seemed to be a somewhat belated test, for 6357 had been inspected with the 160 hp engine as early as 21 February, 1916, the month in which deliveries of the new engine began. General use of the 160 hp Beardmore followed in the summer of 1916 as more engines became available, and F.E.2bs built by Weir and Boulton & Paul emerged with the 160 hp engine in increasing numbers as 1916 wore on. No.6357 itself went to France and was one of the first four 160 hp F.E.s to be allocated to No.11 Squadron on 2 June, 1916. As a 'C' Flight aircraft it lasted a few weeks until it was shot down in July.

By that time No.11 Squadron had become disenchanted with its 160 hp F.E.s. On 9 July 2/Lt Lionel Morris wrote in his notebook:

'The F.E.s are turning out very dud. The engine seems to go dud after about five hours flying.'

Five days later he wrote:

'We have only four or five machines serviceable in the squadron. It transpires that the engines they have sent out [160 hp Beardmore] have all got soft gudgeon pins and their bearings are bedded in all wrong. Consequently they develop trouble after a few hours flying.'

No.11 Squadron was not alone in its troubles with the 160 hp Beardmore, and the engine was, at least at first, much less reliable than the 120 hp original.

Although the F.E.2b was initially employed as a reconnaissance aircraft it began to be used as a bomber in the weeks preceding the Battle of the Somme, and instructions were issued on 10 June requiring F.E.2bs to be fitted with bomb

Installation of R.A.F.5 engine in F.E.2b 6360; photograph dated 10 March, 1916. (*RAE 1229; Crown copyright*)

ribs under the lower centre-section. Throughout the long agony of the Somme the F.E.s and their crews flew and fought with surprising success and outstanding gallantry, and by November some of them began to operate as night bombers. The F.E.2b proved to be, by the standards of its time, effective as a night bomber. It gave its crew exceptionally good forward vision, could lift a fair load of bombs, and under cover of darkness its lack of performance mattered little. Thus when the first British squadron to be formed as a night bombing unit came into being in February 1917 its aircraft were to be F.E.2bs. The personnel of No.100 Squadron, RFC, went to France on 21 March and received their twelve F.E.2bs at St-André a week later. The aircraft were fitted with bomb racks for 25-lb and 112-lb bombs and began their nocturnal raids from Izel-le-Hameau on the night of 5/6 April, 1917. By that time the F.E.2b had been outclassed on daylight operations, for it

Installation of a small searchlight on an F.E.2b, presumably No.4928, made at Farnborough in October 1916. This official photograph is dated 10 October, 1916. (*RAE 1812; Crown copyright*)

A more ambitiously warlike installation of a searchlight was made in the F.E.2b A781 in March 1917. A French Sautter-Harlé searchlight was coupled to two Mark II Lewis guns, the necessary power being provided by a wind-driven generator, and there were four small landing lamps under each lower wing. A781 was later used by No.199 Night Training Squadron. (*RAE 2275; Crown copyright*)

had never been a match for the agile, twin-gun fighters that the Germans had introduced in the autumn of 1916. On 3 April, 1917, IV Brigade orders stated that F.E.2bs were no longer to be used without escort on offensive patrols, and thereafter F.E.2bs photographing, bombing or patrolling normally had an escort of single-seat fighters.

Much of the history of the F.E.2 variants is inevitably bound up in the weapons they carried. The F.E.2b was used by Home Defence squadrons, sometimes modified in various ways to be a somewhat ponderous single-seater but usually armed with a single Lewis gun. Various armament installations were tried on F.E.2bs. These included double-yoked twin Lewis guns, a Vickers pom-pom (first fitted to 6377 in May 1916), a 0·45-in Maxim gun, as on 7001 in August 1916, an ingenious but cumbersome combination of two Lewis guns and a Sautter-Harlé searchlight on A781 in March 1917 (perhaps inspired by the smaller searchlight installation on 4928 in October 1916), the Royal Aircraft Factory gyroscopic bomb sight tested on 4256 and 6360 in the summer of 1916, and various attempts to adapt the F.E. to carry larger and heavier bombs.

Orfordness tried a 336-lb R.A.F. bomb on a 160 hp F.E.2b in October 1917 but this seems not to have found operational employment. In the Field, No.100 Squadron experimented with the carrying of a 230-lb bomb under the nacelle and reported success on 1 June, 1917. Two days later the First Brigade were instructed to send the modified aircraft to No.2 A.D, where six more bomb racks and fittings were to be made. These proved successful, and on 6 October, 1917, RFC Headquarters sent to M.A.2 a telegram:

> We want FETUBIS* to carry 230-lb bomb. They will not do this with Oleo undercarriage and must be fitted with Vee undercarriage.

And that is why the night-bomber F.E.2bs came to have the V-strut undercarriage as standard equipment: it was impossible to fit the 230-lb bomb and its rack when the F.E.2b had the more complex oleo undercarriage.

The F.E.2b bombers were given a modest increase in range by replacing the cylindrical 24-gallon petrol tank with the rectangular 36-gallon main tank of the F.E.2d. On 1 November, 1917, No.2 A.D. was instructed that all F.E.2bs issued to No.100 Squadron thereafter were to have the long-range tanks fitted.

The growing use of the F.E.2b led to the resurgence of its production, particularly in 1918, when the totals of completed aircraft for the four quarters of the year were respectively 15, 140, 185 and 202. On 31 October, 1918, the Royal Air Force had 563 F.E.2bs and 2cs on charge, a remarkable statistic in view of the type's origins dating back to the first few weeks of war.

To add to the destruction caused by their bombs, No.100 Squadron also used a few F.E.2bs armed with Vickers 1-pounder quick-firing guns. Firing trials of such a gun installed in a 120 hp F.E.2b flown as a single-seater had been conducted at Orfordness in August 1916. The weapon had a 40-round belt and the F.E. had a CFS Landing Searchlight Mark II fixed parallel to the axis of the gun and moving with it. By March 1917, No.51 (Home Defence) Squadron had five F.E.2bs armed with the Vickers pom-pom, and Brooke-Popham asked that two be sent to France, specifying that they should have the 160 hp engine and asking for 1,000 rounds of ammunition for their guns. These were sent to No.100 Squadron on 7 April and one of them fired 20 shells in anger on the night of 17/18 April, when trains in the Douai area were attacked. The pom-pom-carrying F.E.2bs had extensively modified nacelles to accommodate the heavy guns and ammunition.

Although it was intended to replace the F.E.2b by Handley Page O/400s and Vimys, this process had made little headway by the time of the Armistice, and the night-bomber version of the design was still in service with Squadrons 38, 83, 101, 102, 148 and 149 of the RAF; some were also with 'I' Flight at Erre.

The F.E.2c variant re-appeared in France by accident. As early as October 1915 the Superintendent of the R.A.F. had been authorised to complete eight aircraft of the batch 6328—6377 as F.E.2cs, but it seems that only 6370 and 6371 were so made. They had been completed by 20 March, 1916; 6370 remained at Farnborough for experimental purposes, but, in the belief that it was an F.E.2b, 6371 was mistakenly sent to France on or about 22 April. Trenchard immediately protested to D.A.E. but when A.D.A.E. asked for the aircraft to be returned Brooke-Popham replied that this was impossible as there were not enough aircraft to meet requirements even with the F.E.2c included.

Thus did 6371 come to have an operational career, for it was allotted to No.22 Squadron and subsequently went to No.25 Squadron on 19 June. It did not last

*Telegraphic abbreviation for F.E.2bs.

F.E.2b of No.100 Squadron modified to accommodate a Vickers one-pounder quick-firing gun. (*Ministry of Defence H1767*)

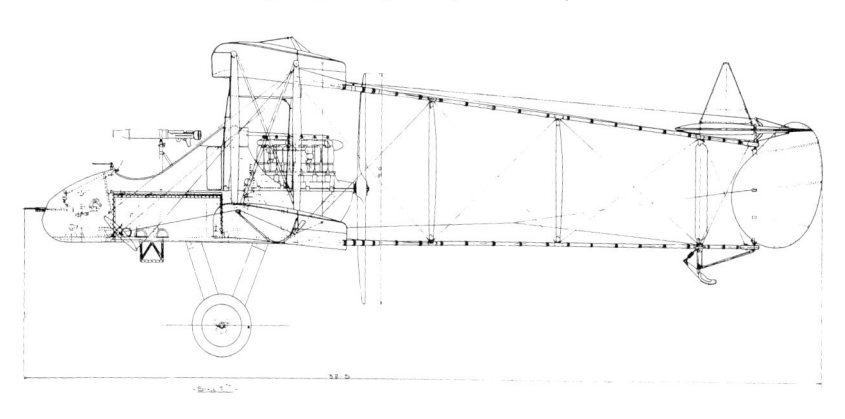

Royal Aircraft Factory drawing of the original F.E.2c design. (*RAE; Crown copyright*)

The F.E.2c No.6370 photographed at Farnborough. The presence of an instrument panel on the 'goal-post' mounting between the seats suggests that the aircraft was being used for experimental purposes. As the photograph is dated 24 July, 1916, the S.E.4a seen on the left is probably No.5611. (*RAE; Crown Copyright*)

long, for it crashed on 17 July, 1916, and was struck off charge, its flying time being recorded as 105 hr 47 min.

That was not the end of the F.E.2c story, however. Late in 1917 six F.E.2bs were converted by the AID at Farnborough to have modified nacelles providing the reversed seating arrangement of the F.E.2c and were so designated. These were A5744, B434, B445, B447, B449 and B450; all six were officially allotted to the RFC in France on 28 November, 1917, and some had reached No.1 A.S.D. by 26 December, 1917, at which time instructions were given for them to be fitted with the F.E.2d main fuel tank. One went to No.100 Squadron in January 1918 and was an instant success. On 17 January Maj J. E. A. Baldwin of the 41st Wing wrote informally to Capt E. R. L. Corballis at RFC Headquarters:

'No. 100 Squadron swear by the F.E.2c and are all very keen to get this machine, if possible. Apparently its performance is very much superior to any other Beardmore F.E. Tempest [Maj W. J. Tempest, DSO, MC, O.C. No.100 Sqn] thinks it is ideal. If there are any more available, is it possible to

F.E.2c night-bomber conversion with the pilot in front. This was a No.100 Squadron aircraft. (*R. C. Bowyer*)

get them allotted to No.100, as vacancies fall due? At the present moment they have got one in the Squadron. Apparently a regular fight ensues as to who is to fly it.'

By that time the VIII Brigade, the forerunner of the Independent Force, was in being, and indicated that they would like to have as many F.E.2cs as possible. On 12 February, 1918, Brooke-Popham asked D.A.E. to ensure that 25 per cent of F.E.s delivered would be F.E.2cs, but it proved impossible to translate this into production aircraft until July 1918. On 13 July, E7112 had been completed as an F.E.2c but crashed a week later on arrival in France; the next production F.E.2c, E7113, was not ready until 28 August and its movements have yet to be traced. It was intended that a part of the deliveries from the batch of 50 (H9913—9962) ordered from Ransomes, Sims & Jefferies in September 1918 should be delivered as F.E.2cs, and apparently it was eventually decided that the first 24 aircraft off

The first F.E.2h, A6545, at Martlesham Heath. (*Ministry of Defence H1786*)

the contract should be completed in F.E.2c configuration but it is doubtful whether any was delivered. Nevertheless, at least 18 of the F.E.2bs called for by the contract (H9937—9954) are known to have been completed by 23 January, 1919.

Various alternative engine installations were made in or designed for the F.E.2b. The 150 hp R.A.F.5 engine was fitted to 6360 in February 1916 and to 4256 in April. This engine was a pusher version of the R.A.F.4a, which fact may account for a reference to an F.E.2b with R.A.F.4a that was used at Orfordness to test the R.A.F. Gyroscopic Bombsight: as earlier noted, both 4256 and 6360 were used in this capacity. References to experimental installations of the 170 hp R.A.F.5b* and 200 hp R.A.F.3a are to be found in some official documents but it is not known whether these were ever more than projects.

The last known engine installation was of a 230 hp Siddeley Puma, with which the aircraft was designated F.E.2h; it was, of course, hoped that performance would be substantially improved. This was initiated on 21 November, 1917, when official approval was given to allotting one F.E.2b to Ransomes, Sims & Jefferies of Ipswich, who were to undertake the conversion. The selected airframe was A6545 and the basic work had been completed by 20 February, 1918, when the translation of the aircraft to Martlesham Heath for erection was imminent. A6545 arrived there on 27 February, 1918, and at once embarked on a strenuous test

*A development of the R.A.F.5 with bore increased to 105 mm: it is doubtful whether this engine was ever completed.

programme; it was tested with both the oleo and V-strut undercarriages. Unfortunately, the F.E.2h's performance was very little better than that of the F.E.2b, and although A6545 went to Orfordness on 17 May it had already been decided that the Puma installation was not sufficiently successful to warrant its adoption. Nevertheless, three more F.E.2hs were created by converting the F.E.2ds A6501—6503; again the work was done by Ransomes, Sims & Jefferies, but it included modification of the nacelles to carry a 6-pdr Davis gun at a near-vertical downward angle, obviously with a view to attacking ground targets. Perhaps this was inspired by the use of the Vickers 1-pdr gun on some of No.100 Squadron's F.E.2bs, but inevitably it proved to be impracticable. The three later F.E.2hs were sent to Grain with the new identities of E3151—3153, and their Davis guns were tested there.

F.E.2b, 2c and 2h

120 hp Beardmore, 160 hp Beardmore, 150 hp R.A.F.5. The F.E.2h had the 230 hp Siddeley Puma

Two-seat fighter-reconnaissance, two-seat bomber, two-seat or single-seat Home Defence fighter.

Span 47 ft 9 in; length 32 ft 3 in; height 12 ft 7½ in; wing area 494 sq ft.

Weights and Performance:

	F.E.2b 120 hp	F.E.2b 160 hp	F.E.2b 160 hp with V under-carriage	F.E.2h Oleo under-carriage	F.E.2h V under-carriage
Weight empty, lb	1,993	2,061	2,042	2,280	2,255
Weight loaded, lb	2,967	3,037	3,036	3,355	3,190
Maximum speed, mph					
at sea level	80·5	91·5	90·1	—	—
at 6,500 ft	—	81	—	—	—
at 10,000 ft	72	76	—	87	89
Climb to					
3,000 ft, min and sec	9 50	7 24	7 24	— —	— —
6,500 ft, min and sec	— —	— —	18 43	12 20	11 5
10,000 ft, min and sec	51 45	39 44	39 44	23 10	20 0
Service ceiling, ft	9,000	11,000	—	14,000	14,500
Endurance, hr	3	—	—	—	—

Armament: One, two, or occasionally three 0·303-in Lewis machine-guns; a few F.E.2bs had one Vickers 1-pdr quick-firing gun or one 0·45-in machine-gun. Various combinations of 25-lb, 40-lb, 112-lb and 230-lb bombs, to a total of about 350 lb.

Manufacturers: Royal Aircraft Factory; Barclay, Curle; Boulton & Paul; Richard Garrett; Ransomes, Sims & Jefferies; G. & J. Weir (with some assembly contracts undertaken by Alexander Stephen & Sons and by Barclay, Curle). One F.E.2b (A8950) built by Blackburn.

Service use: *Western Front*: Sqns 6, 11, 12, 15, 16, 18, 20, 22, 23, 25, 58, 83, 100, 101 and 102; Special Duty Flight. *Home Defence*: Sqns 33, 36, 38, 51, 58 and 76. *Training Duties*: Reserve/Training Sqns 9, 10, 19, 27, 46; Night Training Sqns 188, 191, 192, 199 and 200; School of Aerial Gunnery, Hythe. *Australia*: A778 to Australian Central Flying School, Point Cook, where it was numbered CFS 14. *India*: A790

Serials: **F.E.2b**—4256—4292, 4838—5000, 5201—5250, 6328—6377, 6928—7027, 7666—7715, A778—877, A5438—5487, A5500—5599, A5600—5899, A6571—6600, A8950, B401—500. C9786—9835, D3776—3835, D9081—9230, D9900—9999, E6687—6736, E7037—7136, F2945—2994, F3498—3547, F3768—3917, F4071—4170, F9296—9395, H9937—9962, J601—650.

F.E.2c—6370, 6371, A5744, B434, B445, B447, B449, B450, E7112, E7113, H9913—9936.

F.E.2h—A6545, E3151—3153 (ex A6501—6503).

Rebuilds: A8895.

B704, B4005, B7779, B7782, B7788, B7794, B7795, B7800, B7808, B7809, B7813—7816, B7836—7841, B7843, B7847, B7848, B7856, B7872, D9740—9789 erected from spares, F5852 (ex A5658), F5853, F5854, F5856, F5858, F5859, F5861, F5862, F5863, F6071, F6080, F6081, F9550, F9566, H6888 (ex C9829), H7145 (ex A5783), H7176 (ex A6565), H7178 (ex A6600), H7179 (ex D9759), H7180 (ex D9776), H7228 (ex A6504) H7229 (ex D9757), H7230 (ex A5607), H7231 (ex B460), H7233 (ex A5789).

F.E.2d

On 22 January, 1916, the Deputy Director of Military Aeronautics wrote to Brig-Gen Trenchard, sending design data of a development of the F.E.2b powered by a 250 hp Rolls-Royce engine and stating that it was hoped to provide a squadron of the new type in April 1916. That hope was somewhat optimistic, for the prototype of the Rolls-Royce powered F.E. was only submitted for its pre-flight inspection on 4 April. Designated F.E.2d, it first flew on 7 April with Frank Goodden at the controls and was sent next day to have its cockpit (presumably that of the pilot) cut away. By 20 April it had received the serial number 7995. Its engine was the 31st of the 250 hp Rolls-Royce Mark I batch, and it is clear that the War Office must have secured a promise of a substantial number of Rolls-Royce engines, for a batch of forty F.E.2ds was ordered from the Royal Aircraft Factory early in April 1916 under War Office authority.

The prototype was tested at CFS on 7 May, 1916. It was reported to be superior to the F.E.2b in speed and climb, but somewhat slower to manoeuvre and heavier to land; in flight it was nose-heavy and heavy on the controls. The engine installation and lubrication system were criticised, and the structural adequacy of the airframe was questioned. All of this was a little late in the day, however, for the first production aircraft, A1, was almost completed; it was with the AID at Farnborough on 12 May, and was officially allotted to the Expeditionary Force on 1 June, 1916.

Production of the first batch of forty F.E.2ds was completed by the end of August. The first 30 had the 250 hp Rolls-Royce Mark I as original fit, but in A31—40 the Mark III engine was installed. A38—40 were allotted to the Expeditionary Force on 2 September, but on 7 October A40 was re-allotted to the Royal Aircraft Factory to be tested to destruction. The production F.E.2ds did not perpetuate the stack-type exhaust manifolds that had originally been fitted to the prototype, but had short horizontal manifolds with forward outlets.

Early F.E.2d with rectangular radiator. The Lewis gun is on the middle one of three separate Anderson mountings in the nose, and there is a straight telescopic pillar on an Anderson arch at the back of the observer's cockpit.

Remarkable as it may now seem, the F.E.2d was initially seen as nothing more than a stop-gap type. On 29 May, 1916, the Director of Aeronautical Equipment had written to Trenchard:

> '. . . it is not at present proposed to perpetuate the type of the F.E.2d . . . It is considered that the 160 hp F.E.2b will prove a better machine from the point of view of manoeuvring power. But as the 160 hp engine is only beginning to come forward, and as the performance of the F.E.2d is considerably superior to that of the 160 hp or 120 hp F.E.2b, the use of this machine as a stop-gap to fill the demands you have made for this class of machine cannot possibly be neglected.'

In reply, Brig-Gen Brooke-Popham advised D.A.E. on 2 July that the F.E.2ds would be used to replace the F.E.2bs in No.20 Squadron, the latter type then to be used to replace wastage and complete establishments in other squadrons. Little time was lost in getting the F.E.2ds to the squadron: A1 was with No.20 by 17 June, 1916, by which date the first 12 aircraft had been officially allotted to the Expeditionary Force. On 27 June the 11th Wing reported that 7995, the prototype, and A13 had been taken over from No.1 Aircraft Depot.

Unfortunately, one of the first F.E.2ds to cross the Channel was delivered virtually intact to the enemy. A5 left Farnborough on 1 June, 1916, bound for St-Omer; its pilot was 2/Lt S.C.T. Littlewood, whose total flying time was 32 hours and who had neither flown across the Channel nor in France. His passenger was Lt D. Lyall Grant, who likewise was unfamiliar with northern France. Littlewood misinterpreted his map, crossed the lines at Armentières, and came down on the aerodrome of F1 Abt 292 at Haubourdin, near Lille. His F.E. stood on its nose but sustained little damage; in particular its Rolls-Royce engine, No.1/250/89, was intact and the F.E.2d's secret was out before one had fired a shot in combat. A5 was a typical early-production F.E.2d, having a low-cut coaming about the pilot's cockpit, a telescopic gun-mounting between the cockpits, and three individual Anderson gun mountings round the forward periphery of the front cockpit.

No.20 Squadron had thirteen F.E.2ds on 1 July, 1916, and was the only operational unit using the type until No.57 Squadron arrived in France with a full complement of F.E.2ds. No.20 therefore had to cope with all the problems associated with a new type, and to report on its performance and qualities. On 28 July, 1916, it was decided that all wing panels on F.E.2ds were to be internally cross-braced with swaged tie-rods instead of the piano wire used on the F.E.2b, and on 15 August the 2nd Brigade was instructed to apply the Trafford Jones modification to the undercarriage and report on its suitability for the F.E.2d. The undercarriage modification was adopted for the F.E.2d with effect from 24 September, and on 15 October RFC Headquarters requested officially that all F.E.2ds sent to France should have the modified oleo undercarriage. Appropriately, No.20 Squadron had modified 7995 as the test aircraft by 2 September.

Also on 2 September, Brooke-Popham issued an order specifying the standard gun-mountings for the F.E.2d as the Clark swivelling pillar (No.4 Mark IV) and the telescopic Anderson mounting between the cockpits. On 28 September the RFC in France was advised that F.E.2ds from A1932 would be fitted with the Anderson arch mounting (No.10 Mark II) between the cockpits. By early 1917 the F.E.2d's armament was augmented by fitting a fixed Lewis gun fired by the pilot. This was on a Dixon-Spain mounting* within the partly cut-away decking

*Presumably invented by Capt G. Dixon-Spain of No.20 Sqn.

The luckless F.E.2d A5, delivered direct to the Germans on 1 June, 1916, by 2/Lt S. C. T. Littlewood. (*Egon Krueger*)

The nacelle of F.E.2d A5, which had the large rectangular radiator, three Anderson mountings forward and straight telescopic pillar on Anderson arch at the rear of the observer's cockpit. (*Egon Krueger*)

between the cockpits, usually to starboard; some F.E.2ds had two Lewis guns mounted in this way. Double Lewis guns were tried for a time on the forward Clark mounting but were not popular with observers.

In response to RFC Headquarter's request for an assessment of the F.E.2d, Maj W. H. C. Mansfield, O.C. No.20 Squadron, submitted a long and objective report on 13 August, 1916, to the 11th Wing. Brig-Gen T. I. Webb-Bowen sent it to HQ that day, adding his own comments:

> 'I consider the F.E.2d a most efficient machine for an Army Wing whose principal duties are long reconnaissance, patrols, and escort duty; it has not been used for Artillery work. As a fighting machine it is in my opinion only beaten by the Nieuport Scout; it can outfly a de Havilland [*i.e.*, D.H.2], and is the only machine in this Brigade which can dive with engine all out at really steep angles.
>
> It is open to improvement as regards the radiator which is far too large, and the undercarriage; this latter I believe can be converted into a simple Vee without front wheel and should be a marked improvement.
>
> The Rolls-Royce engine is running well and has not shown any serious mechanical troubles as yet. Plugs and air pressure being the only difficulties encountered.
>
> The best testimonial this machine can receive is the fact that enemy machines will never engage it if they can possibly avoid it.'

The radiator troubles were caused by the excessive size of the rectangular radiator that was initially fitted. Apparently the addition of partial shutters did not improve matters sufficiently, but it was found that adequate cooling was provided by using the same radiator as used on the F.E.2b but mounted externally ahead of the engine. Shutters were later added when the winter of 1916–17 came.

It may have been intended to build relatively few F.E.2ds and perhaps to confine such limited production to the Royal Aircraft Factory, where 85 were built (A1—40, A1932—1966 and A5143—5152). The later aircraft had later Marks of Rolls-Royce, the 250 hp Mark IV engine being fitted to thirteen aircraft of the second batch and seven of the third; the others had Mark I or Mark III engines, but A5151 had the 275 hp Rolls-Royce Mark I and was apparently the first F.E.2d to have this engine, which was later named Eagle V. To provide additional aircraft to maintain the operational squadrons, some of the F.E.s ordered from Boulton & Paul under Contract No.87/A/658 were built as F.E.2ds and some of them had the 275 hp Mark II (Eagle VI) engine. In service the later Marks of engine occasionally replaced earlier versions when engine changes were made.

The F.E.2d gave a good account of itself over the Western Front, mostly on fighter-reconnaissance duties but latterly undertaking bombing with Cooper bombs. The only Victoria Cross to be won by a non-commissioned officer of the RFC was posthumously awarded to Sgt Thomas Mottershead of No.20 Squadron for his action on 9 January, 1917, when, flying A39, he brought the F.E. down in flames from 9,000 ft to save the life of his observer, Lt W. E. Gower. Mottershead was trapped in the F.E. when it crashed, but Gower was thrown clear and survived.

As late as April 1917, No.25 Squadron was re-equipped with F.E.2ds but by that date the type was outclassed; moreover, the advent of the D.H.4 demanded that the excellent Rolls-Royce engines be put to more effective use in the new aircraft. It is a matter of history that many Rolls-Royce engines intended for F.E.2ds were taken back and converted for D.H.4s, consequently F.E.2d production could no longer be sustained. The main Boulton & Paul batch was completed with deliveries as F.E.2bs, and it is likely that some F.E.2ds were converted to F.E.2bs. In France No.57 Squadron began to re-equip with D.H.4s in June 1917, and No.25 Squadron received D.H.4s in July. With No.20 Squadron the F.E.2d remained in service until the autumn of 1917, and aircraft of this unit shot down two of Germany's leading fighter pilots that summer: Karl Schaefer on 5 June, and Manfred von Richthofen on 6 July. Although Schaefer died, von Richthofen recovered from his wound and returned to combat.

This photograph of F.E.2d A27 in German hands clearly shows the Beardmore-type radiator fitted with partial shutters. The Lewis gun is on a telescopic pillar mounting attached to an Anderson arch. A27 originally had a 250 hp Rolls-Royce Mark I engine and went to France on 3 August, 1916. It joined No.20 Squadron on 2 September but returned to No.1 A.D. five days later for repair. Re-issued to No.20 Squadron on 30 December, 1916, with a 250 hp Mark IV engine, it lasted until 17 March, 1917, when Lt Anderson and Lt D. B. Woolley were shot down by Feldwebel Hippert of Fliegerabteilung 227.
(*Egon Krueger*)

Capt F. D. Stevens and Lt W. C. Cambray, MC, with their F.E.2d A6516 of No.20 Squadron at Ste-Marie Cappel. The forward Lewis gun is on a No.4 Mark IV (Clark) mounting, the pilot's fixed Lewis on a Dixon-Spain mounting, and the rear gun on a telescopic pillar on an Anderson arch. This F.E.2d had a 275 hp Rolls-Royce engine and was first allotted to the E.F. on 8 June, 1917.

Some F.E.2ds were allocated to Home Defence squadrons, twenty-three in 1917 and two in 1918, but the type's performance was not good enough to counter the types of German airships then in use and it saw only limited employment in this way. The H.D. aircraft had their losses too: on 31 March, 1917, during an air-raid alert, A12 of No.33 Squadron crashed, and the same squadron lost A6351 on the night of 14/15 August. On 25 September, 1917, A6461 of No.36 Squadron failed to return from air-raid patrol. On 13 March, 1918, 2/Lt E. C. Morris of No.36 Squadron, with 2/Lt R. D. Linford as his observer, coaxed his F.E.2d up to 17,300 ft to attack a Zeppelin over Hartlepool but could not climb close enough to attack effectively. Both officers fired at the airship and pursued it 40 miles out to sea before losing it in cloud.

The F.E.2d saw very little use with training units, presumably because its Rolls-Royce engines were in such great demand for D.H.4s that few F.E.2d airframes retained their full identities. In all, only 40 were issued to training units, 14 in 1916, 23 in 1917 and three in 1918. One of these was probably the prototype, 7995, which survived operational use with No.20 Squadron and was allotted to the Training Brigade on 27 November, 1916.

In April 1917 the design department of the Royal Aircraft Factory was working a seaplane conversion of the F.E.2d. The intended operational application of this variant is unknown, and it is doubtful whether any F.E.2d was so converted.

F.E.2d

250 hp Rolls-Royce Marks I, III and IV; 275 hp Rolls-Royce Marks I and II. (After the introduction of the name Eagle for this family of engines, these became respectively the 225 hp Eagle I, 284 hp Eagle III, 284 hp Eagle IV, 322 hp Eagle V and 322 hp Eagle VI.) Two-seat fighter-reconnaissance.

Span 47 ft 9 in; length 32 ft 3 in; height 12 ft $7\frac{1}{2}$ in; wing area 494 sq ft.

Empty weight 2,509 lb; loaded weight 3,469 lb. Weights with 250 hp Rolls-Royce Mark I.

With 250 hp Rolls-Royce Mark I: Maximum speed at 5,000 ft—94 mph, at 10,000 ft—88 mph; climb to 5,000 ft—7 min 10 sec, to 10,000 ft—18 min 20 sec; service ceiling 17,500 ft; endurance $3\frac{1}{2}$ hr.

Armament: One, occasionally two, fixed 0·303-in Lewis machine-guns; one or two free-mounted Lewis guns; six 20-lb or 25-lb bombs.

Manufacturers: Royal Aircraft Factory, Boulton & Paul.

Service use: Western Front: Sqns 20, 25 and 57. *Home Defence*: Sqns 33, 36, 39 and 78. *Training duties*: No.46 Training Sqn, Bramham Moor; No.59 Training Sqn, Yatesbury; No.98 Depot Sqn, Rochford. *Squadron working up*: No.28.

Serials: 7995, A1—40, A1932—1966, A5143—5152, A6351—6570 (many of this batch completed as or converted to F.E.2b), B1851—1900 (some conversions to F.E.2b).

F.E.4

Work on the design of the F.E.4 began in the summer of 1915. It was designed by S. J. Waters and H. P. Folland and was an ambitious attempt to produce a multipurpose aircraft, for three main functions were envisaged. These were set down in an official document, undated but apparently of late December 1915:

> 'A. Short distance fighter carrying 4 hours fuel, pilot and 2 gunners, 2 guns and mountings, 1,000 rounds. Total weight 5,340 lb.
> B. Short distance bomb dropper. 3 hours fuel, two 599-lb bombs, release gear. Total weight 3,550 lb.
> C. Long distance bomb dropper, as B. Total weight 6,000 lb.
> Two to be built by R.A.F. 100 from Daimler on order. Arrangements now being made for 8 hours petrol at full speed.'

The same document attributed two R.A.F.3a engines to the F.E.4, indicated that a speed of 98 to 104 mph was expected and a climb to 6,000 ft with a load of 2,000 lb in 15 minutes. Furthermore, the aircraft was to be capable of flying on one engine only. Significantly, it was clear that one hundred F.E.4s had been ordered from the Daimler company at that very early stage, long before a prototype had been completed. In fact, the War Office agreement with the Daimler company was formally signed on 30 December, 1915, after weeks of preliminary discussions and preparations.

The reference to modifications providing tankage for 8 hr flying must have been inspired by a memorandum sent to A.D.M.A. by Brig-Gen Trenchard on 11 December, 1915:

'Col Brooke-**Popham** informs me that the F.E.4 is designed for a petrol capacity of 7 hours supply and extra tanks can be put in for a 15 hours supply. In my opinion neither the one nor the other is required. It is 8 hours supply that is wanted with a maximum of 9 hours . . . . 7 hours is not enough and 15 hours is too much.'

The first F.E.4 was submitted for final pre-flight inspection on 8 March, 1916, and flew for the first time under Frank Goodden's pilotage on 12 March. It could be seen to be a very large biplane with unequal-span wings from which the fuselage was underslung. The wide-track undercarriage had substantial oleo legs that were attached to the main spars of the lower centre-section at their outer ends, and there were bumper wheels at the nose of the fuselage to prevent damage when landing on a rough surface. A biplane tail unit enclosing three rudders and a single central fin was fitted. Presumably no examples of the R.A.F.3a engine were available at the time, for the aircraft had two 150 hp R.A.F.5 air-cooled engines (indeed, Napier output of the R.A.F.3a did not begin until July 1916). No great speed was expected, for cooling fans were mounted on forward extensions of the engine crankshafts.

The first F.E.4 with R.A.F.5 engines, photographed at Farnborough with Frank Goodden in the pilot's seat. (*RAE*)

On this F.E.4 the pilot sat in the foremost cockpit and the front gunner sat behind him; inevitably, he would have to stand to make use of his Lewis gun or guns. A cockpit for a rear gunner was provided behind the wings. Dual control was installed in the forward gunner's cockpit.

In this form the F.E.4 was flown several times at Farnborough, having its rear fuselage modified and new two-blade fans fitted; on 26 April, 1916, Frank Goodden flew it before King George V at Farnborough with W. S. Farren and a Mr Pearce as passengers. By that time it had received the official number 7993, and as such Frank Goodden flew it to CFS on 11 May for official trials. At that

The first F.E.4, painted in P.C.10 and with its serial number on the fin, at Central Flying School.

time it had five petrol tanks with a total capacity of 193 gallons, and it was flown with two Lewis guns but with the aft cockpit unoccupied. By then it was proposed to place a gunner's cockpit on top of the upper centre-section to provide an all-round field of fire, possibly because it had been realised that the biplane tail unit obstructed a large part of the field of fire from the rear cockpit.

Performance proved to be lamentable, with a speed of 84·3 mph at ground level, a climb to 6,500 ft in 47 minutes, and a service ceiling of only 5,500 ft. The report indicated that stability was good, apart from a tendency to bear to the left, but included the considerable understatement 'The machine appears to be under engined with two 150 hp R.A.F.5 engines.'

A second prototype was being built, and its conversion to have two 250 hp Rolls-Royce engines had begun as early as 23 March, 1916, possibly because early flights of the first aircraft had made it clear that its performance was unacceptable. Design drawings of the Rolls-Royce F.E.4 are dated October 1916 and depict an elevated gunner's cockpit above the upper centre-section, with the gravity tanks concealed within its forward and rear fairings; there were to be curved skids under the nose in place of bumper wheels, and rectangular radiators, similar to those fitted to the early F.E.2ds. The F.E.4 that had been completed by 3 July, 1916, had no gunner's cockpit, either elevated or in the rear fuselage, but it had bumper

The second F.E.4, with two 250 hp Rolls-Royce engines, photographed at Farnborough. (*Imperial War Museum Q57634*)

wheels as on 7993, and round-top radiators that may well have been of the Beardmore type adopted as standard for the F.E.2d. It had increased fuel capacity provided by two 50-gallon, one 100-gallon and two gravity tanks of 25 gallons each.

The Rolls-Royce F.E.4 was numbered 7994 and made its first flight on 5 July, 1916, piloted by Goodden, who later flew the aircraft on performance tests at Farnborough on 26 July. No document yet found suggests that 7994 ever went to CFS for official trials, but it continued to fly occasionally at Farnborough. It was fitted with bomb-dropping gear early in September but its only recorded bomb test did not take place until 6 November, 1916. On 23 October Goodden had taken up Brig-Gen W. Sefton Brancker as passenger in 7994. At that time Brancker was Director of Air Organisation and a member of the Air Board; as such he was much concerned with the aircraft proposed for the operational use of the RFC.

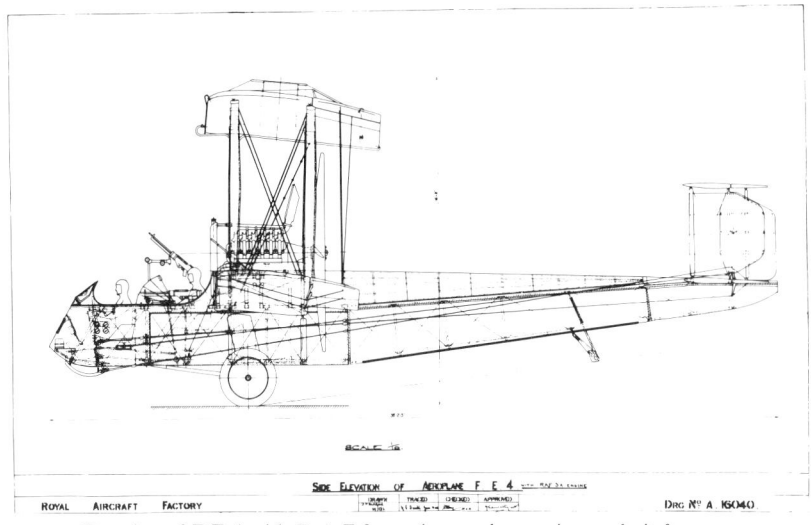

Drawing of F.E.4 with R.A.F.3a engines and overwing cockpit for gunner.
(*RAE; Crown copyright*)

Detailed general-arrangement drawings of the version to be powered with the R.A.F.3a engine had been completed early in June 1916. These had probably provided the basis for the later drawings of the Rolls-Royce version, for they incorporated the overwing gunner's cockpit, nose skids and all other details. This may have been intended to be the production version of the F.E.4, for the Daimler contract included a statement to the effect that the War Department would supply either R.A.F.5 or R.A.F.3a engines for the aircraft at their discretion. It appears that, at the time of the production contract, the Daimler company were expected to be contractors for the R.A.F.3a engine, but in fact only the Armstrong Whitworth and Napier companies produced the engine. The serial numbers A2020—2119 were allotted on 11 July, 1916, for the production aircraft.

The performance of the Rolls-Royce F.E.4 was somewhat better than that of 7993, but it was still not good enough to meet the changing conditions of the autumn of 1916. The prospects of obtaining 200 Rolls-Royce engines were remote, and with the R.A.F.3a the F.E.4's performance could only be substantially worse than that of 7994. Perhaps this was realised in time; perhaps Brancker himself concluded that the aircraft would be elephantine and operationally useless; certainly the Daimler contract was cancelled before any production F.E.4 was completed.

The designation F.E.4a has been associated with the Rolls-Royce version of the type, and although no official document or drawing yet found confirms this, a letter of 24 December, 1915, written by D.G.M.A. to the Daimler company refers to 'the proposed Contract for F.E.4B aeroplanes'. This may have been meant to signify the version with R.A.F.3a engines, but this, too, remains unconfirmed.

In July 1917, fleeting consideration was given to resuscitating 7993 and 7994, which were then lying engineless at the Royal Aircraft Factory. At the Progress and Allocation Committee meeting of 10 July, 1917, General Brancker, referring to the F.E.4s,

'. . . stated there were two old machines at the R.A.F. which might be used for night work, but they required engines. Col Whittington was asked to look into the matter and see what engines could be spared.'

Next day the colonel reported that the only engines available for the F.E.4s were 220 hp Renaults or 190 hp Rolls-Royces, and the Committee was told that only the latter were suitable. Nothing came of this, doubtless because the 190 hp Rolls-Royce was wanted for the Bristol Fighter, and the F.E.4 faded into well-deserved obscurity.

F.E.4

Two 150 hp R.A.F.5; two 250 hp Rolls-Royce Mark I

Three-seat multi-purpose aircraft.

Span 75 ft 2 in (upper), 62 ft 6 in (lower); length 38 ft 2½ in; height 16 ft 9 in; wing area 1,032 sq ft.

R.A.F.5 engines: Empty weight 3,754 lb; loaded weight 5,988 lb.

Rolls-Royce engines: Loaded weight 7,825 lb.

R.A.F.5: Maximum speed at sea level 84·3 mph; climb to 3,000 ft—14 min 40 sec, to 6,000 ft—38 min; service ceiling 6,500 ft.

Rolls-Royce: Maximum speed at 5,000 ft—88 mph, at 10,000 ft—73 mph; climb to 10,000 ft—30 min 5 sec; absolute ceiling 12,000 ft.

Armament: Up to three 0·303-in Lewis machine-guns in fighter version; approximately 1,200 lb of bombs in bomber version.

Manufacturers: Royal Aircraft Factory, Daimler.

Serials: 7993—7994, A2020—2119 (cancelled).

The first prototype F.E.8, No.7456, in its original form with spinner and remotely-controlled Lewis gun in the nose of the nacelle. (*RAE 974; Crown copyright*)

F.E.8

The Royal Aircraft Factory had always been in the forefront in the matter of arming military aircraft, and it is not surprising that its design staff should, early in the war, explore the possibility of building a single-seater armed with a machine-gun. Work on such a design began at Farnborough under the leadership of J. Kenworthy in May 1915, some weeks before the D.H.2 first appeared at Hendon. At that time no British gun-synchronising mechanism of any kind existed, and the adoption of the pusher configuration was more or less inevitable.

Designated F.E.8, the new aircraft was the first true single-seat fighter to be designed at the Royal Aircraft Factory. Two prototypes, 7456 and 7457, were built at Farnborough, the first being completed on 15 October, 1915. It was a classic nacelle-and-tailbooms pusher powered by a 100 hp Gnome Monosoupape rotary engine driving a four-blade propeller; its clean little nacelle was virtually an all-metal component, having a welded steel-tube frame clad in sheet aluminium. The wings were conventional surfaces with wooden spars and fabric covering; their centre-sections were of wide span, and the outboard panels were rigged with marked dihedral. The armament was a single Lewis gun which, on the prototypes, was mounted low down in the nose of the nacelle and had limited movement, controlled by the pilot who operated a sighting bar connected to the gun by linkage. Additionally, the gun could be fixed and aimed by aligning the whole aircraft on the target.

The first flight made by 7456 was on the morning of 15 October, 1915, with Frank Goodden at the controls. It went to CFS for official assessment on 8 November, and in the test report its flying qualities were praised; not surprisingly, the low and inaccessible position of the gun was criticised. Unfortunately, on its return flight to Farnborough on 15 November it crashed badly on landing while being flown by Lt B. C. Hucks and required extensive repairs. The second

The second prototype F.E.8, with remotely controlled Lewis gun in extreme nose of nacelle.

prototype was by then well advanced and was completed by 2 December, identical in all respects with its predecessor. This F.E.8, No.7457, was flown to France by Goodden on 19 December for evaluation by operational pilots of the RFC. By then the Fokker monoplane fighters had become a serious menace on the Western Front, so it is not surprising that the F.E.8 was well received by those RFC pilots who flew it: in the circumstances, it must have looked like a promising counter weapon to the Fokker, and it was slightly faster than the D.H.2.

No.7457, the second prototype F.E.8, in France, 31 January, 1916, when it was with No.5 Squadron. The nose aperture for the Lewis gun had been roughly faired over and the gun placed on a D.H.2-type mounting. External magazine containers had been fitted. In the cockpit is Capt. F. J. Powell. *(Public Record Office AIR 1/942)*

The business end of 6407, an F.E.8 that joined No.40 Squadron on 25 August, 1916. It lasted a month, for it crashed on 27 September, 1916, and was conveyed to No.1 A.D. two days later.

It appears that the F.E.8 had, in any case, been virtually ordered off the drawing board: the first contract for 100 production aircraft was No.87/A/179, placed with The Darracq Motor Engineering Co Ltd of Fulham; the serial numbers 6378—6477 were allotted as early as 11 October, 1915, four days before 7456 made its first flight. The choice of contractor was perhaps unfortunate insofar as Darracq had not previously built aircraft. Numerous difficulties beset and delayed production of the F.E.8, which is perhaps why 50 were ordered from Vickers in February 1916; with the serial numbers 7595—7644, they were built at Weybridge.

Aspects of the production F.E.8 were affected by operational experience gained with 7457 in France. This second prototype went to No.5 Sqn on 26 December, 1915, where it prompted requests for increased petrol tankage, a more practical form of gun mounting, and the removal of the spinner. These were easily accomplished, and in January 1916 this F.E.8 had been fitted with a gun mounting of the type standardised for the D.H.2. It was then virtually in the form of the production version of the design, and went on to do a considerable amount of flying in No.5 Sqn.

The first F.E.8, No.7456, was rebuilt in somewhat slow time, and flew again on 9 April, 1916. It may have been the F.E.8 that was tested at CFS in April 1916, powered by a 110 hp Le Rhône; certainly by 18 May it had been fitted with a 110 hp Clerget 9Z. These alternative engines were most probably tried because the Monosoupape was a somewhat capricious engine, and the supply of French engines for the RFC was usually a source of anxiety. It is known, however, that late in 1916 No.7637 had the 110 hp Le Rhône engine. Various other alternative installations were designed at Farnborough.

No.6390 was a standard Darracq-built F.E.8 that was at Central Flying School in October 1916.

Darracq-built F.E.8s began to come forward in May 1916. No.6378 was at Farnborough on 24 May, whither it was flown to France early in June and was allotted to No.29 Sqn. At least four other F.E.8s were used by that unit, but all had brief careers. Several of the earliest production aircraft were delivered with the conical spinner that had been fitted to the prototypes, but on 12 June, RFC HQ asked the A.D.A.E. to ensure that future F.E.8s would go to France without spinners. Deliveries of Vickers-built F.E.8s began late in June and the type then became available in appreciable numbers.

No.40 Sqn had formed at Gosport in February 1916, and in July its equipment with F.E.8s began, making it the first unit to be wholly equipped with the type. Six

Production F.E.8 with gun at full elevation. (*Ministry of Defence H1672*)

F.E.8s of 'A' Flight of No.40 Sqn went to France on 2 August, 1916, and 'B' and 'C' Flights followed on 25 August. Two days earlier, at Farnborough, Frank Goodden performed the remarkable series of tests that established the standard method of spin recovery. Spinning was not understood at that time, and there had been spinning accidents with some of the earliest F.E.8s. With cold analytical courage Goodden deliberately spun an F.E.8, recovered safely, and encapsulated his actions in Reports and Memoranda No.168. Once pilots knew what to do, they came to like the F.E.8, for it had pleasant flying qualities.

No.40 Sqn had to overcome in the Field the various snags that arose with a new type. The rubber bungee cords on the upper ailerons were replaced by a spanwise balance cable, and the fuel system was modified by the incorporation of a gravity tank (devised by Sgt Ridley of No.40 Sqn) in the centre-section. On October 15 No.41 Sqn arrived in France, similarly equipped with F.E.8s. That date contains virtually all the reasons why the F.E.8 could never hope to duplicate the success of the D.H.2, for it was then precisely a year to the day since 7456 had flown. The F.E.8 was in fact well into obsolescence by the time it reached the front in significant numbers.

The pilots, of course, had to make do with what they had, and right gallantly did they acquit themselves, not a few enemy aircraft falling to the ill-mounted Lewis guns of the little pushers. The most successful F.E.8 pilot was Lt E. L. Benbow of No.40 Sqn, who won eight confirmed victories on the type.

Further production was ordered. An intention to have 25 aircraft built at the Royal Aircraft Factory was reflected in the allocation of the serial numbers A41 65 in April 1916 for such a batch, but none was built. In October Darracq were asked to build a further one hundred and twenty F.E.8s (A4869—4987 and A5491) and deliveries of these began in February 1917.

Corrosion problems discovered in the autumn of 1916 led to the temporary adoption of modified D.H.2 elevators for some F.E.8s, but all-steel elevators and rudders were quickly made to replace the original steel-and-duralumin surfaces. Some aircraft had modified tailskids.

Throughout the winter of 1916–17 the F.E.8s fought doggedly on and into the early spring of 1917. By then they were utterly outclassed, a fact that was emphasised on 9 March, 1917, when ten F.E.8s of No.40 Sqn encountered a strong formation of German fighters led by Manfred von Richthofen and were severely handled. Fortunately for No.40 Sqn its re-equipment with Nieuports was imminent, but No.41 had to battle on until its D.H.5s began to arrive in July 1917. Latterly the F.E.8s of No.41 Sqn undertook ground-attack missions during the Battle of Messines, hazardous work for which its good field of view made it suitable. To the F.E.8 belongs the dubious distinction of being the last operational single-seat pusher of the 1914–18 war, and it continued to serve for a time with training units after its withdrawal from the front line.

F.E.8
110 hp Gnome Monosoupape, 110 hp Le Rhône 9J, 110 hp Clerget 9Z

Single-seat fighter.
Span 31 ft 6 in; length 23 ft; height 9 ft 2 in; wing area 214 sq ft.
Empty weight 895 lb; loaded weight 1,346 lb.
Maximum speed 79 mph at 6,000 ft, 69·5 mph at 10,000 ft; climb to 6,000 ft—9 min 28 sec, to 10,000 ft—23 min 42 sec; ceiling 15,210 ft.
Armament: One 0·303-in Lewis machine-gun.

Manufacturers: Royal Aircraft Factory, Darracq, Vickers.

Service use: *Western Front*: Sqns 5, 29, 40 and 41. *Home Defence*: Two F.E.8s allotted to H.D. unit(s) in 1917. *Training duties*: Reserve Sqns 1 and 10, and at Lympne and Marske. *Other units*: Armament Experimental Station, Orfordness.

Serials: 7456—7457, 6378—6477, 7595—7644, A41—65 (not built), A4869—4987, A5491.
Rebuild: A8894.

F.E.9

The F.E.9 was designed in the summer of 1916 and was intended to be a high-performance replacement for the F.E.2b as a fighter-reconnaissance two-seater. Like its predecessor it was a nacelle-and-tailbooms pusher but of more compact design and configured to give its crew a good field of view and of fire. It appears to have been ordered off the drawing board, for the preliminary drawings were dated 25 and 26 September, 1916, and serial numbers for a batch of twenty-seven (A4818—4844) were allotted under War Office authority on 5 October.

No doubt good performance was expected from the use of the 200 hp Hispano-Suiza engine. The original design had the nacelle faired into the upper centre-section, and the fin was to be a narrow, triangular surface of high aspect ratio; doubtless cooling considerations prompted the revision of the nacelle to place the radiator between the rear centre-section struts, and a greatly enlarged fin was fitted to A4818 when it was built. Otherwise this first F.E.9 faithfully reproduced the unequal-span configuration of the original conception, with widely raked upper wingtips and enormous triangular horn-balance areas on the ailerons. Single-bay bracing created very long extensions on the upper wing, and these

The first F.E.9, A4818, at Farnborough; a photograph dated 15 May, 1917.
(*RAE; Crown copyright*)

were braced from inverted-V kingposts. The observer was given a peculiar and limited form of dual control, provided by two sticks, one of which actuated the elevators, the other the rudder: there was no dual control of the ailerons.

The first F.E.9 had been completed by 4 April, 1917. On the previous day it had been inspected by Maj J. T. C. Moore-Brabazon (later Lord Brabazon of Tara), who had been sent specially from France to report on its suitability for oblique photography. He found the F.E.9 unsuitable for such work, but observed that a standard camera could be mounted in the nose of the nacelle.

Another inspection of the F.E.9 was made on 15 April by Capt B. M. Jones and Capt G. H. Norman, both of the Orfordness Experimental Station and men of some experience in aircraft armament. They found that the aircraft was equipped with two rocking telescopic pillar gun mountings, and recommended that the armament should consist of three Lewis guns, two manned by the observer, one for firing forward, the other for firing to the rear; and a third gun fixed on the

The F.E.9 A4818 in P.C.10 finish, photographed at Hounslow, possibly when on its way to or from France. (*P. H. T. Green*)

starboard side of the nacelle to be fired forward by the pilot. Their recommendations included several detail modifications, but it is not known whether all or any of them were ever made.

On test the F.E.9's performance proved to be disappointing; in particular its climbing performance was markedly worse than had been estimated. Its handling characteristics were in some respects alarming. Much of its testing was done by Capt G. T. R. Hill, MC, BSC, who wrote (in Reports & Memoranda No.728):

> 'The aeroplane can be banked over very quickly for a two-seater, though it is harder to take it off the bank than to put it on. This is to a certain extent true of all aeroplanes; however, on left-hand turns on the F.E.9 it is very marked, and in order even to continue steadily on a steep left-hand turn, great pressure has to be exerted on the stick towards the right and also on the right of the rudder bar. The aeroplane feels as if it wants to put its nose down and turn over on its back. If a steep turn to the left is started at 55 to 60 mph and 1,600 rpm, the speed cannot be kept below about 75 mph after a few seconds, and on each turn I tried I was forced to push the stick to the right with both hands and the rudder bar with my right foot with very nearly my whole strength in order to come off the turn'.

Hill considered that this phenomenon was largely attributable to overbalancing of the ailerons, partly to the rudder, which was thought to be insufficiently powerful. Right-hand turns were less hazardous because they were made against the torque of the propeller.

It seems that no major modifications of the F.E.9 were made before it went to France on 6 June, 1917, having left Farnborough on the previous day. It was flown at No.1 Aircraft Depot, St-Omer, on 6, 7 and 8 June, apparently with Captain Hill

The second F.E.9, A4819, had two-bay bracing, modified ailerons and redesigned fin and rudder. This photograph bears the date 3 November, 1917. (*RAE 2855; Crown copyright*)

The end of A4819 in a comprehensive crash at Biggin Hill. (*RAF Museum*)

at the controls. He had made it clear that the F.E.9 was 'very tricky and liable to get out of control', and Brig-Gen Brooke-Popham instructed Brig-Gen Higgins, commanding 3rd Brigade RFC, 'not to send up one of his pilots to fly it'. On 7 June, Hill took up Lt J. W. G. Clark, an experienced observer of No.13 Sqn, who had been briefed to report on the aircraft. Clark found the unrestricted field of fire highly commendable, flew the F.E.9 briefly on its dual controls, finding it 'quite easy to fly', and concluded:

> 'From an observer's point of view, the range of vision, field of fire and manoeuvring capacities are so good, that I consider the machine excellent for Artillery Observation work'.

Brig-Gen Higgins had doubts about the adequacy of the F.E.9's performance, and Maj-Gen Trenchard had no reservations when he reported on 9 June:

> 'I have inspected the F.E.9 and although the view is excellent and exactly what is required, I consider that this machine is of no use and that all experiments on it should be stopped forthwith. The machine is a year out of date, its performance is considerably below what is required for a new type of machine, and to utilise the 200 hp Hispano for it would be a pure waste of a really good engine'.

So the F.E.9 returned to Farnborough on 14 June (via No.56 Sqn at Liettres on 13 June) with its tail between its legs, doubtless to the disappointment of the R.A.F. design staff. Nevertheless, in the second half of August 1917 there was still an official intention to use the F.E.9s in Home Defence squadrons or with the Middle East Brigade.

Work on the second and third F.E.9s must have been fairly well advanced at that time, and nacelle nose fairings were in production for nearly all of the full production batch. At least 24 such fairings were made, the last being sent for inspection on 14 July, 1917. The Farnborough designers were determined to resolve the F.E.9's control problems, and A4818 went on flying at the R.A.F.

with progressively modified vertical tail surfaces: by 19 September it had used and discarded a high-aspect-ratio balanced rudder, 12 sq ft in area, and had acquired a squat, ugly rudder that was balanced, had no fin, and was 18 sq ft in area. The aileron balance areas were also twice modified, some considerable success being realised with the third (Type C) surfaces in October.

By then the second aircraft, A4819, had been completed. It differed from A4818 in having two-bay wing bracing that dispensed with the overwing kingposts, and had, at least at some time, the Type B ailerons and the 12 sq ft rudder. A4819 was the only F.E.9 to go to an operational squadron. On 15 December, 1917, it was handed over to No.78 (Home Defence) Sqn, RFC, but what it did while with that unit remains a mystery. All that is known is that it crashed at Biggin Hill and was totally wrecked during that winter.

The third aircraft, A4820, also had two-bay interplane bracing, but the precise combination of ailerons and rudder fitted to it remains unknown. It underwent final inspection on 1 November and was flying on 7 November. By 20 November it had been fitted with new ailerons and a new gravity tank, and was given a new undercarriage in January 1918. It was last mentioned on 23 January, 1918, when its engine was tested.

A4818, the first F.E.9, survived at least until the end of March 1918, when it was still doing useful work as a test vehicle in experiments on slipstream effects on controls. Yet another form of fin and rudder (respectively 5 sq ft and 13 sq ft) was tried, and the report of these experiments was printed as Confidential Information Memorandum No.705 of 15 February, 1918.

F.E.9
200 hp Hispano-Suiza

Two-seat fighter-reconnaissance.

Span 40 ft 1 in (upper, with Type A ailerons), 37 ft 9½ in (upper, with Type C ailerons), 26 ft 6 in (lower); length 28 ft 3 in; height 9 ft 9 in; wing area 365 sq ft.

Loaded weight 2,480 lb.

Maximum speed at sea level 105 mph, at 10,000 ft—100 mph, at 15,000 ft—88 mph; climb to 10,000 ft—21 min 20 sec, to 15,000 ft—48 min 30 sec; service ceiling 15,500 ft.

Armament: As designed, two 0·303-in Lewis machine-guns.

Manufacturer: Royal Aircraft Factory.

Service use: Home Defence: A4819 to No.78 Sqn.

Serials: A4818—4844.

R.E.1

The design of a new experimental reconnaissance biplane began to take shape under the original designation B.S.2 in the sketch books of H. P. Folland, probably towards the end of 1912. At an early stage (possibly when the B.S.1 was renamed S.E.2) the two-seater was given the new designation R.E.1, stamping it as the first of the Reconnaissance Experimental series of designs. Nothing yet found suggests that the R.E.1 was regarded as of special importance, which is perhaps surprising at a time when the only military use for aeroplanes that was considered seriously was the reconnaissance role. Nevertheless, the work subsequently done on the two R.E.1s could all be said to have as its object the improvement of their usefulness as reconnaissance vehicles.

The first R.E.1 was completed in July 1913. Mervyn O'Gorman, the Superintendent of the R.A.F., in his 1913 Report on Full-Scale Work (R & M No. 86) introduced the R.E.1 thus:

> 'The experimental reconnaissance armoured aeroplane R.E.1 was designed for the same purposes as B.E.2, as a refinement of it, utilising the same engine, and employing the most recent knowledge to improve the results.'

As initially built, the R.E.1 was not fitted with armour plate, but the fact that the Factory design staff had made provision for it is significant. The aircraft was a neat single-bay biplane with staggered wings, warping for lateral control and a 70 hp Renault engine driving a four-blade airscrew. In general appearance the R.E.1 looked somewhat like a single-bay B.E.2 but was generally better proportioned. The rudder, tailplane and elevators resembled those of the B.E., but a triangular fin was mounted in front of the rudder. The top decking on the fuselage was somewhat deep and housed the two petrol tanks, one between the cockpits, the second ahead of the front cockpit. Structurally the fuselage embodied a modest proportion of steel-tube members, but the patons and longerons were of wood. The employment of a new aerofoil section in the mainplanes allowed a deeper-section rear spar to be used, thus enhancing the strength of the wing truss.

The first R.E.1 in its original form with warping wings and full stagger. This photograph is dated 22 July, 1913. (*RAE 114; Crown copyright*)

It will have been observed that the military mind of the Military Trials period was much occupied with the desirability of Army aircraft being easily road-transportable. Evidently this carried over into the design of the R.E.1, for its mainplanes had special detachable spar fixings incorporating quick-release bolts at the root ends, together with spring pins at the interplane strut sockets. These made it possible for two men to dismantle the mainplanes in about five minutes; the wing panels could then be secured to the fuselage flanks, and the R.E.1 was thereby reduced to transportable proportions.

The second R.E.1, No.608, with warping wings and overwing fins. In the cockpit is Edward Busk, who made extensive use of the R.E.1 in his experiments in stability. (*Imperial War Museum Q66022*)

The first R.E.1 was handed to Flying Department of the Royal Aircraft Factory in July 1913, the second on 12 September, 1913. It is not known whether the two were identical when first completed, but at various times they presented differing appearances. At an early stage the wing span was increased by some two feet, and the second aircraft may have had the longer wings from the outset. The second R.E.1 may have undergone early modification, for a note in surviving Farnborough papers states that on 25 August R.E.1 No.2 was being rebuilt.

As reconnaissance aircraft were expected at that time to be as stable as possible in order to permit the pilot to undertake observation duties without the distraction of having to control his aircraft, it is not surprising that the R.E.1s were used extensively by E. T. Busk in his experiments aimed at producing a completely stable aeroplane. By 6 November, 1913, the first had been fitted with four overwing fins on the upper mainplane, and turning tests were made on that date. The second R.E.1 was similarly equipped but then had reduced stagger and

The first R.E.1 in modified form with ailerons and reduced stagger; its tail unit was also modified. (*RAE*)

a larger balanced rudder without any fin; both R.E.1s retained warping for lateral control while the overwing fins were fitted.

At some time, probably late in 1913, the two R.E.1s were numbered 607 and 608 respectively. It will be convenient to refer to them in this way hereinafter. In the course of the stability experiments Busk had the first R.E.1 modified repeatedly. By 25 November, 1913, when he conducted rolling stability experiments with the aircraft, ailerons had replaced the wing warping, and the dihedral angle was 1° 22′ 30″. By 2 December dihedral had been increased to 2° 42′ 30″; finally, by 10 March, 1914, with dihedral of 3° 12′ 30″, wash out of 0° 40′, ailerons left free and elevator locked, the aircraft was flown in squally conditions without use of either control. 'The flying was very comfortable', recorded R & M No.133, 'and the pilot considered that reconnaissance under these conditions would be considerably easier for a pilot alone'.

Later in March and April 1914, No.607 was extensively repaired and modified. It was fitted with a new rectangular tailplane, and the mainplanes were rigged with 4 in less stagger and given additional lift cables; new undercarriage skids were fitted and the mainplanes covered with new fabric. No.607 went on flying during 1914, having to undergo repairs from time to time.

Its sister aircraft, with ailerons and reduced stagger, yet still retaining its entire original tail unit, was physically handed over to the Military Wing of the RFC on 19 May, 1914, and was allotted to No.6 Sqn. Its transfer to the Military Wing had been effected, on paper at least, at a much earlier date, for on 6 March, 1914, Lt-Col F. H. Sykes, the Officer commanding the Military Wing, informed ADMA:

> 'Aeroplane No.608, type R.E.1 (No.2) having been handed over to the Military Wing as a service machine has now been allotted the new number 362'.

Second R.E.1 with ailerons.

This R.E.1 had this secondary identity for only twelve days. On 18 March, Sykes advised the Chief Inspector of the AID (Maj J. D. B. Fulton) that the War Office had decided that the aircraft was to retain its original number, consequently the allotment of 362 was cancelled and the number was again available (it was in fact subsequently allotted to a Sopwith Tabloid).

During the brief period of the aircraft's existence as 362 it was the subject of a short report, signed by Geoffrey de Havilland as Inspector of Aeroplanes and

Maj Fulton as Chief Inspector. This recorded that the petrol tap was not easily accessible from the pilot's seat, that the aluminium cowling panels about the engine were crudely made and should be improved, and, most remarkable of all, that the throttle lever was mounted on the wing-warping wheel, an arrangement that was rightly felt to be inconvenient.

On the outbreak of war No.6 Sqn was deprived of its aircraft and personnel in order to allow Squadrons 2, 3, 4 and 5 to join the BEF at full strength. This may be why 608 was back at Farnborough by 11 August, 1914. On that day it was flown by Frank Goodden for 35 minutes. When the four operational squadrons of the RFC had been barely a week in the Field a request was sent from the RFC in France on 22 August, 1914, for any spare aeroplanes to be sent over as replacements. Of the five aircraft that were sent in response one was 608, which was flown to France by 2nd Lt C. Gordon Bell and went to No.2 Sqn on 23 August. It lasted just over a week, and was struck off the squadron's charge on 1 September; apparently it was not returned to the Aircraft Park, so it seems likely that 608's career ended then in France.

Not so 607, which was still flying at Farnborough on 25 February, 1915. On that day it was twice flown by W. E. Stutt, in the morning on photography tests and in the afternoon on wireless tests.

Of the R.E.1's intended armour little was heard. This is perhaps not surprising, for it seems that it was of only 1-mm (0·04-in) thickness. On 29 September, 1914, Mervyn O'Gorman wrote to ADMA about a new armoured seat that the R.A.F. had fitted to a B.E.2, and his memorandum included the following passage:

> 'I may say that acting on the 1 millimetre theory I originally armoured R.E.1 in February last and showed it to S. of S. when down here—but the 1 mm idea was knocked out as I said above'.

In view of 607's involvement with Busk's stability experiments it seems likely that the armoured R.E.1 may have been 608.

An R.E.1 was reported to be still in use at the Experimental Station Orfordness as late as 1917. It was used as a target-towing aircraft at that time, employment that suggests that the aircraft concerned might have been an R.E.7 rather than an R.E.1.

R.E.1
70 hp Renault

Two-seat reconnaissance aircraft.
Span (original form) 34 ft; wing area (original form) 316 sq ft.
Empty weight 1,000 lb; loaded weight 1,580 lb.
Original form: Maximum speed 82·8 mph; initial rate of climb 600 ft/min.
Armament: None.

Manufacturer: Royal Aircraft Factory.

Service use: Prewar: No.6 Sqn: *Western Front*: No.2 Sqn. Flown experimentally at R.A.F., 1913–15; Experimental Station, Orfordness (unconfirmed and unlikely).

Serials: 607, 608; the latter was temporarily numbered 362 for a brief period in March 1914.

The seventh R.E.5 exemplified the equal-span form of the design and is here seen at Farnborough in a photograph dated 17 July, 1914. (*RAE 448; Crown copyright*)

R.E.5

The history of the R.E.5 is surprisingly long and somewhat complex. Of the types that followed the R.E.1 in the R.E. series, H.R.E.2 was supplied to the RNAS as No.17 and R.E.3 was likewise a singleton; the R.E.4 would have been a two-seater powered by a rotary engine, rather like a two-seat S.E.4 in nasal configuration, but it did not get beyond preliminary sketches. The R.E.5 was designed in 1913 as an enlarged development of the R.E.1.

Late in October 1913 it was decided that the small non-rigid airships and man-lifting kites operated by No.1 Sqn, RFC, should be transferred to the Naval

On the sixth R.E.5 Norman Spratt attained an altitude of 18,900 ft on 14 May, 1914. This photograph is said to record the preparations for his take-off. (*RAE 340; Crown copyright*)

Wing. A financial consequence of this decision was the transfer to the War Office of a credit, reported to be of £25,000, which Col J. E. B. Seely, Secretary of State for War, promptly used to order twenty-four R.E.5s from the Royal Aircraft Factory.

Obviously no time was lost in producing the R.E.5s, for the first was handed over to Flying Department at the R.A.F. as early as 26 January, 1914, the second on 6 February, the third and fourth on 13 and 14 February respectively. These were all standard two-seaters with equal-span 45 ft 4·34 in two-bay wings having ailerons for lateral control. Power was provided by a 120 hp Austro-Daimler engine that had a curious hooded cowling and a large flat radiator carried internally behind the engine. The Austro-Daimlers for the R.E.5s were made by Arrol-Johnston Ltd of Dumfries. The fuselage was, for the time, remarkably deep, and the crew sat high in commodious cockpits, the pilot in the rear; the undercarriage cross-axle was carried on a twin-skid, four-strut structure typical of R.A.F. practice of the period. Structurally the R.E.5 was of unusual interest because extensive use was made of steel tubing in its fuselage and tail unit. This last incorporated a triangular fin, to which the large plain rudder was hinged.

R.E.5s Nos.5 and 6, which were handed over to Flying Department on 5 and 12 March, 1914, respectively, differed from their predecessors in being single-seaters with an extended upper wing, the span of which was 57 ft 2·39 in. These were intended for high-altitude flights, and on 14 May Norman Spratt flew No.6 to a record height of 18,900 ft. This R.E.5 was given the official serial number 380 and was allotted to No.6 Sqn, whose C.O., Maj J. H. W. Becke, made repeated attempts to better Spratt's altitude record during the summer of 1914.

Throughout that summer production continued more or less steadily, and at least fifteen of the R.E.5s had been completed by the beginning of the war. Latterly, delays in delivery occurred because trouble was experienced with the crankshafts of the Austro-Daimler engines, which had to be replaced by strengthened shafts. R.E.5s Nos.11 and 13 were delivered as long-range aircraft with additional tankage, the latter having the 57-ft span upper wings; these two were handed to the R.A.F. Flying Dept on 21 May and 20 July respectively.

The sixth R.E.5 had the long-span upper wing and was given the serial number 380. Maj J. H. W. Becke used this aircraft in several attempts to establish a new altitude record. No.380 went to the Aircraft Park at St-Omer on 28 September, 1914, and was issued to No.2 Squadron on 1 November. It was reported as wrecked on 14 December, 1914, and was apparently struck off charge.

No.16 was apparently delayed because in mid-July it was being fitted with a device known as the Forbath Stabilizer, while No.15, which was being fitted with airbrakes on 28 August, was not handed to Flying Department until 16 September. For all Col Seely's haste to order the R.E.5s, their delivery to the RFC was slow and did not keep pace with initial output. It seems likely that only five (334, 335, 361, 380 and 382) went to the Military Wing, mostly to No.6 Sqn, before the war, but possibly the crankshaft troubles delayed the handing over of others. Of these, No.361 is known to have been in service with the Military Wing as early as 24 March, 1914, and was probably the first to see use with the RFC. It seems that in some details, such as the arrangement of instruments on the dashboard, it was not typical. Like the R.E.1, it had its throttle lever mounted on the centre of the pilot's control wheel. As late as 30 November, 1914, R.E.5s Nos.11, 21, 22 and 23 were all at Farnborough without RFC serial numbers, in company with 382, 617, 631 and 667.

The R.E.5 No.631 went first to CFS on 17 August, 1914, and was taken to France by No.7 Squadron on 8 April, 1915. It remained with No.7 until it was struck off squadron charge on 29 July, being subsequently struck off the charge of the RFC on 1 August, 1915. It is here seen in France with its engine cowling removed. (*RAF Museum*)

It seems likely that by that time No.18 had been abandoned without receiving an official number. This R.E.5 had been completed with the Royal Aircraft Factory's oleo undercarriage (as later fitted to the R.E.7 and F.E.2b) and the 57-ft upper wing. It was handed over to Flying Department on 27 October, 1914, and advantage was immediately taken of its large wing area and sturdy undercarriage, for it was used that day in an experiment in dropping what was then a very heavy load. A cast-iron weight of 10 cwt was suspended under the fuselage and released from 1,000 ft to bury itself in the ground. This was successfully accomplished but immediately after the drop the R.E.5's pilot, Frank Goodden, found that some of the fuselage woodwork had been ignited by the exhaust. He succeeded in landing the aircraft before it was seriously damaged on that occasion, but was less fortunate on 31 October when, in the course of further weight-carrying experiments, No.18 was wrecked. One other R.E.5 did not go to the RFC. This was No.14, which was flown to Hendon on 2 September, 1914, by Geoffrey de Havilland to become 26 in RNAS service. It survived at least until 23 November, 1915, when it was conveyed to the White City Depot.

The R.E.5s Nos.20 and 21 were fitted with wireless, and the former had an enlarged fin, possibly of the curved shape, 11 sq ft in area, standardised for the R.E.7. No.22 was renumbered 24 for no apparent reason, and was designated for experimental purposes on its completion at the end of December 1914. By mid-April 1915 it had been fitted with an oleo undercarriage and a jettisonable petrol tank, and early in June it was re-engined with a 140 hp R.A.F.4; a change to another R.A.F.4 had been made by 3 September, the dates being such that these engines, numbered 2497 and 5669, must have virtually been prototypes, for no production R.A.F.4a was delivered before December 1915. The R.A.F.4a installation was designed with an inverted-Y exhaust manifold like that of the later B.E.12 prototype.

Another earlier experimental engine installation that was made in an R.E.5 airframe in July 1914 was of an eight-cylinder Sunbeam, presumably a 150 hp Crusader. Test records relating to this combination have yet to be found, and it remains uncertain whether this R.E.5 ever flew with the Sunbeam engine. Other experiments included the fitting of a plough brake at the mid-point of the axle of an oleo undercarriage; this was done in mid-September 1914. In March 1915 an R.E.5 with the oleo undercarriage was being used as a test vehicle for an experimental 500-lb bomb designed at the R.A.F.

The fuselage of R.E.5 No.651 immediately after being off-loaded in France. It went to No.2 Squadron on 27 September, 1914, was transferred to No.16 Squadron on 10 February, 1915, and was struck off charge on 14 February.

As noted elsewhere, No.6 Squadron was broken up at the start of the war to provide personnel and equipment for the four squadrons that accompanied the BEF to France. None of the R.E.5s joined these squadrons, however, and it is known that 334 and 361 returned to the Royal Aircraft Factory on 2 September, 1914. On 27 September, Nos.613, 651, 659 and 660 joined No.2 Sqn in France, and on the following day the single-seat 380 arrived at the Aircraft Park at St-Omer; on 1 November it, too, went to No.2 Sqn, perhaps as replacement for 613 and 660, wrecked on 1 and 15 October respectively. In the time between its arrival in France and its allocation to No.2 Sqn it may have been modified to become a standard two-seater, for under 8 October, 1914, the records of the Aircraft Park

This R.E.5 had the 57-ft upper wing and is believed to have been a No.7 Squadron aircraft. (*R. C. Bowyer*)

contain a reference to an unidentified R.E.5 single-seater noted: 'To be converted to two-seater'. In its turn 380 was wrecked on 14 December, and two weeks later 659 left the squadron, returning to England via the Aircraft Park on 29 December, and leaving No.2 Sqn with one R.E.5 only, 651, until 745 arrived on 7 January, 1915. On 10 February, 651 and 745 were transferred to No.16 Sqn but were struck off strength almost immediately, 745 on 11 February, 651 three days later. From then until 8 April the RFC had no operational R.E.5s in France, but on that date No.7 Sqn arrived, bringing with it seven R.E.5s: these were 361, 617, 631, 674, 677, 678 and 737.

Thereafter No.7 Sqn continued to be the RFC's largest user of the R.E.5. On 29 April, 674 left in order to be reconstructed, returning to the unit on 11 June. By then 2458 was with the squadron having arrived on 14 April, and 2457 came on 15 June, direct from No.1 Aircraft Park, where it had arrived from England that day. No.7 used its R.E.5s as bombers and reconnaissance aircraft and it seems likely

Equal-span R.E.5 with oleo undercarriage, photographed at Central Flying School. Its rudder is marked with the Union Flag. Beyond it is the D.H.1 prototype. (*RAF Museum*)

that most, if not all, of them had the 57-ft span upper wing. On 26 April, 1915, two of No.7's R.E.5s and seven B.E.2cs of No.8 Sqn took off from St-Omer laden with 20 lb bombs to attack various targets. One of No.7 Sqn's R.E.5s, 2457, was the aircraft flown by Capt J. A. Liddell with 2/Lt R. H. Peck as his observer on 31 July, 1915, when, severely wounded in combat near Bruges, he lost consciousness but recovered in time to regain control of his damaged aircraft and bring it to a safe landing near Furnes. Liddell was awarded the Victoria Cross for his outstanding gallantry, but died of his wounds a month later.

A camouflaged equal-span R.E.5 captured more or less intact by the Germans and put on display in an exhibition of captured war material. (*A. E. Ferko*)

The R.E.5s of No.7 Sqn began to be withdrawn that July, 674, 678 and 631 being struck off strength on the 3rd, 25th and 29th of the month respectively, while 677 was returned to the Aircraft Park on the 10th. The others went in September, 617 was struck off on 28 September, while 2458 had returned to the Aircraft Park on the 11th. On 30 September, 2457 was transferred to No.12 Sqn, but returned to the Aircraft Park, presumably for repair, on 18 October. On 30 October, 2458 was sent to No.12 Sqn and was joined on 26 November by the reissued 2457. The latter went back to the Aircraft Park on 12 December, 1915, leaving 2458 as the last R.E.5 in the Field. It lingered on until 20 February, 1916, when it, too, went back to the Aircraft Park. It is known that 2457 returned to England on 30 December, 1915, and was subsequently used for training purposes by No.7 Reserve Sqn at Netheravon. It is doubtful whether any of the trainees who flew in 2457 in the evening of its career knew of its more glorious past as Liddell's aircraft.

Although the R.E.5 was never a great military aeroplane it contributed a good deal towards later developments, notably in the delivery of heavy bombs. It would probably have been almost helpless in combat, for the only known gun mounting designed for it provided only for the rearward firing of a single Lewis gun, manned by the observer, who occupied the forward cockpit. Its flying characteristics were probably not endearing, but structurally it seems to have been sturdy. Of it Oliver Stewart wrote, in *The Clouds Remember*: 'The R.E.5 resembled a blowzy old woman, and floundered about the sky in a safe if unattractive manner'.

R.E.5
120 hp Austro-Daimler

Two-seat reconnaissance and bomber.
Original span 45 ft 4·34 in. Some R.E.5s had an upper-wing span of 57 ft 2·39 in. Wing area 498·347 or 569·107 sq ft.
Maximum speed 78 mph at sea level.
Armament: One 0·303-in Lewis machine-gun; makeshift combinations of pistols and rifles; small load of 20 lb Hales bombs.

Manufacturer: Royal Aircraft Factory.

Service use: Prewar: No.6 Sqn. *Western Front*: Sqns 2, 7, 12 and 16. *Training duties*: No.7 Reserve Sqn, Netheravon.

Serials: 334, 335, 361, 380, 382, 613, 617, 631, 651, 659, 660, 674, 677, 678, 688, 737, 745, 2456—2459, 2461.

R.E.7

The ultimate form of the R.E.5 was a far cry from the original design, having the extended upper wing, enlarged and rounded fin, the R.A.F. oleo undercarriage, and provision for carrying the R.A.F. 336-lb bomb. This version, with some major modifications, went into large-scale production under the new type number R.E.7. The aircraft that is named in contemporary R.A.F. documents as the prototype R.E.7 still had the elliptical wingtips and upper centre-section of the R.E.5, and was flying at Farnborough late in March 1915.

Early in 1915 contracts for a total of 133 R.E.7s were distributed to the Austin Motor Co (33), Coventry Ordnance Works (50) and D. Napier & Son (50), quickly followed by an order for 100 more, given to the Siddeley-Deasy Motor Car Co. The production R.E.7 differed markedly from the prototype, having an extensively redesigned wing structure of longer span; the upper wing was made in two halves that met at a central cabane and the wingtips were of revised planform. A rectangular long-span tailplane replaced the R.E.5 pattern used on the prototype. The initial power unit was the 120 hp Beardmore, cowled as on the R.E.5 and with the radiator behind the engine.

The earliest recorded note of a production R.E.7 relates to the Napier-built 2287, which was at Farnborough by 3 July, 1915. It seems that subsequent deliveries came along several weeks later, and it was not until 28 September, 1915, that an R.E.7 (2235) went to France to join No.12 Squadron at St-Omer. No.12 was the first unit to have the type and had four on its strength on 31 December, 1915 (these were 2287, 2289, 2290 and 2458). These early R.E.7s all had the 120 hp Beardmore and in consequence were underpowered. This may have been why No.12 Sqn used the type as an escort for bombers if occasion

The prototype R.E.7 at Farnborough, a photograph dated 26 March, 1915. The bomb-like object under the fuselage is a jettisonable petrol tank that had both nose and tail fuses and could be dropped as a missile, either to explode on contact or not, the dropping mechanism being so designed that the pilot could, when necessary, withdraw a safety pin to activate the explosive charge. (*RAE 191*)

demanded. On one such mission, 2235, being flown by Capt G. A. K. Lawrence, was attacked by four German aircraft but managed to regain the sanctuary of the Allied lines with no fewer than 97 bullet holes.

Combat presented serious problems for the R.E.7, in which the observer occupied the front seat and was obstructed by the converging cabane struts. Various gun mountings were produced, among them an arrangement first of two,

Demonstration of a singularly unpromising gun mounting on the prototype R.E.7. This photograph, dated 4 June, 1915, illustrates the capaciousness of the pilot's cockpit. (*RAE 860; Crown copyright*)

subsequently three, brackets for a Lewis gun on the forward cabane struts. There was also a remarkable arrangement of a gun above the upper wing controlled by the observer via a system of linkage actuated by a dummy gun and pistol grip, but it seems unlikely that this found any operational acceptance. A rearward-firing Lewis gun could be mounted on a bracket secured to the rear cabane struts; this was first tried on the prototype in June 1915. The most desperate arrangement consisted of drastic modification to the root ends of the upper mainplane to create a central aperture through which the observer stood with his head and shoulders above the wing, in which position he was able to operate a barbette-type mounting for a Lewis gun. This had, in theory at least, a 360 deg field of fire above the aircraft, but its effect on the R.E.7's dismal performance must have been considerable.

Standard production R.E.7, C.O.W.-built, with 120 hp Beardmore engine. This aircraft went to No.21 Squadron on 24 January, 1916. (*National Aeronautical Collection, Ottawa*)

The overwing barbette was first tried on 2185 on 29 December, 1915. This was regarded as being essentially successful, but an improved gun mounting was called for, and the observer's platform was too weak. The report did not seem to impress Trenchard, however, and it seems unlikely that any R.E.7 so armed went to France.

The intention to send out a squadron equipped throughout with R.E.7s was known by December 1915, for on 12 December Trenchard advised ADMA:

'I do not propose to keep up the R.E.7 machines in No.12 Sqn after the first squadron of R.E.7s has arrived.'

The unit in question materialised as No.21 Sqn, which went to France on 23 January, 1916, equipped with R.E.7s powered by the 120 hp Beardmore. Two days previously the Royal Aircraft Factory excitedly reported that removal of the silencer and substitution of six individual exhaust stubs gave the R.E.7 more power and improved its climb. This was passed on to RFC Headquarters but drew only this blunt rejoinder from Trenchard on 6 February:

'Reference your M.A./R.E.7/39 (M.A.26) of the 2nd instant, the R.E.7 is underpowered to such an extent that the effect of 5 per cent extra horse power will be negligible and it is therefore not proposed to take any action in the matter.'

Operational experience with No.21 Sqn did nothing to improve Trenchard's opinion of the R.E.7, and on 8 March he wrote to ADMA:

'. . . . I am of the opinion that the R.E.7 with 120 hp Beardmore engine is useless in the field. The lifting power of the machine rapidly decreases, and although the engines appear to be running now better than they ever did and every care has been given to looking at the propellers, the fact remains that not half of the machines can get off the ground if it is at all sticky with full load, nor can they climb to 8,000 ft. Therefore, I shall be glad to have these machines replaced at an early date.'

R.E.7 No.2348 with R.A.F.4a engine; a photograph dated 8 January, 1916. This aircraft was one of Farnborough's hacks. By August 1916 it had become a three-seater with a Beardmore engine. (*RAE 1049; Crown copyright*)

Replacements began to arrive in April in the form of more R.E.7s, but this time powered by the 140 hp R.A.F.4a, an air-cooled V-12 engine. The operational début of the new engine was not auspicious, serious lubrication problems affecting serviceability so badly that on 7 May Maj R. Campbell-Heathcote, O.C. No.21 Sqn, reported that he had only two out of seven R.E.7s serviceable.

Despite these difficulties No.21 Sqn pressed home several quite significant bombing raids during the Battle of the Somme. On 2 July, 1916, six R.E.7s dropped six 336-lb R.A.F. bombs on a German infantry headquarters and on ammunition dumps with telling effect, and other targets were to suffer from similar attentions in the ensuing weeks. Nevertheless, the R.E.7 was never an asset to the RFC, even with the R.A.F.4a engine; not only were its lubrication problems insoluble in the absence of a dual oil-pump system, still some way in the future, but the aircraft, having little more power than a Tiger Moth, was still underpowered. On 23 July, 1916, Brooke-Popham wrote to the DAE:

'It has been decided that the R.E.7s with R.A.F.4a engine are useless for

R.E.7 No.2299 was used at Farnborough as a flying test-bed for the Rolls-Royce engine that was later named Falcon. This official photograph is dated 24 July, 1916.
(*RAE 1609; Crown copyright*)

work in the Field, until the engines are fitted with all the latest improvements, including a satisfactory double oil pump system and extra air inlet. It is considered quite impossible to carry out this work in the Field; it is therefore intended to replace the R.E.7s now in No.21 Sqn with B.E.12s.'

The logic of that decision is hard to detect, but the B.E.12 was, at the time, an unknown quantity in operational terms. No.21 Sqn's R.E.7s were quickly flown back to England and the type's operational career was ignominiously terminated.

It was not quite the end of the road for the R.E.7, however, for the RFC continued to use a few in France as target tugs at the Aerial Musketry Range at Berck-sur-Mer. Initially 120 hp Beardmore aircraft were used but were eventually replaced by R.E.7s with the R.A.F.4a engine, remaining in use for this duty well into 1918. Numbers of R.E.7s were used by training units at home, and the Royal Aircraft Factory made extensive use of the type for experimental purposes.

R.E.7 with 225 hp Sunbeam Mohawk engine.

The R.E.7 No.2348 after extensive modification became a three-seater with a Beardmore engine. Its load in this photograph, dated 8 August, 1916, optimistically includes a 336-lb bomb. (*RAE 1642; Crown copyright*)

Other engines tried in R.E.7s were the 160 hp Beardmore, 190 hp Rolls-Royce (Falcon), 200 hp R.A.F.3a, 220 hp Renault, 225 hp Sunbeam and 250 hp Rolls-Royce Mark III (Eagle III). At one time the formation of a squadron of R.E.7s with the 220 hp Renault was briefly considered, but the idea came to naught.

A remarkable variant was the three-seat R.E.7 of which at least two were converted from standard aircraft. No.2348, which in January 1916 had an R.A.F.4a engine, had by August become a three-seater with a 160 hp Beardmore, and No.2299 was similarly modified but had the considerable advantage of a 250 hp Rolls-Royce Mark III engine (it had earlier flown as a two-seater fitted with the first 190 hp Rolls-Royce (Falcon) engine). No.2299 was complete by

R.E.7 No.2299 also underwent transformation into a three-seater with a 250 hp Rolls-Royce Mark III engine. In this form it went to No.20 Squadron in France on 30 September, 1916, and was used operationally to some purpose by that unit until it crashed on 31 January, 1917. Its remains were crated and returned to England on 3 February, 1917. (*Imperial War Museum Q63815*)

mid-August 1916, and was soon fitted with the overwing barbette gun mounting for the front-seat observer. It was flown to France on 11 September, 1916, by Lt Oliver Stewart and was extensively tested at No.2 Aircraft Depot by Lt Vernon Busby, who was later to lose his life in the crash of the first Handley Page V/1500.

Although the R.E.7 three-seater was not enthusiastically reported on, it was not condemned out of hand; indeed, certain duties were considered to be within its compass. In particular, No.20 Sqn thought it might be useful for night work, and RFC HQ authorised the 2nd Brigade to use it across the lines at night. HQ also approved suggestions for fitting a Sopwith gun mounting on the overwing barbette (this was in fact a Scarff mounting), and for fitting a synchronised Vickers gun for the pilot, for which a B.E.12 interrupter gear was suggested. How the pilot was to aim this weapon did not seem to be considered, but the installation was made.

Just what the three-seat R.E.7 did while with No.20 Sqn has never been chronicled. One who knew it at the time was Air Vice-Marshal H. G. White who, writing in 1954, recalled:

> 'Its armament consisted of two free Lewis guns and one fixed Vickers gun firing through the propeller—including, on occasions, the blades. When in action the front gunner had to perch himself precariously on a Scarff ring fitted in the top centre section, but all too frequently came adrift when rotating the gun mounting and invariably finished up in the pilot's cockpit.
>
> The ingenious R.A.F. periscope bomb-sight was fitted for use by the pilot, but since not one of the many 500-lb [sic] bombs dropped from this aircraft was ever seen to explode, this excellent piece of apparatus proved to be entirely superfluous.'

R.E.7s saw some employment as target tugs, notably at the Aerial Musketry School, Berck-sur-Mer. Here 2256 is seen with an unusually detailed flag target; the aircraft bears the presentation inscription *Accra II*.

The operational career of the three-seater was relatively brief, and on 30 January, 1917, RFC HQ wrote to the D.A.E. that the aircraft was being returned to England '. . . as it is the only one of this type out here, and as there are not likely to be any more Rolls-Royce engines available for this type of machine for some time to come'. This was a pity in a way, for 2299 was probably the only R.E.7 that was of any real use. However, it crashed on the very day of its official withdrawal, 31 January, 1917, and its remains were returned to England in a transit case on 3 February.

R.E.7
120 hp Beardmore, 140 hp R.A.F.4a, 160 hp Beardmore,
190 hp Rolls-Royce Mark I, 200 hp R.A.F.3a,
220 hp Renault, 225 hp Sunbeam Mohawk,
250 hp Rolls-Royce Mark III

Two-seat reconnaissance-bomber.
Span 57 ft (upper), 42 ft (lower); length 31 ft 10½ in; height 12 ft 7 in; wing area 548 sq ft.
140 hp R.A.F.4a: Empty weight 2,170 lb; loaded weight 3,449 lb.
Three-seater: Loaded weight 4,309 lb.
120 hp Beardmore: Maximum speed at sea level 82 mph, at 5,000 ft—73 mph; climb to 5,000 ft—30 min 35 sec.
140 hp R.A.F.4a: Maximum speed at sea level 84·9 mph; climb to 6,500 ft—37 min 50 sec; service ceiling 6,500 ft; endurance 6 hr.
Three-seater: Maximum speed at sea level 94 mph, at 5,000 ft—80 mph, at 10,000 ft—72 mph; climb to 5,000 ft—13 min 15 sec, to 10,000 ft—32 min 10 sec.
Armament: *Two-seater*: one 0·303-in Lewis gun, one 336-lb R.A.F. bomb or equivalent; *Three-seater*: one 0·303-in Vickers machine-gun, two 0·303-in Lewis machine-guns, one 336-lb R.A.F. bomb or equivalent.

Manufacturers: Austin, Coventry Ordnance Works, Napier, Siddeley-Deasy.

Service use: Western Front: Nos.12 and 21 Sqns. Three-seater used only by No.20 Sqn. *Training duties*: CFS; Reserve/Training Sqns 28, 35 and 42; Squadrons mobilising— Nos.19, 42, 52, 62 and 82; No.98 Depot Sqn; No.1 (Auxiliary) School of Aerial Gunnery, Hythe; Aerial Musketry Range, Berck-sur-Mer, France; Wireless School, Brooklands; Australian Flying Corps Training Wing, Harlaxton. *Other units*: Wireless Flight, Biggin Hill.

Serials: 2185—2234, 2235—2236, 2237—2266, 2267, 2482—2336, 2348—2447, 6016—6115, 6828—6927, 7545—7594. (The last three batches, 6016—6115, 6828—6927 and 7545—7594, were cancelled).
Rebuild: A5156.

R.E.8

In the autumn of 1915 the RFC in France formulated a request for a new type of aircraft to replace the B.E.2c and its immediate derivatives for corps reconnaissance and artillery spotting duties. The requirement specified that the new type must be capable of defending itself. To provide this replacement the Royal Aircraft Factory designed the R.E.8, a tractor two-seater in which the observer occupied the rear cockpit, with a good field of fire for his free-mounted Lewis gun; the engine was the 140 hp R.A.F.4a that, in its early form, had proved to be a sore trial in the R.E.7 and B.E.12.

Basic design work had been completed by early March 1916, the original design having a large fin with constant-chord rudder; for want of a British interrupter gear at that time, the pilot's armament was to be a fixed Lewis gun and there were to be bullet-deflecting plates on the four blades of the airscrew. Two prototypes, 7996 and 7997, were put in hand; these flew on, respectively, 17 June and 5 July, 1916. Both had a completely redesigned vertical tail assembly in which the fin area was drastically reduced; although the pilot's gun in these prototypes was apparently a Lewis there were no deflectors on the airscrews.

On 16 July, 1916, Capt Frank Goodden flew 7997 to France. He had already written a lengthy report on the first prototype, criticising many details of the design, and the R.E.8 was subjected to unusually thorough scrutiny in France. On 18 July, Goodden took up Brig-Gen Brooke-Popham in the observer's cockpit; in a report to Trenchard Brooke-Popham listed points of criticism, but concluded:

'The flying qualities of the machine are splendid. The air speed indicator shows 103 mph full out, flying level, and Goodden flew for at least 5 minutes at 47 on the speed indicator, the engine revs 1125, doing turns, without losing height at all, and the machine perfectly under control the whole time. It is very handy and easily manoeuvred, lands very slowly, and pulls up quickly on the ground.'

Next day instructions were issued requiring certain experienced officers, pilots and observers to go to No.2 Aircraft Depot to examine and assess the R.E.8. All reported on the aircraft, from Col C. A. H. Longcroft down to pilots and observers of several squadrons. Numerous details were criticised, but there was a fair measure of unanimity in liking the R.E.8 as a flying machine, though Longcroft thought it somewhat heavy on the controls and not easy to turn quickly.

By 26 July, Brooke-Popham was able to record a considerable number of modifications that had been made at the Aircraft Depot and to list others that were still required. Specifically he stated that the pilot's weapon should be a Vickers gun mounted externally on the port side. On 11 August he wrote at length to D.A.E. listing twelve modifications made at the Aircraft Depot and enumerating twelve more that were considered necessary. Next day Capt R. H. Mayo flew 7997 back to Farnborough.

The first production R.E.8, A66, without Vickers gun but with early ring mounting for the observer's Lewis gun. This aircraft had been completed by 13 September, 1916, and was sent to Orfordness, where this photograph was taken. It is known to have crashed there.

The incidence of allotment of serial numbers suggests that the initial production batch of R.E.8s, fifty numbered A66—A115, had been ordered under War Office authority from the Royal Aircraft Factory early in April 1916—that is, before either of the prototypes had flown. Deliveries of this batch did not begin until mid-September, possibly because of the need to incorporate the modifications that had been made to 7997 after its return from France. These were submitted for inspection by the AID on 31 August, and the first production R.E.8, A66, was similarly submitted on 13 September.

A78 photographed at No.2 A.D., Candas, on 4 November, 1916, shows the original shape and size of the fin area on the earliest R.E.8s. This aircraft also has the early type of mounting for the Lewis gun. It was one of the R.E.8s that first equipped No.52 Squadron, and was transferred to No.34 Squadron on 14 February, 1917. It was lost exactly two months later, when Lts H. R. Davies and J. R. Samuel were reported missing on 14 April, 1917.

By that time large-scale orders had been placed with several contractors; at the end of September no fewer than 1,200 were under order, a very large commitment to an aircraft still to be used operationally.

The earliest production aircraft resembled the prototypes in having small fin surfaces and a somewhat rudimentary rotary mounting for the observer's gun. On the first few R.A.F.-built R.E.8s, the pilot's Vickers gun was mounted internally, firing through a port situated behind and below the engine. Some may have had the Arsiad interrupter gear, for the use of this mechanism was at least contemplated, but the majority had the Vickers-Challenger gear, characterised by its long external tappet rod. The internal Vickers installation was short lived, and by early November 1916 work was well in hand to transfer the guns to the external position. A later proposal to reposition the Vickers on top of the fuel tank was quickly abandoned before the end of November 1916.

It is clear that from an early stage it was expected that the Hispano-Suiza engine would replace the R.A.F.4a and on 5 December, 1916, Brooke-Popham specified that on the Hispano version the Vickers gun was to be mounted centrally, as on the Spad 7. An installation of the 200 hp Hispano-Suiza engine was made in A95 in December 1916, in which form the aircraft was designated R.E.8a. Surviving drawings of this variant depict a flat frontal radiator and the pilot's Vickers gun installed, as on the S.E.5, in the port shoulder of the forward decking. As it proved difficult to obtain enough Hispano-Suiza engines for the S.E.5 production programme none was ever available for the R.E.8a and no production ensued.

The first RFC squadron to receive the R.E.8 was No.52, a new unit that went to France on 21 November, 1916. Its R.E.8s came from the initial R.A.F.-built batch. Early operational experience with the R.E.8 was unhappy: many accidents occurred, in most of which the aircraft spun in from low heights, and fire frequently followed when the engine was pushed back into the fuel tanks. Morale suffered so badly that No.52 Squadron exchanged its R.E.8s for the B.E.2es of No.34 Sqn in January and February 1917.

The R.E.8's spinning tendencies were investigated and the aircraft itself was virtually exonerated in a report dated 5 March, 1917, concerning comparative tests that had been conducted with three R.E.8s, one standard, one with reduced upthrust on its engine, and one with an enlarged fin. In his conclusions on the report Lt-Col F. C. Jenkins wrote:

> 'In carrying out these tests, I came to the conclusion that this machine can only be spun by determined effort, or by inexperienced flying. In the latter case, it might be due to such an elementary error as taking off with insufficient speed and attempting to turn without sufficient bank'.

It therefore seems clear that the basic cause of most of the R.E.8 crashes lay in the woefully inadequate training given to the RFC's pilots at that time, aggravated no doubt by the frequent failures of the R.A.F.4a engine, and perhaps even by the original but ambiguous wording of the instructions about tail trimming painted on the aircraft's sides.

The Royal Aircraft Factory worked assiduously on the R.E.8's controls and performance, fitting a variety of enlarged fins, some with horn-balanced rudders. The fin area of the production aircraft was increased in two stages, first by enlarging the upper fin and then the lower. This evidently did not satisfy No.42 Sqn, for in 1917 that unit made a much enlarged upper fin by bolting a standard B.E.2e fin to the stern post and taking the leading edge up to the top of the post by a length of steel tubing. Although this fin was not adopted for operational R.E.8s, it was widely used on aircraft of training units. Correspondence on these fins went to and fro (curiously by mid-May 1918 RAF HQ was asking for six specimen large fins to be sent to France for evaluation) until on 22 May the Air Ministry, with faintly perverse logic, ruled that training-unit R.E.8s were to revert to having standard fins.

As 1917 passed, more and more squadrons were equipped with the R.E.8. Various modifications were introduced throughout the year, one of the most important being the early introduction of the Scarff ring mounting for the observer's Lewis gun. The use of Prideaux links instead of a webbing belt doubled

R.E.8 A3432 was built by Siddeley-Deasy and had a Scarff ring mounting on the rear cockpit. On 7 February, 1917, it was at Coventry, allotted to the E.F. for 'A' Replacement Squadron (presumed to be No.21 Squadron), but two days later it was re-allotted to the Experimental Flight at Orfordness, where this photograph was taken.

the Vickers' ammunition capacity from 250 to 500 rounds, and the fixed gun's functioning was improved by the use, from August 1917, of the Constantinesco synchronising gear in place of the Vickers-Challenger mechanism.

At about the same time a wooden undercarriage replaced the original unsatisfactory faired-steel-tubing V-struts. Less obvious but no less important was the replacement, from July 1917, of the original Claudel carburettor by the Brown and Barlow carburettor; this improved engine performance considerably and was greatly liked in the squadrons.

In that summer of 1917, A4600 was fitted at Farnborough with a set of B.E.2d mainplanes and was extensively flown in this form. Structurally the equal-span and two-bay wings were more satisfactory than the R.E.8 configuration, and this modification was developed into two two-bay developments of the R.E.8. One of these had wings of equal chord and was designated R.E.9; the other had mainplanes of unequal chord with the upper wing rather lower down and closer to the fuselage than on the R.E.8 and was named R.T.1; both had enlarged fins, horn-balanced rudders and narrow-chord elevators.

A typical operational R.E.8 of No.52 Squadron, with the enlarged fins that were introduced to minimize spinning proclivities. A4267 was built by The Austin Motor Co and was at Coventry, allotted to the E.F., on 25 April, 1917. It went first to No.5 Squadron and was with No.52 Squadron early in 1918 until it was destroyed by fire on 6 March, 1918.

Although a contract for 525 R.T.1s was proposed in October, these two derivatives of the R.E.8 were abandoned before the end of 1917, but in December the idea of fitting B.E.2d wings to the R.E.8 was revived. On 29 December D.D.A.E. informed RFC HQ that drawings were then available for issue to contractors for this drastic modification to be made to production aircraft, and asked HQ whether it would be advisable to proceed. D.D.A.E. pointed out that it would be some time before the modification could become effective and that he did not propose to proceed with the matter unless RFC HQ desired him to do so. Evidently HQ did not so desire, for no more was heard of the idea, possibly because there was still hope that the replacement of the R.E.8 by Hispano or Arab-powered Bristol Fighters might be brought forward. It never was; the long-deferred expectation that replacement would begin in September 1918 was not realised; and the R.E.8 had to remain in service beyond the Armistice.

After the abortive attempt to employ the 200 hp Hispano-Suiza in the R.E.8 airframe, a few alternative power units were tried or suggested. In May 1917 a 200 hp R.A.F.4d was installed in A3406; the second prototype engine was used, for the first Daimler-made R.A.F.4d was not delivered until August 1917.

A3902 displays the greatly enlarged fin fitted to many training R.E.8s.

Production ceased when only 16 had been delivered of the 500 ordered, ruling out the engine as an alternative power unit. When the seventy-five R.E.8s numbered D4811—4885 were ordered from D. Napier and Son in October 1917, the Rolls-Royce Eagle, presumably of one of the earlier Marks, was specified. With the potential need for Eagles measured against the low output of these engines the R.E.8s stood no chance of being powered in this way, and as far as is known all of the 75 aircraft concerned were delivered as standard R.E.8s.

Official statistics indicate that 2,262 R.E.8s were delivered to the RFC up to 31 March, 1918. Production had still to reach its maximum then, and over 1,800 more were still to be delivered to the Royal Air Force before the end of 1918. The R.E.8 was therefore widely used on the Western Front, and in Italy, Palestine, and Mesopotamia, as a reconnaissance, artillery-spotting and bomber aircraft. Its successes were modest and little known, its failures progressively more numerous as the war went on; yet experienced and determined crews proved that the R.E.8 could defend itself pugnaciously in combat.

R.E.8
140 hp R.A.F.4a

Two-seat reconnaissance.

Span 42 ft 7 in (upper), 32 ft 7½ in (lower); length 27 ft 10½ in; height 11 ft 4½ in; wing area 377·5 sq ft.

Empty weight 1,803 lb; loaded weight 2,869 lb (with two 112-lb bombs).

Maximum speed 98 mph at 6,500 ft, 92·5 mph at 10,000 ft; climb to 6,500 ft—21 min, to 10,000 ft—39 min 50 sec; service ceiling 11,000 ft. Performance with two 112-lb bombs.

Armament: One 0·303-in Vickers machine-gun; one, occasionally two, 0·303-in Lewis machine-gun(s); bomb load up to 224 lb.

Manufacturers: Royal Aircraft Factory, Austin, Coventry Ordnance works, Daimler, Napier, Siddeley-Deasy, Standard.

Service use: *Western Front*: Sqns 4, 4(A), 5, 6, 7, 9, 12, 13, 15, 16, 21, 34, 42, 52, 53 and 59; No.69 (Australian) Sqn, RFC, later No.3 Sqn, Australian Flying Corps; GHQ Communication Flight; GOC 3rd Brigade Communication aircraft attached No.56 Sqn; Aerial Ranges, France. *Italy*: Sqns 34 and 42. *Home Defence*: Sqns 50, 76 and 77. *Mesopotamia*: Sqns 30 and 63. *Palestine*: Sqns 14, 113 and 142; No.67 (Australian) Sqn, RFC, later No.1 Sqn AFC. *Training duties*: Training Sqns 7, 8, 9, 13, 15, 16, 17, 20, 23, 25, 26, 31, 35, 38, 39, 42, 46, 51, 52, 53, 57, 59, 60, 64, 65, 66 and 110; Training Depot Stations Nos.2, 5 and 35. *Sqns mobilising*: Nos.62, 82 and 83; Aerial Observation School, Heliopolis; Wireless & Observers School, Brooklands.

Serials: 7996—7997
A66—115, 3169—3268, 3405—3504, 3506—3530 (not built), 3531—3680, 3681—3830, 3832—3931, 4161—4260, 4261—4410, 4411—4560, 4564—4663, 4664—4763.
B2251—2300, 3401—3450, 5001—5150, 5851—5900, 6451—6480, 6481—6624 (B6625—6630 delivered as R.T.1s), 6631—6730, 7681—7730.
C2231—3080, 4551—4600, 5026—5045, 5046—5125.
D1501—1600, 3836—3910 (cancelled), 4661—4810, 4811—4885, 4886—4960, 6701—6850.
E1—300, 1101—1150, 1151—1250.
F1553—1602, 1665—1764, 3246—3345, 3548—3747.
Known rebuilds: B730, 732, 734, 736—738, 741, 742, 750, 755—760, 764, 765, 780—784, 786, 787, 791—793, 798, 810, 811, 814, 820—825, 830, 832—837, 840—846, 853—855, 876, 1498, 4021, 4026—4036, 4038—4040, 4045—4048, 4050, 4051, 4053—4057, 4059, 4060, 4062, 4065, 4067, 4068, 4069, 4075, 4086, 4087, 4089, 4090, 4093, 4094, 4097, 4098, 4100, 4101, 4103—4106, 4109, 4118, 4134, 7734, 7738—7740, 7754, 7761, 7802—7805, 7827, 7834, 7853, 7887, 7888, 7893, 7917, 8045, 8097, 8798, 8872, 8874—8900, 8902—8907, 8909, 9473, 9997.
C3381, 3433, 4282, 4294.
D4966, 4967, 4970, 4972, 4973, 4975, 4977, 4980, 4981, 4996, 4998, 9737—9739, 9790—9799.
F662, 663, 666, 669, 672, 675—677, 681—685, 687, 689, 690, 694, 699, 700, 5871, 5872, 5874—5883, 5885, 5891, 5892, 5894—5897, 5899, 5901, 5902, 5904—5909, 5976, 6007, 6009, 6010, 6012—6019, 6044—6050, 6085, 6091, 6097, 6203, 6204, 6218, 6270, 6273, 6277, 6279, 6299.
H6843, 6845, 6849, 6857, 6865, 6870, 6879, 6883, 6896, 6900, 7017—7038, 7040—7057, 7136—7143, 7182—7193, 7207, 7262—7268, 7340, 8121.

S.E.2

The history of the S.E.2 is rendered somewhat complex by the fact that it existed in three successive and distinctly different forms. In the summer of 1912 work began at the Royal Aircraft Factory on the design of a high-speed single-seat biplane, powered by a 100 hp fourteen-cylinder Gnome rotary engine and initially designated B.S.1, or Blériot Scout No.1. Principal design responsibility was assumed by Geoffrey de Havilland, and much of the detail work was done by H. P. Folland. Their creation earned a unique place in aviation history as the first single-seat scout ever made and the primary ancestor of tens of thousands of single-seat fighters.

In the B.S.1, Geoffrey de Havilland was determined to create an aircraft capable of exceeding 90 mph, and the little biplane was both compact and unusually clean for its time. Its basic outlines were scaled down from those of the B.E.3, but the B.S.1 was characterised by a fine semi-monocoque fuselage, circular in cross-section. Lateral control was by wing warping and a twin-skid undercarriage was fitted.

The B.S.1 as it first appeared at Farnborough. It had been redesignated S.E.2 before the end of March 1913. (*RAE 71197; Crown copyright*)

The B.S.1 was completed early in 1913 and made its first flight with Geoffrey de Havilland at the controls. Progressive testing established that its average maximum speed was 91·7 mph, its minimum speed 51 mph, and its initial rate of climb 900 ft/min. These figures vindicated de Havilland's hope of achieving at least 90 mph, and were all the more creditable because the Gnome's actual output was about 82 bhp. With somewhat optimistic foresight the engine mounting had been made large enough and strong enough to take a 140 hp Gnome. Geoffrey de Havilland found the aircraft light on the controls but thought the rudder was too small. He designed a larger rudder and it was put in hand in the workshops, but while awaiting its completion he continued to fly the aircraft with the original surface. At this time a punctured tyre led to the replacement of the wheels by a larger pair, thus increasing the side area forward of the c.g. On 27 March, 1913, de Havilland attempted a sharp turn and the undersized rudder could not overcome the effect of so much forward side area: the aircraft went into a spin and crashed, injuring its distinguished pilot.

This rear view of the B.S.1 at Farnborough admirably reveals its outstandingly clean lines. (*Science Museum*)

A surviving official report makes it clear that the aircraft had been redesignated S.E.2 before this mishap occurred, and it was as such that the aircraft was rebuilt. The Superintendent of the Royal Aircraft Factory at the time was Mervyn O'Gorman, who immediately proposed to reconstruct the aircraft with an 80 hp Gnome and to build a new type, the S.E.3, with a 100 hp single-row Gnome. O'Gorman's proposals and associated estimates were officially approved on 25 April, 1913; the S.E.2 was rebuilt, but the S.E.3 got no further than preliminary sketches.

The reconstructed S.E.2 emerged with wings and undercarriage little altered and fuselage slightly modified. Its tail unit was completely redesigned to include two small triangular fin surfaces above and below the fuselage, together with a high-aspect-ratio rudder that was shod on its lower contour in lieu of a tailskid. Although it was hoped to complete the reconstruction by 3 August, 1913, it seems likely that it was not finished until October; Geoffrey de Havilland flew it during that month. Early in November it was flown by Capt Charles Longcroft, RFC, at O'Gorman's invitation. Although O'Gorman then offered to allow RFC pilots to fly the S.E.2 when it was not required by the Factory, he virtually had to be instructed by Maj Sefton Brancker, on 23 December, to issue the aircraft to the Military Wing. The S.E.2 was duly handed over on 17 January, 1914.

S.E.2 after its first reconstruction, about October 1913. (*RAE 207; Crown copyright*)

It was flown by four officers of No.5 Sqn, who generally thought well of it, and it was subsequently sent to No.3 Sqn. By that time it must have acquired the official serial number 609, and as such in April 1914 it was flown by Maj J. F. A. Higgins back to Farnborough for modification. O'Gorman had foreshadowed improvements that he had in mind as long ago as 5 November, 1913, including the replacement of the monocoque fuselage by a fabric-covered built-up structure. The modifying of the S.E.2 occupied the summer of 1914, and when it reappeared early in October it had a new rear fuselage constructed as O'Gorman had proposed. More significantly, it had greatly enlarged fin and rudder surfaces, together with a new tailplane and elevators. A new engine cowling of individual aspect was fitted, and the two-blade airscrew had a small spinner; the mainplanes and undercarriage differed little from those of the basic B.S.1 design, but all external bracing was by the new streamline-section Rafwires. In this form the reconstructed S.E.2 was flown on 3 October, 1914, by Frank Goodden, and he and Geoffrey de Havilland flew it a few times more before it was sent to France to join the RFC in the Field.

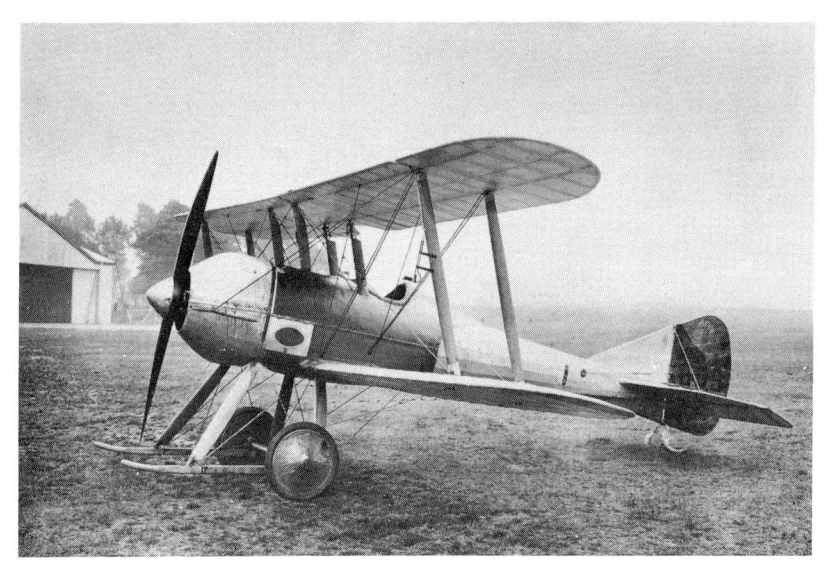

S.E.2 after its second reconstruction, numbered 609, and photographed at Farnborough shortly before going to join the RFC in the Field. (*RAE 614; Crown copyright*)

The S.E.2 joined No.3 Sqn at Moyenneville on 27 October. Its initial stay with that unit was brief, for it was sent to the Aircraft Park four days later, presumably for repair. On 18 November the S.E.2 was re-allocated to No.3 Sqn, in which there was an Air Mechanic James T. B. McCudden who later won conspicuous success as one of the war's most brilliant fighter pilots. In his book *Five Years in the Royal Flying Corps* he later wrote:

> 'I forgot to mention that an S.E. joined No.3 Squadron at Moyenneville and was fitted with two rifles in the same way as the Bristol Scout . . . . This S.E.4 [*sic*] was the first machine on active service to be fitted with the RAF streamline wire. It was fitted with an 80 hp Gnome, and was a little faster than the Bristols with the same engine, but did not climb quite as well.'

Evidently the S.E.2 did quite well, by the standards of the time, for it remained with No.3 Sqn until March 1915, apparently without having to return to the Aircraft Park. It might have remained operational longer, but on 12 March it was damaged by the explosion of a bomb on No.3 Sqn's aerodrome, and two days later it and its log-book were handed over to the Aircraft Park. It seems that it did not physically leave No.3 Sqn until 16 March; it was sent back to England on 24 March and struck off the strength of the RFC in the Field. Its subsequent history is not known.

It is doubtful whether the S.E.2 saw any aerial combat during its operational life, for its armament was less than practical. At that period of the war it must, almost certainly, have been looked upon as a high-speed scout or reconnaissance aircraft.

S.E.2 data overleaf.

S.E.2
100 hp Gnome, 80 hp Gnome

Single-seat scout.

Span 27 ft 6¼ in; length in original form 20 ft 5 in, in first reconstruction 20 ft 10 in; height in original form 8 ft 4½ in, in first reconstruction 9 ft 3$\frac{11}{64}$ in; wing area 188 sq ft.

Empty weight 850 lb (original), 720 lb (first reconstruction); loaded weight 1,230 lb (original), 1,132 lb (first reconstruction), 1,246 lb (second reconstruction).

Speed at sea level 92 mph (original), 91 mph (first reconstruction), 96 mph (second reconstruction); endurance 3 hr (first reconstruction).

Armament: Two 0·303-in rifles.

Manufacturer: Royal Aircraft Factory.

Service use: *Prewar*: Sqns 5 and 3; *Western Front*: No.3 Sqn.

Serial: 609.

S.E.4a

The designation S.E.3 was applied to a 1913 design for a high-speed single-seat biplane; provision for its construction was made in April 1913, an estimate of £2,050 for that purpose being approved on 25 April. Some design work was done by H. P. Folland, his ideas including several very advanced proposals. The S.E.3 was never built, but Folland evolved from it the high-performance S.E.4 that was built in 1914 and attained the phenomenal speed of 135 mph. Some structural work was done on a second S.E.4, but was apparently suspended when the first crashed irreparably on 12 August, 1914. It seems that a short-lived tentative attempt to resume work on the second S.E.4 was made on 19 November, but this was stopped two days later.

The first S.E.4a embodied many aerodynamic refinements not found on the three succeeding aircraft. (*RAE 822; Crown copyright*)

A new design had been prepared during the summer of 1914. This was for a structurally conventional single-seater using a tailplane and elevators similar to those of the S.E.4 but otherwise not obviously related to the earlier type. Nevertheless, it was designated S.E.4a, possibly because it embodied variable-camber mainplanes similar in principle to those of the S.E.4. The equal-span mainplanes carried on their trailing edges full-span control surfaces that acted as ailerons in response to side-to-side movement of the control column, but could be raised or lowered simultaneously by the rotation of a handle mounted on the control column.

In general appearance the S.E.4a was a neat and attractive little single-bay biplane with staggered wings and a normal pair of parallel interplane struts on each side. The upper wing was made in two halves brought together on the centre line of the aircraft at a trestle-like cabane structure; there was no centre-section.

The S.E.4a has always been regarded, with some justification, as an experimental aircraft from which it was hoped to gather information on stability and manoeuvrability, but there are grounds for believing that production was contemplated. On 3 May, 1915, the Assistant Director of Military Aeronautics (Lt-Col D. S. MacInnes) wrote to the Officer Commanding the Administrative Wing, RFC:

'I am to inform you that for purposes of official nomenclature the Royal Aircraft Factory designed machines F.E.2b and S.E.4 now being put out to contract, will be known as 'Fighter Mark I' and 'Scout Mark I' respectively, and will be referred to as such in all communications concerning them'.

The 'S.E.4' mentioned therein can hardly have been, at that late date, the true original S.E.4; certainly the S.E.4a was more of a production possibility for contractors to undertake.

However all that may be, only four S.E.4as were built at the Royal Aircraft Factory and were allotted the serial numbers 5609—5612. The first was completed in June 1915 and was submitted for final pre-flight inspection at Farnborough on 23 June. It seems likely that, with the S.E.4a at least, it was intended to obtain data on performance as well as on stability and manoeuvrability, for it had a number of refinements that were not applied to the other three aircraft. Its fuselage was carefully faired throughout its length, form being provided by stringers and formers, and the airscrew was fitted with a large spinner that blended well with the gently cambered engine cowling but left only a narrow annular slot for the admission of cooling air. To augment this limited flow of air the large spinner had a circular frontal opening and an internal fan. The stub-wings to which the lower mainplanes were attached had full aerofoil sections, the aileron cables were led internally through the lower wings, and the head fairing behind the cockpit had increased height.

This first S.E.4a made its initial flight on 25 June, 1915, piloted by Frank Goodden. It was last recorded at Farnborough on 6 September, 1915, by which time it had done a fair amount of flying in the hands of Goodden and W. Stutt, and had had several engine changes. As far as is known, this first S.E.4a was normally fitted with the 80 hp Gnome.

The second S.E.4a, 5610, was inspected for flight on 12 July, 1915, and made its first flight on 21 July, piloted by Stutt. It visited CFS at Upavon on 9 August and was last reported at Farnborough on 20 August. Like its immediate predecessor, 5610 had an 80 hp Gnome, but was built to the basic design and lacked the refinements of 5609. There was no spinner, the engine being housed in a fully cambered circular cowling; behind this cowling there were only rudimentary flank fairings and the fuselage was otherwise flat-sided. The lower wings were attached to simple tubular spanwise members, the resulting gaps between wing roots and fuselage giving the pilot some downward view. Aileron cables ran externally and the head fairing was not so high as on 5609.

By 23 July the third S.E.4a was virtually complete, and it made its first flight on 27 July. This aircraft, 5611, was originally fitted with an 80 hp Le Rhône, but this was replaced by an 80 hp Gnome by 2 September. The whereabouts of this S.E.4a between December 1915 and June 1916 are unknown, but it is known to have been back at Farnborough by 26 June. A further engine change had been made by 11 October, 1916, when the aircraft was inspected following the installation of an 80 hp Clerget engine. How long it retained the Clerget is not known, but 5611 survived at least until September 1917.

No.5611 was typical of the other three S.E.4as, with its flat-sided fuselage and deeply cambered engine cowling, which in this case housed an 80 hp Le Rhône engine. This photograph is dated 29 July, 1915. (*RAE 841; Crown copyright*)

Last of the S.E.4as, 5612, was given its final inspection on 9 August, 1915; apparently it had an 80 hp Clerget engine. It did very little flying at Farnborough, being last recorded there on 17 August, and it may have been one of the two S.E.4as that were used at Home Defence stations. The other may have been 5610, which was at Joyce Green by 8 September, 1915.

An unidentified S.E.4a was stationed as a Home Defence aircraft at Hounslow. On 24 September, 1915, while being flown by Capt Bindon Blood, it spun into the ground and its pilot was fatally injured.

For Home Defence duties a machine-gun mounting had been designed for the type. The first S.E.4a had overwing mountings above the cabane, possibly for a single Lewis gun, and there seems no reason why later S.E.4as should not have been similarly armed. Provision of armour plate was also designed, but confirmation that any was ever fitted to an S.E.4a has yet to be found.

The S.E.4a looked promising, but it was underpowered and overweight, and was not developed. It was found that depressing the wing flaps by 12 deg reduced the stalling speed from 45 to 40 mph, while reflexing them through 5 deg added 1 or 2 mph to the aircraft's top speed. Trim was not noticeably affected by the use of the flaps, and control was not appreciably reduced. The flap gear was not liked by the Royal Aircraft Factory pilots and was not properly tested until the summer of 1916. Despite its shortcomings the S.E.4a acquired something of a reputation as an aerobatic aircraft, hence its manoeuvrability must have been good.

S.E.4a
80 hp Gnome, 80 hp Le Rhône, 80 hp Clerget
Single-seat experimental aircraft and fighter.
Span 27 ft 5·2 in; length 20 ft 11½ in; height 9 ft 5 in.
Armament: One 0·303-in Lewis machine-gun.

Manufacturer: Royal Aircraft Factory.

Service use: Home Defence: S.E.4as stationed at Joyce Green and Hounslow.

Serials: 5609—5612.

S.E.5 and S.E.5a

The excellent Hispano-Suiza engine designed by Marc Birkigt made its initial test runs at the Barcelona works of the Hispano-Suiza company in February 1915. The French Government was quick to obtain examples of the new engine; these were tested rigorously at Chalais-Meudon in July. Lt-Col H. R. M. Brooke-Popham saw the engine and immediately recognised its outstanding qualities. He recommended that it should be ordered for the RFC, and a British order for 50 was placed in August 1915.

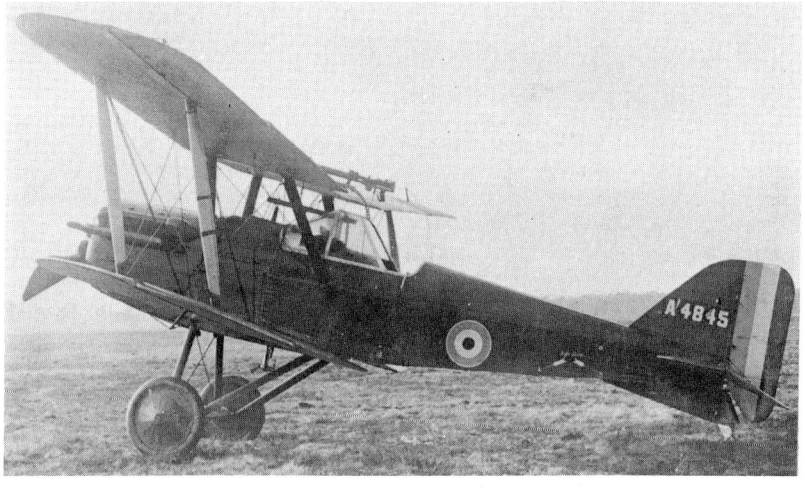

The first production S.E.5, A4845, photographed at Martlesham Heath. When thus recorded it had the original voluminous windscreen irreverently dubbed The Greenhouse by the pilots of No.56 Squadron. That the pilot sat high behind this transparent structure is made clear by the high situation of the Aldis optical sight. (*T. Heffernan*)

Capt Albert Ball running up the engine of his S.E.5, A4850, at London Colney. His aircraft had been extensively modified by the removal of the windscreen and addition of a head fairing; the seat had been lowered and a downward-firing Lewis gun had replaced the Vickers. Curiously, this S.E.5 was briefly allotted to the Training Brigade in April 1917, but an amendment dated 11 April ensured its retention by No.56 Squadron. Ball was flying A4850 when he met his death near Annoeullin on 7 May, 1917.

Lt Cecil Lewis, author of the classic *Sagittarius Rising*, stands beside S.E.5 A4855 at London Colney shortly before leading No.56 Squadron to France on 7 April, 1917. A4855 was flown back to England on 21 June, 1917.

Although the RFC's earliest intention was to fit the Hispano-Suiza to B.E.2cs, a more potent demonstration of the engine's capabilities was provided by the appearance of the French Spad 7.C 1 single-seat fighter in April 1916, a matter of weeks after Maj Gen Trenchard had submitted his specifications for the new aircraft that the RFC would require by the onset of winter in 1916; these included requirements for a new fighter.

The precise sequence of events at the Royal Aircraft Factory at about that time will probably never be known. From the testimony of the late Sir William Farren it is known that the Factory was invited by Gen Brancker to design a single-seat fighter with the Hispano-Suiza engine; and Dr A. P. Thurston stated in March 1921 that just such an aircraft had been designed by Frank Goodden, and that the design was taken up by the R.A.F. and turned over to H. P. Folland. Whether Goodden's design preceded or followed Brancker's invitation remains unknown, but there is no doubt that overall design responsibility rested with Folland.

The new fighter was designated S.E.5, and its basic design had been settled and drawn up by mid-June 1916. As originally conceived it was to be armed with a

No.56 Squadron had to return to England in June 1917 to augment London's defences against aerial attack; the squadron was based at Bekesbourne, where this photograph was taken. S.E.5 A8913 was Lt Keith Muspratt's aircraft at the time, having been allotted to the E.F. on 10 May, 1917. It flew its first patrol with No.56 Squadron on 23 May, subsequently saw service with Nos.60 and 40 Squadrons, and was issued to No.2 Squadron, Australian Flying Corps, on 9 March, 1918.

single Lewis gun firing forward through a hollow airscrew shaft; to this end the engine was to have a spur reduction gear, a fact that suggests that the aircraft was probably intended to have the 200 hp geared Hispano-Suiza. First thoughts on the tail unit were for rather small fin surfaces above and below the fuselage, and a rudder of parallelogram form with the tailskid on its lower edge.

Three prototypes were ordered as A4561—4563, and an initial production batch of 24 (A4845—4868) was ordered about seven weeks before the first prototype was ready and flown. It was complete by 20 November, 1916, when it could be seen to be a single-bay biplane of classic simplicity, conforming closely to the original design. It differed primarily in having a 150 hp direct-drive Hispano-Suiza and a vertical tail assembly adapted from that of the contemporary project,

The aircraft of the second production batch of S.E.5s initially had the large windscreen but introduced the modified wingtips that reduced the span. This photograph of A8904 on the compass base at Farnborough is dated 1 May, 1917, although the aircraft actually went to France on 30 April. It flew its first patrol with No.56 Squadron on 6 May and went or was sent to No.2 A.S.D. on 19 May, 1917. By September 1917 A8904 was with No.60 Squadron and in February 1918 was with No.1 Squadron. It was finally struck off charge on 26 March, 1918. (*RAE 2370; Crown copyright*)

the F.E.10. The airframe was a typical fabric-covered, wire-braced wooden structure, and the equal-span wings had widely raked tips. No armament was fitted, perhaps because the direct-drive engine made the original Lewis installation impossible.

This S.E.5, A4561, made its first flight on 22 November, 1916, piloted by Frank Goodden. Next day Capt Albert Ball flew it for 10 minutes, but he decided, on that fleeting acquaintance, that he did not like the aircraft. The second prototype, A4562, was well advanced at that time, and underwent its final inspection on 27 November. Goodden flew it on 4 December, and it subsequently underwent modification to production standard. The armament that was then fitted consisted of a fixed Vickers gun and a Lewis gun on a Foster overwing mounting. The Vickers was mounted at an upward angle of 5 deg in the port shoulder of the forward decking and was synchronised by the new Constantinesco C.C. mechanism. A very large windscreen was fitted and was designed to protect the pilot when he had to deal with a jammed Vickers gun. The fuel system was modified, and an overwing gravity tank replaced the earlier leading-edge tank built into the port upper wing.

On 24 December, 1916, Goodden flew A4562 to France for evaluation by the RFC. Reaction was generally favourable, and a number of modifications were made or recommended. The aircraft returned to Farnborough on 4 January and presumably was modified as requested, for it did not fly again until 26 January. Two days later, when Goodden was flying it, its port wings folded up at 1,500 ft and Goodden was killed. Following exhaustive investigation by Lt A. P. Thurston, structural modifications to the wing structure were made.

The third prototype, A4563, passed its pre-flight inspection on 12 January, 1917, and had been flown by Goodden that same day. It had a 200 hp geared Hispano-Suiza engine and an overwing gravity tank but otherwise closely resembled A4561. Goodden flew it up to 26 January, whereafter it underwent extensive modifications.

Production of the initial batch was under way at Farnborough, and the first production S.E.5, A4845, was completed by 1 March, 1917. It resembled the modified A4562, having a windscreen that was, if anything, more voluminous than that of the prototype. This S.E.5 was tested at Martlesham Heath, where its poor lateral control was criticised; its performance was markedly inferior to that of A4562. Deliveries of the first 24 production S.E.5s were completed by 30 March.

Most of them went to No.56 Sqn, then forming at London Colney, with Albert Ball as one of its Flight Commanders. His dislike of the S.E.5 was undiminished and he set about modifying his aircraft, A4850, by removing the cumbersome windscreen and replacing the Vickers gun by a useless downward-firing Lewis; his aircraft also had a new leading-edge gravity tank built into the centre-section. A4853 was also modified in several ways at London Colney.

The Squadron went to France on 7 April, 1917, and was immediately visited by Brig-Gen Brooke-Popham, who next day issued instructions for several modifications to be made to the S.E.5s; of these the first was the removal of the large windscreen. Hence the squadron remained grounded for two weeks, and did not fly its first offensive patrol until 22 April when Ball led six S.E.5s with strict instructions not to cross the lines. On 26 April Ball had a hard but successful combat on A4850 and this appeared to improve his opinion of the S.E.5.

Further production was under way, and efforts had been made to improve lateral control. The span had been reduced by shortening the rear spars of the wing extensions; this was primarily done for strength reasons and blunted the wingtips appreciably. At the same time the action of the aileron controls was improved by shortening the levers on the ailerons. It is possible that it was these modifications that led to the revised designation S.E.5a, but their introduction virtually coincided with the fitting of the 200 hp Hispano-Suiza to production aircraft, and this more distinctive change has come to be associated with the later designation. The second production batch of 50 aircraft (A8898—8947) was ordered late in December 1916 and deliveries began in mid-April. Fifteen of these were delivered with the 200 hp Hispano-Suiza engine, and all subsequent S.E.5as were powered with this engine or British-made variants thereof. The 150 hp aircraft were gradually replaced in the summer of 1917; some were modified in the Field to take the 200 hp engine, and as from 11 July, 1917, all S.E.5 fuselages overhauled at No.2 Aircraft Depot were converted to take the later and more powerful engine.

A4563, the third prototype, was the first S.E.5 to have the 200 hp geared Hispano-Suiza engine. It is here seen at Martlesham Heath in May 1917, embodying all major features of the 200 hp production S.E.5a. It had a long operational career, joining No.56 Squadron on 11 June, 1917, and seeing combat before being sent, badly shot about, to No.1 A.D. on 23 September, 1917. After later service with No.84 Squadron it was returned, wrecked, to No.2 A.D. on 24 February, 1918, and was struck off charge next day, having flown a total of 129 hr 35 min. (*A&AEE; Crown copyright*)

Large-scale production of the S.E.5a had been decided upon as early as January 1917, and in February the first major contracts were given to the Martinsyde and Vickers companies, each being required to build 200 aircraft. Deliveries from Vickers began early in July, from Martinsyde in August, the aircraft conforming closely to the modified form in which the third prototype, A4563, reappeared late in May 1917. A4563 itself went to France in June 1917 and began a long operational career by being first allocated to No.56 Sqn.

In service much difficulty was experienced with the 200 hp Hispano-Suiza engine, especially those made by the Brasier company, in many of which the

S.E.5a A8941 at Central Flying School, fitted with a 200 hp 1170-geared Hispano-Suiza but retaining the L-shaped exhaust manifolds more usually associated with the 150 hp engine. (*Wing Commander G. H. Lewis, DFC, from Peter Liddle's 1914–18 Personal Experience Archives, presently housed within Sunderland Polytechnic*)

reduction gears and airscrew shaft were imperfectly hardened. The Wolseley-built geared Hispano-Suiza (Wolseley W.4B Adder) was no better, and the basic engine design suffered from lubrication problems. To add to the spares problems associated with this difficult engine the geared version had no fewer than five reduction-gear ratios, and there were minor variations in crank-pin lengths and diameters and in reduction-gear housings. These difficulties were not confined to the RFC: in November 1917 Colonel Duval, Chef du Service aéronautique au Grand Quartier Général, reported that 200 hp Spad aircraft in the French Aviation militaire were grounded on two days out of three because their Hispano-Suiza engines were unserviceable.

The 200 hp Sunbeam Arab was tried as an alternative to the Hispano-Suiza as early as November 1917. Although it was hoped to adopt the Arab as a standard power unit for the S.E.5a, the engine could never be persuaded to run satisfactorily, for it was beset with problems of its own. Because expectations of

A few S.E.5as had elevators of reduced chord. One such was B4890, built at the Royal Aircraft Factory and first allotted to the RFC with the E.F. on 25 October, 1917. It was brought down intact on 29 November, 1917, and its pilot Lt Dodds was made a prisoner of war. Here photographed at Boistrancourt, the base of Jasta 5, this S.E.5a's defeat has been variously attributed to Vizefeldwebel Josef Mai of Jasta 5 and to Schubert of Jasta 6. (*Egon Krueger*)

477

B4891 also had narrow-chord elevators and was used by No.56 Squadron, in which unit it was flown by Capt J. T. B. McCudden. Originally powered by a Peugeot-built Hispano-Suiza, it was first allotted to the E.F. on 21 November, 1917, and went to No.56 Squadron early in December. McCudden fitted it with the spinner from an L.V.G. C V that he had shot down on 30 November, and went on to fly 124 hours on it. This S.E.5a had the so-called three-strut undercarriage. (*C. C. H. Cole*)

B189, aircraft 'S' of No.40 Squadron, was wearing additional markings on its elevators when this photograph was taken. It was one of fourteen Martinsyde-built S.E.5as, B187—B200, that were allotted to the E.F. on 20 February, 1918; all were powered by Peugeot-built Hispano-Suiza engines. (*K. M. Molson*)

S.E.5as of No.111 Squadron in Palestine. Both apparently have the 1170-geared engine; the nearer has the late all-wood undercarriage and a load of Cooper bombs in the rack under the fuselage; its serial number is B139. That S.E.5a was allotted to the Middle East on 13 February, 1918, but this photograph probably dates from the post-RFC period. (*RAF Museum*)

S.E.5a B4875 with experimental installation of the Eeman triple gun mounting, with three Lewis guns firing at an upward angle of 45 deg. The gravity tanks for fuel and water were transferred from the leading edge of the centre section to the starboard upper wing, and the installation was made in October 1917. An earlier fitting of the Eeman mounting had been made in B542, which was allotted to the E.F. on 3 September, 1917, at which time the Eeman installation was required to be made by the Technical Department. (*RAE 133195*)

Typical installation of 200 hp geared (1170 gearing) Hispano-Suiza in S.E.5a B536, Capt K. W. Junor's *Bubbly Kid II*, of No.56 Squadron. This S.E.5a was complete at Weybridge on 11 August, 1917, when it was allotted to the E.F. It went first to No.60 Squadron but was badly damaged following engine failure on 8 December. When repaired, it was issued to No.56 Squadron on 24 February, 1918, (*K. M. Molson*); and (*right*) standard armament installation on S.E.5a with ring-and-bead sights on the Vickers gun. The Aldis optical sight is not in place but the circular mounting brackets for it can be seen. Also visible on the instrument panel is a container for a spare 97-round magazine for the Lewis gun. The S.E.5a seen here is B4899, on which the second installation of a Wolseley Viper engine was made.
(*T. Heffernan*)

C6481 was just too late to see service with the RFC, but it illustrates perfectly the standard Viper engine installation, though its retention of the steel-tube undercarriage is remarkable. In its batch, S.E.5as up to C6419 had been allotted to the E.F. up to the 31st of March, 1918, most of those from C6402 onwards having the Viper engine. (*R. C. B. Ashworth*)

Arab deliveries were not realised, about four hundred S.E.5a airframes were languishing in store engineless in January 1918. The combination of engine difficulties seriously retarded the formation and re-equipment of squadrons with the S.E.5a, and until November 1917 only three squadrons were flying the type in France. These were Nos.56, 60 and 84; Nos.40 and 41 completed their re-equipment in that November, and next month No.68 Sqn exchanged its D.H.5s for S.E.5as, while No.24 Sqn received its first S.E. on Christmas Day.

By October 1918 all but two of the S.E.5a squadrons then in France had aircraft powered by the 200 hp Wolseley Viper. This direct-drive engine had been created by the Wolseley company as a result of an error of interpretation of an order for 400 direct-drive engines of the basic 150 hp Hispano-Suiza 8Aa type already built in England as the Wolseley Python. The production Viper had a compression ratio of 5·3 to 1 and other modifications. It had its own share of troubles, which had to be overcome, and much experimental and development work went into achieving a satisfactory radiator installation.

Apart from engine troubles the S.E.5a had structural problems of the fin and mainplane trailing portions, and there were difficulties in the field with the inadequate steel-tubing undercarriage struts. Appropriate modifications cured these, and the S.E.5a went on to acquire an excellent reputation for strength and for stability as a gun platform, a reputation emphasised by the many victories secured by the leading fighter pilots who flew the type. These included Mannock, Bishop, McCudden, Beauchamp Proctor, McElroy, Hazell, Jones, Dallas, and an impressive list of their gallant contemporaries.

Various attempts were made during the war to augment the S.E.5a's armament and to improve its manoeuvrability, but despite installations of twin overwing Lewis guns, an Eeman triple-gun mounting, horn-balanced rudders, twin-fin assemblies and narrow-chord elevators and ailerons, the S.E.5a that fought on to victory at the end of 1918 was essentially the same as the first aircraft of the hard-fought battles of 1917, and not significantly different from the original design by Folland—or was it Goodden?

S.E.5 and S.E.5a

150 hp Hispano-Suiza 8Aa, 150 hp Wolseley Python I,
180 hp Hispano-Suiza 8Ab, 180 hp Wolseley Python II,
200 hp Hispano-Suiza 8BCa, 8BCb, 8BDa, 8BDd, 8BEa, 8BEb, 8Bd, 200 hp Wolseley Adder I, II or III,
200 hp Wolseley W.4A* Viper, 200 hp Sunbeam Arab I or II,
220 hp Hispano-Suiza 8Bc or 8Be

Single-seat fighting scout.

Span 27 ft 11 in (originally), later 26 ft 7·4 in; length 20 ft 11 in; height 9 ft 6 in; wing area 249·8 sq ft (originally), later 245·8 sq ft.

Empty weight 1,399 lb (150 hp Hispano-Suiza), 1,531 lb (200 hp Adder); loaded weight 1,935 lb (150 hp Hispano-Suiza), 2,048 lb (200 hp Adder), 1,980 lb (200 hp Viper).

Speed at 10,000 ft—114 mph (150 hp Hispano-Suiza), 126 mph (200 hp Adder); speed at 15,000 ft—98 mph (150 hp Hispano-Suiza), 116·5 mph (200 hp Adder), 117·5 mph (200 hp Viper).

Armament: One fixed 0·303-in Vickers machine-gun, one 0·303-in Lewis machine-gun on Foster mounting, four 25-lb Cooper bombs.

Manufacturers: Royal Aircraft Factory, Austin, Blériot & Spad, Martinsyde, Vickers, Wolseley.

Service use: Western Front: Sqns 1, 24, 32, 40, 41, 56, 60, 64, 68 (No.2 Sqn, Australian F.C.), 74 and 84. Home Defence: Sqns 37, 39 and 61. Macedonia: Sqns 17 and 47. Mesopotamia: No.72 Sqn. Palestine: No.111 Sqn. Squadrons mobilising: Nos.74, 85, 92 and 94. Training duties: CFS; Training Squadrons Nos.3, 27, 28, 54, 55, 56, 58, 60, 72, 74 and 189; No.7 Training Depot Station; No.1 School of Aerial Fighting, Ayr; School of Military Aeronautics, Reading; No.5 School of Military Aeronautics, Denham; School of Technical Training, Reading; School of Inspection, Watford. Other duties: Administrative Wing Experimental Squadron, Royal Aircraft Factory. Wireless Experimental Establishment, Biggin Hill. Experimental Station, Orfordness.

Serials: A4561—4563, 4845—4868, 8898—8947.
B1—200, 501—700, 4851—4900, 8231—8580.
C1051—1150, 1751—1950, 5301—5450, 6351—6500, 8661—9310, 9486—9635.
D201—300, 301—450, 3426—3575, 3911—4010, 5951—6200, 6851—7000, 7001—7050, 8431—8580.
E1251—1400, 3154—3253, 3904—4103, 5637—5936, 5937—6036.
F551—615, 851—950, 5249—5348, 5449—5698, 7751—7800, 7951—8200, 8321—8420, 8946—9145.
H674—733.
Also allotted for S.E.5as for RFC/RAF but cancelled were B1001—1100, C7901—8550 and H5291—5540.
Known rebuilds: B733, 848, 875, 891, 7733, 7735, 7737, 7765, 7770, 7771, 7786, 7787, 7796, 7824, 7830, 7831, 7832 (renumbered D7017), 7833, 7850, 7870, 7881, 7882, 7890, 7899, 7901, 7913, 8791, 8932.
F4176, 5910, 5912, 5924, 5969, 6060, 6276, 9568.
H7072 (ex E3182), 7073 (ex E3981), 7074 (ex E3962), 7161 (ex C9491), 7162 (ex E5703), 7163 (ex E5693), 7164 (ex E5704), 7165 (ex E5705), 7166 (ex E5692), 7181 (ex D4946), 7247—7254 (previously numbered F6420—6427 in error), 7256—7261 (previously numbered F6428—6433 in error).

One of the Short School Biplanes that were used at Central Flying School, Upavon.

Short School Biplane

The earliest aeroplanes built by Short Brothers to their own design showed a certain amount of Wright influence, perhaps inevitably in view of their initiation into heavier-than-air flying machines having been the building of six Wright biplanes. In 1910 they turned to the Sommer configuration, their earliest known biplane in this form being Short No.18, which had been built for G. C. Colmore. Later that year came the better-known S.26, S.27 and S.28, and July 1911 saw the début of S.32, an open Farman-type pusher with 70 hp Gnome, extensions on the upper wings, two side-by-side seats and full dual control.

This aircraft had been built for Frank McClean, who offered to provide flying training on S.32 and the similar S.33 to members of the London Balloon Company, R.E., T.F. These biplanes proved well suited to flying training, and two more were ordered by the War Office for the RFC. Built as S.43 and S.44, they were allotted to the Military Wing on 27 June and 4 July, 1912, respectively, and were delivered to Upavon in July 1912 for use by CFS. One of them was in use

In this line-up of aircraft at CFS in January 1913, the biplane third from the right is one of the early Shorts used as primary trainers at that time.

Short School Biplane No.402 after a mishap, a photograph reported as having been taken in November 1912. (*RAF Museum*)

as early as 15 July, 1912, when, with the aircraft wrongly reported as a Short-Wright biplane, Capt P. W. L. Broke-Smith flew from Upavon to Larkhill with Capt E. L. Gerrard, RMLI, as passenger. Broke-Smith then flew the Bristol Boxkite F8 from Larkhill to Upavon, followed by Gerrard on the Short.

On 22 July, S.44 flew repeatedly, its pilots being Capt J. D. B. Fulton and Lt A. M. Longmore, RN, and went on flying throughout that week. Both S.43 and S.44 were flying on 9 August, and on 17 August the first CFS course began, the instructors being Broke-Smith, Fulton, Gerrard and Longmore. When the official system of numbering came into force in August 1912, S.43 and S.44 became 401 and 402. Thereafter the two Shorts rendered sterling service at Upavon throughout the remainder of 1912 and 1913, providing basic flying training to considerable numbers of embryo RFC pilots. Among the names of those who learned and those who taught on the sturdy Shorts were many who were later to win renown in the RFC and Royal Air Force.

For part of 1913, No.401 apparently did rather more flying than 402, but both were reported to be still in use late in November 1913. Precisely when and how their careers ended has yet to be discovered, but they were no longer on the strength of CFS at the end of May 1914.

School Biplane
70 hp Gnome

Two-seat elementary trainer.
Span 46 ft 5 in; length 42 ft 1 in; wing area 517 sq ft.
Empty weight 1,100 lb; loaded weight 1,540 lb.
Speed 45 mph.

Manufacturer: Short Brothers.

Service use: Central Flying School.

Serials: 401, 402.

Short Tractor Biplane

A series of sensible tractor biplanes built by Short Brothers began with S.36, completed late in 1911 and subsequently lent by Frank McClean to the Naval Flying School at Eastchurch. S.41 and S.45 were generally similar and were used by the Naval Wing of the RFC, having both wheel and float undercarriages. Three further tractors were ordered for the War Office for use as trainers at Central Flying School, and were built as S.49—S.51. These had a form of top decking about the cockpits, double-acting ailerons, kingpost bracing for the extensions of the upper wings, and wheel undercarriages.

The first of the three to be delivered was at Farnborough on 1 October, 1912, when it was flown on several circuits of the aerodrome by Mr T. O. M. Sopwith. It was again flown on 8 October, this time by Gordon Bell, and late next day Maj E. L. Gerrard set out for Upavon but was obliged to return by growing darkness. He succeeded in flying the Short to Upavon on 10 October, its arrival being recorded under its official identity of 413. Next day the new aircraft was flown by Gerrard and by Capts J. M. Salmond and J. D. B. Fulton.

Thereafter 413 flew quite regularly at Upavon with a variety of pilots. Its greatest moment occurred on 28 November, 1912, when Lt R. R. Smith-Barry took it to a height of 7,000 ft, establishing a new CFS height record. On 3 December, while being flown by Lt T. O'Brien Hubbard with Stoker Edwards as pupil, the Short 'came down at a very steep angle of descent and was badly smashed up'. Neither was badly hurt but No.413 was a write-off.

The War Office's other two Short tractor biplanes were evidently delivered to Farnborough in December 1912, for there on 10 December Gordon Bell made a one-hour flight on a 'new Short tractor biplane', and flew it again on speed tests two days later. On 17 December he was reported to be 'out morning and afternoon putting two new Short tractor biplanes through tests, flying with usual skill in very tricky winds'. For undetermined reasons the Shorts were not delivered to CFS for two months; they acquired the official numbers 423 and 424 while at Farnborough.

Maj. E. L. Gerrard flew 424 from Farnborough to Upavon on 17 February, 1913, but made an inauspicious arrival. He brought off a good landing in a gale, but while he was taxi-ing towards the sheds a sudden gust blew the Short over. Damage to the aircraft was said to be slight, but it is doubtful whether 424 ever flew again. Gerrard flew 423 to Upavon without incident on 22 February, and this aircraft was in brief use for a few instructional flights on 25 February.

Thereafter virtually nothing is known about the two Shorts, but it is probable that they were returned to Farnborough. When Their Majesties King George V and Queen Mary visited Farnborough on 9 May, 1913, the airships *Beta* and *Gamma* and eighteen aeroplanes made demonstration flights, one of the eighteen being an unspecified Short that could, of course, have been one of several Short types then extant. Towards the end of 1913 the Short tractors' serial numbers 423 and 424 re-appeared on two B.E.8s, a clear indication that, whatever official subterfuge and engine transfers there may have been, the Military Wing had unquestionably done with the Short Tractors.

It has been said that the two engineless airframes of the two Shorts were transferred to the Admiralty in August 1914 for use by the RNAS; repaired by that service and fitted with 100 hp Clerget engines they were reported to have served at Eastchurch as 1268 and 1279. If true, this is a remarkable story.

Short Tractor Biplane, S.50, which was allotted the serial number 424, later transferred to a B.E.8 (*RAE; Crown copyright*)

The serial numbers 1268—1279 were not officially listed until December 1914; these numbers were then allotted for a complete batch of twelve Short Tractor Seaplanes 126 Type that were intended to be gun-carrying aircraft. Evidently some misprint or other error had occurred, for Short No.126 was the solitary S.81 gun-carrying pusher seaplane: possibly, therefore, 1268—1279 were intended to be pushers, or alternatively 126 may have been an incorrect rendering of 166. The numbers 1268—1279 did not appear at all in a September 1915 list; the Air Department's *Disposition of Aircraft Lists* of 21 June, 1915, and 2 October, 1915, list all the aircraft then at Eastchurch and all other RNAS stations, yet 1268 and 1279 are conspicuously missing from these lists.

The March 1916 *List of H.M. Naval Aircraft built, building and under repair* not only fails to include the vital numbers but lists all aircraft transferred from and to the RNAS. Nowhere do two Short tractor landplanes of War Office origin appear. However, a 1917 list states that the block of ten aircraft 1269—1278 was cancelled, leaving 1268 and 1279 as Short Tractors of indeterminate type, apparently ordered under Contract C.P.01509/15. This same 1917 record lists 'Seaplanes: obsolete types, of which none remain in commission', and there the elusive 1268 and 1279 appear, certainly with 100 hp Clerget engines attributed to them, but indubitably listed as seaplanes. It seems, therefore, that any connection between the erstwhile 423 and 424 on the one hand and 1268 and 1279 on the other can be discounted.

Tractor Biplane
70 hp Gnome

Two-seat trainer.
Span 42 ft; length 35 ft 6 in; height 11 ft 6 in; wing area 450 sq ft.
Empty weight 1,080 lb; loaded weight 1,500 lb.
Speed 60 mph.

Manufacturer: Short Brothers.

Service use: Central Flying School.

Serials: 413, 423, 424.

Short S.62

The Short pusher biplane S.38 began life as a simple boxkite substantially similar to S.43 and S.44; in this form it earned a unique place in history as the aircraft on which Lt C. R. Samson, RN, made the first successful take-off from a British warship on 10 January, 1912. S.38 was wrecked on 9 July, whereafter it was rebuilt and redesigned, re-emerging with extended wings, smaller front elevator, reduced gap, a nacelle with two seats in tandem, and an entirely new tail unit embodying an enlarged tailplane, enlarged elevators and twin rudders.

In this form the reconstructed S.38 became the prototype of a small Short-built batch of generally similar biplanes, most of which went to the Naval Wing of the RFC. Their Short series numbers and official identities have been incorrectly attributed and related in the past, but there is no doubt that the only example of the S.38-type biplane to go to the Military Wing was S.62.

It seems likely that S.62 was completed in March 1913. On Saturday 22 March, a very stormy day, Gordon Bell flew the Short from Eastchurch to Hendon with his mechanic as passenger. While the aircraft was on the ground a violent gust of wind blew it on to its back with such force that one blade of the propeller was embedded 14 inches in the ground. The biplane was extensively damaged.

After being rebuilt, S.62 was acquired by the RFC, possibly because its full dual control suggested that it would be a useful trainer. The timing of its availability also suggested that it might have been hastily bought to be yet another aircraft needed to bolster Col Seely's claims relating to the RFC's aircraft strength. S.62 was officially allocated to the Military Wing on 3 July, 1913; with the serial number 446 it was flown from Farnborough to Upavon by Lt F. V. Holt on 19 July, 1913, and thereafter was used by CFS. It was flown by Maj J. D. B. Fulton, Capt T. I. Webb-Bowen, Maj E. L. Gerrard, Air Mech F. Dismore, Lt F. J. L. Cogan, Lt G. N. Humphreys, and Maj G. C. Merrick, DSO.

The flying career of No.446 was brief and ended tragically. On 3 October, 1913, Maj Merrick was descending steeply to land when, at about 300 ft, he apparently slipped forward on to the controls: he was not strapped in although the Short had safety belts. The Short's dive steepened so quickly that Merrick was thrown out and fell to his death; the aircraft became inverted, righted itself, but turned over on striking the ground. It was not repaired and was written off.

The Short S.38-type biplane S.62, photographed at Hendon on 22 March, 1913, shortly before it was blown over by a violent gust of wind. After reconstruction it was acquired by the RFC's Military Wing in July 1913 as No.446.

The wreck of Short No.446 after Maj Merrick's fatal accident on 3 October, 1913. (*RAF Museum*)

S.62
50 hp Gnome

Two-seat trainer.
Span 52 ft; length 36 ft; wing area 500 sq ft.
Empty weight 1,050 lb; loaded weight 1,500 lb.
Speed 48 mph.

Manufacturer: Short Brothers.

Service use: Central Flying School.

Serial: 446.

Short 827

Early in 1917 the Fairey-built Short 827 seaplane 8560 was transferred from the RNAS to the RFC for use at the School of Aerial Gunnery at Loch Doon, where the RFC operated Farman seaplanes, F.B.A. flying-boats, and a modified B.E.2c with centre-float undercarriage. For RFC service the Short was re-numbered A9920.

As noted in the text, the RFC's solitary Short 827 came from the Fairey-built batch 8550—8561, which included both the original configuration, as exemplified here by 8556, and the three-bay version. (*K. M. Molson*)

The Fairey-built three-bay Short 827 8561, the aircraft that followed 8560 which was transferred to the RFC. No.8561 was in use at Calshot in 1918. (*W. Evans*)

A photograph of this unique Short 827 has yet to be found, consequently it is not possible to be certain of its configuration. The standard Short 827 had two-bay wings with long wire-braced extensions on the upper mainplanes, but a second version existed in which the lower mainplanes were greatly lengthened and almost equalled the upper surfaces in span. This variant had full three-bay interplane bracing, and was rather more numerous than has hitherto been believed. Known 827s that had the three-bay wings were 8231, 8233, 8237 and 8561.

The last of these was, of course, the immediate production successor to 8560, and it is relevant to record that the Fairey company prepared their own structural drawing (No.141) of the extended lower wing. This does not represent confirmation of the use of this wing structure on 8560, however, for 8556 is known to have had the standard unequal-span wings.

Nothing is known about the activities of A9920, but it survived in the Loch Doon area until the end of the war. When the Armistice was signed it was in store at Bogton, the aerodrome near Dalmellington that was available to the landplanes of the Loch Doon school, for want of the intended landing ground at Loch Doon itself which, according to local belief, was never completed.

Short 827
150 hp Sunbeam Crusader

Two-seat bomber-reconnaissance seaplane.

Span 53 ft 11 in; length 35 ft 3 in; height 13 ft 6 in; wing area 506 sq ft on standard version.

Empty weight 2,700 lb; loaded weight 3,400 lb.

Maximum speed 61 mph; climb to 3,000 ft—16 min.

Armament: Normally four 20-lb or two 65-lb bombs and one 0·303-in Lewis machine-gun.

Manufacturer: Fairey Aviation.

Service use: School of Aerial Gunnery, Loch Doon.

Serial: A9920 (ex 8560).

Short Bomber

The Short Bomber came into being as a stop-gap pending the production of the Handley Page O/100, for the RNAS did not wish to delay its strikes against German naval installations. To this end the Admiralty issued a requirement for a single-engine bomber capable of carrying eight 112 lb bombs, and to meet it The Fairey Aviation Company designed the F.1 with the new and untried 200 hp Brotherhood engine, The Grahame-White Aviation Company built their Type 18 with a 285 hp Sunbeam Maori engine, Hewlett & Blondeau designed and began to build an ungainly biplane intended for the 250 hp Rolls-Royce, and Short Brothers put a Short 184 (255 hp Sunbeam Mohawk) airframe on a four-wheel land undercarriage. (The Wight Bomber N501 came along later, offered as an alternative to the Short).

In its original form the Short aircraft was exactly a Short 184 on wheels, but subsequent development saw modified wing structures and seating arrangements before the three-bays-plus-extensions and pilot-in-front configuration was adopted. Despite its unprepossessing appearance and dismal performance, the Short could lift a worthwhile load and had the advantage of limited commonality of minor components with the Short 184. For better or worse it was selected for production: small wonder that J. Samuel White thought they could do better with a design of their own when invited to tender for the Short.

Production was shared by Shorts with four other contractors, and the first production aircraft, still with the original length of fuselage, became available early in 1916. Longitudinal stability was so poor that the fuselage had to be lengthened by 8 ft 6 in, a modification that must have delayed production.

The original No.3 Wing, RNAS, was disbanded in December 1915, shortly before the end of the Dardanelles campaign. In May 1916 the Wing began to re-form under Capt W. L. Elder, RN, with the intention of operating as a strategic bombing force (the first ever formed) based at Luxeuil; its equipment was to be Sopwith 1½-Strutters and Short Bombers. At that time, however, Maj-Gen Trenchard was so concerned over the RFC's deficiencies in face of the need to prepare for the Battle of the Somme that he was compelled to appeal for help in making up the RFC's strength, which he estimated as being short of no less than

Phoenix-built Short Bomber A3932 (ex 9833) photographed at Farnborough.

twelve squadrons. A review of all RFC units at home produced a list of only twelve aeroplanes of adequate performance, leaving only the Admiralty as a source of aircraft. Worthwhile assistance could only be offered at the expense of No.3 Wing, but the Admiralty's response was generous and practical. Naval Sopwith 1½-Strutters were transferred to the RFC, and offers of Short Bombers, Bristol Scouts and Nieuport 12 two-seaters followed.

Twenty Shorts went to the RFC, apparently in ones and twos, starting in September 1916 with the Phoenix-built 9833, which went direct to the Southern Aircraft Depot at Farnborough on or about 23 September. It received the new identity of A3932, and was quickly followed by 9325, which became A4005. The transfers continued haphazardly and seemingly at short notice. No.9319 went from RNAS Eastchurch to the Isle of Grain for training duties on 13 October but little more than a week later it was allotted the RFC serial number A5155. Last of the Shorts to forsake the RNAS for the RFC were 9316, 9480 and 9481, which became A5157—5159 late in November 1916.

The transfers of the Shorts to the RFC greatly hampered the proper development of No.3 Wing RNAS, for no fewer than twelve of the twenty transferred Short Bombers had originally been allotted to No.3 Wing.* Doubtless the Admiralty had to be careful to send only Shorts that had the excellent Rolls-Royce engine: those built by the Sunbeam Company had the 225 hp Sunbeam Mohawk engine, which was unknown in the RFC. The twenty transferred bombers comprised five built by Short Brothers, one by Parnall & Sons, two by The Phoenix Dynamo Manufacturing Co, and twelve by Mann, Egerton & Co.

Despite the urgency of Trenchard's appeal that had secured the transfer of so many RNAS aircraft to the RFC, none of the Shorts went to the RFC in France, and as far as is known none found any operational employment with that Service. This is hardly surprising, for all transfers were made long after the Somme carnage had begun; indeed, by the time when the Shorts arrived for the RFC the enemy had his fleet new Albatros and Halberstadt fighters in service, and they would have made short work of the lumbering, helpless Shorts by daylight. But if the RFC found no operational use for the Bombers there can be little doubt that their twenty excellent Rolls-Royce engines found employment in other airframes.

Short Bomber
250 hp Rolls-Royce

Two-seat bomber.
Span 85 ft (upper); length 45 ft; height 15 ft; wing area 870 sq ft.
Loaded weight 6,800 lb.
Maximum speed at 6,500 ft—77·5 mph; climb to 6,500 ft—21 min 25 sec, to 10,000 ft—45 min; service ceiling 9,500 ft; endurance 6 hr.
Armament: Four 230-lb or eight 112-lb bombs; one 0·303-in Lewis machine-gun.

Manufacturers: Short Brothers; Mann, Egerton; Parnall; Phoenix Dynamo.

Service use: None known.

Serials: A3932 (ex 9833), A4005 (9325), A5153 (9484), A5154 (9483), A5155 (9319), A5157 (9316), A5158 (9480), A5159 (9481), A5170 (9478), A5171 (9772), A5173 (9485), A5179 (9482), A5180 (9477), A5181 (9488), A5182 (9476), A5203 (9315), A5214 (9320), A5489 (9479), A5490 (9487), A6300 (9832).

*These were 9315, 9316, 9319, 9476, 9477, 9478, 9480, 9481, 9482, 9485, 9772 and 9832.

Sopwith 80 hp Biplane No.319 saw brief service with No.5 Squadron in 1914. It crashed in July and was struck off charge on 21 August, 1914.

Sopwith 80 hp Biplane

By mid-1912 Thomas Octave Murdoch Sopwith had made a considerable name for himself as a highly competent pilot. Early that year he set up a flying school at Brooklands, and early in July his workshop mechanics completed a three-seat tractor biplane embodying wings of Wright inspiration and powered by a 70 hp Gnome. In the autumn of 1912 Sopwith decided to pursue aircraft design and construction in preference to flying tuition. That October he offered this first tractor biplane to the RFC, and on 17 October Maj F. H. Sykes, then commanding the Military Wing, wrote to the Director of Fortifications and Works recommending its purchase for £650 but wisely suggesting that its wings should be tested at the Royal Aircraft Factory before acceptance. In the event, however, the aircraft was bought by the Admiralty for the Naval Wing of the RFC and was flown to Eastchurch by Harry Hawker on 22 and 23 November, 1912.

Early in 1913 The Sopwith Aviation Company was in being and working in a disused roller-skating rink in Kingston-on-Thames. The first all-Sopwith design was built there and made its first flight at Brooklands on Friday, 7 February, 1913; Tom Sopwith was the pilot, Harry Hawker his passenger. The aircraft was a well-proportioned two-bay biplane powered by an 80 hp Gnome and characterised by having three large transparent panels let into the fuselage sides to improve downward view from the cockpits. The forward cockpit could accommodate two passengers seated side by side; the pilot occupied the rear cockpit.

This biplane was specifically built for the Admiralty, who had asked for an aircraft with higher performance than the early Sopwith tractor. It was exhibited at the 1913 Olympia Aero Exhibition in company with the first Sopwith Bat Boat, the two aircraft making a considerable impression on the aviation world. The three-seater was flown at Brooklands on 1 March, 1913, was accepted by Lts Spenser Grey and L'Estrange Malone of the Naval Wing, and was flown by them to Hendon. On 8 May Hawker flew a second example of the type to Farnborough where it underwent what *The Aeroplane* chose to describe as 'The War Office tests for climbing and rolling'; during these it climbed to 1,000 ft in 2 min and 22 sec. Next day, 9 May, 1913, in the speed tests its recorded fast speed was

73·6 mph, its low speed 40·6 mph. From Farnborough Hawker flew the aircraft to Hendon, via Brooklands, and won the altitude contest in the Fifth London Aviation Meeting by climbing to 7,400 ft in 15 min. On 31 May he bettered this performance substantially by climbing to 11,450 ft, a new British altitude record.

This second aircraft had an enlarged and balanced rudder of completely new design and slightly modified windows in the fuselage sides; it retained the wing warping and double tailskid of the first aircraft. Precisely how development and modification proceeded, or how many of the excellent tractor biplanes were built, remain uncertain, but it is known that by the end of June 1913 an example was at Brooklands fitted with ailerons for lateral control. This was apparently a separate and different aircraft from the altitude-record machine.

Another view of Sopwith No.319, probably photographed at the RFC Concentration Camp at Netheravon in the summer of 1914.

The fine performances of the Sopwith had not escaped the attention of Maj F.H. Sykes, and on 20 June, 1913, he sought War Office permission for a Military Wing officer to fly the type. Six days later permission was granted '. . . on the understanding that no expense connected therewith falls on the public'. It therefore seems likely that Capt A. G. Fox must himself have paid for the privilege of flying the aileron-equipped aircraft at Brooklands on 29 June. He first flew it solo for 20 minutes, then took it up again with two passengers, who must have had a heart-stopping experience when the engine failed at 100 ft above Brooklands' notorious sewage farm. Fortunately the aircraft's docility and controllability enabled Fox to get away with what no pilot should ever do: he about-turned and landed downwind. Not surprisingly, he was impressed, and reported:

> 'The machine glides well and is very easy to handle and I consider it a better machine than the B.E. (Gnome type)—in fact I like it better than any machine I have flown.'

Sykes sent a copy of Fox's full report to the Director of Artillery, recommending that nine of these Sopwith biplanes should be ordered for the Military Wing. Remarkably, this was evidently accepted and acted upon, for nine Sopwiths were eventually delivered to the Military Wing. The first of these was being tested at Brooklands early in November 1913; the second was expected to be ready about 15 November and subsequent deliveries were to be at the rate of one a week.

Before the type went into Squadron service, No.243 was tested to destruction in the Royal Aircraft Factory. The other eight were all used, at one time or another, by No.5 Squadron, seemingly without distinction. Two were destroyed on 12 May, 1914, when 324 and 325 collided in flight; Capt E. V. Anderson and Air Mech Carter, the occupants of 324, were killed; Lt C. W. Wilson was injured but survived.

No.5 Squadron's use of the type seemed not to inspire the unit to want to take its Sopwiths to France when the four RFC squadrons joined the BEF. By then few of the Sopwiths remained, for 315 and 319 had crashed, and only 246, 247 and 333 are positively known to have survived. The first two of these were concealed at Farnborough in August 1914, subsequently being struck off charge on 24 September; while 333, which had seldom been serviceable in the squadron, was transferred to CFS on 6 August. The only operational Sopwiths of the type were the few that went to Europe with the RNAS.

80 hp Biplane
80 hp Gnome

Two-seat tractor biplane.
Span 40 ft; length 29 ft 9 in; height 10 ft 4½ in; wing area 365 sq ft.
Empty weight 1,060 lb; loaded weight 1,810 lb.
Speed 73·6 mph; climb to 1,000 ft—2 min 22 sec; endurance 2½ hr.

Manufacturer: Sopwith.

Service use: No.5 Sqn; CFS.

Serials: 243, 246, 247, 300, 315, 319, 324, 325, 333.

Note: There appears to have been a tenuous, and possibly only temporary, association of the serial number 360 with a Sopwith aircraft having an 80 hp Gnome, but the precise type remains unidentified. The number 360 was apparently allotted at different times to two separate Maurice Farman Longhorns.

Sopwith Tabloid

Late in November 1913 the Sopwith company completed a compact little biplane powered by an 80 hp Gnome. Its design was something of a team effort by the staff of the Sopwith works, but primary credit has been given to Harry Hawker. How much inspiration may have been provided by the earlier Royal Aircraft Factory S.E.2 will never be known, but the Sopwith S.S. was essentially similar in conception, being an equal-span single-bay biplane with staggered wings and warping for lateral control.

In appearance the little Sopwith was less refined than the more elegant S.E.2, and it was structurally simpler and lighter; it could, moreover, squeeze a suitably compact or pliable passenger into the cockpit beside the pilot. In terms of performance the Sopwith was the equal of the S.E.2, for it underwent official assessment at Farnborough on 29 November, 1913, when it proved to have a maximum speed of 92 mph, a minimum speed of 36·9 mph, and an initial rate of climb of 1,200 ft/min.

The unfortunate arrival of the Military Wing's second Sopwith Tabloid, No.326, at Farnborough, 6 May, 1914.

Within two weeks of establishing these impressive performance figures and making a sensational appearance at Hendon on the little Sopwith, Harry Hawker was on his way back to his native Australia, taking the prototype S.S. with him. He intended to demonstrate the aircraft, possibly in the hope of securing some orders from the Australian Government.

The Sopwith S.S. had, almost as soon as it appeared, been nicknamed Tabloid. Although the word was then, as now, a perfectly ordinary English noun, it was also used by the manufacturing chemists Burroughs, Wellcome & Co. as a trade name, and they contemplated legal action for infringement. Wisely they did not pursue litigation, but their concern was sufficient to make the name inseparable from the aircraft.

In spite of the frequently repeated allegation that the War Office was incapable of looking beyond the Royal Aircraft Factory for the aircraft that the RFC's Military Wing required, that Department lost commendably little time in ordering a batch of Tabloids for the Military Wing. It should be noted particularly

A Tabloid after arrival at Farnborough but before being numbered.
(*RAE 0136; Crown copyright*)

that no order had been placed for production of the S.E.2. No fewer than nine Tabloids were ordered by the War Office at a price of £1,075 each under Contract No.A/2368 dated 18 December, 1913, a substantial order at that time. Three more were ordered on 14 March, 1914. The official intention was to use these compact and speedy aircraft as fast scouts and dispatch carriers, and it was from this proposed employment—so very seldom enacted in reality—that the later fighting single-seaters retained the name of scouts.

Delivery of the first production Tabloid had been promised for 27 February, 1914, but in fact it was not flown at Brooklands until 11 April, 1914; its pilot was Howard Pixton. He flew the aircraft again next day, as did Vickers' pilot Harold Barnwell, who seized the opportunity to execute the first loops made by a Brooklands aviator. Pixton's induction flights were timely, for a few days previously he had been unable to become satisfactorily airborne in the special floatplane version of the Tabloid that was to be the Sopwith company's entry in the 1914 Schneider Trophy contest at Monaco. When Pixton flew the seaplane at Monaco on 20 April, 1914, he won the contest handsomely, covering the 280 km (174 miles) in 2 hr and $13\frac{2}{5}$ sec, an average speed of 86·8 mph. In its power unit the Schneider victor differed from the standard Tabloid by having a 100 hp Gnome Monosoupape instead of the 80 hp Gnome.

On 22 April, 1914, the first production Tabloid was flown to Farnborough, where it received the official number 378 and its airframe was tested to destruction in the Royal Aircraft Factory. The second production aircraft arrived at Brooklands on 24 April and was flown to Farnborough by Pixton on 6 May, but it turned over on landing on rough ground and sustained damage. Deliveries continued more or less regularly: by 26 May six Tabloids had been completed, and Pixton delivered aircraft to Farnborough on 13 May and 2, 3 and 9 June. The production Tabloids differed from the prototype in having a modified engine installation and a revised tail unit in which a plain rudder was hinged to a triangular fin.

Pixton's mishap of 6 May led to modifications of the undercarriage, and some of the later Tabloids, exemplified by 394, were delivered with an additional strut in the undercarriage on each side. Early in July 1914, Sykes wrote to the War Office recommending that the strengthened undercarriage should be fitted to all of the Sopwiths before they were taken into squadron service; he also gave instructions that the Tabloids were not to be flown until this was done. On 22 July the D.G.M.A. wrote to Sopwiths, seeking confirmation that the firm were prepared to fit the additional struts and stronger wheels to all aircraft already delivered.

The Schneider Trophy aircraft was fitted with a wheel undercarriage soon after its return to England, but in this case the wheels were borne on a simple structure of two conventional V-struts separated laterally by parallel spreader-bars having the half-axles pivoted at their mid-points. In this form the 100 hp Schneider Tabloid appeared at Brooklands on 20 May, and was later flown at Farnborough, whence Hawker flew it back to Brooklands on 16 June, 1914. Hawker made all subsequent deliveries of RFC production Tabloids to Farnborough; these occurred on 23 and 29 June, 7 and 24 July, and 4 August.

The first Tabloid to be issued to an RFC unit was 381, which was flown from Farnborough to Netheravon by Maj J. F. A. Higgins on 30 June, 1914, its allocated destination being No.5 Squadron. Maj Higgins reported on the Tabloid's handling qualities on 1 July. He found that it had a pronounced tendency to climb on full throttle and constantly flew left wing low; considerable

Tabloid No. 326 at Farnborough after repair, a photograph dated 23 May, 1914. *(RAE 369; Crown copyright)*

force was required to keep the aircraft level in the rolling and pitching planes. Certain controls and instruments were inconveniently placed, and the undercarriage was much too weak. (At that time, of course, 381 had the original unmodified undercarriage.) In Higgins's opinion the S.E.2 was more pleasant to fly.

An official statement of machines and engines serviceable on Friday 31 July, 1914, conspicuously excludes all mention of any Tabloids, doubtless because the necessary undercarriage modifications had not been completed.

Harry Hawker made a second delivery flight to Farnborough on 4 August, 1914, the aircraft concerned being the Tabloid prototype that he had brought back from Australia. By then its rear fuselage had been stripped of fabric (in order, so it was earnestly said, to improve looping performance), and its wings were rigged with marked dihedral on the lower surfaces, nil on the upper—a prophecy, perhaps, of the later Camel. The Tabloid prototype had been impressed for the use of the RFC, having been purchased for £900, and was given the official serial number 604. It was initially allotted to Lt C. C. Barry (a puzzling allocation: Barry did not receive his aviator's Certificate until 2 September, 1914), and was apparently intended to join the squadrons with the BEF. As it had a V-strut undercarriage by August 1914 it was presumably exempt from the instruction grounding Tabloids with unmodified undercarriages, but it did not go to France. Instead, as related later, it was transferred to the RNAS in September.

Possibly owing to concern about the undercarriage's robustness, none of the four RFC squadrons took any Tabloid to France, but four crossed the Channel in packing cases with the Aircraft Park that arrived at Boulogne on 18 August, 1914. These were 362, 386, 387 and 611, of which the last two were sent to No.3 Squadron on 24 August. They did not last long: 387 was struck off the squadron's strength on 2 September, 611 next day; both were reported to be wrecked. It was of this precise period that Louis Strange recalled in his book *Recollections of an Airman* how 2nd Lt Norman C. Spratt '. . . flew a Sopwith Tabloid and forced the enemy [an Albatros] to land by circling round above him and making pretence to attack him'. McCudden, in *Five Years in the R.F.C*, recorded Spratt's action in greater detail and dated it on 28 August; he thought Spratt had 'a handful of fléchettes' in his Tabloid. Three days later Spratt again rose to attack an enemy aircraft, this time taking a revolver from which he fired 30 rounds into his adversary without success. He landed to make a home-made weapon consisting of a hand grenade on a length of wire cable; his intention was to fly above the enemy aircraft and try to entangle the cable-borne grenade in the enemy's airscrew. At that time Spratt was one of four flying officers attached to the Aircraft Park, yet his log-book records that, when 'chasing German aeroplanes' on 29, 30 and 31 August, 1914, he was flying 387, which was supposed to be on the strength of No.3 Squadron.

Not until 26 December, 1914, was 386 issued to No.4 Squadron. It lasted a little longer than 387 and 611, for it was not struck off until 22 January, 1915. It seems likely that the RFC decided that the Tabloid was useless as an operational aircraft, for 362 was sent back to England from the Aircraft Park on 4 February, 1915, together with 654, which had arrived at the Park on 15 January.

The RFC apparently discarded the Tabloid thereafter, and it was in the hands of the RNAS pilots that the type performed its solitary action of any consequence. The Naval service, in the person of Sqn Cdr Spenser Grey, approached the War Office on 8 September, 1914, to ask whether the Admiralty might be allowed to take over three or four Tabloids from the RFC. It was felt that they might be useful in the kind of work that Naval pilots were then undertaking in

Tabloid No.394 with the strengthened undercarriage mentioned in the text. It has been reported that it saw brief service with RFC Squadrons Nos.7 and 6, but it was sold to the RNAS, being renumbered successively 904 and 167, and was flown by Lt-Cdr Spenser Grey on 9 October, 1914, in his attempted bombing attack on German airship sheds at Cologne. (*RAE*)

Flanders. Surprisingly, the War Office agreed immediately to release three Tabloids and, by 10 September, Nos.394 and 395 had been handed over direct to what was still referred to in official documents as the Naval Wing; No.604 had been handed over to the Sopwith company for the Naval Wing. The first two were sold to the Admiralty for the price of £1,075 each, while 604 went for £800 (having only recently been bought by the War Office for £900 on impressment); with the addition of Departmental Expenses the total cost to the Admiralty was £3,154.5s.0d.

The RNAS allotted to 394 and 395 the new identities of 904 and 905, but it was found that these had already been taken up by, respectively, the Short S.58 landplane and Short S.80 seaplane. The three ex-RFC Tabloids were then renumbered 167–169, of which 169 was the former prototype, 604 during its brief sojourn in the RFC.

Tabloid
80 hp Gnome, 100 hp Gnome Monosoupape

Single-seat scout.
Span 25 ft 6 in; length 20 ft 4 in; height 8 ft 5 in; wing area 241·3 sq ft.
Empty weight 730 lb; loaded weight 1,120 lb.
Maximum speed 93 mph.
Armament: Temporary and makeshift use of revolver and fléchettes has been reported.

Manufacturer: Sopwith.

Service use: *Prewar*: No.5 Sqn; *Western Front*: Sqns 3 and 4.

Serials: 326, 362, 378, 381, 386, 387, 392, 394 (later 904 and 167), 395 (later 905 and 168), 604 (later 169), 611, 654.

Sopwith LCT, the 1½ Strutter

In December 1914 The Sopwith Aviation Co designed a small two-seat biplane that had an 80 hp Gnome and an unusual arrangement of central bracing that supported the upper wings. In its day this aircraft was known as the 'Sigrist bus', and in the hands of Harry Hawker it set up a British altitude record of 18,393 ft on 6 June, 1915. The W-form cabane bracing reappeared on a new and slightly larger military two-seater that was completed in December 1915, and the unusual strut arrangement led to the aircraft's unofficial name, the 1½ Strutter.

To the Sopwith drawing office the handsome new aircraft was the LCT, these initials probably signifying Land Clerget Tractor, for the engine was the 110 hp Clerget 9Z. At the time the Sopwith company was primarily an Admiralty contractor, and it was to the RNAS that the prototype LCT went with the official serial number 3686. From the outset the new Sopwith was regarded as a fighter yet, as originally designed, it was to be armed only with one Lewis gun fired by the occupant of the rear cockpit. The cockpits were widely separated, and it was evidently expected that the location of the gunner relatively far aft, combined with a rotating and elevating mounting for the gun, would enable him to fire almost straight ahead over the upper wing. Thus it appears that the philosophy underlying the design must have been that the superior performance to be expected from a tractor configuration outweighed the sacrifice of the unobstructed field of fire forward provided by a pusher. This was bold thinking with only 110 hp available.

An early batch of ex-RNAS 1½ Strutters of No.70 Squadron, RFC, at Farnborough. Second from the left is 5720, (ex 9387); all the aircraft have partly faired-in Vickers guns and Nieuport-type (Etévé) mountings for the Lewis guns. (*RAF Museum*)

The 1½ Strutter introduced a new concept of the military aeroplane, for it was small and unusually compact. It was innovative, too, for it introduced the British flying Services to the variable-incidence tailplane controlled by the pilot. In anticipation of the flat glide that its clean design could be expected to induce, the aircraft was fitted with air-brakes in the trailing portions of the lower centresection.

It is greatly to the credit of the Air Department of the Admiralty that the staff there unhesitatingly ordered the 1½ Strutter for the RNAS, a batch of fifty two-seaters being closely followed by a further order for 100 aircraft, many of which were completed as single-seat bombers.

The prototype 1½ Strutter was tested at CFS in January 1916. Its speed trial was flown on 24 January, 1916, when its maximum speed at ground level with the Lewis gun strapped to the circular rail of the rear cockpit proved to be 105·1 mph. The narrative portion of the official report contained alarmed references to the single-bay interplane bracing, the lack of flying wires under the lower wings, and the aft location of the undercarriage wheels. More complimentary observations were made on the aircraft's flying qualities, but the distance needed for landing was considered abnormal, even with the use of the air-brakes. These could reduce the landing run by 100 yd but caused 'a good deal of vibration' when applied.

No.7762, the first Sopwith 1½ Strutter built by Ruston, Proctor & Co. It had been completed by 29 June, 1916, when it was officially allotted to the E.F. It went to France via the AID at Farnborough, where it was reported on 20 July, 1916, and two days later it was issued to No.45 Squadron from No.2 A.D. It returned to England and was again allotted to the E.F. on 24 March, 1917; its only other known movement was to No.1 A.D. on 19 April, 1917.

This report was telegraphed to the G.O.C. Royal Flying Corps on 27 January, 1916, consequently the RFC knew something of the potential of the new Sopwith at an early stage. Whatever reservations there may have been about the aircraft's structure were probably outweighed by its good performance, and the RFC did not take long to order the 1½ Strutter. With the Sopwith company's production capacity fully taken up by Admiralty orders, the War Office had to look elsewhere for contractors, and in March 1916 Contract No.87/A/368 was given to Ruston, Proctor & Co for the construction of 50 Sopwith two-seaters; the serial numbers 7762—7811 were allotted on 27 March, 1916.

High-level consideration of the Allied military operations of 1916 began in December 1915, and it became apparent that the most propitious time for a British offensive in France would be in July. Much happened to cause reconsideration, most of all the German offensive of 21 February, 1916, that began the martyrdom of Verdun; but it was not until Fort Vaux fell on 7 June that it was finally agreed that the Somme offensive should be launched on 1 July. Some months earlier, Trenchard, seriously concerned at the RFC's lack of the squadrons that would be needed to provide effective support to the Army, sent an

One of the 1½ Strutters that were transferred from the RNAS to the RFC was 9381, which became 7942 in the RFC. It was with the AID at Farnborough on 14 June, 1916, on which date it was formally allotted to the E.F., and it saw operational service with No.70 Squadron.

urgent request for more aircraft. All that RFC resources in England could muster was a total of twelve aircraft of adequate performance; no additional airframes or engines could be expected from France; there was only one course open to the War Office, namely an approach to the Air Department of the Admiralty. This won immediate recognition and prompt assistance, and the RNAS agreed to transfer to the RFC a number of 1½ Strutters from their Sopwith contracts.

Transfers of two-seat 1½ Strutters began in April 1916, when 9386, 9387 and 9389 were sent to Farnborough to assume the RFC identities of 5719—5721 respectively. The matter was formally recorded in an RNAS memorandum dated 15 May, 1916, in which it was proposed that one-third of currently outstanding Sopwith production, to a total of forty 1½ Strutters, should be transferred to the RFC. Eventually at least seventy RNAS 1½ Strutters were transferred to the RFC.

Another 1½ Strutter that was transferred from the RNAS was A1914, which was originally 9716. It was at Farnborough on 9 September, 1916, having been allotted to the E.F. as early as 31 August. Apparently it still had not gone to France by early November, but it evidently fell into German hands before the end of 1916. Surprisingly, for that relatively late date, it still had a Nieuport (Etévé) gun mounting on the rear cockpit.

Another 1½ Strutter that was captured intact by the Germans was A993, which was at the Southern Aircraft Repair Depot, Farnborough, allotted to the E.F. on 28 March, 1917. On going to France it was issued to No.43 Squadron but came down in German-held territory on 28 April, 1917. Its pilot was 2/Lt C. M. Reece, its observer 2AM A. Moult, and it came down at Roncy.

The earliest 1½ Strutters delivered to the RNAS had no Vickers gun for the pilot. At least two mechanical interrupter gears were then under development: the Scarff-Dibovsky for the RNAS and the Vickers-Challenger potentially for the RFC. These became available at just about the time when production of Vickers guns increased sufficiently for some to be made available for use in aircraft. Precisely when or by whom it was decided to arm the Sopwith 1½ Strutter with a fixed, forward-firing Vickers gun seems not to have been recorded, but it must have been in March or April 1916. It was a decision of historic significance, for it made of the 1½ Strutter the first expression of the two-seat fighter arrangement that established a pattern for the later and greater Bristol Fighter.

A8226 was a 130 hp 1½ Strutter that was allotted to the E.F. on 25 April, 1917, and became one of the Sopwiths of 'C' Flight, No.45 Squadron, at Ste-Marie Cappel. On 27 May, 1917, it was shot down near Ypres by Offizierstellvertreter Max Müller of Jasta 28 as his 13th victory. The 1½ Strutter's pilot and observer, respectively Capt L. W. McArthur, MC, and 2/Lt A. S. Carey, were killed. (*RAF Museum*)

In the RFC the first squadron to have the Sopwith two-seater was No.70, which was formed at Farnborough on 22 April, 1916. Its three Flights were made up progressively as 1½ Strutters became available, each going to France as it attained full strength. 'A' and 'B' Flights went to Fienvillers on 24 May and 29 June respectively, their Sopwiths armed with Vickers guns and Vickers-Challenger gear; those that had a rotatable mounting for the observer's Lewis gun had the Nieuport form of mounting. The Sopwiths of 'C' Flight came from the RNAS fully equipped with both the Scarff-Dibovsky interrupter gear and the Scarff ring mounting on the rear cockpit. This Flight arrived in France on 30 July, 1916, and its aircraft's armament made a considerable impression on the RFC's high command.

The Scarff-Dibovsky gear was expected. On 3 June Brooke-Popham had reported to D.A.E. that none of the Sopwiths then in France was serviceable. The

The 1½ Strutter A5252 was built by the Wells Aviation Co and became aircraft A of 'A' Flight of the Wireless Experimental Establishment, Biggin Hill. It had three spanwise aerials along the underside of the upper wing and others under the lower.

drive for the Vickers-Challenger gear was taken off the Clerget's pump spindle, which also drove the rpm counter; the pump spindle was too weak and broke regularly. Moreover, the cutting of the cowling panels to take the Vickers gun allowed oil to pass through, blinding the pilot and reaching the magneto. Immediate action was taken in England to ensure that all subsequent Sopwith two-seaters sent to the RFC would have the Scarff-Dibovsky gear.

As soon as the Scarff ring mounting was seen, its superiority over the cumbersome Nieuport mounting was at once evident, and Trenchard asked that all 1½ Strutters supplied to the RFC should be fitted with it. In practice it proved somewhat difficult to comply with this request, for as late as 1 December, 1916, an instruction had to be given that 1½ Strutters received with a Nieuport mounting would be issued in that form and the receiving squadron would have to replace it with a Scarff mounting. On A1903 (ex-9695), issued to No.70 Squadron on 11 September, 1916, the Lewis gun was carried on a Vickers ring mounting. This was not developed, possibly because A1903 was damaged in combat on 15 September and struck off charge at No.2 A.D. on 26 September. The aircraft was repaired in England and returned to France early in 1917 to fly with No.43 Squadron, but it is doubtful whether it retained its Vickers rear-gun mounting.

Deliveries of 1½ Strutters from War Office contractors began in June 1916, on the 29th of which month 7762 was ready at Lincoln. The first four Ruston, Proctor Sopwiths, 7762—7765, were all allotted to No.45 Squadron on 22 July, but not until 12 October was the squadron able to go to France. It quickly found that the 1½ Strutter was already outclassed, an experience repeated by No.43 Squadron which, also equipped with 1½ Strutters, went to France on 17 January, 1917. All three squadrons used their Sopwiths as fighter-reconnaissance aircraft, but the difficulties increased as time went on. A small increase in performance was achieved by fitting the 130 hp Clerget 9B in place of the 110 hp Clerget 9Z, but in the RFC this modification came too late. Although A956 was tested with the 130 hp engine in January 1917 it was mid-April 1917 before production 1½ Strutters could be so equipped in reasonable numbers, and as late as 16 May, 1917, RFC Headquarters instructed No.1 A.D.:

> 'As far as possible, only Navy Sopwith two-seaters with 130 hp Clerget engines should be issued to No.70 Squadron.
> This squadron will take preference over Nos.43 and 45 Squadrons as regards the issue of 130 hp Sopwiths.
> 110 hp Clerget Sopwith two-seaters when available should be issued to Nos.43 and 45 Squadrons.'

Although the great majority of 1½ Strutters sent to France thereafter had the 130 hp engine, the Sopwith was by then hopelessly outclassed.

Vickers-built 1½ Strutters became available from late July 1916, while the Fairey and Hooper factories started to deliver in October. Not until April 1917 did deliveries from the Wells Aviation Co begin, but very few were allotted to the Expeditionary Force.

As both the late Wing Commander Norman Macmillan and Marshal of the RAF Lord Douglas of Kirtleside have indicated in their published works, the 1½ Strutter was not the most tractable of aeroplanes. Part of its slow response was attributable to its enormous 13 ft 6 in tailplane, and word of a smaller empennage for the type had reached Maj G. A. K. Lawrence, O.C. No.70 Squadron, in mid-October 1916. His request to be given 'three or four experimental small tails' for trial on his squadron's aircraft was turned down by Trenchard. An enquiry sent by Brooke-Popham on 28 October, 1916, elicited the information that a 12-ft tailplane had been tested on a 1½ Strutter, in which form the aircraft's control response had been so sluggish and recovery from a dive so difficult that the modification had been abandoned.

Late in December, lengthened ailerons were introduced and were standardised. Amazingly, it appears that no one troubled to advise the RFC in France that this change, with its substantial implications for spares and maintenance, was to be made.

As early as 20 June, 1916, Brooke-Popham had asked No.1 A.D. to see how bomb racks could best be fitted to the 1½ Strutter to enable it to carry four 20-lb bombs or either one or two 100 lb bombs. On 22 June he instructed No.70 Squadron to have their Sopwiths fitted with a rack for four 20-lb bombs and a bomb sight for the use of the pilot. Unfortunately, the standard bomb sights then in general use could not be suitably installed on the 1½ Strutter, and this problem was never resolved.

It reappeared with the production of the single-seat bomber version of the 1½ Strutter, of which the War Office had ordered 100 (A6901—7000) from Hooper & Co, with a provisional order for 150 more. Additionally, at least some aircraft of the Morgan-built batch A5950—6149 were delivered in the single-seat bomber configuration. It had been agreed at an official conference on 15 December, 1916, that the bomber version was to be built for the RFC and that the aircraft would be kept in store until the Expeditionary Force needed them, but the question of the bomb sight and type of bomb to be used was not decided. The RNAS found it possible to fit their Equal Distance Bomb Sight to the 1½ Strutter, but an official RFC report on the matter, dated 24 January, 1917, concluded:

> '. . . . it is considered impossible to design a satisfactory bombsight for this machine, and that if even moderate accuracy is needed in bombing with this machine, the bombs must be dropped from very low heights.'

Although A.D.A.E. seemed to think that the Equal Distance Bomb Sight should be provided for RFC 1½ Strutter bombers, it never was and the variant saw no operational service in the RFC.

The Hooper-built batch of 1½ Strutters A6901—A7000 were ordered as single-seat bombers but this variant saw no operational employment with the RFC. Some were transferred to the RNAS for conversion to two-seat shipboard configuration, while others were used for training. In the latter category was A6984, which in December 1917 was with No.36 Training Squadron at Montrose. (*R. Gerrard*)

By mid-April 1917 it had been decided to hand over the bombers to the Russians, and the provisional order for 150 Hooper-built bombers was cancelled on 22 June. As at 10 August there were 87 Sopwith bombers in store, but production could not be stopped until it was known whether the Russian Government wanted the aircraft. Evidently the Russians decided to accept some, but in the second half of September returned 50 Sopwiths. Thereafter arrangements were made to transfer a number of the 1½ Strutter bombers to the RNAS, and others were used by the RFC at various training units.

In February 1917 there was an intention to try out a 1½ Strutter two-seater on artillery spotting duties. The necessary trials were to be conducted by No.46 Squadron, and it is likely that the aircraft in question was A882 (ex-9668), which was issued to No.46 Squadron from No.2 A.D. on 14 February, 1917. The nature of No.46 Squadron's report is not known, but the 1½ Strutter was not adopted for artillery work.

It was at about this time that other interrupter gears were introduced on the 1½ Strutter. Just before mid-February 1917 the Ross gear became available for test; it had been designed by Capt Ross of No.70 Squadron and the first

The 1½ Strutter A8778 was equipped for night flying, having navigation lights and flare brackets, and was used for Home Defence duties by—so its markings suggest—No.44 (H.D.) Squadron.

installation was made in A2431 of that unit. It was favourably reported on by the O.C. No.70 Squadron, at that time Capt A. W. Tedder, an officer whose distinction as a great leader of the Royal Air Force was to emerge in the Second World War, and was adopted for the 1½ Strutter. Tedder's report on the Ross gear was dated 18 February, 1917: next day A.D.A.E. wrote to Trenchard to inform him that

> 'A type of interrupter gear known as the Kauper gear has been produced which is especially suitable for use with Clerget engines as on the Sopwith machines.
>
> Either this gear or the Constantinesco, whichever becomes available first, will be provided for fitting to the Sopwith two-seaters in use by you.'

As far as is known, no RFC 1½ Strutter saw operational service with the Constantinesco gear, but some of the later production aircraft had the Sopwith-Kauper mechanism.

For nocturnal combat this 1½ Strutter had a spotlight coupled to its Lewis gun. (*Air Vice-Marshal H. J. Roach, CB, CBE, AFC, from Peter Liddle's 1914–18 Personal Experience Archives, presently housed within Sunderland Polytechnic*)

Sopwith 'Comic' modification of a 1½ Strutter for Home Defence duties. B762 had been reconstructed at the Southern Aircraft Repair Depot and was initially allotted to the RFC in France on 4 August, 1917. It was re-allotted to Home Defence Group on 18 August and went to No.78 Squadron, which may have been responsible for converting the aircraft to the single-seat configuration. (*E. F. Cheesman*)

For Home Defence duties fifty-nine Sopwith 1½ Strutters, three of which were single-seaters, were sent to Home Defence squadrons in the summer of 1917. Some of the two-seaters of No.78 Squadron were converted in a manner devised by Capt F. W. Honnett of 'A' Flight to become single-seaters, the pilot occupying what was normally the rear cockpit. Armament varied, several of these conversions having one Lewis gun or a double-yoked pair on a suitably adapted Foster mounting, firing over the upper wing. These conversions were known irreverently as Sopwith Comics.

On the Western Front the replacement of the gallant old Sopwith two-seaters began on 25 July, 1917, when No.70 Squadron began to re-equip with Camels.

The 'Comic' night-fighter could be armed with the usual fixed Vickers gun augmented by a Foster-mounted overwing Lewis gun, or a pair of overwing Lewis guns. The former arrangement is seen here on A5259, believed to be of No.44 (H.D.) Squadron.

Some of the Camels that might have eased life for Squadrons 43 and 45 were hurriedly allotted to Home Defence units after the Gotha raids on London of 13 June and 7 July, consequently No.45 Squadron took until 1 September to re-equip with the new single-seater, No.43 until 3 October. In fact, all 1½ Strutters had been withdrawn from the squadrons by 26 September.

Thereafter the type served on with training units, in both two-seat and single-seat forms. Some of the two-seaters had dual control fitted, an installation that entailed some structural alterations at the back of the rear cockpit. The 1½ Strutter may have given more varied and versatile service with the RNAS, but nothing it did elsewhere outshone the gallantry of its RFC crews, who battled on through so much of the bitter air fighting of 1917.

1½ Strutter

110 hp Clerget 9Z or 130 hp Clerget 9B. A8194, a two-seater intended for Rumania, was tested with a 110 hp Le Rhône 9J

Two-seat fighter-reconnaissance; single-seat bomber; single-seat fighter (Home Defence).

Span 33 ft 6 in; length 25 ft 3 in; height 10 ft 3 in; wing area 346 sq ft.

Weights and performance:

Date of trial	25.4.16	27.12.16	July-Aug 1917	27.1.17
Aircraft	Possibly 9386 (5719) Two-seater	A2391 Two-seater	A6901 Single-seat bomber	A956
Engine	110 hp Clerget	110 hp Clerget	110 hp Clerget	130 hp Clerget
Weight empty, lb	—	1,259	1,354	—
Weight loaded, lb	2,052	2,149	2,362	—
Maximum speed, mph				
at sea level	104·9	—	—	—
at 6,500 ft	—	100·5	—	—
at 10,000 ft	—	96·5	94	—
Climb in min and sec to				
2,000 ft	3 12	— —	— —	— —
5,000 ft	8 51	— —	— —	4 50
6,500 ft	— —	10 40	13 55	7 15
10,000 ft	— —	20 25	26 55	— —

Armament: One 0·303-in Vickers machine-gun and one 0·303-in Lewis machine-gun; four 25-lb bombs. RFC single-seat bomber version had only the fixed Vickers gun and was expected to carry twelve 25-lb bombs. Home Defence single-seat conversion could have two Lewis guns or a combination of one Vickers and one Lewis.

Manufacturers: Sopwith; Fairey; Hooper; Morgan; Ruston, Proctor; Vickers; Wells.

Service use: *Western Front*: Sqns 43, 45 and 70; one two-seater briefly with No.46 Sqn. *Home Defence*: Sqns 37, 39, 44, 78 and 143. *Macedonia*: Composite (RFC and RNAS) Fighting Squadron at Hadzi Junas. *Training duties*: Training Sqns 1, 6, 10, 17, 18, 28, 30, 36, 40, 45, 54, 62 and 73; No.198 (Night) Training Sqn, Rochford; CFS; School of Special Flying, Gosport. *Squadrons mobilising*: No.6 Sqn, Australian Flying Corps; RFC Sqns 52, 56, 80, 81 and 84. *Other units*: Wireless Experimental Establishment, Biggin Hill.

Serials:
I. *Transfers from the RNAS*
5719—5721 (ex 9386, 9387 and 9389); 7942 (ex 9381); 7998—8000; A377—386; A878—897 (A882 ex 9668, A888 ex 9675, A889 ex 9676, A890 ex 9678, A891 ex 9681, A896 ex 9679, A897 ex 9685); A1902—1931 (A1903 ex 9695, A1904 ex 9696, A1907—1919 ex 9697, 9703, 9705, 9707, 9702, 9713, 9710, 9716, 9719, 9721, 9725, 9728, 9731; A1921—1925 ex 9737, 9740, 9743, 9746, 9749); A2431—2432 (ex 9694 and 9682); A2983—2991 (A2983—2989 ex 9684, 9691, 9688, 9690, 9693, 9687, 9892). A8726—8731 allotted for transfers from RNAS but not used.

II. *Production for RFC*
7762—7811, A954—1053, A1054—1153, A1511—1560, A2381—2430, A5238—5337, A5950—6149, A6901—7000, A8141—8340, A8744—8793, B2551—2600.

III. *Acquired from France by RAF for use as shipboard aircraft*
F2210–2229; former S.F.A. numbers 7034, 7083, 7084, 7096, 7100, 7107—7112, 7119—7121, 7124, 7126, 7098, 7099, 7103, 7104, F7547—7596.

IV. *Rebuilds*
B711, B714, B715, B729, B744, B745, B762, B799, B812, B816, B827, B862, B4016, B4044, B7903, B7914—7916, B7946, B8911, B8912, B9910, C4300.

Note: One official record quotes A162 as a 1½ Strutter. If correct, this number must have been allotted by No.1 A.D.

Sopwith Sparrow

In the autumn of 1915 The Sopwith Aviation Company built a small single-seat biplane to the design of Harry Hawker, the firm's chief test pilot. The aircraft was built to the full-size drawings laid out in chalk on the shop floor by Hawker, who used the little biplane as a personal transport when it was built. Power was provided by a 50 hp Gnome, reputed to have been originally fitted to Tom Sopwith's Burgess-Wright biplane; lateral control was by wing warping. The tips of the mainplanes were distinctively curved and cut back, and the tailplane had corresponding reverse rake on its extremities. In these features and in its general proportions the aircraft established the principal geometry of the later Pup.

The 50 hp single-seater looked a frail little wisp of an aeroplane, but Hawker had no qualms about using it for aerobatics. On Sunday, 14 November, 1915, he flew across to Hendon on it, looped twelve times and performed other manoeuvres to the delight of the onlookers.

Evidently the promise of Hawker's Runabout (as such the 50 hp single-seater was known) was sufficient to justify its development into the enchanting and efficient Pup, the prototype of which was flying early in February 1916. An intriguing fact is that the little 50 hp biplane was not allowed to disappear unrecorded. Although the original drawings to which it was built had been made in chalk on the floor of the Sopwith Experimental shop, a full general-arrangement drawing was prepared in the Sopwith drawing office. It bore the drawing number 1719, and it is remarkable for the fact that it was drawn on 12 September, 1916, nearly a year after the original aircraft was built. The date on which it was traced, 18 November, 1916, was later still. By way of designation all that appears on this drawing is 'Type 50 hp Gnome'.

Sopwith Sparrow with national markings as applied at the factory. The origin of the striped finish is not known.
(*The late Air Chief Marshal Sir John Whitworth-Jones, GBE, KCB, via E. A. Harlin*)

This is of interest because of two facts. The first is that a single-seat, warping-wing biplane bearing a very close resemblance to the Runabout was acquired in 1925 by Mr R. C. Shelley of Billericay. On this aircraft's instrument panel was a brass plate inscribed:

        'Machine type   SL.T.B.P.
        Engine   50 h.p. Gnome
        No   476'

At that time its engine was in fact No.683—3831, which had apparently been rebuilt by F. W. Berwick & Co Ltd in August 1916. The designation SL.T.B.P. has never been satisfactorily interpreted.

The second relevant fact is that the official serial numbers A8970—8973 were allotted in May 1917 for four Sopwith aircraft enigmatically described in the official record as:

'Sopwith (small) scouts. No R.N.A.S. [numbers] have been allotted. Warping wings fitted for 50 Gnomes [*sic*]. At Sopwiths, Brooklands, ready but without engines.'

It is unconfirmed that these four scouts were similar to Hawker's Runabout, and interpretation of the above quotation in the absence of supplementary information must to some extent be conjectural. Nevertheless, it is interesting to speculate as to why a G.A. drawing was found to be necessary as late as September 1916—perhaps to enable the four 'small scouts' to be built? Why, if the seeming designation SL.T.B.P. was in fact contemporary with the original Runabout of 1915, did it not appear on the drawing dated September 1916? And could Mr Shelley's acquisition of 1925 have been one of the mysterious quartet of small 50 hp Scouts?

A possible explanation lies in the reports of the 92nd and 93rd meetings of the Progress and Allocation Committee held on, respectively, 11 and 12 June, 1917.

Unfortunately, the report of the 76th meeting held on 22 May, 1917, does not appear to have survived. Paragraph 9 of the report of 11 June is headed *Sopwith Sparrows with 50 hp Gnome* and reads:

'Lt Col Whittington referred to the statement made on the 22nd May that an officer of the Technical Department would inspect these machines and say whether they were worth buying, and asked if any decision had been come to, as he had heard nothing since.

Col Beatty arranged to report on this.'

Under the same heading the report of the following day's meeting ran:

'General Pitcher stated that he sent a report on this to D/D.G.M.A. about three weeks ago, advising strongly that they should not be bought, as they were very old machines and to make a thorough examination would entail taking them down.'

There the matter terminated, as far as the Progress and Allocation Committee was concerned.

These references leave little doubt that the aircraft in question became (probably for write-off purposes) A8970—8973, and that they were a small production batch of the Hawker Runabout biplane. The early use of the name Sparrow adds both piquancy and an element of confirmation: hitherto it has always been believed that the Sparrow was a single prototype of a radio-controlled pilotless aircraft powered by a 35 hp A.B.C. Gnat engine. This Sparrow's airframe clearly owed a great deal to the Runabout, and it now seems likely that the basic Gnome-powered design was named Sparrow, the Gnat-powered radio-controlled aircraft being merely a later modification, naturally retaining the name. It may even have been a conversion of one of the four earlier biplanes.

The modified Sparrow with 35 hp A.B.C. Gnat engine, intended to be the prototype of a pilotless radio-controlled aircraft.

It further seems equally likely that the Gnome-powered single-seater acquired by R. C. Shelley from Mr Mouser in 1925 was one of the four 'production' Sparrows. What appears to have been another similar aircraft was converted into a two-seater with ailerons on the lower wings. Neither Shelley's aircraft nor the two-seater seems to have borne any identifying markings, but this is not surprising: in the circumstances it is unlikely that anyone would trouble to paint the numbers on A8970—8973.

Sparrow
50 hp Gnome

Single-seat scout.
Span 26 ft 9½ in; length 19 ft.

Manufacturer: Sopwith.

Serials: A8970—8973.

Sopwith Pup

From Harry Hawker's little single-seat Runabout the Sopwith company's design office quickly developed the prototype of a single-seat fighter powered initially by an 80 hp Le Rhône, and armed with a single fixed and synchronised Vickers machine-gun. This had been completed by 9 February, 1916, on which date it was passed for flight by the firm's experimental department. It was essentially similar to the Runabout, but the fuselage structure was revised internally to take the more powerful engine (from the beginning the 100 hp Gnome Monosoupape was intended to be an alternative power unit) and the use of a wide-span centre-section necessitated divergent supporting struts. Ailerons replaced the wing-warping of the Runabout.

The Sopwith company were Admiralty contractors at the time, and the prototype was evaluated by the RNAS. A copy of the report was sent to RFC

A Pup of the first Standard-built batch, A648 was allotted to No.54 Squadron on 15 December, 1916, and returned to England with the squadron on 10 July, 1917.

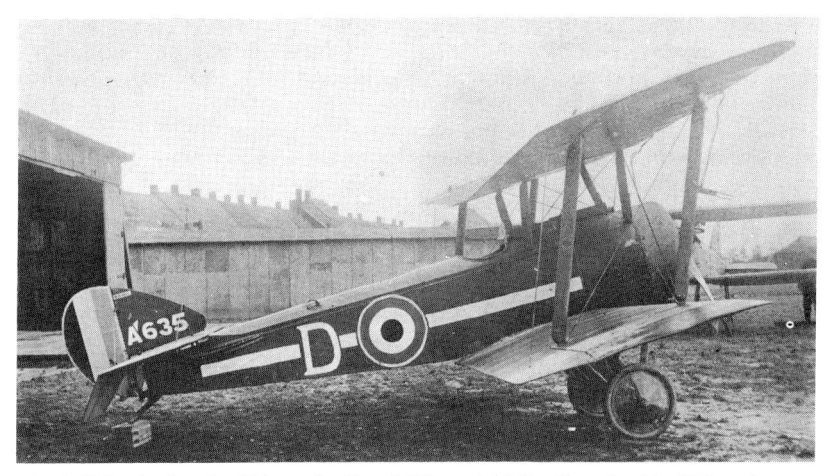

Also built by the Standard Motor Co, Pup A635 was initially allotted to No.54 Squadron on 15 November, 1916, and it was issued to that unit from No.1 A.D. on 12 January, 1917. It was later used by No.66 Squadron and was brought down in German-held territory in October 1917 while being flown by 2/Lt M. Newcomb, who was made prisoner.
(*Egon Krueger*)

Headquarters in France, where Maj-Gen Trenchard instantly recognised the aircraft's potential and asked for a squadron of the new Sopwiths. At an early stage the aircraft was dubbed Sopwith Pup by the flying Services. Although both the RFC and RNAS officially eschewed this frivolous name, it became inseparable from the type. (In official RFC parlance the new single-seater was the Sopwith Scout; to the RNAS it was the Sopwith Type 9901 with wheel undercarriage, Type 9901a when fitted with flotation gear, central cut-out in the centre section, and tripod mounting for an upward-firing Lewis gun.) Before the end of April 1916 an official tender for 50 aircraft to be built for the RFC by the Standard Motor Co had been put out, whereas details of the first RNAS order for the type were not recorded until mid-June.

Nevertheless, the first deliveries of production Pups were made to the RNAS early in September 1916. These preceded by some three weeks the delivery of the

A7311, a Standard-built Pup, was Capt J. T. B. McCudden's personal aircraft, which he first flew at Hounslow on 7 May, 1917, and subsequently at various training units.
(*RAF Museum*)

first Standard-built Pup, A626, which was tested at CFS in mid-October and went to France on 4 November. Its initial allocation was to No.70 Squadron, RFC, but towards the end of December it went to No.8 (Naval) Squadron, then one of the RNAS units giving invaluable support to the RFC on the Western Front. For a few days it was on the strength of 'B' Flight of Naval Eight, but on 4 January, 1917, it was brought down intact by Leutnant Friedrich Mallinckrodt of Jagdstaffel 10, and the Germans thereby gained a sound specimen of the latest British fighter.

Production of the Pup for the RFC was substantially expanded in the autumn of 1916, when contracts were proposed for manufacture by The Darracq Motor Engineering Co of Fulham, the Wells Aviation Co of Chelsea, and Whitehead Aircraft of Richmond. The Wells contract was cancelled within a few days, and the Darracq order was never confirmed, but the Whitehead contract for 100 Pups (A6150—6249) went ahead, being dated 16 October, 1916. The Whitehead company did well to be able to start deliveries before the end of January 1917, A6150 having reached Farnborough by 27 January, followed next day by A6151.

By that date the Pup had been on an operational footing in France with the RFC for about a month. No.54 Squadron had received its first Pup, the Standard-built A627, on 22 October, 1916, and was fully equipped with twenty Pups by 20 December; this famous fighter squadron went to France on 24 December. As more Pups became available some were delivered in March 1917 to No.66 Squadron, then forming at Filton, and the unit was able to go to France on 12 March, 1917. Next month No.46 Squadron began to exchange its Nieuport two-seaters for Pups and became operational as a single-seat scout squadron in May.

Pup B2166, built by Whitehead Aircraft, was used for training at Scampton, and is seen there in August 1917.

Operating the lightly built Pup in the Field was not without difficulties. The initial production aircraft had a variable-incidence tailplane similar in design and operation to that of the $1\frac{1}{2}$ Strutter, but as early as January 1917 later Pups were being delivered with fixed tailplanes. As the incidence mechanism necessitated its own form of fin as well as tailplane none of the fixed tail surfaces were interchangeable, and problems of supply and store-keeping arose at once. The fixed tailplane was standardised by the end of January 1917.

514

The cockpit of A6174, a Whitehead-built Pup that was allotted to the RFC in France on 6 March, 1917. It went to No.54 Squadron, and on 9 May, 1917, 2/Lt G.C.T. Hadrill was obliged to come down in German-held territory with his Pup but little damaged.

The Monosoupape-powered Pup had a distinctive engine cowling with cut-away lower segment and peripheral slots. This one may have been photographed at London Colney. (*Air Chief Marshal Sir Hugh Saunders, CB, KBE, MC, DFC, MM, from Peter Liddle's 1914–18 Personal Experience Archives, presently housed within Sunderland Polytechnic*)

B1807 was one of the few Pups to acquire a civil registration after the war. In this photograph it is seen at a training unit, armed and with a Monosoupape engine, with an Armstrong Whitworth F.K.3 in the background. After the war it became G-EAVX and had an 80 hp Le Rhône engine.

At an early stage stronger centre-section struts were introduced, and interplane struts proved to be of varying lengths from different contractors. The ammunition stowage had to be modified when the Prideaux disintegrating-link belt was introduced in February 1917. In the autumn of 1916 the CFS Test Committee had expressed doubts as to the adequacy of the Pup's structure, following which A631 was tested to destruction by the Royal Aircraft Factory. Factors of safety proved to be 4·46 on the front spars and 5 on the rear, values that were acceptable at the time. Nevertheless, the strains of combat flying so affected the mainplane panels of the Pup that they had to be replaced after about 40 hours. It was found that this deterioration applied to all Pups irrespective of constructor or whether used by RFC or RNAS.

A tendency to tail-heaviness was increased by the adoption of ash longerons instead of spruce, and this was ultimately cured by increasing the incidence of the tailplane. Interestingly, it had been found that the tail-heaviness could be eliminated by making a small rectangular cut-out in the centre-section ahead of the rear spar. In operational use the heavily padded windscreen was found to obscure a vital part of the field of aim and was often replaced by simpler screens. Almost everyone who flew the Pup in combat soon felt a need for heavier armament, and various attempts were made to install an overwing Lewis gun but it seems that none was truly successful.

Despite its modest power unit the Pup was able to hold its own in combat because of its exceptional ability to maintain its height in all the gyrations of dog-fighting. Even in November 1917, long after the Pup had been outclassed in France, RFC formations of Pups, Bristol Fighters and D.H.5s were so disposed that the Pups held the uppermost level at about 15,000 ft. Nevertheless, it was expected that more power would improve performance, and the first installation of the 100 hp Gnome Monosoupape engine was made in A653 in April 1917. Modifications to the cowling followed, and substantial numbers of Pups were built with the engine. Operational use of the Monosoupape-powered aircraft appears to have been confined to Home Defence squadrons, possibly because its performance was not significantly better than that of the standard Pup.

In the late summer of 1917 several installations of the 110 hp Le Rhône were made in Pups. Tests of A6220 with this engine at Dover and Croydon showed some improvement in performance and satisfactory handling, but the greater weight of the engine made it impossible to get the tail down into the three-point attitude on landing. In France B5908 was tested with the 110 hp Le Rhône and went briefly to No.66 Squadron on 14 October, 1917. Trenchard had doubted the adequacy of the Pup's strength for the more powerful engine, and indeed static-loading tests of B5941 at Farnborough proved that the load factor of the 110 hp Le Rhône version was appreciably lower* than that of the 80 hp Pup. On 6 November, 1917, the fitting of the 110 hp engine was officially abandoned, but it is interesting that this work was still in progress at a time when the Camel and S.E.5a were in operational use.

*It proved to be 6·6 as opposed to the approved figure of 7·95 of the original 80 hp Pup.

B5259 was one of the very few Pups that had the 110 hp Le Rhône engine, which probably explains the large-diameter cowling seen here. On 9 November, 1917, it was reported to be with the E.F. but allotted to the Training Division, an allotment that was amended to leave the aircraft with the E.F. At some time this Pup did in fact turn to training duties, for these photographs were taken at London Colney, when the aircraft had a Lewis gun above the centre-section and a head fairing behind the cockpit. (*Alex Revell/RAF Museum*)

Some Pups of training units had the 80 hp Gnome engine.

The last operational Pups in the RFC were those of No.54 Squadron, which replaced them with Camels in December 1917. Thereafter the RFC made extensive use of the Pup for training purposes, and many of these training aircraft had the 80 hp Gnome; it was intended that training Pups in the Middle East were to have the 80 hp Clerget, but this engine saw limited use in some Pups in England. Some Pups had a strengthened undercarriage. To the end the Pup remained immensely popular with all who flew it.

Pup
80 hp Le Rhône 9C, 80 hp Gnome, 80 hp Clerget,
100 hp Gnome Monosoupape, 110 hp Le Rhône 9J

Single-seat fighting scout.
Span 26 ft 6 in; length 19 ft 3¾ in; height 9 ft 5 in; wing area 254 sq ft.
Weights and performance:

	80 hp Le Rhône	100 hp Monosoupape	110 hp Le Rhône
Empty, lb	787	856	—
Loaded, lb	1,225	1,297	—
Maximum speed at sea level	111·5 mph	—	—
at 5,000 ft	105 mph	—	110 mph*
at 10,000 ft	102 mph	104 mph	87 mph*
	min sec	min sec	min sec
Climb to 5,000 ft	6 25	5 12	— —
to 10,000 ft	16 25	12 24	8 55
Service ceiling	17,500 ft	18,500 ft	—
Endurance	3 hr	1¾ hr	—

*Indicated air speed not corrected to standard atmosphere.

Armament: One 0·303-in Vickers machine-gun, occasionally supplemented or replaced by one 0·303-in Lewis machine-gun; some aircraft carried a small load of 25-lb Cooper bombs.

Manufacturers: Sopwith, Standard Motor, Whitehead.

Service use: *Western Front*: RFC Sqns 46, 54 and 66; one Pup (A626) briefly with No.70 Sqn; at least one (B2188) with No.101 Sqn, February 1918; Special Duty Flight. *Italy*: One Pup with No.66 Sqn. *Home Defence*: Sqns 37, 46, 50, 61 and 112. *Training duties*: Reserve/Training Sqns 3, 6, 7, 10, 11, 18, 23, 28, 30, 36, 40, 42, 43, 55, 56, 63, 67, 73, 81, 89, 188, 189 and 195; Training Depot Stations Nos.3, 4 and 7; Central Flying School, Upavon; School of Special Flying, Gosport; No.1 School of Aerial Fighting, Turnberry; No.3 School of Aerial Fighting, Bircham Newton; No.4 Auxiliary School of Aerial Fighting, Marske; School of Aerial Fighting, Heliopolis, Egypt; Scout School at No.2 Aircraft Depot, Candas, France; No.84 Wing, Vendôme, France; Squadrons working up—Nos.56, 64, 65, 74, 84, 87, 89 and 94.

Serials: For production aircraft—A626—675, A5138—5237, (Wells contract; cancelled, numbers re-allotted), A6150—6249, A7301—7350, B1701—1850, B2151—2250, B5251—5400, B5901—6150, B7481—7580, C201—550, C1451—1550, C3707—3776, D4011—4210. The numbers A8732—8737 were allotted for Pups (N5193, N5194, N5190, N5183, N5182 and N5192) to be transferred from the RNAS but cancelled.

Rebuilds: B735, B803—805, B849, B1499, B4082, B4128, B4131, B4136, B7752, B8064, B8784—8786, B8795, B8801, B8821, B8829, B9440, B9455, B9931, C3500—3503, C4295, C8653—8654, C9990, C9991, C9993, E9996, F4220.

Sopwith Triplane

As noted in the history of the Sopwith Pup, the RFC acted swiftly to order that type within days of the appearance of the prototype. The prototype of the almost contemporary Sopwith triplane was passed by the Sopwith experimental department on 28 May, 1916; like the Pup, the triplane was a single-seat fighter armed with a single Vickers gun, but it had the more powerful 110 hp Clerget 9Z rotary engine. Again the RFC lost no time in demonstrating its interest in a new type, for a War Office contract for the production of triplanes under the reference MA/Aeros 1377 had been drawn up as early as 7 June, 1916. This was still under consideration at the end of June, but surviving official documents contain no indication as to the intended contractor.

The official records remain enigmatic about triplane production. The prototypes were delivered to the RNAS and wore naval serial numbers; forty were ordered by the Admiralty from Clayton & Shuttleworth with the identities N5350—5389, and deliveries began on 2 December, 1916; similarly N5420—5495 were ordered from Sopwith for the RNAS; and N5910—5934, originally ordered from Oakley of Ilford as 1½ Strutters, were changed to triplanes about the end of December 1916.

It was at precisely that time that the apparently non-Naval serial numbers A9000—9099 were allotted for Sopwith triplanes, and according to the official register these were to be built by Clayton & Shuttleworth under Admiralty contract. Approximately three weeks later, in about the third week of January 1917, the serial numbers A9813—9918 were allotted for 106 Sopwith triplanes, also to be built by Clayton & Shuttleworth. This time there was no mention of an

Admiralty contract, so it is possible that A9813—9918 were intended to be the RFC's triplanes.

Early in 1917 forward planning for the equipment of RFC squadrons provided for Sopwith triplanes to form the operational equipment of No.65 Squadron, RFC. That unit would have been the first RFC squadron to have the triplane had not fate decreed otherwise.

September 1916 had seen the menacing advent on the Western Front of new types of German fighters superior to anything that the RFC then had. On 29 September Trenchard advised the War Office that he intended to send, via Field Marshal Sir Douglas Haig, a request for substantial increases in the numbers of fighting squadrons attached to each British army. Haig wrote on 30 September, setting out the situation clearly and cogently. Unfortunately, Haig's letter arrived at a time when relations between the War Office and Admiralty were somewhat strained, because the Admiralty had sought and obtained Treasury approval for their independent purchase of aircraft and aero-engines to the value of about £3,000,000 (a sum that, in those days, would buy considerable quantities of aircraft and engines). This was contrary to the arrangement under which the Air Board was responsible for organising and co-ordinating the supply of aviation material in order to prevent undesirable and wasteful competition between the Admiralty and the War Office. This responsibility had been laid upon it by the War Committee at its meeting of 11 May, 1916; hence the Air Board's displeasure was understandable and justified. This situation may have been responsible for the fact that the Air Board did not discuss Haig's letter until 11 December, 1916: the lapse of time of nearly $2\frac{1}{2}$ months at a period of crisis for the RFC now seems scarcely credible, but there had been bitter inter-Departmental strife since October.

N5430 was the only Sopwith Triplane to be used by the RFC (*A&AEE; Crown copyright*)

Certainly in that time matters did not improve on the Western Front and the Air Board belatedly recognised that something had to be done, and done urgently. Clearly, no miracle of aircraft production could be conjured up and there was only one obvious source of supply. Whether any pride had to be swallowed, principles abandoned or concessions made is not now of moment: what happened was the calling of a conference at the Admiralty on 14 December,

1916, with Gen Brancker representing the War Office. It was on this occasion that the Admiralty agreed to hand over to the RFC half of the production of Spads provided for under their current contracts. This was stated to amount 'to 62 machines all of which would be complete with engines, the first 12 to be handed over in March'. This figure was obviously based on the fact that at that time a total of 125 Spads were on order for the RNAS, 75 to be built by Mann, Egerton & Co, the other 50 by the British Nieuport concern. The latter contract was subsequently amended and finally cancelled, but 20 more Spads were ordered from Mann, Egerton. A further conference held on 26 February, 1917, 'agreed that all the Army Sopwith triplanes should be handed over to the Navy and that the Navy would then hand over all their Spads to the Army instead of half.'

In fact it is doubtful whether any handover to the RNAS of a triplane built for the RFC ever took place, even on paper, for the total number of triplanes delivered to the RNAS was much too small to have incorporated any aircraft intended for the RFC. Nevertheless, it is tempting to speculate about the six built by Clayton & Shuttleworth with twin Vickers guns and taken on by the RNAS as N533—N538. Were they the odd six of the 106 apparently ordered for the RFC? And what of the ten or so triplanes apparently supplied to the French Aviation maritime?

One Naval triplane, N5430, was evaluated by the RFC at Martlesham Heath and Orfordness; it had the 130 hp Clerget 9B and was the subject of the official test report No.75, prepared in December 1916. At that time the RFC was still expecting to receive triplanes for some of its squadrons, but N5430 apparently remained with the RFC after the Spad-exchange decision had been reached. While at Orfordness in July 1917 it was flown by Capt Vernon Brown (later Air Commodore Sir Vernon Brown, CB, OBE, MA, FRAeS) against the formation of Gothas that attacked London in daylight on 7 July; machine-gun trouble frustrated Brown's gallant attempt to defend the capital.

This triplane was still active late in 1918, for on 1 October, 1918, it visited Farnborough from Orfordness, returning whence it had come on the same day. It could not possibly, as has been alleged, have gone to Russia.

Triplane
130 hp Clerget 9B

Single-seat fighting scout.
Span 26 ft 6 in; length 18 ft 10 in; height 10 ft 6 in; wing area 231 sq ft.
Empty weight 1,103 lb; loaded weight 1,543 lb.
Maximum speed at 6,500 ft—112·5 mph, at 10,000 ft—106·5 mph, at 15,000 ft— 95 mph; climb to 6,500 ft—6 min 30 sec, to 10,000 ft—11 min 50 sec, to 15,000 ft— 22 min 20 sec; service ceiling 20,500 ft; endurance 2¾ hr.
Armament: On N5430, one fixed 0·303-in Vickers machine-gun; on six of the Clayton & Shuttleworth aircraft, two fixed 0·303-in Vickers guns.

Manufacturer: Sopwith.

Service use: N5430 flown experimentally at Martlesham Heath and Orfordness.

Serials: A9813—9918 were apparently ordered for the RFC. N5430 lent or transferred to RFC from RNAS, and later incorrectly repainted as A5430.

The Camel N6332 was transferred from the RNAS to the RFC on 25 May, 1917, and is seen in this photograph at No.2 A.D., Candas, on 26 May. It retained its Naval serial number with the RFC, and was sent to No.70 Squadron on 28 June, 1917. On 17 July, 1917, it was shot down near Waterdamhoek by Vizefeldwebel Franke of Jasta 8 as his 4th combat victory; its pilot, Lt W. E. Grosset, was taken prisoner.

Sopwith F.1 Camel

The earliest recorded reference to a Sopwith single-seat fighter biplane with the 110 hp Clerget engine is to a proposed version of the Sopwith 9901 (Pup) for which the Admiralty serial number N503 was allotted in August 1916. This project is last mentioned in an official return of 4 December, 1916, and it seems that no version of the Sopwith Pup was in fact built with the 110 hp Clerget.

However, on 22 December, 1916, the experimental department of the Sopwith company passed the design of the new Sopwith F.1, a single-seat fighter powered by the 110 hp Clerget 9Z. It had originally been intended to give the F.1 equal dihedral on upper and lower wings, but, in the interests of speedy production, Fred Sigrist, Sopwith's Works Manager, had decided to make the upper wing in one piece with straight spars; as rough compensation the dihedral of the lower wing was doubled and rigged at 5 deg.

The F.1 bore a strong resemblance to the Pup, but its fuselage was deeper and the main masses of engine, weaponry, pilot and fuel were more closely grouped together, thus giving the aircraft a squat, hunched appearance. This arrangement unfortunately placed the pilot under the trailing portion of the upper wing, where he had no view forwards and upwards.

As far as can be determined, two prototypes numbered N517 and N518 were supplied to the Admiralty, and another, identified only as Sopwith F.1/3, was evaluated for the RFC. N517 was soon in France; it was recorded as having been tested on 28 February, 1917, and subsequently as being at the RNAS Depot, Dunkerque, on 1 March, 1917. Even earlier, on 20 February, 1917, Brig-Gen D. le G. Pitcher, Technical Director General, wrote to inform Trenchard that it was proposed to send him 'for inspection a sample Sopwith F.1 single-seater fighter (130 Clerget)'. Additionally, RFC Headquarters were in touch with RNAS Dunkerque at this time and before the end of February had received a report of unknown origin (only part of it has survived) on the new Sopwith. This makes it clear that the early F.1s were very tail heavy, so much so that it was proposed and officially agreed to re-rig the aircraft with the upper wing 3 inches farther aft. Confirmation that this was ever done, even experimentally, is lacking, however.

It is of interest to note that, whereas N517 had no central cut-out between the spars of the centre-section, the sample aircraft supplied to the RFC had a central rectangular hole roughly 18 in by 8 in. Early in March 1917, the RFC and RNAS agreed that the cut-out should be made standard, but the reasons for doing so seem not to have been recorded. It will be recalled that a few early RNAS Pups (which basically gave their pilots a much better upward field of view) had their entire centre-sections covered with transparent material: more significantly, it was found that the Pup's tail-heaviness was alleviated by cutting a small rectangular hole in the centre-section. The cut-out in the Camel's centre-section was too small to improve the pilot's view greatly, and it was not until 3 July, 1918, that modification F.1/78.II introduced on production Camels a much wider cut-out that really did give the pilot some upward view. It therefore seems possible that the absurdly small cut-out that was adopted as standard in March 1917 may have had as its primary object the reduction of the Camel's tail-heaviness.*

The sample F.1 sent to the RFC was tested at No.1 A.D. on 2 March, 1917, by 2/Lt K. L. Caldwell of No.60 Squadron and on 4 March by Capt A. M. Lowery of No.70 Squadron. Both reports related to the weapon installation, for both the RNAS and RFC had been swift to point out that the close-fitting decking over the breech mechanisms of the twin Vickers guns made it impossible to rectify stoppages of the starboard gun. Both guns had right-hand feed—inevitably, that being the only type of Vickers gun then available—and this meant that normal remedial action could not be taken on the starboard gun as there was insufficient room to operate the cocking handle or feed-block pawls.

Attempts were made to produce left-hand feed blocks but were curiously slow to succeed. Only thirteen had been delivered to the Sopwith company by 16 July, 1917; one was fitted to the production Camel B3761, another to a Dolphin. It is doubtful whether any other Camel was thus equipped, and a decision communicated to RFC Headquarters on 18 August, 1917, virtually made it certain that Camels would not receive guns with left-hand feed, whereas the available production of such guns would be allotted to Dolphins. In operational service, therefore, squadron Camels had the forward decking cut away over the starboard gun's breech mechanism to give the pilot ready access for rectifying stoppages, and this apparently continued to be done until the end of the war, only the French-based night-fighting Camel squadrons being allowed to retain the full cockpit decking in the interests of pilot comfort. The approved method for RFC Camels was devised by No.2 A.D. and was standardised on 28 July, 1917. This modification led to difficulties in providing and fitting an adequate windscreen and many shapes and sizes were tried at different times.

*The date is of considerable interest if only because the first 130 hp Clerget F.1 Camel did not arrive at Martlesham Heath for official trials until 24 March, 1917. This was the F.1/3, and it had been preceded on 15 March by N5, the prototype 2F.1.

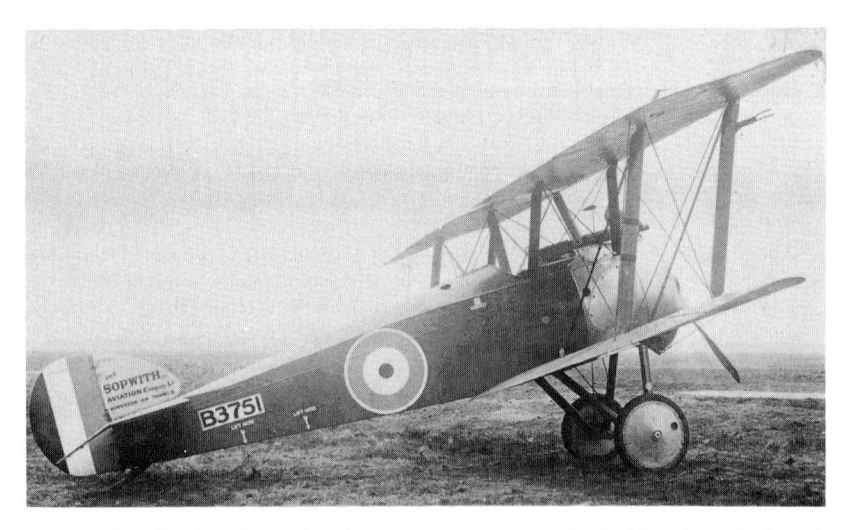

The text describes how it was found necessary to cut away the decking about the breech mechanism of the starboard gun to enable stoppages to be cleared in combat. This may be why B3751, the first Camel to be built for the RFC by Sopwith, appeared at Martlesham Heath with the forward top decking modified to expose the guns completely. It was reported to be undergoing tests of guns and gun gear before 7 July, 1917, but did not stay long at Martlesham. It returned early in January 1918 for tests with a B.R.1 engine and had gone back to the Sopwith works to be armoured by 12 January, 1918. (*T. Heffernan*)

The unusual gun installation on Camel B3751 (*T. Heffernan*)

524

The RFC ordered substantial numbers of Camels, the first contract being for 250 (B2301—2550) ordered from Ruston, Proctor; the serial numbers were allotted in mid-April 1917. It was closely followed by a contract for 200 Sopwith-built Camels (B3751—3950) and 50 from Portholme Aerodrome Ltd (B4601—4650). For the Ruston, Proctor aircraft the 130 hp Clerget was specified, for Portholme Camels the 110 hp Le Rhône, and the Sopwith-built aircraft could have either. The 150 hp Bentley B.R.1 was only fitted to RNAS Camels at that time.

An official assessment of the situation was made on 26 May, 1917, and it was recognised that the shortage of Le Rhône engines made unlikely the early availability of that power unit for Camels in 1917. It was proposed to have six Camel squadrons operational in France by the end of the year, three of them created by re-equipping the three 1½ Strutter squadrons and thus diminishing the increased need for Clerget engines. Even the supply of Clergets proved inadequate; although all available engines were allotted to Camel contractors from mid-July 1917, some aircraft were expected to leave the factories engineless, and on 18 July the Director of Aeronautical Equipment asked whether the RFC in France could provide Clerget engines for these Camels.

B6234 was a Sopwith-built Camel that was first recorded at Kenley Aircraft Acceptance Park on 23 August, 1917, when it was officially allotted to the E.F. It was captured intact by the Germans on 5 December, 1917, when Vizefeldwebel Konnecke of Jasta 5 brought down Lt L. J. Nixon of No.3 Squadron. Nixon became a prisoner. (*Egon Krueger*)

The first official allotment of a Camel to the RFC in France was made on 15 June, 1917; the aircraft was B2302,* the second Camel built by Ruston, Proctor. Next day the Sopwith-built B3813 and B3814 were similarly allotted. The RNAS was some way ahead of the RFC in terms of Camels: two production aircraft had reached Dunkerque by 17 May, 1917; on 7 June there were five Camels with No.4 (Naval) Squadron, and at the Dunkerque Depot were eight Bentley Camels and nine Clerget Camels. A week later No.4 (Naval) had nine Camels and N517 was with No.12 Squadron, while by 21 June No.6 (Naval) Squadron was an all-Camel unit with ten aircraft and No.3 (Naval) had acquired N6377; thereafter more

*In fact this Camel was damaged and its allotment to France was cancelled on 14 July. It seems that B2301 went to France, however, for it was tested at No.2 A.D. with a modified cowling on 19 June, 1917.

Camels were added to the various RNAS squadrons. On 27 June, eight more Camels were allotted to the RFC, bringing the total of allotments for the month to eleven. The production aircraft had modified cowling panels and the upper wing was made in three pieces.

The RFC's first production Camel had in fact come from the RNAS as early as 25 May, 1917, when N6332 was transferred. It was tested at No.2 A.D. on 30 and 31 May and was sent to No.70 Squadron on 28 June. No.70 was the first RFC squadron to be equipped with the Camel and had its full complement by the end of July. During that month No.45 Squadron also began to exchange its gallant old 1½ Strutters for Camels, and by the end of 1917 there were nine RFC squadrons of Camels operational in France, and four RNAS squadrons operating with the RFC also flew the type.

This was not achieved without problems. From the outset it was known that the forward top decking would have to be modified to provide access to the breech mechanism of the starboard gun, and before the end of June 1917 instructions were given to both A.D.s that they would have to convert the interrupter mechanisms on the earliest Camels from the Mark II to the Mark III version of the Sopwith-Kauper gear, which was the standard interrupter gear on the Clerget Camel. This gear was tested on N6332 on 26 June, 1917, when it was found that it needed constant adjustment and would wear rapidly.

Re-equipment of squadrons was retarded by the compulsion to improve Britain's air defences following the German air attacks on London of 13 June and 7 July, 1917. On 8 July, A.D.A.E. (Lt-Col C. W. Whittington) wrote to Trenchard:

> 'I am to inform you that instructions have been issued that the first ten Sopwith Camels becoming available this month are to be handed over to Home Defence Group, and that altogether 24, out of the 54 promised to you during the month, are to be handed over to Home Defence Group.'

And on 10, 11 and 12 July, 1917, nine Camels were allotted to the Home Defence Wing, two to the RFC in France, and four to the Training Brigade.*

*These were B3752, B3763—3765, B3767, B3776, B3815, B3816 and B3827 to H. D. Wing; B2303—2304 to the E.F.; and B3847—3850 to the Training Brigade.

B6345, a Clerget Camel, was aircraft F of No.28 Squadron, and saw service on both the Western and the Italian Fronts.

B2443 was built by Ruston, Proctor and was at Lincoln Aircraft Acceptance Park on 12 October, 1917, when it was allotted to the RFC in France. It went to No.45 Squadron and became aircraft K of that unit. (*RAF Museum*)

The earliest assessments of the Camel given by fighting pilots enthusiastically recognised that the aircraft's good combat qualities outweighed such disadvantages as the severely limited upward view from the cockpit and the difficulties of clearing gun jams. Unfortunately, early experience with production Camels was disappointing because in too many cases performance fell far short of the figures obtained on official trials. The RFC had expected their aircraft to reach the ceiling of 21,000 ft attained by the test machines, but on 14 August, 1917, Brooke-Popham wrote to D.A.E.:

> 'The ceiling of the Sopwith Camel with the 130 Clerget engine is given in the table of performances, issued by the Air Board and dated June 30th, as 21,000 ft.
> We are unable to get machines out here over 18,000 ft.
> The G.O.C. directs me to point out that this matter is of the utmost importance and that he wishes to know whether any possible explanation can be given.'

Much correspondence and activity ensued. On 24 August, 1917, Capt H. T. Tizard (the late Sir Henry Tizard) admitted in a letter to Brig-Gen Brooke-Popham:

> 'With regard to the report of the bad performance of Sopwith Camels in the Field, we have found, firstly, that the Sopwith Camels fitted with English Clergets and single ignition are considerably inferior in performance when new to 130 French Clerget Camels, and, also, that the performance of the latter appears to fall off considerably after the machine has been flown for some time. . . .
> Our conclusion that the Clerget engines are at the bottom of the trouble appears to be borne out by reports from Dunkerque, to the effect that B.R. Camels keep their performance while Clerget Camels fall off.'

Tizard advised Brooke-Popham that Capt R. H. Carr was about to go to France with a B.R.1 Camel, N518, and various items of test equipment. After Carr's investigations Trenchard himself sent the following despairing note to D.G.M.A. by King's Messenger on 3 September, 1917:

Clerget Camels of No.46 Squadron in the snow. The Camel at right is B5419, which was due to go to a Repair Park to have its Clerget replaced by a 110 hp Le Rhône in February 1918.

'I am sending the results of eight tests that were carried out by Captain Carr when he was out here.

It will be seen that the only machine with a satisfactory result was the Sopwith Camel that Captain Carr brought over with him which is fitted with an A.R. engine.

This satisfactory result, however, was gained by the extra horse power and not by the machine itself.

Will you please say what is proposed to do, as the position is now very unsatisfactory, but *do not stop* the Camels coming out as they are much better than $1\frac{1}{2}$ Strutters or nothing.'

Martlesham took very seriously the business of evaluating Camels' performance, comparing the effects of aircraft and engine life, and of different airscrews. The results of these tests were recorded in Confidential Information Memorandum No.47 (later identified as Air Publication 1011) of November 1917. This concluded:

'In general, we consider the bad performance of the Camels to be due to low engine power and in particular that there are indications that a French engine, when new, is of considerably greater effective horse-power at great heights than the corresponding English engine, and also that the power at great heights of both engines appears to fall off disproportionally to the power on the ground with time. . . .

From the experience at this Station, it is considered that of all rotary engines the Clerget engines are the worst for losing power after running for some time. . . . It is necessary to go over the valve timing of a Clerget engine practically after every flight. . . .'

In addition to Martlesham's investigations, Lt T. C. Thrupp and 2/Lt J. H. Ledeboer of the Technical Department went to France on 12 November, 1917, and visited, successively, Squadrons No.3, 46, 43 and 70, of which No.46 had only received their first five Camels a few days previously. In his report dated 20 November, 1917, Ledeboer included these observations:

'Generally speaking, at these three squadrons [*i.e.*, No.3, 43 and 70] the main ground of complaint is the unsatisfactory performance of the Camel with 130 Clerget engine. . . . Up to 10,000 or 12,000 feet the performance of the Camel is satisfactory; thence upward the climb falls off rapidly and at 15,000 feet quick manoeuvring becomes impossible. From 10,000 feet upwards the average German machine walks away from the Camel both in speed and climb. None of these squadrons have had 110 Le Rhônes or 150 B.R.1. . . .

Another general complaint was the marked inferiority from the flying point of view of machines built by Ruston, Proctor as compared with those built by Sopwith and other firms. Machines built by Hooper and Boulton & Paul are reported good as also one machine made by Portholme. Ruston's machines are reported heavy and sluggish and exceptionally tail heavy. . . .

Squadrons are experiencing considerable trouble with the Sopwith-Kauper gun gear, propellers being frequently shot through. The guns also have a tendency to "run away" . . .'

By the time when C.I.M. 47 was circulated, the RFC in France was beginning to acquire operational experience with Camels powered by the Le Rhône engine. No.3 Squadron had received its first three Camels on 5 October, 1917, and was fully equipped with the type by 12 October. In view of Ledeboer's report these cannot have been Le Rhône Camels, and it seems that No.3's equipment with that variant began early in December. By 9 December, 1917, Trenchard had had enough of the Clerget, and on that date he wrote to D.A.E.:

'. . . I prefer the 110 hp Le Rhône to the 130 hp Clerget and wish to have as many Camel squadrons as possible equipped with the former engine. I am now equipping one Squadron with Le Rhône Camels . . .'

He went on to ask that No.80 Squadron should have Le Rhône aircraft when it came out in January 1918, and to require three Clerget Camel squadrons to be re-equipped with Le Rhônes by the end of February 1918. These proposals were

B5423 of No.54 Squadron in German hands after its capture in January 1918. Its pilot was Lt F. M. Ohrt, an American serving with the RFC. This was a Le Rhône Camel, allotted as such to the E.F. on 7 November, 1917

Le Rhône Camels of 'C' Flight, No.3 Squadron, at Inchy. (*D. R. Neate*)

made in the expectation that the long-stroke Clerget 9Bf and the B.R.1 would become generally available in March or April. To all of this D.A.E. (Brig-Gen W. B. Caddell) agreed on 12 December, 1917.

As soon as practicable, Camels allotted to France were mostly those that had either the 110 hp Le Rhône or the 140 hp Clerget 9Bf; a few with the 130 hp Clerget 9B were sent across the Channel, but most aircraft with this engine went either to training units or into store. For example, official allotments of Camels to France in February 1918 were as follows:

(a) With 110 hp Le Rhône—98; (b) With 140 hp Clerget 9Bf—74; (c) With 130 hp Clerget 9B—30; (d) With 130 hp Clerget 9B to store—33.

Additionally, some of the Clerget-powered Camels in France were converted at Aircraft Depots from February 1918 onwards to have the Le Rhône engine (for example, B2451, 2453, 2522, 2524, 3814 and 5419). From April 1918 a conversion kit was provided by No.1 Aircraft Depot.

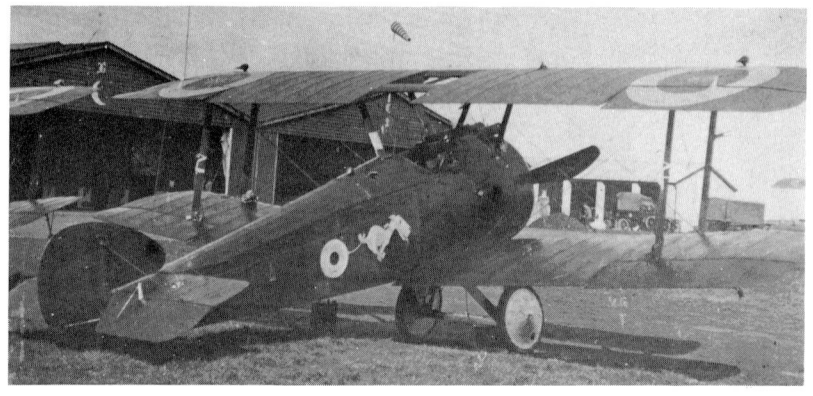

Standard Camel equipped as a night fighter, with navigation lights above the upper wing and white omitted from the roundels. No.44 Squadron, Hainault.

Le Rhône-powered night-fighting Camels of No.78 Squadron. C1555 was built by Hooper & Co and was completed early in January 1918. (*K. M. Molson*)

The Le Rhône version of the Camel had the further advantage of being equipped with the Constantinesco C.C. synchronising mechanism for its guns. One or two Clerget Camels had been experimentally fitted with the Constantinesco gear: on 22 July, 1917, No.1 A.D. was asked to fit the gear to a newly arrived Camel, and in August 1917 Martlesham tested B3851 with Clerget engine and Constantinesco mechanism. Neither installation had any sequel, possibly because the demand at that time for Constantinesco gears for Bristol Fighters, D.H.4s and S.E.5as left no surplus for use on other types.

As already noted, the Camel had an early introduction to Home Defence duties. Initially, Clerget-powered aircraft were supplied to the Home Defence

B2402 typifies the special night-fighting conversion of the Camel with cockpit moved aft and twin overwing Lewis guns. This one was built as a standard Camel by Ruston, Proctor and subsequently converted; it was used by No.44 (H.D.) Squadron at Hainault Farm.
(*C. H. Barnes*)

B3811 was a Sopwith-built Camel that was fitted with a 100 hp Gnome Monosoupape engine at a very early stage, having been completed by 14 July, 1917. It arrived at Martlesham for testing on 26 July, was extensively flown and went to Orfordness on 5 September. In fire-proofing experiments it was tested with a fuel tank packed in felt soaked in carbon tetrachloride, and was flown on 23 and 24 October, 1917. When it was shot at on 24 October it ignited and presumably was not repaired or reconstructed.
(*Aeromodeller*)

squadrons but eventually the Le Rhône was standardised for these duties. Despite the Camel's sensitivity on the controls it was successfully flown at night, but it was found that pilots lost their night vision when the guns were fired, being momentarily blinded by the muzzle flash. This led to the creation of a special

Camel B6329 was tested with the 150 hp Gnome Monosoupape engine and arrived at Martlesham Heath for that purpose on 15 November, 1917. Performance tests were followed by engine tests, and the aircraft went to Orfordness on 25 January, 1918. It was supposed to return to Martlesham for further running tests after the engine had reached 50 hours but it seems not to have done so. A further Camel with the 150 hp Monosoupape, B2541, went from Martlesham to Orfordness on 26 April, 1918, presumably to continue the development of the installation, and by that time the first production Camels with this engine were expected.

night-fighter variant, armed with two overwing Lewis guns and with the cockpit moved aft. Use of this, the so-called Sopwith Comic, version of the Camel was limited, and by the summer of 1918 its use was being discontinued.

The Camel served in most theatres of war and in several capacities, and was used by Belgian, Greek and American units as well as the RFC and RNAS. In combat and in the right hands it gave an excellent account of itself in spite of some undeniable shortcomings, and it was in the forefront of much of the horrifyingly costly ground-attack work that began in earnest during the Battle of Cambrai. Casualties in the low-flying fighters then averaged 30 per cent and led to the development of the Sopwith T.F.1, an armoured version of the Camel with two downward-firing Lewis guns. This idea proved to be unpractical and the T.F.1 was abandoned.

Other engines fitted to some Camels included the 100 hp Gnome Monosoupape and the 150 hp Monosoupape. Neither installation was developed for the RFC, but Camels with the 150 hp engine were later supplied to the US Air Service.

In training units the Camel acquired a sinister reputation, for there were many accidents involving the type. Not only was flying training at that time dangerously rushed and inadequate, but training in handling the sensitive rotary engines was also sketchy. Inexperienced pupils were unable to react quickly enough when the fine-adjustment fuel control had to be altered soon after take-off; the faltering engine's loss of power, if not instantly corrected, could lead to a stall and spin at a level too low to permit recovery of control. Eventually a two-seat version of the Camel was evolved for training purposes but the earliest examples were introduced only a few days before the formation of the Royal Air Force.

The Camel fought on until the end of the war, at its best in its B.R.1-powered form that the RFC never knew, tenacious and magically manoeuvrable when flown with understanding of its fiercely pronounced idiosyncrasies, savagely unforgiving of incompetent handling.

F.1 Camel

130 hp Clerget 9B, 140 hp Clerget 9Bf, 110 hp Le Rhône 9J,
100 hp Gnome Monosoupape, 150 hp Gnome Monosoupape.

Single-seat fighter.

Span 28 ft; length 18 ft 9 in (Clerget), 18 ft 8 in (Le Rhône); height 8 ft 6 in; wing area 231 sq ft.

Weights and performance:

Aircraft	B2312	F.1/3	B3829	B3811	B6329
Engine	130 hp Clerget	140 hp Clerget	110 hp Le Rhône	100 hp Mono	150 hp Mono
Weight empty, lb	962	—	—	882	930
Weight loaded, lb	1,482	1,452	1,422	1,387	1,441
Maximum speed, mph					
at 6,500 ft	108	—	—	—	—
at 10,000 ft	104·5	—	—	110·5	—
at 15,000 ft	97·5	113·5	111·5	102·5	113
Climb to (min and sec)					
6,500 ft	6 40	5 0	5 10	5 30	5 5
10,000 ft	11 45	8 30	9 10	9 50	8 50
15,000 ft	23 15	15 45	16 50	20 0	16 5
Service ceiling, ft	18,500	24,000	24,000	18,500	22,000
Endurance	—	—	—	$2\frac{3}{4}$ hr	$2\frac{1}{4}$ hr

Armament: Two 0·303-in Vickers machine-guns; four 25-lb Cooper bombs. The converted night-fighting version had two 0·303-in Lewis machine-guns.

Manufacturers: Sopwith; Boulton & Paul; British Caudron; Clayton & Shuttleworth; Hooper; Marsh, Jones & Cribb; Nieuport & General; Portholme; Ruston, Proctor.

Service use: Western Front: Sqns 3, 28, 43, 45, 46, 54, 65, 70, 71 (Australian), 73 and 80. Italy: Sqns 28, 45 and 66. Home Defence: Sqns 37, 44, 50, 78 and 143. Macedonia: Sqns 17 and 47. Training duties: Training Sqns 1, 3, 6, 10, 11, 18, 31, 36, 40, 42, 55, 62, 63, 65, 67, 70, 71, 73 and 89; Training Depot Stations—No.7; Central Flying School; School of Special Flying, Gosport; Scout School at No.2 A.S.D.; Advanced Air Firing School, Lympne; No.1 School of Air Fighting, Ayr. Other units: Wireless Experimental Establishment, Biggin Hill. Squadrons working up—No.94.

Serials: Contract-built production aircraft: B2301—2550, 3751—3950, 4601—4650, 5151—5250, 5401—5450, 5551—5650, 5651—5750, 6201—6450, 7131—7180, 7181—7280, 7281—7480, 9131—9330.
C1—200, 1551—1600, 1601—1700, 3281—3380, 6701—6800, 8201—8300, 8301—8400.
D1776—1975, 3326—3425, 6401—6700, 8101—8250, 9381—9530, 9531—9580, 9581—9680.
E1401—1600, 4374—4423, 5129—5178, 7137—7336.
F1301—1550, 1883—1957, 1958—2007, 2008—2082, 2083—2182, 3096—3145, 3196—3245, 3918—3967, 3968—4067, 4974—5073, 5174—5248, 6301—6500, 7144—7146 (cancelled), 8496—8595, 8646—8695, 9446—9495 (cancelled), 9496—9695 (cancelled).
H734—833, 2646—2745, 3996—4045, 7343—7412.
J651—680, 681—730.
Renumbered aircraft: B381, B3977 (ex N6338), B9990 (ex N6344).
Rebuilds: B778, 802, 847, 885, 893, 895, 898, 900, 3977 (ex N6338), 3980, 7732, 7743—7746, 7756, 7757, 7760, 7769, 7772, 7776, 7777, 7783—7785, 7790, 7791, 7793, 7797, 7807, 7817, 7820, 7821, 7829, 7835, 7859, 7860, 7862—7864, 7867—7869, 7874, 7875, 7883, 7896, 7905, 7932, 7968, 8025, 8108, 8155, 8187, 8201, 8205, 8212, 8217, 8220, 8830, 8921.
E9964—9983, 9990.
F2189—2208, 4175, 4177, 4178, 4183, 4184, 4187, 4193, 4194, 4199—4201, 4204, 4213, 4214, 5914—5921, 5923, 5925—5928, 5930—5932, 5936, 5938—5948, 5950, 5951, 5953—5960, 5966—5968, 5972, 5981, 5983, 5985, 5987, 5989—5994, 6022—6039, 6052, 6053, 6058, 6061, 6063, 6064, 6067, 6076, 6082—6084, 6086—6090, 6100, 6102, 6106, 6107, 6109—6111, 6117, 6122, 6123, 6126, 6132, 6135, 6138, 6147, 6149—6153, 6155, 6157, 6169, 6175, 6176, 6180, 6184, 6185, 6189—6194, 6197—6201, 6210, 6211, 6216, 6219—6221, 6223, 6225, 6226, 6228, 6240, 6245, 6249—6252, 6254, 6257—6259, 6261, 6264, 6268, 6269, 6271, 6281, 6282, 6285, 6292, 6294, 6295, 6300, 9409, 9410, 9413, 9509, 9548 (ex B5556), 9575, 9579, 9591, 9599, 9623, 9624, 9628—9632, 9634, 9635, 9637, 9695.
H6844 (ex D9427), 6847 (ex D6615), 6848 (ex D8196), 6850 (ex F1538), 6851 (ex E5178), 6852 (ex F5947), 6853 (ex D3415), 6854 (ex F2154), 6855 (ex E5177), 6856 (ex D6627), 6860 (ex F6184), 6861 (ex D9492), 6862 (ex D6442), 6863 (ex F5989), 6864 (ex E4400), 6867 (ex C3353), 6868 (ex F6197), 6869 (ex C6723), 6871 (ex F1899), 6872 (ex D9486), 6874 (ex F5946), 6875 (ex F5931), 6876 (ex B7896), 6877 (ex E1547), 6878 (ex D8171), 6884 (ex E4403), 6886 (ex F1917), 6889 (ex C3302), 6890 (ex D9653), 6891 (ex F6210), 6892 (ex F6147), 6897 (ex F2139), 6898 (ex D9669), 6899 (ex C8382), 6901 (ex H7276), 6902 (ex F6216), 6903 (ex D1942), 6904 (ex C3351), 6993—7016—renumbering of previously wrongly numbered rebuilt Camels, 7077—7092—renumbering of previously wrongly numbered rebuilt Camels, 7097, 7098, 7110 (ex D6644), 7111 (ex D1887), 7112 (ex B7769), 7113 (ex F1413), 7114 (ex D9567), 7115 (ex D9588), 7116 (ex F5981), 7117 (ex F1962), 7151 (ex E7173), 7205 (ex F1415), 7206 (ex D9598), 7207 (ex F1312), 7208 (ex E5173), 7209 (ex F2129), 7210 (ex D9599), 7211 (ex falsely numbered 'F6328'), 7212 (ex B9149), 7213 (ex D6442), 7214 (ex D6596), 7215 (ex D8162), 7216 (ex F5955), 7217 (ex C3355),7218 (ex E5156), 7219 (ex F3938), 7220 (ex F5936), 7221 (ex F1928), 7222 (ex F2144), 7223 (ex D1851), 7224 (ex D9634), 7225 (ex E7182), 7226 (ex D9438),

7232 (ex B7817), 7234 (ex B3786), 7235 (ex D9402), 7236 (ex B895), 7237 (ex D6638), 7238 (ex E1483), 7239 (ex D6446), 7240 (ex F6053), 7241 (ex F3108), 7242 (ex D6194), 7255 (ex falsely numbered 'F6337'), 7269—7289 (renumbering of previously wrongly numbered rebuilt Camels). 8253, 8258—8262, 8264, 8291, 8292.

Note: Additionally, Lt C. E. F. Arthur, a test pilot at No.2 A.S.D. in 1918, included in his log book rebuilt Camels F6055 and F6172, serial numbers that elsewhere have been attributed to rebuilt D.H.9s.

Sopwith 5F.1 Dolphin

Before the RFC began to receive production Camels a new Sopwith single-seat fighter was in the air. It was powered by a 200 hp Hispano-Suiza engine and underwent initial maker's trials on 22 May, 1917, when its indicated air speed at 5,000 ft was recorded as being 116—118 kt; when corrected and converted this came out as 143·88—146·18 mph. These startling figures were included in a brief description sent to RFC Headquarters in France; on the document Trenchard pencilled: 'This is very interesting. I wonder what it is like.'

The first prototype Dolphin takes an airing at Brooklands in the company of Camel N6336.

In some respects the new Sopwith 5F.1 was unlike anything the RFC had previously had. Few other aircraft of the 1914–18 war had such a radical configuration imposed primarily in the interests of giving the pilot the best possible outlook from the cockpit. Starting from a determination to place the pilot with his head in the middle of the open centre-section frame of the upper wing, the designer (Herbert Smith) found himself obliged to adopt negative stagger on the mainplanes. The pilot had a superlative view, possibly equalled only by that of the Bristol M.1C, in the most vital directions, and his two Vickers guns were immediately in front of, and perilously close to, his face. On the first prototype the engine had a single tall frontal radiator; the rear top decking was, at the cockpit, level with the rear spar member of the centre-section frame; and the tail unit had the general appearance and proportions of that of the Camel.

The first Dolphin went to Martlesham Heath early in June 1917, and its trial report records a speed of 123·5 mph at 10,000 ft and a service ceiling of 21,500 ft. Thereafter it went to France on 13 June, flown by Capt H. T. Tizard, and was greeted by a barrage of anti-aircraft fire from British guns, whose gunners did not recognise the Dolphin's unfamiliar silhouette. Next day it was flown by Capt W. A. Bishop, DSO, MC, of No.60 Squadron, who reported thus:

> 'I have the honour to submit the following report on a test flight of the Sopwith Dolphin.
> *View*—The view is exceptionally good, the pilot being able to see everything in front and behind him, as well as below him in front. He also has a good view above behind.
> *Handiness*—The machine is extraordinarily quick on turns and very handy.
> *Guns*—The guns are in a position where the pilot can easily work at them, *i.e.*, when correcting jams, etc.
> *Speed*—With the engine only giving 2,000 revolutions the speed at 2,000 ft was 106 knots =122 mph.'

The first Dolphin at Martlesham Heath, where it arrived for testing early in June 1917. It went to France on 13 June. (*T. Heffernan*)

Trenchard promptly sent a telegram to D.G.M.A. on 14 June, giving his opinion that the S.E.5 should stand as the RFC's primary fighting scout. He went on

> '. . . the Dolphin . . . as seen by me yesterday is a good machine provided certain mechanical defects in radiator and a few other small defects are got over. But the view must not be altered. I prefer this machine to the 200 French Hispano-Suiza Spad provided that the delivery is not later than the 200 Spad would be.'

Perhaps the unconventional configuration of the Dolphin made Trenchard hesitate to prefer it to the S.E.5, for the 200 hp S.E.5a was still something of an unknown quantity in mid-June 1917, but those responsible for ordering the RFC's aircraft had so much confidence in the Dolphin that they awarded a contract for 500 (C3777—4276) to the Sopwith company before the end of June. Further contracts were placed with Hooper (28 June), Darracq (13 July, 1917, 8 June, 1918, and 28 September, 1918) and Sopwith (29 November, 1917, 13 March, 1918, and 6 April, 1918), with E9997 being specially allotted for an

The second prototype Dolphin, here seen at Brooklands, had two small radiators let into the underside of each upper wing root, and there were cut-outs in the lower wings to give the pilot some downward view. The fin was originally very small and roughly rectangular, the rudder being horn-balanced.

experimental Dolphin on 27 April, 1918, but possibly being a belated allotment for the then-defunct fourth prototype.

Martlesham had criticised the efficiency of the Dolphin's radiator, had found the aircraft nose heavy, and had considered that improvement in rudder control was needed. These criticisms led to design development through three further prototypes before the production Dolphin evolved, having flank radiators, tapered nose, enlarged vertical tail surfaces that incorporated a horn-balanced rudder, and the addition of two upward firing Lewis guns mounted on the forward spanwise member of the centre-section frame. The second Dolphin arrived at Martlesham on 27 July, 1917, the third on 18 October; the latter was the first Dolphin to have the additional Lewis guns, and it went to Orfordness for evaluation on 25 November. It was apparently back at Martlesham early in December, but was grounded for weeks with engine trouble.

The fourth prototype, which represented the production form of the Dolphin in all essential particulars, was awaited by Martlesham from 6 October but was

The second prototype Dolphin with flank radiators and redesigned fin and balanced rudder. This photograph was taken at Martlesham Heath, where the aircraft had arrived on 27 July, 1917, it returned to its makers on 26 September. (*A&AEE; Crown copyright*)

The third Dolphin prototype, fully armed, at Martlesham Heath. It reached the Testing Squadron there on 18 October, 1917, but engine difficulties held up its trials for a time. It went to Orfordness on 25 November for a few days but more engine trouble occurred after its return to Martlesham, and by 26 January, 1918, its engine was out. Apparently the airframe went to Farnborough in March 1918.

held up at the Sopwith works by engine trouble. Apparently it was officially regarded as a production aircraft but it seems that it was not tested at Martlesham, though photographic evidence indicates that it went there. Contract-built production Dolphins were then imminent, and C3777 arrived at Martlesham on 21 November, 1917, followed by C3778 on 23 November and C3779 seven days later.

There can be no doubt that the fourth prototype was the unnumbered Dolphin that went to France early in November 1917. It was tested at No.1 Aeroplane Supply Depot on 4 November, and on 15 November it was flown from No.1 A.D. to No.19 Squadron at Bailleul by the squadron's C.O., Maj W. D. S. Sanday, DSO, MC. The new aircraft was flown extensively by pilots of No.19 Squadron,

The fourth prototype Dolphin evidently appeared fleetingly at Martlesham Heath, but went to France and, as related in the text, was used operationally by No.19 Squadron, acquiring the locally allotted serial number B6871.

among them Maj A. D. Carter, who first flew the Dolphin on 19 November. He was later to become one of the leading exponents of the type.

On 7 December Maj Sanday reported on the Dolphin:

> 'The machine itself seems to handle perfectly at all heights, and does not lose any of its controllability at 15,000 ft. It has not been over this height in this Squadron, but there is every reason to think that it would be equally as good at 20,000 ft.
>
> It is very stable and can be flown hands off for lengthy periods. It was, however, found to feel the effects of a very strong rough wind rather more than a Spad during the first 2,000 ft.
>
> The view is perfect and the machine easy to fly. Twelve pilots of this Squadron have flown it; they all handled it satisfactorily and were delighted with the machine; the weather has not been very suitable for lengthy tests.'

In addition to this highly complimentary report, a number of structural and mechanical points were criticised. The difficulty experienced in changing the engine's oil filter led to a modification in the design of the cross-bar of the engine mounting and took some time to resolve. After further experience the squadron recommended other modifications, but continued to report on the Dolphin in glowing terms.

Before the Dolphin saw combat it was decided that only one Lewis gun would be carried on operational aircraft, and on 23 December, 1917, Brooke-Popham instructed both A.D.s accordingly, stipulating that only the starboard Lewis was to be mounted.

The fourth prototype remained with No.19 Squadron, and on 28 December production Dolphins C3820, C3789 and C3794 were collected by squadron pilots at Marquise, with C3788 following next day. The prototype had come close to being involved in combat on 12 December when, in the words of the Squadron Record Book:

> 'At 2.7 p.m. several Gothas were seen approaching the aerodrome. Four Spads and one Dolphin left the ground. Capt. Huskinson was catching up the Gothas near the lines but did not follow up the chase and he was not allowed to cross the lines on this machine.'

Re-equipment of No.19 Squadron proceeded until by 9 January, 1918, all the Spad 7s had been replaced by Dolphins. In this process the fourth prototype remained with the unit and was allotted the serial number B6871, doubtless to put it on the same official footing as the production aircraft. It survived until 26 February, 1918, when it was recorded as missing; its pilot was 2/Lt J. L. McClintock.

Although No.19 Squadron's pilots apparently welcomed the Dolphin with rapture, many of their contemporaries had misgivings inspired by the exposed position of the pilot's head and the risk of serious injury in the event of overturning on landing. On 31 January, 1918, D.A.E. wrote to the G.O.C., RFC, informing him that a quick-release device was to be incorporated in the side cross-bracing of the centre-section to enable the pilot to disconnect the bracing wires and escape through the fuselage side in the event of an inverted landing. This was built into later production Dolphins to provide emergency egress on the starboard side, but in practice this proved impracticable owing to the presence of the top gun fitting, and in May 1918 the quick-release arrangement was re-designed for the port side.

Production Dolphin with all guns mounted.

Of even greater significance was the hurried strengthening of the front main spar butt attachments in the wings. Before January 1918 was out strict instructions were issued to contractors that no Dolphin was to go to France without this modification; on 26 January, No.2 A.S.D. was instructed that no Dolphin was to be flown until the appropriate fittings had been reinforced. This followed the breaking-up of an aircraft in flight owing to failure of a lower-wing attachment. As far as is known, the Dolphin was never structurally suspect after the modification was made.

A closer look at a four-gun fighter, 1917 model. The cockpit of the Dolphin was not over-large at best and the risk of facial and cranial injury to the pilot in the event of overturning was obvious to all who had to fly it.

Combat experience led to difficulties with the flank-mounted radiators, for they were frequently damaged by empty cartridge cases and ejected links, which cut the tubes and caused leaking. Wire-mesh screens were fitted, but in some cases this was too fine and led to boiling; by mid-April 1918 it was found that ordinary rabbit wire-netting, suitably mounted, was entirely adequate. For the radiators themselves, a sensible (indeed, obvious) modification made it possible for the pilot to open or close both shutters simultaneously.

Further to improve pilots' chances of surviving overturning on the ground, the Technical Department evolved a cabane structure that served as a crash pylon and as the mounting for the Lewis gun. One such cabane was sent to France on 2 April, 1918, and was quickly followed by the production Dolphins C4044—4046, all equipped with the cabane. Examples were sent to Squadrons No.19 and 79 but pilots did not favour the modification because it made it much more difficult to use the Lewis gun effectively, and it was thought that the gun would be much more likely to injure the pilot in the event of overturning. The cabane was therefore not adopted.

On 7 December, 1917, the Sopwith-built Dolphin C3828 was at Kenley Aircraft Acceptance Park and on that day was allotted to the RFC with the Expeditionary Force in France. By February 1918 it was with No.19 Squadron as aircraft P. Here it is seen at Bailleul just after being ferried to the squadron by Maj W. D. S. Sanday, DSO, MC, the Commanding Officer. By the airscrew is Flt Sgt J. M. Flynn. C3828 was shot down in May 1918; its pilot, Capt Chadwick, was made a prisoner.

No.79 Squadron had arrived in France equipped with Dolphins on 18 February, 1918, thus preceding the re-equipment of No.23 Squadron by a few days. It had been intended to replace No.23's Spad 13s with Dolphins from 31 January but this was postponed until 28 February. In fact, however, No.23 did not lose its Spads until late April, and was not fully equipped with Dolphins until 4 May. Although it was originally intended to equip twelve squadrons with Dolphins, the RAF was to have only one more Dolphin squadron (No.87 which arrived in France on 26 April) before the war ended. No.91 Squadron should have gone to France as a Dolphin unit in November 1918, and Nos.93 and 81 were planned for December, with two further squadrons following in January 1919. Replacement by Dragons/Nighthawks and Buzzards was expected to start in April 1919.

In operational use the Dolphin proved to be an excellent fighter, especially at higher altitudes, and all recorded reports speak of its performance and handling qualities in highly complimentary terms. These reports leave little doubt that the

Dolphin was superior to the S.E.5a with the same engine, but it has always been consistently under-rated and has never received well-deserved credit for its fine qualities, probably because so many pilots were daunted by its reverse stagger and exposed pilot's position. What could be done with the Dolphin was clearly demonstrated by such pilots as Maj A. D. Carter, who achieved no fewer than nine combat victories on the type; while Capt F. W. Gillett, DFC, an American pilot of No.79 Squadron, attained the even higher score of 14 enemy aircraft and three kite-balloons brought down.

Inevitably, the Dolphin suffered from the same engine troubles that bedevilled the S.E.5a and Spad 13, and these were accentuated by chronic oil-tank leaks, at least during its service with the RFC. In the history of the S.E.5 and S.E.5a something has been said about the difficulties experienced with faulty reduction gears on Brasier-made Hispano-Suiza engines. Eventually someone thought of removing the reduction gear from some of these engines, thus converting them into direct-drive engines that did not differ essentially from the Wolseley Viper that came to be used in the later S.E.5as. This de-geared engine was tried in a Dolphin, C8194, which was sent to Martlesham on 22 September, 1918, for official trials. Its performance was somewhat poorer than that of Dolphins with the geared engine, but the RAF in France evidently considered this price worth paying in order to have reliability. On 9 February, 1919, Brooke-Popham asked that all Dolphins sent to France thereafter should have either the direct-drive converted Hispano or the Wolseley Viper, and on 28 February D.A.E. agreed that future deliveries would be confined to Dolphins with ungeared French Hispano-Suiza engines. This variant was named Dolphin III, the designation Mk II having been allotted to the French version of the Dolphin with the 300 hp Hispano-Suiza.

That final form of the Dolphin did not appear until long after the RFC had lost its separate identity, but the history of the Dolphin cannot be closed without mention of its brief and limited use on Home Defence duties. When Herbert Smith designed the Dolphin he was well aware of the need for effective night fighters, and this influenced his thinking. Early in 1918 seven Dolphins were

C3854 with only one of the Lewis guns on the centre-section frame, the officially adopted operational armament for the type. This Dolphin was used at No.2 School of Aerial Fighting and Gunnery, Marske.

One of the three Dolphins that were fitted with a cabane structure as a form of crash pylon. Operational pilots did not like this well-meant protective structure, however, and it was not adopted. (*Aeromodeller*)

allotted for Home Defence duties; one (C3862) was tried by No.78 Squadron at Suttons Farm and the others apparently went to No.141 Squadron at Biggin Hill. The first of these was soon crashed by Lt Langford-Sainsbury (later Air Vice-Marshal T. A. Langford-Sainsbury, CB, OBE, DFC, AFC), but the arrival of the others enabled the squadron to create a Flight of Dolphins. On the nocturnal duties of the unit the Dolphin was not popular, for the Hispano-Suiza was temperamental and the risk of overturning greater than in daylight. In February 1918 the Sopwith company modified C3858 by adding two half-hoops of steel tubing above the upper wing in line with the inboard interplane struts. This Dolphin also had a variable-incidence tailplane. No development ensued, doubtless because the Dolphins were soon withdrawn from Home Defence duties.

To this day the Dolphin remains little known and seriously underrated. One who knew it well was Wing Commander W. M. Fry, MC, who recalled:

'Looking back, the Dolphin was a splendid machine, strong, easy on the controls, no flying vices, and owing to its large wing surface it had an excellent performance high up. In fact, I should think at the time it had a better performance at, say, over 15,000 ft, than any other British or German machine. Its armament was also first-class for those days.'

There seems little doubt that the Dolphin was, in truth, the best operational fighter of its period. It was a better aircraft than the mediocre Snipe, which might not have been developed had the Dolphin not suffered from the shortcomings of its Hispano-Suiza engine.

Final proof of the Dolphin's quality is provided by its adoption by the French with the 300 hp Hispano-Suiza engine; it was agreed as early as 27 November, 1917, to supply a Dolphin, and it seems that this was done almost immediately. This may have been the Dolphin that had the first installation of the 300 hp engine and crashed fatally late in April 1918. D3615 later had the same engine, and production was initiated by the S.A.C.A. concern. With the high-powered engine the Dolphin would have been a fighter to reckon with in 1919.

5F.1 Dolphin

200 hp Hispano-Suiza 8Ba or Bb, 220 hp Hispano-Suiza 8Bc or Be, 200 hp Wolseley W.4B Adder

Single-seat fighting scout.

Span 32 ft 6 in; length 22 ft 3 in; height 8 ft 6 in; wing area 263·25 sq ft.

Weights and performance:

Aircraft	First prototype	Fourth prototype with one Lewis gun	Third prototype with cabane	C3777
Weight empty, lb	1,350	—	—	—
Weight loaded, lb	1,880	—	1,959	2,003
Maximum speed, mph				
at 10,000 ft	123·5	108	121·5	128
at 15,000 ft	116	98	114	119·5
Climb to (min and sec)				
6,500 ft	5 30	— —	7 5	6 25
10,000 ft	9 25	9 30	12 5	11 0
15,000 ft	17 20	16 0	23 0	20 12
Service ceiling, ft	21,500	—	20,000	—
Endurance	2¼ hr	—	—	—

Armament: Two fixed 0·303-in Vickers machine-guns; one (occasionally two) semi-free 0·303-in Lewis machine-gun; four 25-lb Cooper bombs.

Manufacturers: Sopwith, Darracq, Hooper.

Service use: *Western Front*: Sqns 19, 23 and 79. *Home Defence*: Sqns 78 and 141. *Training*: No.23 Training Wing; Training Sqns 11, 60 and 70; No.1 School of Special Flying, Gosport. *Other units*: Squadrons working up, Nos.85, 87, 90 and 92.

Serials: *Fourth prototype*: B6871.
Production aircraft: C3777—4276, C8001—8200, D3576—3775, D5201—5400, E4424—4623, E4629—5128, E9997, F7034—7133, J1—150, J151—250.
Known rebuilds: B7849, B7851, B7855, B7861, B7876, B7877, B7927, B7928, B7937, B7953, B7955, B7978, B8189, F5961, F5962, F6020, F6144, F6145, F6146, H6866 (ex D3532), H7243—7246.

Sopwith 7F.1 Snipe

In the spring of 1917 the RFC's armoury of operational fighter aircraft still included D.H.2s, F.E.2ds and F.E.8s; a few squadrons of Pups were supplemented by the triplanes of the RNAS fighter squadrons attached to the RFC; the S.E.5, Camel and Bristol Fighter were still only in prospect and unknown quantities. It was at this time that the Air Board drew up specifications for twelve types of future aeroplanes; of these, Specification Type A.1(a) set out the requirements for a single-seat fighter for the RFC. These called for a speed of 135 mph at 15,000 ft, an average rate of climb of 1,000 ft/min above 10,000 ft, a ceiling of at least 25,000 ft, all with fuel for 3 hr and three machine-guns with 1,000 rounds of ammunition (the third gun, to be pivoted to fire upwards, was optional).

The first Snipe prototype at a very early stage with Camel-type tail unit and small dihedral angle on the wings.

It is hard to believe that this specification was not supplied to the Sopwith company, yet their new rotary-powered fighter designed in the summer of 1917 did not seem to have been intended to comply with the official requirement. Its basic layout had been determined by mid-August 1917 as an unambitious development of the Camel, designed to take only the 150 hp A.R.1, 150 hp Monosoupape, 130 hp Clerget or 110 hp Le Rhône—three of which were already standard power units of the Camel. Also like the Camel, the new fighter was armed with twin 0·303-in Vickers guns, their original installation closely resembling that of the earlier type.

In its geometry the new Sopwith 7F.1 was a single-bay biplane with equal dihedral on upper and lower wings. The narrow centre-section was supported on four vertical struts, and was virtually an open frame, a considerable concession to pilot view. In the fuselage design further consideration was given to the pilot's outlook by making the fuselage deep enough for his seat to be placed so high that

First Snipe prototype with increased dihedral.

his eyes were almost level with the rear-spar member of the centre-section frame. This was perhaps the detail in which the 7F.1 improved most upon the Camel, from which the pilot's field of view was, by the standards of the time, atrocious. The tail unit did not differ appreciably from that of the Camel, and its vertical surfaces looked disproportionately small. From the beginning the Sopwith company named their new creation Snipe. Not unnaturally, when the first prototype was built they chose to power it with the most powerful rotary engine then available, the 150 hp Bentley B.R.1.

At a very early stage it was found necessary to increase the dihedral angle of the mainplanes, and a different airscrew was fitted. The Snipe was still without an official identity, for it had been built as a private venture under Licence No.14. Indeed, its progress might well have gone no further had not W. O. Bentley designed the B.R.2 rotary engine, which underwent its first bench runs early in October 1917. It was an immediate success, and a large-scale production programme for the new engine was quickly drawn up. A second Snipe prototype, identical with the modified form of the first, was fitted with the first B.R.2 engine. Soon afterwards the aircraft was given official recognition in the form of Contract No.A.S.31668, which provided for six prototypes: for these the serial numbers B9962—9967 were allotted on 10 November, 1917. The prototype with the B.R.2 became B9963 and went to Farnborough with that identity on 23 November.

The second Snipe prototype resembled the first, but had a 230 hp Bentley B.R.2 engine in place of the B.R.1.

Surviving official documents suggest that of the six aircraft only B9965—9967 were officially regarded as effective prototypes, and subsequent events make it clear that the Snipe was considered as a contender under the Specification for Type A.1(a). Official insistence on high-power engines is indicated by the fact that the 200 hp eleven-cylinder Clerget 11E was, early in November 1917, envisaged as an alternative power unit. The first Clerget 11E had been completed in mid-1917, the second was nearly ready by 18 July, and the engine was evidently expected to deliver 260 hp. It is therefore not surprising that the Air Board ordered substantial numbers of the Clerget; known orders were for 700 from the parent Clerget-Blin company and 1,300 from Delaunay-Belleville, and there may

546

The second Snipe prototype was eventually numbered B9963 but retained the original inadequate fin and rudder. (*RAF Museum*)

have been a further proposed order, for in June 1918 it was expected that 2,330 would be delivered against British orders. Development of the Clerget proved to be protracted, however, and only 27 had been delivered under British contracts by the end of 1918. Its hoped-for output of 260 hp was never realised; the best it could do was 215 hp.

Late in November 1917 the Snipe prototype B9965 was expected at Martlesham Heath, but a crash at Brooklands on 19 November necessitated the aircraft's return to Kingston for repair. It arrived at Martlesham on 18 December and was tested; it crashed five days later and was sent back to the Sopwith works for further repairs. At that time B9965 was a single-bay biplane but differed from the earlier prototypes in having fully-faired sides on the fuselage, a wider-span

The third Snipe prototype, B9965, as it first appeared at Martlesham Heath on 18 December, 1917, with fully-faired fuselage, revised wing and centre-section, and new form of fin and rudder. By February 1918 it had been fitted with two-bay wings. (*RAF Museum*)

centre-section and modified upper wings, and a new fin and rudder assembly that looked no more adequate than the original.

Early in January 1918, B9966 was in prospect, but it was B9965 that again went to Martlesham on 25 January. In the course of being repaired it had been fitted with new two-bay wings spanning 30 ft, possibly to improve the Snipe's chances of fulfilling the requirements of the new specification Type I, *Single-seat Fighter (high altitude)*. As called for in that specification, B9965 had a Lewis gun on the centre-section. The aircraft was hustled back to Martlesham before it had been fully adjusted, for it was nose-heavy on turns and, not surprisingly, lacked adequate rudder control.

Snipe B9965 in its two-bay form, with Lewis gun on the centre-section, February 1918.

Competitive with the Snipe were the Austin A.F.T.3 triplane (later named Osprey), Boulton & Paul Bobolink, and Nieuport B.N.1. The four were subjected to a comparative assessment based on their Martlesham test reports and, although the B.N.1 appeared to be the best all-round fighter, the Snipe was selected for no clearly identifiable reason. B9965 flew to France on 11 March, 1918, for evaluation, first of all at No.1 Aeroplane Supply Depot, St-Omer, subsequently by pilots of Nos.65 and 43 Squadrons, RFC. Comments were not too unfavourable, but the inadequate rudder and heavy ailerons were criticised.

On 20 March contracts for the large-scale production of the Snipe were given to seven contractors, the orders totalling 1,700 aircraft. All subsequent development was conducted by the Royal Air Force, notably with B9966, which first went to Martlesham on 18 May, 1918. It was on this prototype that the Snipe's considerable control problems were investigated until the aircraft at last had a fin and horn-balanced rudder that looked adequate, and its upper ailerons were given inverse taper and overhanging horn balances. Much was obviously expected of the Snipe development powered by the 320 hp A.B.C. Dragonfly radial engine; the production of hundreds with the name Dragon was intended, but the engine had incurable faults and no Dragons saw service.

Another variant of late 1918 was the 7F.1A Snipe Ia, a long-range version intended for escort duty, especially for the bombers of the Independent Force. A prototype and 50 production Snipe Ias were built, but structural deficiencies imposed such limitations on the flying of the aircraft that it could not possibly have been used effectively in its intended capacity.

The Snipe was perhaps fortunate that it went into operational service so late in the war that the German Jagdgeschwader were in operating difficulties and the German industry was unable to bring forward new types in any numbers. In practical terms the Snipe was not much of an improvement over the Camel, in spite of its long period of development, yet in the postwar period it was to be preferred to the incomparably superior Martinsyde Buzzard as the Royal Air Force's standard single-seat fighter.

7F.1 Snipe
230 hp Bentley B.R.2

Single-seat fighting scout.

Span, with original plain ailerons, 30 ft; length, with original rudder, 19 ft 2 in; height 9 ft 6 in; wing area 256 sq ft.

Empty weight 1,212 lb; loaded weight 1,992 lb. Figures for B9965 with two-bay wings.

Maximum speed at 10,000 ft—124·5 mph, at 15,000 ft—113·5 mph; climb to 10,000 ft—8 min 50 sec, to 15,000 ft—17 min 35 sec; service ceiling 19,500 ft; endurance $2\frac{1}{4}$ hr. Figures for B9965.

Armament: Two fixed 0·303-in Vickers machine-guns; B9965 briefly had one 0·303-in Lewis gun on overwing mounting.

Manufacturers: Sopwith; Boulton & Paul; Coventry Ordnance Works; Kingsbury Aviation; Marsh, Jones & Cribb; Napier; Portholme; Ruston, Proctor.

Service use: In the RFC, prototypes B9963 and B9965 were flown at Farnborough and Martlesham Heath for test purposes; in France B9965 was flown for evaluation by Sqns 43 and 65, and at No.1 Aeroplane Supply Depot, St-Omer. All subsequent service with RAF units.

Serials: (being all known numbers allotted for Snipes, whether built or not):
B9962—9967, E6137—6536, E6537—6686, E6787—6936, E6937—7036, E7337—7836, E7987—8286, E8307—8406, F2333—2532, F7001—7030 (delivered as Dragons), F9846—9995, H351—650, H4865—5064, H8513—8662 (changed to Nieuport Nighthawks), H8663—8762, H9964—9966 allotted for 7F.1a Snipe Ia prototypes, J301—400, J451—550, J2392—2541, J2542—3041, J3042—3341, J3617—3916 (built as Dragons), J3917—3991 (possibly deliveries began as Nighthawks), J4092—4591, J6493—6522.
Rebuilds: H6846 (ex E8035), H6880 (ex E8024), H6894 (ex E8032), H6895 (ex E8060), H7149 (ex E8075), H7150 (ex E8107), H7152 (ex E8049), H7153 (ex E8130), H7154 (ex E7341), H7227 (ex E8109).

On 9 May, 1918, at Brooklands, Capt H. Robin Rowell prepares to take off for France in the first Salamander, E5429. The Naval officer in the picture is Lt Denis Allen.
(*Sir Robin Rowell, from Peter Liddle's 1914–18 Personal Experience Archives presently housed within Sunderland Polytechnic*)

Sopwith T.F.2 Salamander

The RFC began to use tractor fighters systematically on ground-attack duties in August 1917 during the Battle of Ypres and, more significantly and at greater cost, in the Battle of Cambrai. By that time D.H.5s and Camels were the RFC's most successful ground-attack aircraft, but their losses averaged 30 per cent for each day on which they were used in this way. In November 1917, before the Battle of Cambrai ended, the RFC submitted a request for a single-seat aircraft specifically for ground-attack duties. The armament specified by the RFC was:

> 'One or two of the guns to be so fitted to fire downwards at an angle of 45° to the line of flight; if possible these guns to be variable over 20°, *i.e.*, depression to be variable from 35° to 55°. One gun on the top plane firing straight ahead and upwards.'

In response to this, Wg Cdr Alec Ogilvie of Section T.6 of the Technical Department arranged for the Sopwith company to fit a Camel with suitable armour, and the experimental station at Orfordness was asked to determine how a Camel could be fitted with the downward-firing armament that the RFC wanted, and how sighting arrangements could be provided. Before February 1918 was out, Camel B9278 had been fitted with unhardened steel plates along the underside of the forward fuselage; further armour protected the pilot's back, the carburettor and petrol pipes. Two Lewis guns were arranged to fire through the cockpit floor at a 45 deg angle, and a third Lewis gun was carried above the centre-section on a Bowan & Williams mounting. At the same time a second Camel, B6218, was fitted with a periscopic sighting arrangement intended to facilitate the aiming of downward-firing guns.

550

The armoured Camel was given the new Sopwith type number T.F.1 and first flew in its revised form on 15 February, 1918. Accompanied by B6218, it flew to France on 7 March, 1918. It took the RFC only a few days to recognise that the downward-firing guns were unpractical, and on 13 March Maj-Gen J. M. Salmond reported:

> '. . . with regard to Sopwith Camels Nos.6218 and 9278. It is not considered that either of these machines are of any practical value from the point of view of firing into enemy trenches or at hostile parties on the ground. The present Sopwith Camel is considered more efficient in every way for this purpose.'

On the same day it had been decided, in a discussion at the War Office, to provide two squadrons of Camels armoured like B9278 but retaining the standard installation of twin Vickers guns. In fact all that happened was that slabs of 11-mm or 12-mm armour plate were fitted on the undersides of the seats in some Camels, and no externally armoured Camel ever equipped any squadron in France.

The RFC had drawn up a comprehensive table of the types of aircraft wanted for 1918. Type 2 on this list was to be a single-seater for attacking ground targets, to be armed with two Vickers and one Lewis guns with 1,200 rounds and capable of carrying four 20-lb bombs. In January 1918 the Technical Department approached the Sopwith company and invited design proposals. By 25 January an outline proposal envisaged the use of a 230 hp B.R.2 engine, armament consisting of one gun firing ahead and two firing downwards at 45 deg with ±10 deg variation in the angle; 1,200 rounds of ammunition and four 20-lb bombs were to be carried. Armour plate was to enclose the entire forward portion of the fuselage, and the speed at sea level was to be at least 120 mph. A week later Sopwiths' proposal had taken more positive shape as a Snipe derivative in which the forward fuselage was made solely of the armour plate and dispensed with basic structure and bracing; the armour was estimated to weigh 605 lb. The downward-firing guns were to be Lewises, while the forward-firing weapon could be either a Lewis on the upper wing or a synchronised Vickers. One fuselage was already being erected at that time (1 February, 1918).

On 18 February it was recommended that six prototypes should be ordered; some ten days earlier the Vickers gun had been chosen as the forward-firing weapon, and the thicknesses and disposition of the armour plate determined. All the armour plate had been received by the Sopwith company by 22 February, by which date the aircraft had received the Sopwith designation T.F.2. Early in March it was provisionally arranged that three of the prototypes should have two conventionally mounted and synchronised Vickers guns in place of the downward-firing Lewises, a sensible precaution in view of the possibility that the RFC might decide against the downward-firing guns then being evaluated in France on the T.F.1. It was intended that the fourth, fifth and sixth T.F.2s would have the twin Vickers installation, but on 11 March, knowing that the downward-firing guns had proved to be unacceptable after all, Wg Cdr Ogilvie wrote to Tom Sopwith, asking that all six aircraft should have the twin forward-firing Vickers. At that stage the use of an overwing Lewis was intended.

The decision of 13 March, 1918, to create two squadrons of armoured Camels almost stopped further development of the Sopwith T.F.2. The memorandum of the high-level discussion held on that date recorded:

> 'Work should be stopped on such of the Sopwith T.F.2 machines as were

not in too far advanced stage of completion, as it was considered that the weight and loss of manoeuvrability of those machines did not compensate for the extra protection. The first one or two machines, however, would be completed and sent to E.F. for inspection, having two Vickers guns firing straight ahead and no gun on the top plane.'

It seems that in fact nothing was done to retard or reduce work on the T.F.2s, but none was completed before the RFC lost its separate identity on its assimilation into the Royal Air Force on 1 April, 1918.

The first prototype, E5429, made its initial flight at Brooklands on 27 April and was flown to France on 9 May. It went to Squadrons No.73, 3 and 65, and was flown by several pilots before crashing on 19 May. The RAF's high command was lukewarm towards the Salamander, as it was named, but the pilots' reports were, on the whole, favourable, though all regarded the lateral control as poor. Maj-Gen J. M. Salmond reported back and asked for balanced ailerons to be fitted. An initial order for 500 Salamanders was given to the Sopwith company in May, and later contracts were given to other contractors.

Other prototypes followed E5429, and various modifications were introduced or considered. These included the balanced upper ailerons and enlarged fin and rudder of the late production Snipe, but the Salamander surfaces differed in structural strength. The 200 hp Clerget 11E was to have been an alternative power unit, and disruptive camouflage areas were painted on some aircraft. Salamander squadrons No.96 and No.157 were forming at the time of the Armistice, and forward planning envisaged eight B.R.2-Salamander squadrons and five with Clerget-Salamanders in France by the end of May 1919.

The signing of the Armistice prevented the movement to France of any Salamander unit, however, but the type remained in the RAF's inventory until 1922. Production eventually totalled 526.

T.F.2 Salamander
230 hp Bentley B.R.2, 200 hp Clerget 11Eb

Single-seat armoured ground-attack aircraft.

Span, with original ailerons, 30 ft 1½ in; length 19 ft 6 in; height 9 ft 4 in; wing area 266·5 sq ft.

Empty weight 1,844 lb; loaded weight 2,512 lb.

Maximum speed at 500 ft—125 mph, at 6,500 ft—123·5 mph; climb to 6,500 ft—9 min 6 sec; service ceiling 13,000 ft.

Armament: Two fixed 0·303-in Vickers machine-guns; four 25-lb bombs.

Manufacturers: Sopwith, Air Navigation Co, Glendower Aircraft, Palladium Autocars, Wolseley, National Aircraft Factory No.1.

Service use: None with RFC.

Serials: E5429—5434, F6501—7000, F7601—7750, F7801—7950, J5892—5991, J5992—6091, J6092— 6491.

The first Spad 7 to be delivered to the RFC received the official serial number A253 and is here seen at No.2 A.D., Candas, shortly after delivery. It was issued to No.60 Squadron on 20 September, 1916, returned to No.2 A.D. on 17 October and was sent to England on 23 October. In July 1918 it was still in use with No.56 Training Squadron.

Spad 7

The first Spad 7 made its initial flights in April 1916 and was immediately recognised by the RFC as a fighter of great potential. It was the first single-seat fighter to be powered by Marc Birkigt's brilliant new 150 hp Hispano-Suiza engine, and was armed with a single Vickers machine-gun that was synchronised by a special mechanism also designed by Birkigt. Only days after the prototype first flew, Trenchard wrote to Col Régnier, Directeur de l'Aéronautique militaire, to ask whether the RFC might order three examples of the new type. Early in May, Régnier agreed in principle but it was not until early September that S.126 was delivered as the first RFC Spad. Nevertheless, this was a remarkable feat of procurement by the British Aviation Supplies Depot, for the aircraft was delivered less than three weeks after the first production aircraft for the French air service was taken over by Armand Pinsard. S.126 was flown to No.2 A.D. at Candas by Adjudant Strohl on 9 September, 1916, and became A253 in the RFC.

It went to No.60 Squadron on 20 September but stayed barely four weeks, returning to No.2 A.D. on 17 October and being sent to England six days later. By that time the RFC had two more Spad 7s, S.140 and S.143, that had been flown to No.2 A.D. on 27 and 28 September, becoming respectively A262 and A263. Not surprisingly, the RFC wanted more Spads, and soon received French permission to order 30 more. These were built by Blériot Aéronautique, and deliveries began on 10 November, 1916.

Re-equipment of No.19 Squadron with Spads began on 9 October, 1916, when A263 was delivered to the unit, but early deliveries were slow, and it was February 1917 before No.19 was fully equipped with the new type. Likewise, No.23 Squadron could only begin to replace its F.E.2ds with Spads early that February and took until April to complete its re-equipment. Early in January 1917 fourteen

B1627 was a French-built Spad 7 that was issued to No.19 Squadron on 14 May, 1917. Five days later, while being flown by Lt S. F. Allabarton, it was brought down by anti-aircraft fire at Harnes. It sustained some damage in landing. (*A. E. Ferko*)

Spad 7s had been delivered to the RFC and an additional quota of 50 had been approved by the French authorities. In response to a plea for more Spads put forward by Trenchard personally, the French transferred ten from their own Réserve Générale de l'Aviation late in January. Further production Spad 7s for the RFC were built by the Avionnerie Kellner et ses Fils, with whom an order for 120, plus spares equivalent of 30 more, had been placed by the RFC early in 1917.

The first Kellner-built Spad 7 was completed in April but had to remain at the Kellner works to serve as a model, consequently the first delivery was not made until 21 May. By 21 July a total of 30 had been delivered from Kellner. Fortunately, in April the French authorities had released another 30 to the RFC from the R.G.A., and these sufficed to keep the two RFC Spad squadrons going.

By June 1917 the RFC was receiving some Spads fitted with the 180 hp Hispano-Suiza 8Ab engine, which was basically the same as the 150 hp Hispano-Suiza 8Aa but had its compression ratio increased. With this engine the Spad had a significantly improved performance. In the Spad 7 the 8Ab engine required

B3508 was another 19 Squadron Spad 7 that fell into German hands and was flown by German pilots. It was at No.1 A.D. on 4 August, 1917 and was recorded as missing on 6 October, 1917; 2/Lt G. R. Long, its pilot, was wounded and had to land behind the German lines. (*A. E. Ferko*)

certain installation modifications. At an earlier stage, late in 1916, the gun installation had to be modified. Originally the webbing ammunition belt was carried wound on a large transverse drum ahead of the cockpit and was fed, empty, on to a similar drum installed longitudinally behind the cockpit. This imposed considerable loads on the gun, which had to pull the belt drum round as it fired, and led to double feeds when the rotational momentum of the drum forced the belt on when firing ceased. The advent of the Prideaux disintegrating-link belt in November 1916 enabled an ammunition box to replace the drum, and this eventually cured the problem.

It had become clear in 1916 that the RFC's need for fighting scouts of the calibre of the Spad 7 was so great that adequate supplies might not be forthcoming from native French production, and steps were taken to initiate production in England. Both the RFC and RNAS were interested in the type and both wanted

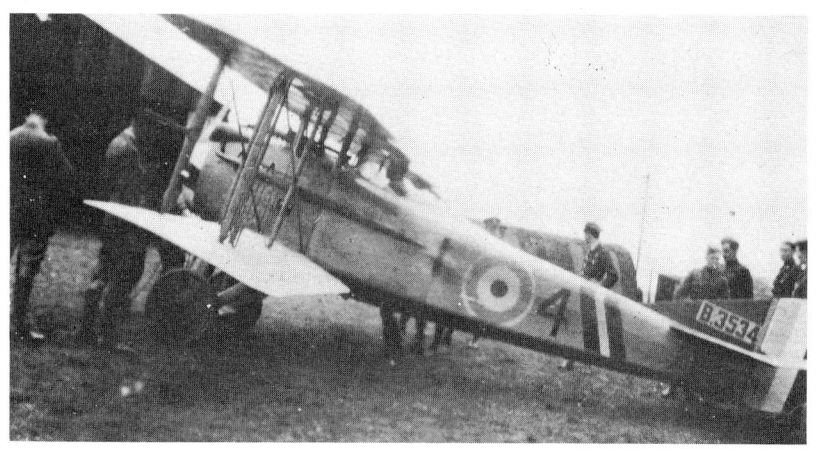

B3534 was a No.23 Squadron Spad 7 that was with the unit in August 1917. It was returned to No.1 A.D. for repair on 18 August. (*RAF Museum*)

it to be built by British firms. The RNAS received the Spad 7 S.211 at Dunkerque on 23 November, 1916, and the aircraft briefly bore the unexplained number N3399 for about a week before receiving the official RNAS serial number 9611. It seems likely that this Spad had been supplied to serve as a model or prototype.

In mid-December 1916 the RNAS ordered one hundred Spad 7s, fifty each from Mann, Egerton and the British Nieuport company, but quickly ordered 25 more from the former. A sample Spad, possibly 9611, was sent to Mann, Egerton's Norwich works, and production began. At that time the RFC was in process of seeking help from the RNAS, and at a conference held on 14 December, 1916, the Admiralty agreed to transfer to the RFC half of its Spads then on order. For the 62 aircraft thus promised the RFC numbers A9100—A9161 were allotted early in January 1917. However, at a subsequent conference on 26 February, 1917, the Admiralty agreed to hand over all of its Spads in exchange for all the Sopwith triplanes then on order for the RFC. By then the British Nieuport order had been changed to S.E.5s, consequently only Mann, Egerton's output was involved. Eventually one hundred and twenty Spad 7s built by that company were delivered to the RFC, while 9611 was also transferred, acquiring the new identity of B388 on 2 March, 1917.

The RFC's choice of British contractor was L. Blériot (Aeronautics) of Brooklands, from whom one hundred Spad 7s were ordered. It seems that instructions to proceed with this order must have been given early in November 1916, but the serial numbers A8794—8893 were not officially allocated until 29 December. Some delay occurred initially for the surprising reason that the parent French firm supplied incomplete drawings, and a sample Spad (possibly A253) had to be completely dismantled to enable the firm to make adequate drawings. This re-drawing may have been one reason why the British Blériot concern elected to fit a large opaque housing over the gun breech. No doubt this was intended to provide the pilot with protection when clearing gun faults, but all it did was to mask totally the pilot's view in that most vital area about the gun sight.

A8794 was the first Spad 7 made at Addlestone by Blériot & Spad Ltd. It was at Farnborough on 19 May, 1917, and was issued to No.19 Squadron from No.1 A.D. on 20 July, 1917. (*RAF Museum*)

Some of these British Blériot built Spads went to the RFC in France. Almost immediately Brooke-Popham ordered the removal of the housing over the gun, and such aircraft as went to France were thus modified. In service, however, it was found that the British-built Spads had a poorer performance than the French-built aircraft. They were also nose-heavy in flight and were not popular with RFC pilots. The RFC therefore sought to do without these aircraft, and one official document dated 3 August, 1917, went so far as to say that 'the English Spads have been cancelled'. This was not so, but thereafter the RFC in the Field did not exert itself to obtain British-made Spads for the squadrons, despite the fact that estimates of wastage and availability showed that the squadrons would cease to exist by February 1918. In the event, No.19 Squadron began to receive Dolphins in November 1917, and later that month Kellner-built Spad 13s began to come forward, enabling No.23 Squadron to be re-equipped with the more powerful twin-gun type. From the end of November 1917 the reconstruction of Spad 7s by Aircraft Depots ceased, and the type was gradually phased out from the Western Front.

Thereafter a few remained in use with Squadrons Nos.30 and 63 in Mesopotamia and with No.72 Squadron in Palestine, but most of the others were sent from France to England for service with training units.

The Spad 7 was, in the RFC, never quite the success that it was in French service. It seems that it was not regarded by the RFC as a particularly good dog-

Spad 7 A8812, another Addlestone product, was used by No.30 Squadron in Mesopotamia. It had an overwing Lewis gun, and is here seen at Faluja.
(Peter Liddle's 1914–18 Personal Experience Archives, presently housed within Sunderland Polytechnic)

The overwing mounting for the Lewis gun on A8812 of No.30 Squadron. The Spad 7 retained the large hooded windshield.
(Peter Liddle's 1914–18 Personal Experience Archives, presently housed within Sunderland Polytechnic)

fighting aircraft, yet in the hands of individualistic French pilots it proved to be an outstandingly effective weapon. The RFC regarded it as more of a diving-attack aircraft than a patrol or close-combat type; the view from its cockpit was poorer than that from the S.E.5a and this made formation flying more difficult on the Spad. Nevertheless it remains one of the great classic types of the war and in many respects a better weapon than the Spad 13 that should have replaced it but never quite did.

Spad 7
150 hp Hispano-Suiza 8Aa, 180 hp Hispano-Suiza 8Ab, 150 hp Wolseley W.4A Python I, 180 hp Wolseley W.4A Python II

Single-seat fighting scout.

Span 7·822 m (25 ft 8 in) (upper), 7·573 m (24 ft 10⅛ in) (lower); length 6·08 m (19 ft 11⅜ in); height 2·2 m (7 ft 2⅜ in); wing area 17·85 sq m (192·137 sq ft).

Empty weight 500 kg (1,102 lb); loaded weight 705 kg (1,554 lb).

150 hp: Maximum speed at sea level 193 km/h (120 mph), at 2,000 m—187 km/h (116 mph), at 3,000 m—180 km/h (112 mph), at 4,000 m—174 km/h (108 mph); climb to 2,000 m—6 min 40 sec, to 3,000 m—11 min 20 sec; service ceiling 5,500 m (18,045 ft).

180 hp: Maximum speed at 2,000 m—212 km/h (132 mph), at 3,000 m—204 km/h (127 mph), at 4,000 m—200 km/h (124 mph); climb to 2,000 m—4 min 40 sec, to 3,000 m—8 min 10 sec, to 4,000 m—12 min 49 sec; service ceiling 6,553 m (21,500 ft).

Armament: One 0·303-in Vickers machine-gun; some French Spad 7s could carry two 10-kg Anilite bombs on racks mounted on the rear undercarriage legs; it has been reported that some aircraft of No.19 Sqn, RFC, had two 25-lb Cooper bombs in internal racks behind the cockpit.

Manufacturers: SPAD; Blériot Aéronautique; Kellner; L. Blériot (Aeronautics); Mann, Egerton.

Service use: *Western Front*: Sqns 19, 23 and 60. *Mesopotamia*: Sqns 30 and 63. *Palestine*: No.72 Sqn. *Training duties*: Training Sqns 27, 31, 54, 56 and 60; 17th Wing, Beaulieu; 18th Wing Fighting School. Sqns working up: Nos.72 and 81.

Serials: (a) *French production*: A253, 262, 263, 310, 312, 6627, 6633, 6634, 6640—6642, 6649, 6654, 6661, 6662, 6663, 6681—6683, 6685, 6687, 6688, 6690, 6695—6697, 6703—6706, 6709—6714, 6746—6749, 6753, 6759, 8965.
B388 (ex 9611), 1524—1538, 1560—1565, 1573, 1580, 1581, 1586—1589, 1591—1593, 1620, 1622, 1623, 1627, 1628, 1653, 1660, 1661, 1663, 1664, 1667, 1669, 1695—1698, 3457, 3460, 3464, 3471, 3472, 3475, 3488—3493, 3498, 3501—3510, 3515, 3516, 3519, 3520, 3523, 3524, 3528—3535, 3537—3539, 3550—3553, 3556, 3557, 3559, 3560, 3562—3576, 3615, 3616, 3618—3620, 3638—3642, 3645, 3646, 6758, 6761, 6762, 6772, 6773, 6775—6777, 6780, 6787, 6794, 6795, 6796, 6802, 6805, 6816, 6817.
(b) *British production*: A8794—8893, A9100—9161, B1351—1388, B9911—9930.
(c) *Rebuilds*: B861, B879—881, B887, B9969.

A typical Spad 12.Ca 1, S.459. (*Musée de l'Air STAé 02335*)

Spad 12

The Spad 12.Ca 1 was a late-1916 design, created originally by Louis Béchereau of the S.P.A.D. at the request of Georges Guynemer, who asked for a fighter capable of carrying the 37-mm Hotchkiss shell-firing gun. Béchereau designed a single-seater very similar in structure and appearance to the Spad 7 but powered by the Hispano-Suiza 8C engine, a variation of the 200 hp geared Hispano-Suiza 8B in which advantage was taken of the configuration of the spur reduction gear to install a 37-mm Hotchkiss gun between the cylinder banks, its shortened barrel aligned with a hollow airscrew shaft. The completed aircraft was somewhat larger than the Spad 7, had slight stagger, and its shapely engine cowling enclosed the power unit without having to incorporate the protuberant fairings over the camshaft covers that appeared on the Spad 7. A single Vickers machine-gun was mounted in a trough on top of the fuselage, offset slightly to starboard.

According to French sources, 300 of the cannon-armed fighter were ordered for the Aviation militaire with the official S.T.Aé. designation Spa.12.Ca 1, but the type did not see much operational use. Guynemer had his Spad 12 by July 1917 and is reported to have won his 49th, 50th, 51st and 52nd victories on it. René Fonck also flew a Spad 12 and won eleven of his 75 combat victories on it, and the type was flown by Albert Deullin, Fernand Chavannes, Lionel de Marmier and, doubtless, others. No escadrille was ever completely equipped with the type, however: by 1 October, 1918, only eight were operational, yet the Spad 12 was officially regarded as a standard fighter until the end of the war.

An aircraft of such potent striking power could not escape the attention of the RFC. Its weapon installation had been the subject of early RFC interest, for British Requisition No.141 of 19 July, 1917, sought the acquisition of

> '2 Hispano-Suiza engines, of the type known as Modèle Canon, fitted with Spad Gun as well as 37 mm Canon.'

Appropriate instructions were sent to the Officer Commanding the British Aviation Commission in Paris, but confirmation that these engines were ever delivered has yet to be found.

On 10 September, 1917, British Requisition No.182 called for the acquisition of 'one 200 hp Spad'. This may have been retrospective documentation relating to the Spad 13 S.505 that had reached England by 8 September and became B9445. It might equally well have related to a Spad 12, for in its weekly report dated 5 January, 1918, the B.A.S.D. reported to RFC Headquarters: 'One 200 hp "Cannon" Spad machine is being tested and tuned up at Buc and, weather permitting, should be delivered during the forthcoming week.' Carburettor troubles ensued, however, and the aircraft, S.449, could not be delivered until 9 March, when it was flown to No.2 A.D. by Adjudant de Courcelles.

Despite the fact that RFC Headquarters had asked for the aircraft to be flown there, its arrival at Candas seems to have been unexpected (nevertheless, it was promptly given the RFC serial number B6877). On 12 March, 1918, Capt E. R. L. Corballis wrote to Col Blackburn: 'There is a Cannon Spad at No.2 Repair Park, No.6877. Do you know what Brooke-Popham wants done with this machine, and are you dealing with this?' A terse pencilled note was written on this communication, stating: 'G.O.C. is seeing it today, then it is to go to England. 2 A.S.D. are told.' England, in the shape of M.A.2 section of the Air Ministry, emphatically wanted the aircraft and sent the following telegram to RFC Headquarters on 13 March: 'The Spad fitted with 37-mm gun should be flown to Martlesham Heath as soon as possible.'

On 18 March, 1918, Flight Sergeant Piercey flew B6877 to England. It was evaluated at Martlesham Heath, but no document yet found contains an official test report: all that is known is that one of the Martlesham pilots who flew it described it as a 'soggy, nose-heavy job'. From Martlesham the only British Spad 12 was despatched to the Isle of Grain on 4 April, 1918, but crashed en route. As no further mention of the aircraft appears subsequently it must be concluded that the Royal Air Force did not attempt to repair B6877 or to secure a replacement for it.

Spad 12
220 hp Hispano-Suiza 8Cb, No.9523

Single-seat fighting scout.

Span 8 m; length 6·4 m; height 2·55 m; wing area 20·2 sq m.

Empty weight 587 kg; loaded weight 883 kg.

Maximum speed at 2,000 m—203 km/h, at 4,000 m—190 km/h, at 5,000 m—177 km/h; climb to 2,000 m—6 min 3 sec, to 3,000 m—10 min 2 sec, to 4,000 m—15 min 42 sec, to 5,000 m—23 min 13 sec; ceiling 6,850 m; endurance 1¾ hr.

Armament: One 7·7-mm Vickers machine-gun and one 37-mm Hotchkiss shell-firing gun.

Manufacturer: SPAD.

Service use: Flown on test at Martlesham Heath.

Serial: B6877.

B9445 was a Spad 13 that was apparently supplied to the RFC as a sample aircraft. (*K. M. Molson*)

Spad 13

On 20 February, 1917, Maj F. L. Scholte of the British Aviation Commission reported to RFC Headquarters that the S.P.A.D. were building 20 new single-seaters powered by the 200 hp geared Hispano-Suiza 8B engine and armed with twin synchronised Vickers guns. The first of these was reported to be ready by 13 March and only awaiting favourable weather for its trials. Thus did the RFC first learn of the Spad 13, which in appearance closely resembled the earlier Spad 7, but could be distinguished by its more substantial fuselage, the curved trailing edge to its rudder, and the left-handed airscrew of its geared engine.

By 27 April, 1917, at least one Spad 13 was operational at La Bonne Maison aerodrome near Fismes. On the instructions of Brig-Gen Brooke-Popham, Capt W. J. C. K. Cochran Patrick visited La Bonne Maison on 29 April to inspect it and report. Patrick gave a good description of the Spad 13, quoting a speed of 190 km/h at 4,000 m, which altitude was reached in 11 minutes. He also reported that 'The French pilots were very enthusiastic over the 200 hp geared Hispano-Suiza Spad.' Nevertheless, the introduction of production Spad 13s in significant numbers proved to be slow and protracted, and in service the type was severely handicapped by the problems that plagued its 200 hp Hispano-Suiza engine. Later aircraft had the 220 hp Hispano-Suiza 8Bc or 8Be with increased compression ratio and improved performance.

As with the Spad 7, so with the Spad 13 was the B.A.S.D. in Paris commendably quick off the mark. Within a month of Patrick's report its officers had secured for the RFC a specimen of the new Spad; this was S.498 which, delivered by road to No.2 A.D. at Candas on 1 June, 1917, became B3479 in the RFC. Under test it returned spectacular performance figures, and RFC Headquarters naturally wanted Spad 13s rather than the Spad 7s then on order. B3479 went to No.19 Squadron on 9 June and flew operationally with that unit; in various combats it registered several victories, notably in the hands of Capt F. Sowrey and Lt G. S. Buck. This Spad 13 subsequently went to No.23 Squadron, with which it remained in operational use until 23 March, 1918, when it was destroyed, its total flying time being 85 hr 19 min.

For production quantities of Spad 13s, the RFC turned to the Kellner company. The existing contract for one hundred and twenty Spad 7s plus the equivalent of 30 more in spares was altered to provide for 60 of the aircraft to be Spad 13s. Additionally the RFC gave a written undertaking to order a further 100. The first Kellner-built Spad 13s for the RFC began to come forward in November 1917 and it was hoped that 60 would be delivered by the end of the year. However, production was delayed by shortages of materials and diminished by the unwarranted insistence of the French authorities on taking half of Kellner's output, even of aircraft built to meet RFC contracts. Not until the week ending 23 March, 1918, did Kellner deliveries to the RFC total 61. By that time, official documents were mentioning a second contract for only 70 aircraft, and even that was reported to be reduced to 25.

A Spad 13 for the RFC at No.2 A.D, Candas. (*RAF Museum*)

The great majority of the Spad 13s delivered to the RFC had the original fully rounded wingtips and early centre-section bracing. One or two, such as B6882 (S.4563) had the later blunt-tipped wings, but did not arrive until the type was on the point of withdrawal. It had been intended that all Spads delivered under the later Kellner contract would have the modified wings.

In the RFC only No.23 Squadron was fully equipped with the type. Replacement of that unit's Spad 7s began in December 1917, and certain operational snags arose during that winter. The aircraft were, for instance, given S.E.5-pattern wheels because the original French wheels frequently buckled on landing. For a time it seemed that No.23 would go on flying its Spad 13s until June 1918 in order to avoid delaying the arrival in France of No.85 (S.E.5a) Squadron and other Dolphin squadrons. However, No.23 itself was completely re-equipped with Dolphins by 4 May, 1918, and the Spads were returned to the Depot.

The Spad 13s of No.23 Squadron at La Lovie. (*Ministry of Defence H1122*)

Although the Spad 13 has over the years acquired a reputation grossly inflated by American pulp magazines, it was never the operational success it deserved to be. In French service the Spad 7 had to remain in use well into 1918 to make good the gaps that would otherwise have appeared in the Escadrilles de Chasse if reliance had been placed solely on the Spad 13. The RFC claimed no spectacular successes with the type during its four months of service with No.23 Squadron.

The Spad 13 was not built outside France, but production in England may have been considered. Before Kellner-built Spad 13s became available to the RFC in France a specimen of the type was delivered and was sent to England. This was S.505, which became B9445 in the RFC; it had reached Brooklands by 8 September, 1917. It was expected to go to the Aeroplane Experimental Station at

Officers of 'C' Flight, No.23 Squadron, in front of a Spad 13 of the squadron, its rounded wingtips clearly visible, at Ham near Peronne, February 1918. Left to right: Lt Keary, Lt Fielder, Capt W. M. Fry (Flight Commander), Lt Stringer and Lt Trudeau. (*Wing Commander W. M. Fry, MC*)

Martlesham Heath but spent the remainder of September with the Aircraft Inspection Department. As far as can now be determined B9445 never did go to Martlesham, and its ultimate fate is unknown.

Spad 13

200 hp Hispano-Suiza 8Ba, 8Bb, 8Bd, or 220 hp Hispano-Suiza 8Bc or 8Be Single-seat fighting scout.

Span 8·25 m (original), later 8·08 m; length 6·25 m; heignt 2·6 m; wing area 21·11 sq m (original), later 20·2 sq m.

Empty weight 601·5 kg; loaded weight 856·5 kg

Maximum speed at 2,000 m—218 km/h, at 3,000 m—214 km/h, at 5,000 m—203 km/h; climb to 2,000 m—4 min 40 sec, to 3,000 m—7 min 50 sec, to 5,000 m—18 min 30 sec; ceiling 6,850 m; endurance 1 hr 40 min. Performance figures with 220 hp engine.

Armament: Two 7·7-mm Vickers machine-guns.

Manufacturers: SPAD, Kellner.

Service use: *Western Front*: Sqns 19 and 23.

Serials: B3479, B6731—6739, B6835, B6838—6862, B6864—6867, B6872—6875, B6878—6886, B9445.

Vickers Boxkite

Aviation at Brooklands began in September 1907 when Alliott Verdon Roe started his experiments with his first primitive biplane. Others followed him, their names virtually a roll-call of Britain's aviation pioneers, and in 1910 the first flying school to set up in business at Brooklands was opened by Mrs Hilda Hewlett and Gustav Blondeau, whose partnership later progressed to building aircraft. When 1911 began, the Hewlett & Blondeau flying school had been joined by the Avro, Bristol, Deperdussin, Hanriot, and Spencer schools; and early in 1912 Vickers Ltd established a flying school.

The original equipment of the Vickers school consisted of the Vickers monoplanes Nos.2, 3, 4 and 5, but these were not entirely suitable for ab initio instruction and in mid-July the Vickers school acquired a so-called racing Farman biplane built by Hewlett & Blondeau. Presumably this was a Farman boxkite

A Vickers-Farman biplane near the Vickers sheds at Brooklands, some time before the 1914–18 war. (*RAF Museum*)

biplane in which the lower wings had been shortened by one bay. On 16 October, 1912, an aircraft described as 'No.2 Farman' was reported to be flying with the Vickers school, and next day 'No.1 Farman' was reported wrecked.

Sunday 1 December, 1912, saw the first reported appearance of a biplane described as the new Vickers-Farman, a designation that, at the time, need not have connoted anything more than its connection with the Vickers school: the identity of its makers was not recorded. Similarly, on 25 February, 1913, a 'new school biplane' was reported in use; on 10 May, Capt H. F. Wood was testing a new biplane fitted with a new Vickers seven-cylinder radial engine; and next day Archibald Knight was testing yet another new biplane powered by a 70 hp Gnome. This last joined the Vickers school, but the radial-powered biplane evidently did not.

On Monday 9 June, 1913, Knight was instructing Maj W. Sefton Brancker and 2nd Lt C. F. Beevor on 'No.19 biplane', while Harold Barnwell was flying biplane No.20. Next month a further new biplane, No.21, was reported, on 10 July. These numbers were evidently Vickers numbers, suggesting that these three aircraft at least had been built by Vickers themselves. In mid-September there appeared a further new biplane, No.26, which had a Vickers radial engine; this aircraft apparently saw some use as an instructional machine. It may have been the so-called Vickers Pumpkin, which was a Farman airframe to which was added a corpulent rudimentary nacelle inadequately housing two side-by-side seats that were staggered laterally, the port one lying slightly behind the one on the starboard side, and the flying controls being the responsibility of the occupant of the starboard seat. The Pumpkin itself was an equal-span biplane of typical Farman configuration; its power unit was described as a 50 hp Vickers radial engine, which was in fact the Vickers-Boucier. The possibility that the Pumpkin was later than and different from No.26 cannot be ruled out, for the Pumpkin itself was described as 'the new Vickers School biplane' in *The Aeroplane* of 22 January, 1914.

Clearly the Vickers school went on using its Farman-type biplanes until the outbreak of war, and may have added to their number in 1914. The foregoing paragraphs show that the school used considerably more than three Farmans during its existence, and a Vickers advertisement of the time shows that at least one of them was a two-bays-plus-extensions biplane. The company probably modified the aircraft to suit their own requirements and apparently built at least three themselves. On the outbreak of war the Vickers school closed, and two of its Farman-type biplanes, described, as they frequently were, as Vickers Boxkites, were assimilated by the RFC as 639 and 642 on 17 August, 1914. They were allotted to No.1 Squadron for training purposes but did not last long: 642 was struck off charge on 14 October, 639 on 20 October.

Boxkite
50 hp Gnome, 70 hp Gnome

Two-seat primary trainer.

Manufacturers: Some may have been made by Hewlett & Blondeau, and it seems that at least three of all those used by the Vickers school were built by Vickers.

Service use: *Training duties*: No.1 Sqn.

Serials: 639, 642.

No.649, one of the pre-production Vickers Fighting Biplanes delivered to the RFC in 1914. (*RAF Museum*)

The early Vickers Gun-carrying Biplanes

The first Vickers gun-carrying biplane to be built was designed to meet an Admiralty request for such an aircraft. It was exhibited at the Olympia Aero Show that opened on 14 February, 1913, being a pusher biplane of unequal span and unprepossessing appearance. Its armament was a single belt-fed Vickers machine-gun with limited movement, mounted on the nose of the nacelle. Although grandiosely named *Destroyer* it crashed on its first attempted take-off.

This biplane was at some time retrospectively designated E.F.B.1. Other designs followed, all powered by the 100 hp Gnome Monosoupape rotary engine. Those that came to be known as E.F.B.2 and E.F.B.3 were virtually contemporary early in 1914. E.F.B.2 had warping wings and first flew at Bognor in December 1913; E.F.B.3 had ailerons and an impossible installation of its Vickers gun. Both were tested at Farnborough but neither, it seems, went to the RFC. Here it should be noted that at some time between 1 April and 31 August, 1913, a Vickers biplane of unspecified type was ordered by the Director of Army Contracts at a price of £1,800. Perhaps it was hoped that E.F.B.2 or E.F.B.3 might meet the War Office requirement, but confirmation that either ever did has yet to be found.

By early May 1914, E.F.B.3 had been extensively modified, its nacelle having been redesigned and a new fin and rudder replacing the original curved rudder. As far as can be determined, the modified E.F.B.3 did not go to the RFC either.

Precise details of the line of development of the Vickers gun-carrying biplanes remain disappointingly few and obscure as far as this formative period is concerned. What is known is conveyed by two surviving (but undated) Vickers design drawings. No.1991V delineates 'Aeroplane fuselage, F.B.2, 3, 4, 5 and 7'; No.2019V is of 'Aeroplane fuselage, F.B.6'. Both are drawings of basically similar nacelles for two-seat pusher biplanes, each having a pillar mounting for a gun in the forward cockpit. F.B.6 had different petrol and oil tanks, and its pilot was to have a wheel control for the ailerons whereas the nacelle in drawing No.1991V was to have only a conventional universally-jointed stick.

These drawings suggest that Vickers intended to build five aircraft of one form with works identities of F.B.2, 3, 4, 5 and 7 and one of different design as F.B.6. Support for this postulate is provided by surviving photographs of a two-bay equal-span biplane with F.B.4 stencilled in small characters under its wingtips and elevators, and of a generally similar but unequal-span biplane marked in like fashion F.B.6. What F.B.1 may have been can only remain conjectural. Contemporary references in the aeronautical press indicate beyond doubt that F.B.6 was at Brooklands on or about 1 July, 1914, and went to Farnborough on 9 July, returning the same day. On 17 July another new Vickers gun-carrier was flown from Joyce Green to Brooklands; this was evidently one of the five equal-span aircraft. All known aircraft of this group, F.B.6 included, had wooden interplane struts and semi-circular tailplanes.

One of the equal-span aircraft was flown by Harold Barnwell to Farnborough on Monday 20 July, 1914, to be tested by the AID. Next day it was flown by three RFC officers: Maj Robert Brooke-Popham, Capt R. Cholmondeley and 2nd Lt T. O'Brien Hubbard. Opinion was not unfavourable, but Brooke-Popham's thoughtful report contained some reservations. Nevertheless, he recommended that an example of the Vickers should be bought for experimental work with machine-guns. On 23 July, Lt-Col Sykes submitted the three pilots' reports to the D.G.M.A. and recommended the purchase of two aircraft.

The RFC wanted a further test of the Vickers, but on 25 July the aircraft was blown over on to its back and damaged. International events then overtook matters. Within a few weeks of the start of the war, six Vickers pushers were acquired by the RFC with the official serial numbers 649, 664, 682, 686, 704, and 747. It is known that 664 was F.B.4 and 704 F.B.6, consequently it is reasonably certain that the other four were F.B.2, 3, 5 and 7, though not necessarily in that order. At least four of the six had stronger tail-booms than originally fitted; six Vickers guns with pivot mountings and parallel-motion sights were allotted for the aircraft on 7 October, 1914.

Third of the pre-production Vickers Fighting Biplane series was F.B.4, marked as such on the undersides of its wingtips and elevators. It is here seen at Farnborough on 21 October, 1914, fitted with a gun mounting and an air-speed indicator. By that date it had acquired the RFC serial number 664. (*RAE 618; Crown copyright*)

Vickers F.B.6 at Farnborough on 27 October, 1914, with gun mounting installed and wearing the RFC serial number 704. (*RAE 626; Crown copyright*)

Possibly the first Vickers gun-carrier actually to join the RFC was 664, reported as taken on charge on 10 September, 1914. Initially with No.6 Squadron, it went to No.7 Squadron, then about to form, on 24 September. It was joined in No.7 by 649 on 14 October, by 682 next day, and subsequently (but apparently briefly) by 686 and 704. None of these Vickers biplanes went to France with No.7 Squadron on 8 April, 1915; Nos.649 and 664 had been converted to have dual control in January 1915, in which form 649 went to No.11 Squadron in February. No.686 had been struck off charge on 9 November, 1914, 682 on 13 February, 1915.

To No.664 has been attributed the first operational action by a Vickers Fighter. This occurred on Christmas Day 1914, when the aircraft was on stand-by at Joyce Green. 2nd Lt M. R. Chidson, with Corporal Martin as gunner, took off to attack a Taube that had ventured up the Thames. Some of Martin's shots hit the German monoplane, and later events suggested that he had shot it down.

No.747 went to the RNAS at Dover on 9 February, 1915, to be exchanged for a naval Avro 504; five days later it was recorded as being on the strength of No.1 Naval Aeroplane Squadron. On 21 April, 1915, it was renumbered 862, thus assuming the identity of an RNAS Vickers F.B.5 that must have been deleted some time between 12 March, 1915 (when the original 862 was at Dunkerque) and 21 April.

The dual-control conversions saw some service as trainers, for it is known that 649 at least was used by No.2 Reserve Aeroplane Squadron. This early Vickers survived until 2 September, 1915.

Gun-carrying Biplanes
100 hp Gnome Monosoupape
Two-seat fighter-reconnaissance biplane.
Armament: One 0·303-in Vickers machine-gun.
Manufacturer: Vickers.
Service use: Sqns 7 and 11. *Training*: No.2 Reserve Aeroplane Squadron.
Serials: 649, 664, 682, 686, 704, 747. After transfer to the RNAS 747 was renumbered 862 on 21 April, 1915.

Vickers F.B.5

Contemporary with the RFC's early Vickers gun-carriers was a similar two-seat pusher biplane that was delivered to the RNAS in September 1914. In that Service it received the serial number 32, whence it became officially known as the Vickers Type 32. Much of its airframe appeared to be identical with those of the RFC aircraft, but the centre-section struts were of steel tubing and the nose of the nacelle was of different design. No.32 retained the semi-circular tailplane and original short ailerons common to all the equal-span gun-carriers, and it was armed with a Vickers gun on a pillar mounting.

The surviving Vickers works drawing that most closely resembles No.32 bears no type or works number, but this aircraft, more than any other, deserves to be regarded as the prototype of the Vickers E.F.B.5. The RNAS ordered twelve Vickers gun-carriers, to be numbered 861—872, on 14 August, 1914, and the RFC followed suit on 19 September by ordering 1616—1627; these were subsequently augmented by 1628—1639 and 1640—1651. These were two-bay biplanes of equal span, powered by the 100 hp Gnome Monosoupape engine, and of typical pusher configuration with the tailbooms converging in plan to meet on the rudder's axis of rotation.

Deliveries to the RFC began in mid-December 1914, production aircraft being distinguished by nacelles similar to that of No.32 and by redesigned tailplane and elevators of greater area and rectangular planform. The original constant-chord rudder was fitted to the early production Vickers at least up to 1633, but all later aircraft had a greatly enlarged rudder of rounded outline. Although it had been decided in November 1914 that the standard weapon for the RFC's Vickers

No.1616 was the first production Vickers F.B.5 for the RFC, and is here seen at Farnborough on 24 December, 1914, with Frank Goodden at the controls. It went to the 1st Aircraft Park, in France, on 12 April, 1915, and was issued to No.5 Squadron two days later. On 9 June it returned to the Aircraft Park, apparently for repair, and rejoined No.5 Squadron on 6 September. It was again sent back to the Aircraft Park on 18 September and was struck off charge on 20 September. (*RAE 677; Crown copyright*)

No.1621 went to No.2 Squadron on 5 February, 1915, but moved on to No.16 Squadron on 10 February; it was the first Vickers F.B.5 to join the RFC in France, and had the early constant-chord rudder. It came down intact behind the German lines on 2 March, 1915, and in this photograph is seen in German hands, its Vickers gun conspicuously manned.
(Egon Krueger)

Fighting Biplanes was to be the Lewis magazine-fed gun, several of the early production aircraft still had the belt-fed Vickers gun. Problems of mounting the Lewis persisted for some months, and the wheel-traversing Vickers No.2 Mark I mounting, although some improvement upon the No.1 Marks I and II mountings, was far from ideal in combat, and squadrons evolved their own.

Vickers F.B.5 No.2883 at Farnborough in August 1915. It was allotted to No.18 Squadron on 24 August but did not go to France with that unit. In October 1915 it was on the strength of No.10 Reserve Aeroplane Squadron at Joyce Green. (*RAE 896; Crown copyright*)

At some indeterminate time the production aircraft was designated Vickers E.F.B.5, abbreviated to F.B.5. To the RFC the type was the Vickers Fighting Biplane or Vickers Fighter: the name 'Gun Bus', so long associated with the type, appears to have been prewar journalese. It is not to be found in any official document yet discovered; it was never used by those men of the RFC who wrote anything about it; its use is therefore best avoided.

First of the fighting Vickers to go to the RFC in France was 1621. Initially sent to No.2 Squadron on 5 February, 1915, it was transferred to No.16 Squadron five days later. Its career was brief, for it fell into German hands intact on 2 March. The first squadron to operate the F.B.5 in any numbers was No.5, which received a total of nine between 19 March and 21 June, 1915; of these only three remained on the squadron's strength on 30 June.

Production expanded early in 1915. Further orders were placed with Vickers, and in February the firm entered into an agreement with the French Darracq

Maintenance on 1623 of No.5 Squadron in the Field. The operation consisted of replacing the tail-booms, which had been shot through in combat. The combination of rudder stripes and fin roundel is interesting. This F.B.5 joined No.5 Squadron on 1 April, 1915, but was wrecked on 29 June, 1915, when it stalled and crashed, killing Lt Barfield and 1/AM Sutcliffe.

company for the manufacture of Vickers designs in France. The RNAS ordered twelve F.B.5s, the RFC fifty (5454—5503), from Darracq. First delivery to the RNAS occurred on 15 June, 1915, but the RFC had to wait until 19 August for 5454. Darracq deliveries were slow, which may be why the six F.B.5s 3601—3606, intended for the RNAS, were transferred to the RFC without engines in late September and early October, becoming 5074, 5075, 5078, 5079, 5083 and 5084.

On 25 July, 1915, No.11 Squadron arrived in France equipped throughout with Vickers Fighters. It brought eight aircraft with it and acquired three more by the end of the month. Although combat successes were won by the determination of the crews, the Vickers F.B.5 did not really have the performance needed to cope with the better German aircraft, and combat reports of the period repeatedly record the fact that the Vickers' adversaries escaped simply by superior speed. The indicated air speed of an F.B.5 at combat altitude was 50—55 mph, and no doubt even that performance suffered owing to the difficulties experienced with the Monosoupape engine. The situation was not improved by the use of Monosoupapes made in England by Peter Hooker Ltd, for these proved inferior to the French-made engines. A trial installation of a 110 hp Le Rhône in one aircraft produced a slight improvement in performance but was not followed up.

Improvements in the gun mounting were made as time went on, but it appears that the most efficient mounting was that evolved, by mid-October 1915, by No.11 Squadron. To supplement the Lewis gun many observers carried a Lee-Enfield rifle, but in the autumn of 1915 a mounting for a second Lewis was fitted to some F.B.5s. This was carried on a pyramidal mounting about 2 ft high, fitted between the cockpits so that the gun could be fired by either the pilot or the observer. Such aircraft suffered a reduction in performance, and their directional stability was adversely affected. This mounting was apparently abandoned by early March 1916.

By then another Vickers squadron, No.18, was operational, having arrived in France on 19 November, 1915; but by early 1916 the gallant old pusher was outclassed and obsolete. On 16 March Trenchard wrote to D.D.M.A.:

> 'The Vickers Fighters in their present condition are now hopelessly outclassed and must be considered as quite out of date . . . It is essential that these machines be replaced by something better at an early date.'

Although the F.B.9 was by then in prospect, it was not enough of an improvement on the F.B.5 to represent an adequate replacement. F.B.5s were still coming forward, latterly from the second Darracq-built batch, 7510—7519, but few, if any, can have seen service in France.

A squadron-designed gun mounting on a Darracq-built F.B.5.

After the advent of the Vickers F.B.9 some F.B.5s with training units were fitted with V-strut undercarriages similar to that of the F.B.9. One such was 5677.

Most of the F.B.5s were sent to England after withdrawal from operations and gave valuable service with training squadrons. Some of them acquired simple V-strut undercarriages similar to that of the F.B.9.

F.B.5

100 hp Gnome Monosoupape; individual installations of 110 hp Le Rhône 9J and 110 hp Clerget 9Z

Two-seat fighter-reconnaissance.

Span 36 ft 6 in; length 27 ft 2 in; height 11 ft; wing area 382 sq ft.

Empty weight 1,220 lb; loaded weight 2,050 lb.

Maximum speed at 5,000 ft—70 mph; climb to 1,500 ft—5 min, to 5,000 ft—16 min; service ceiling 9,000 ft; endurance 4½ hr.

Armament: One 0·303-in Lewis machine-gun (Vickers gun on a few early aircraft), frequently supplemented by one 0·303-in Lee-Enfield rifle. Some F.B.5s had a second Lewis gun mounted between the cockpits.

Manufacturers: Vickers, Darracq.

Service use: *Western Front*: Sqns 2, 5, 7, 11, 16 and 18. *Training duties*—Reserve/Training Sqns 2, 6, 8, 9, 10, 14, 19 and 24; Central Flying School; Machine-Gun School, Hythe; School of Instruction, Reading.

Serials: 1616—1627, 1628—1639, 1640—1651, 2340—2347, 2462—2467, 2865—2868, 2870—2883, 4736, 5074, 5075, 5078, 5079, 5083, 5084, 5454—5503, 5618—5623, 5649—5692, 5729 (single armoured F.B.5 with 110 hp Clerget), 7510—7519.

The Vickers F.B.7 as it first appeared, with engines wholly exposed. (*K. M. Molson*)

Vickers F.B.7

It would appear that Vickers' sensible belief in the operational usefulness of gun-carrying aircraft was so strong that the firm considered that proportionately greater destruction could be wrought by an aircraft carrying a large-calibre gun. Shortly after the war started Maj H. F. Wood of Vickers engaged L. Howard Flanders to design an aircraft capable of carrying a Vickers 1-pounder quick-firing gun and having some armour for the protection of the gunner.

Under the Vickers type number F.B.7, Howard Flanders designed a large biplane remarkably like a scaled-up Flanders B.2. To obtain the power required to lift the substantial military load, two 100 hp Gnome Monosoupape engines were installed as tractors, driving opposite-handed airscrews. The upper wing had long extensions; these alone had marked dihedral and carried the ailerons, which had pronounced wash-out of incidence. The pilot's spacious cockpit was located behind the wings, a long way aft of the gunner's position in the nose. To support the gun a substantial tripod was bolted to the front cockpit's floor, which was a $\frac{3}{8}$-in thick sheet of extra-hard duralumin, optimistically regarded as bullet-proof. The gunner's seat was fixed to the rotating mounting and traversed with it, his cockpit's ample dimensions permitting him to make a full 360 deg circle.

The F.B.7 first flew in the summer of 1915, and at some time its engine installation was modified. Fore-and-aft mountings replaced the original overhung mountings, and oil-catchment rings were fitted round the engines. Behind these were added bulky nacelles of circular cross-section that must have seriously reduced the crew's field of view.

For official evaluation the F.B.7 went to CFS at Upavon. Among those who flew it while there was Capt (later Air Commodore Sir) Vernon Brown; writing in 1920 he recalled that it was 'a horrid machine with two 100 hp Monos'. Notwithstanding this unfavourable opinion, and despite the F.B.7's undistinguished performance, the War Office ordered a batch of twelve, almost certainly to be numbered 5717—5728. Vickers subcontracted the order to the Darracq company's London works at Fulham, and the design was extensively modified. A completely new fuselage was designed, having a rectangular cross-section throughout its length instead of the more complex pentagonal/triangular section of the prototype. Apparently the crew sat side by side in a wide cockpit ahead of the wings, but it is not known whether this betokened any revision of the armament, for the short, blunt nose revealed no provision for mounting a heavy gun. Both upper and lower wings had dihedral, which now started immediately outboard of the engines. A long-span tailplane of new design with elliptical tips was fitted; it was braced from a pair of kingposts above, producing an untidy complex of bracing struts and cables at the rear of the fuselage. Because Gnome engines were scarce and in great demand, the first production aircraft, 5717, was completed with two 80 hp Renaults. It was designated Vickers F.B.7A.

With power thus reduced by 20 per cent, the F.B.7A's performance was so diminished that Vickers had to recognise that the aircraft could not hope to meet any operational requirement. The firm therefore asked the War Office to reduce the contract to one aircraft only, an action that must have been taken and agreed early in 1916, possibly in a War Office letter dated 12 February, 1916, for Vickers wrote to the Darracq company confirming the cancellation arrangements on 16 February. These provided for Vickers to pay Darracq £1,300 for the first

Vickers F.B.7 with quick-firing gun mounted, cowlings on the engines, and bulky nacelles. (*Imperial War Museum HU.1727*)

The only Vickers F.B.7A to be completed, No.5717. (*British Aircraft Corporation*)

F.B.7A as it stood, and a further £3,550 for work done on the other eleven aircraft.

It is uncertain whether 5717 had been completed at that date. It went to Weybridge and was seen at Brooklands, but was not officially allotted to CFS for trials until 1 June, 1916. Confirmation that it went there has yet to be found, but any official trials that might have been undertaken could only confirm that the F.B.7A was operationally useless.

F.B.7

F.B.7—two 100 hp Gnome Monosoupape
F.B.7A—two 80 hp Renault

Two-seat heavy gun carrier.

F.B.7: Span 59 ft 6 in (upper), 37 ft 6 in (lower); length 36 ft; wing area 640 sq ft.

F.B.7: Empty weight 2,136 lb; loaded weight 3,196 lb.

F.B.7: Maximum speed at 5,000 ft—75 mph; climb to 5,000 ft—18 min; endurance 2½ hr.

Armament: One Vickers 1-pounder automatic gun.

Manufacturers: Vickers, A. Darracq.

Serials: 5717—5728 (5718—5728 cancelled February 1916).

Vickers E.S.1

Shortly after the war started Harold Barnwell, Vickers' chief test pilot, designed a single-seat scout intended to achieve a high performance. It seems that he designed and built this aircraft in secret, obtaining the use of a 100 hp Gnome Monosoupape engine by subterfuge. His aircraft has been described as having a well streamlined fuselage of circular cross-section with short-span unstaggered wings. It is uncertain whether Barnwell's scout ever flew, for accounts of its demise differ, but the aircraft was badly damaged, either by undercarriage collapse or by overturning.

Vickers E.S.1 in its basic form with Monosoupape engine and long-chord cowling.
(*Vickers Ltd*)

Vickers E.S.1 No.7509 at Central Flying School. (*A&AEE; 1643; Crown copyright*)

This accident inevitably disclosed the existence of the Barnwell scout, but its design was regarded as sufficiently promising for its revision to be entrusted to Rex Pierson, who was then a young member of the Vickers drawing-office staff. The modified aircraft was built with the Vickers designation E.S.1, the initials signifying Experimental Scout, and was completed in August 1915. Photographs that have repeatedly been identified as depicting the original E.S.1 show a single-bay biplane with wings of equal span and square-cut tips on a portly fuselage of circular cross-section, fully-faired throughout its length. The Monosoupape engine had a full-circular cowling of unusually deep chord. The cockpit was placed between the rear centre-section struts, in such a position that the pilot's field of view must have been minimal, for it was restricted by the wings and the girth of the fuselage side fairings.

By November 1915 new wings with rounded tips and lengthened ailerons had

Vickers E.S.1 with Clerget engine and without clear-view panel in centre-section.

been fitted, and on 6 November Harold Barnwell flew the E.S.1 over Hendon, creating something of a sensation with a series of climbing loops. The E.S.1 went to No.1 Aircraft Depot at St-Omer on 27 December, 1915, for evaluation by the RFC in France. On arrival it had no official serial number, a deficiency that apparently so offended some tidy official mind that on 2 January, 1916, it was given the number 5127 in the block 5001—5200 that had been allotted to the BEF for aircraft purchased in France.

This was beyond doubt the aircraft about which R. R. Money wrote the following passage in his book *Flying and Soldiering*:

> 'One day, Reggie Howitt brought the first Vickers Bullet, a single-seat tractor biplane with a Monosoupape engine and a deep monocoque [*sic*] fuselage. The depth and bulginess of this fuselage allowed the pilot hardly any view except skywards, and Howitt, who had come over from England in bad weather said: "You can't see anything at all unless you tilt the - - - - - - sideways." In addition, its Monosoupape engine was well cowled in with no gap in the cowling at the bottom to let surplus petrol get away. In the hands of a good rotary pilot like Howitt this gave no trouble; but when others took the machine up, they usually choked the engine, thereby filling the cowling with petrol which immediately caught fire. After several incidents of this nature the machine was eventually crashed by Captain Playfair, who, as a Bristol Scout pilot, had been asked to test the latest effort.'

It seems that the E.S.1 was not too seriously damaged, however: it was reported to have been damaged on 5 January, 1916, but was serviceable again on 7 January. On 10 January Brooke-Popham wrote to ADMA:

> 'The Vickers Scout with Monosoupape engine has been tested out here. It has been decided that with the present engine, it is dangerous to fly owing to the risk of fire and orders have been given for it not to be flown any more.'

Four days later D.D.M.A. asked for the aircraft to be returned to England in a packing case, and this request was implemented on 23 January.

That might have been expected to be the end of the road for the Vickers E.S.1 but, remarkably, this was not so. Six more aircraft were ordered under Contract No.87/A/344, the serial numbers 7509 and 7756—7760 being allotted on 18

Vickers E.S.1 with 110 hp Clerget and transparent centre-section. This is said to have been photographed at Savy in July 1916, with No.11 Squadron, but confirmation is lacking.

The E.S.1 No.7756 is known to have been with No.11 Squadron. It was flown to France on 18 May, 1916, but had a short career, being returned to No.2 A.D. from No.11 Squadron on 7 June, at which date it was reported to be in poor condition.

February, 1916. These differed from 5127 principally in being armed with a single Vickers gun sunk into the top decking just inboard of the port centre-section struts; they also had streamline Rafwires for interplane bracing in place of the wire cables of 5127. In May 1916 CFS tested 7509 with a 110 hp Clerget and 7756 with a 110 hp Le Rhône, each returning excellent performance figures. Flying characteristics were criticised, however, both aircraft being described as tiring to fly and requiring great care on landing. A later note on 7756 said that it was very unstable and tail heavy.

First to go to France was 7756, which was flown to No.1 A.D. at St-Omer by Maj J. E. Tennant on 18 May, 1916. It was sent to No.11 Squadron, which returned it to No.2 A.D. in poor condition on 7 June. Also in France at that time were 5127 and 7758, which were with No.32 Squadron on 4 June, having apparently gone to France with the squadron on 28 May. No.5127 disappears from the record after 7 June, and 7758 was returned to No.1 A.D. on 22 June, the Depot having been instructed by RFC Headquarters to remove and store the engine (provided it had done less than 15 hours), and to keep the gun and its Vickers-Challenger interrupter gear before returning the airframe to England. While with No.32 Squadron, 7758 was flown by the Commanding Officer, Maj L. W. B. Rees.

Surviving official papers indicate that a fourth Vickers Scout had gone to France with No.70 Squadron, obviously with 'A' Flight of that unit, which arrived in France on 24 May, 1916. This must have been 7757, but it, too, was due to be sent back to England before the end of June. Advising D.A.E. on 27 June of the intention to return the Vickers Scouts, RFC Headquarters said '. . . it is found that these machines are not of military value, owing to the difficulty of seeing out of the machine'.

The intention to return the E.S.1s to England may have thwarted the despatch to France of 7509 and 7760. These two had been officially allotted to the RFC in France on, respectively, 17 and 16 June, 1916, but on 19 June RFC Headquarters telegraphed to D.A.D.A.E., subsequently confirming by memorandum that 'Vickers Scouts 7509 and 7760 are not required by the Expeditionary Force, and no more of this type need be allotted.' The allotments were therefore cancelled on 21 June, when both aircraft were allotted to No.50 (Home Defence) Squadron at

The occupant of this Monosoupape-powered E.S.1 is said to have been Maj L. W. B. Rees, VC, which suggests that the aircraft may have been one of the two E.S.1s that were used by No.32 Squadron, probably 5127.

Dover. The only other E.S.1, 7759, was also used by No.50 Squadron for a time, as was the durable 5127, which was with the squadron until 23 March, 1917, when Capt A. J. Capel set out to fly it from Bekesbourne to Farnborough, but had to make a forced landing at Crondall.

The two last aircraft of the batch, 7759 and 7760, differed from all the others in having no belly fairing on the fuselage; they and at least one of the others had a large clear-view cut-out in the centre-section. It is difficult to be positive about

Vickers E.S.1 Mk II No.7759 with No.50 (H.D.) Squadron.
(*Peter Liddle's 1914–18 Personal Experience Archives, presently housed within Sunderland Polytechnic*)

Vickers E.S.1 Mk II No.7760 at Gosport with Capt E. L. Foot in the cockpit.

type designations for the production aircraft: the designation E.S.1 Mk II has been attributed to the aircraft fitted with the 110 hp Clerget engine but probably applied to all the production machines. A designation E.S.2 also gained currency from an early date, and may have related to 7759 and 7760. In the RFC the type was known officially as the Vickers Scout, but the nickname Bullet seems to have had some unofficial usage.

E.S.1
100 hp Gnome Monosoupape, 110 hp Le Rhône 9J,
110 hp Clerget 9Z

Single-seat fighting scout.
Span 24 ft 5½ in; length 20 ft 4 in; height 8 ft; wing area 215 sq ft.
Weights and performance:

	Monosoupape	Le Rhône	Clerget
Weight empty, lb	843	1,052	981
Weight loaded, lb	1,295	1,600	1,502
Maximum speed, mph			
at sea level	118	117·2	112·2
at 5,000 ft	114	—	109
at 8,000 ft	—	109	106
Climb to (min and sec)			
3,280 ft	3 45	— —	— —
5,000 ft	6 40	5 15	6 25
10,000 ft	18 00	14 45	18 00
Service ceiling, ft	15,500	16,500	12,500
Endurance	3 hr	2 hr	3¼ hr

Armament: One 0·303-in Vickers machine-gun.

Manufacturer: Vickers.

Service use: *Western Front*: Sqns 11, 32 and 70. *Home Defence*: No.50 Sqn.
Training: Central Flying School.

Serials: 5127, 7509, 7756—7760.

The prototype Vickers F.B.9, later numbered 7665. This aircraft had a successful operational career with No.11 Squadron. (*Vickers Ltd*)

Vickers F.B.9

Towards the end of 1915 the Vickers design team created a somewhat refined development of the F.B.5 that emerged in December. The new type retained the 100 hp Gnome Monosoupape but had new wings, tailplane and elevators; all had rounded tips, and the span was reduced to 33 ft 9 in. A new design of nacelle was used, having a traversing and elevating gun mounting, and a simplified V-strut undercarriage. All external bracing wires were streamline-section Rafwires.

The performance obtained on maker's trials was regarded as promising, and prompted Maj H. F. Wood of Vickers to write to Brig-Gen Trenchard in optimistic terms. The new prototype was tested at CFS on 5 January, 1916, and its flying qualities were found to be good; the gun mounting and the lack of leg-room for the observer were criticised. Precisely how and when the type became designated F.B.9 is not known; the type number F.B.5A has been associated with the prototype but remains unconfirmed; certainly the production aircraft were designated F.B.9. Equally certainly the prototype was officially numbered 7665 and went to France on 24 January, 1916, where it was taken on strength by No.11 Squadron next day.

Pilots of the squadron regarded 7665 as a considerable improvement on their F.B.5s, but the observers condemned the F.B.9's gun mounting as useless. It seems likely that a squadron mounting replaced the Vickers No.3 Mark I mounting, if only because 7665 was flown operationally with considerable success, most notably by Capt Champion de Crespigny. He was at the controls on 2 April with 2nd Lt J. L. M. de C. Hughes-Chamberlain as observer, and attacked five German aircraft over the lines. One of these was shot down and a second driven off, but the other three shot 7665 about so effectively that de Crespigny had to descend and, with rudder controls and propeller shot away, was unable to avert a crash. The Vickers was damaged beyond repair and was struck off charge on 7 April, 1916.

Production F.B.9 with a later form of gun mounting, possibly a No.3 Mark II.

The F.B.9 was ordered in quantity from Vickers and (in the expectation of slightly earlier deliveries) from the Darracq company. Problems beset Darracq production from an early stage: the design of the nacelle had to be fundamentally changed, after production was well begun, to accommodate the revised gun mounting on which No.11 Squadron understandably insisted; in April the machines that made the streamline-section bracing wires in the Darracq works broke down. Thus it was late April before the first Darracq-built F.B.9, No.7812, was delivered. Even then it was found to be badly rigged, one of its broken Rafwires having been replaced by a cable. It was subsequently found that the Darracq-made Rafwires were faulty, whereupon Vickers offered to send over some spare sets. When these arrived they proved to be of the wrong dimensions, and arrangements had to be made for Vickers to send over wires ready swaged but not threaded; Darracq had then to cut the threads on them and provide the necessary fittings.

No.7812 was with No.11 Squadron by 19 May, 1916, but proved troublesome, for it could not be induced to climb satisfactorily. By 25 June, No.11 Squadron

Darracq-built F.B.9 No.7826, which was issued to No.11 Squadron on 1 July, 1916. It was returned to No.2 A.D. on 20 July. (*Bruce Robertson*)

Vickers F.B.9 with Scarff ring mounting for the Lewis gun.

had four F.B.9s, and at least six of the type saw some service with the squadron during the initial stages of the Battle of the Somme. The F.B.9's operational career was short, however, and the type was withdrawn by 21 July.

Although the batches A1411—1460 and A8601—8625 were ordered from Vickers under Contract No.87/A/485, it appears that responsibility for 50 of these F.B.9s was transferred to the Wells Aviation Co. Official records are not wholly clear about these aircraft, but the official rigging handbook for the F.B.9 states unequivocally that A1411—1460 were built by Wells. It seems that the first Vickers-built F.B.9 was delivered late in February 1917 while Wells output began about a month later, five Wells-built aircraft having been delivered by 24 March. The last of the seventy-five F.B.9s was delivered early in September, and it is possible that at least some of the Wells aircraft may have been assembled from Vickers-made components. At such a late period these F.B.9s can only have been intended for training duties and all may have been built with dual controls from the outset. The dual-control F.B.9 had an extended nacelle nose to give the occupant of the front seat adequate leg room. A few that were used for gunnery training were fitted with a Scarff ring mounting on the front cockpit. The 100 hp Monosoupape remained the standard engine, but at least one F.B.9 had a 110 hp Le Rhône. In the training squadrons the F.B.9 survived into 1918, and A8613 was selected for preservation for museum purposes.

F.B.9
100 hp Gnome Monosoupape, 110 hp Le Rhône

Two-seat fighter-reconnaissance.

Span 33 ft 9 in; length 28 ft 5½ in; height 11 ft 6 in; wing area 340 sq ft.

Empty weight 1,029 lb; loaded weight 1,892 lb.

Maximum speed at 6,500 ft—79 mph, at 10,000 ft—75 mph; climb to 6,500 ft—19 min, to 10,000 ft—51 min; service ceiling 11,000 ft; endurance 5 hr.

Armament: One 0·303-in Lewis machine-gun; occasionally a second Lewis gun was carried.

Manufacturers: Vickers, Wells Aviation, Darracq.

Service use: *Western Front*: No.11 Sqn. *Training duties*: Reserve Sqns 6 and 10; No.188 Night Training Sqn; No.2 Auxiliary School of Aerial Gunnery; CFS.

Serials: 7665, 5271—5290, 7812—7835, A1411—1460, A8601—8625.

Vickers F.B.12

In 1914 C. B. Redrup was a partner in the Hart Engine Company of Leeds. He had been designing aero-engines since 1910, and by 1914 was the patentee of two designs, one for a 150 hp air-cooled radial engine, the other for a seven-cylinder engine of 35 hp. When war began, Vickers recognised that France's aero-engine industry would be fully stretched in meeting the demands of the French aviation services, consequently Britain might be unable to obtain the engines needed for the aircraft of the RFC and RNAS. In September 1914, Vickers entered into negotiations with the Hart company, and agreed to provide financial assistance for the development of the 150 hp engine. Initially, encouraging progress was made and several Vickers aircraft were designed with the 150 hp Hart engine as their intended power unit.

Vickers F.B.12 with 80 hp Le Rhône engine. This was the original form of the design.

One of these was the Vickers F.B.12, a single-seat pusher generally similar to the D.H.2 and F.E.8. It is not known when work on the design began, but it seems logical to assume that it must have been some time before the Vickers-Challenger interrupter gear became available and made possible fixed-gun installations in tractor aircraft. Delay in developing the Hart engine led to some redesign to accommodate an alternative engine, choice falling, faute de mieux, on the 80 hp Le Rhône rotary. With little more than half of its designed power the F.B.12 was seriously underpowered.

As originally built the F.B.12 was a two-bay biplane with generously rounded wingtips; the span and chord of the upper wing were greater than those of the

Vickers F.B.12 with 100 hp Gnome Monosoupape. (*Vickers Ltd*)

lower. The blunt-nosed nacelle had a basic structure of steel tubing, was of circular cross-section, and was mounted in mid-gap; the tall undercarriage was attached to the underside of the lower centre-section.

The F.B.12 was completed in June 1916 and subsequently went to CFS for official trials, the report thereon being dated August 1916. This recorded a speed of 94·5 mph low down and a climb to 10,000 ft in 27 min 35 sec, a respectable performance on a mere 80 hp. Later, a 100 hp Gnome Monosoupape replaced the 80 hp Le Rhône, and in a further modification a new set of mainplanes having increased area and straight raked tips was fitted. The design of the undercarriage was modified at this stage, and the aircraft was given the new designation F.B.12A.

Considerable faith was retained in the Hart engine, for an official contract for 50 production aircraft with that engine was given to Vickers in November 1916. The serial numbers A7351—7400 were allotted on 10 November, and the contract called for a guaranteed speed of 100 mph.

Vickers F.B.12A, a photograph that illustrates the revised design of wings and undercarriage.

At about the same time, the F.B.12A, now effectively the prototype for a production design, was given the serial number A5210. With this identity the aircraft was allotted to the Expeditionary Force on 8 December, 1916, at which time it was at the Southern Aircraft Depot at Farnborough, having been flown there from Joyce Green on 4 December by Lt Oliver Stewart. In his log-book Stewart recorded that the F.B.12A was '*very* stiff on ailerons and requiring *hard* left rudder, also very nose heavy (*awful*)'. While at Farnborough it was damaged, the necessary repairs delaying its flight to France, but eventually it arrived at No.1 A.D. on 24 December, 1916, and it is of interest to note that the aircraft was recorded in No.1 A.D.'s Daily Aircraft Return as 'Vickers Scout (Hart) A5210'. The F.B.12A was flown on test at No.1 A.D. St-Omer on 27 December, 1916, by Capt C. H. Nicholas of No.32 (D.H.2) Squadron and on or about the same date by Lt E. L. Benbow of No.40 (F.E.8) Squadron. Both pilots agreed that the Vickers was significantly faster than the single-seat pusher scouts they were flying operationally, yet neither expressed any enthusiasm or preference for the F.B.12A. It was nose-heavy, very heavy on the ailerons and elevators but sensitive on the rudder, the view from the cockpit was considered unsatisfactory in some respects, the cockpit itself being cramped, the seat too low, and the windscreen opaque and too large, while the gun mounting made it impossible to use the 97-round drum on the Lewis gun. The F.B.12A's lack of an emergency gravity tank was inevitably criticised.

Apparently it was found that the F.B.12A had been badly out of true at the time of these tests. It was re-rigged and tested again by Capt A. M. Lowery, who agreed that it was faster than the F.E.8 but found the controls 'very slow and hard to move'. He was also severely critical of the poor view for landing, and found he had to break the windscreen in order to see well enough to land the aircraft.

Doubtless Brooke-Popham knew that the batch of Hart-powered F.B.12s was already on order, which may be why, on 1 January, 1917, he ordered No.1 A.D. to make certain modifications to A5210 and to complete them by 6 a.m. on 4 January. None of the alterations could have improved the aircraft's handling characteristics but, remarkably, one of them was a requirement to fit a reel for a wireless aerial with appropriate fair-lead and a dummy aircraft tuner set. The work took a few hours longer than Brooke-Popham had specified because the F.B.12A's nacelle was so restricted that only one man at a time could work on it.

On 5 January, Lt Benbow flew the F.B.12A twice and Brooke-Popham wrote to the D.A.E. stating that the aircraft, as fitted with a Monosoupape engine, offered no advantages over the F.E.8 and was not wanted by the RFC with the Expeditionary Force. However, he asked that a Hart-powered F.B.12 be sent to France for evaluation as soon as possible, requesting that the gun mounting thereon should be capable of elevation and that the view from the cockpit should be improved. On 7 January, 1917, the F.B.12A was flown back to England, its mock-up wireless installation being left in place at the request of Maj Wood of Vickers.

Early in 1917 an example of the Hart engine was installed in an F.B.12 airframe that was built by Wells Aviation. This apparently had the steel-tubing nacelle and long-span wings, presumably like those of the F.B.12A; the Hart-powered aircraft was designated F.B.12B. Minor mishaps marred the initial engine runs, but it seems that the aircraft not only flew but went to Martlesham Heath. It is not known whether an official test report was compiled, but the F.B.12B's career was abbreviated by a crash, following which Vickers abandoned the Hart engine. This led to a dispute between Redrup and the company, and eventually litigation followed in which **Redrup** complained that 'The aeroplane was of an experimental pusher type whereas a test of the engine should properly have been made in a standard aeroplane'.

The abandonment of the Hart engine inevitably affected the production aircraft. For these a new nacelle, with a wooden basic structure, and an enlarged fin and rudder had been designed, the modified aircraft being designated F.B.12C. It was intended that production would be undertaken at Weybridge (and it is not impossible that a few F.B.12Cs may have been built there), but the capacity in that factory was wanted for the F.B.14. It has been reported that production of the F.B.12C was transferred to the Wells company, but this now seems less than certain, and official documents are not clear on the point: indeed, one indicates that Wells built seventy-five F.B.9s but no F.B.12s. Possibly the Wells factory was used as an assembly shop for either or both types, using Vickers-made components. The first production F.B.12, A7351, was fitted with a 110 hp Le Rhône, and may therefore have been, strictly speaking, an F.B.12D.

Martlesham Heath was not impressed by A7351 when it was tested there in May 1917. Speed was only 87 mph at 6,500 ft, and it took 18 min 30 sec to climb to 10,000 ft. The report could find nothing complimentary to record:

> 'Insufficient elevator control at low speeds; heavy on lateral controls. The main petrol cock is behind the seat, and inaccessible; the main switch is liable to be knocked off by pilot's arm; and the compass is difficult to read when flying. The gun is too far forward for easy changing of drums. There is only one petrol tank.'

There can be little doubt that this report was the reason for an official visit to Vickers works on 22 May, 1917. Next day Capt Wheatley, who made the visit, reported to the Progress and Allocation Committee that all of the wing surfaces and nearly all spares called for under the contract had been made; twelve of the nacelles had been erected and the remainder were ready for erection. In his view it would be impossible to cancel the contract. Gen Brancker suggested that an Anzani engine might be tried in the aircraft; this was agreed, and a 100 hp Anzani was allotted from Naval resources.

Vickers F.B.12C A7352, one of the production batch, which had a 100 hp Anzani.

This Le Rhône-powered F.B.12C, probably A7351, was photographed at Rochford, the aerodrome of No.61 (H.D.) Squadron and No.198 Depot Squadron. A7351 was with the latter unit and made two operational sorties against German raiders, on 7 July and 12 August, 1917. *(The late Douglass Whetton)*

This engine was fitted to the second production aircraft, A7352, which went to Martlesham Heath in June 1917. Its trials were interrupted for want of necessary engine spares that did not arrive until mid-July. The official trials were concluded during the week that ended on 28 July, and the performance that they recorded was considerably worse than that of A7351, as the appended figures show. The Anzani-powered F.B.12C languished at Martlesham until 1 September, 1917, awaiting disposal; thereafter it disappears from the record.

In late June 1917 there had been a proposal to fit a 125 hp Anzani in one of the F.B.12Cs, but by 7 July Vickers had stated that this engine would not fit the airframe. By 23 July it had been decided that production of the F.B.12C must cease; completed aircraft were to be delivered to store; and eighteen such aircraft were accepted. In addition to A7351 and A7352, thirteen were officially recorded as delivered in July and three in August 1917.

Another proposal for an alternative engine was explored in September and October, despite the decision to terminate production. A 150 hp Smith Static radial became available, apparently because the RNAS did not want it: the Smith engine had been tried by the RNAS in a modified Vickers F.B.5 and the A.D. Navyplane. At its meeting on 13 September, 1917, the Progress and Allocation Committee discussed the possible use of the engine, which was considered to be unsuitable for operational service. Brancker thought it might be used on Home Defence aircraft, and it was decided to make a trial installation in a Vickers F.B.12C. A7353 was allotted to the Technical Department for the purpose, and apparently went to Martlesham Heath to be modified, yet was not mentioned in Martlesham's weekly reports. It was found that the F.B.12C needed too many modifications, however, and the idea was abandoned on 25 September. Production of the Smith engine (by Heenan & Froude of Worcester) continued, however, and the Committee found on 3 October that it had six engines and 20 sets of spares on its hands. Eventually four were taken over by the Technical Department, but it is doubtful whether any practical use was made of them.

The designation F.B.12D was allotted to a sub-type to be powered by either the 110 hp Le Rhône or 100 hp Gnome Monosoupape. Whether this made A7351 an F.B.12D is uncertain, but it seems unlikely that any production F.B.12C ever had a 100 hp Monosoupape.

F.B.12

F.B.12: 80 hp Le Rhône 9C
F.B.12A: 100 hp Gnome Monosoupape
F.B.12B: 150 hp Hart
F.B.12C: 110 hp Le Rhône 9J or 100 hp Anzani

Single-seat fighter.

F.B.12: Span 26 ft (upper), 24 ft (lower); length 21 ft 6 in; height 8 ft 8 in; wing area 204 sq ft.
F.B.12C: Span 29 ft 9 in (upper), 26 ft 9 in (lower); length 21 ft 10 in; height 8 ft 7 in; wing area 237 sq ft.

Weights and Performance:

	F.B.12	F.B.12C (Le Rhône)	F.B.12C (Anzani)
Weight empty, lb	845	927	953
Weight loaded, lb	1,275	1,447	1,473
Maximum speed, mph			
at sea level	94·3	—	—
at 6,500 ft	86	87	86·5
at 10,000 ft	81	81	77
Climb to (min and sec)			
6,500 ft	13 10	9 40	15 25
10,000 ft	27 35	18 30	35 35
Service ceiling, ft	11,500	14,500	10,000
Endurance	3 hr	3¼ hr	3¾ hr

Armament: One 0·303-in Lewis machine-gun.

Manufacturer: Vickers.

Service use: Home Defence—one F.B.12C is officially recorded as having gone to a Home Defence unit in 1917: this probably was A7351 with No.198 Depot Squadron.

Serials: A5210, A7351—7400.

Vickers F.B.14

In the immediate postwar period the British aviation journal *Flight* printed a series of articles under the general heading *Milestones*, each dealing with the products of leading British aircraft manufacturers. In the article on Vickers aircraft published on 12 June, 1919, there appeared this note on the F.B.14:

> 'This machine was a two-seater tractor biplane designed at the War Office request for a 200 hp B.H.P. engine, but, owing to this engine not having emerged from the experimental stage, a request was received to remodel the machine for 160 hp Beardmores.'

For want of any alternative or additional information that statement was for many years accepted by aviation historians (the author included) and was reproduced several times.

Now that official documents are available for study, however, the origins of the F.B.14 seem rather less clear-cut. The initial contract for the F.B.14

Original form of Vickers F.B.14, with 120 hp Beardmore engine. (*W. Evans*)

(No.87/A/453) was given to Vickers on 23 May, 1916, whereas the first prototype of the B.H.P. engine did not begin its initial bench trials until June 1916.

There seems little doubt that the fifty-two F.B.14s (A676—727) provided for under the contract were ordered off the drawing board, and it is not impossible that Vickers might have been given advance drawings of the B.H.P. engine. Nevertheless, not only does it seem unlikely that the entire combination of new airframe and new, untried, engine would be ordered with such abandon, but the original F.B.14 had every appearance of having been designed for the Beardmore engine (almost certainly one of 120 hp) with which it was fitted. It was a single-bay biplane with unequal-span wings that had rounded tips and outward-raked interplane struts. The original fin was small, the undercarriage short, and the engine was surmounted by a long, rearwards-raked exhaust stack; there was no

Modified prototype F.B.14 at CFS, still with elliptical wingtips but with revised engine installation, taller undercarriage and enlarged fin and rudder. (*T. Heffernan*)

evidence of any gun for the pilot, and the observer's Lewis was on a rocking-pillar mounting.

This prototype was tested at CFS in the spring of 1916. A record of the results has still to be found, but performance must have been poor, for the aircraft was seriously underpowered. The design was revised extensively, the second form of the F.B.14 having a 160 hp Beardmore engine in a redesigned cowling, a taller undercarriage, a rounded top decking on the rear fuselage, and a greatly enlarged fin and rudder. The mainplanes still had rounded tips and wooden, cross-braced, cabane struts, but a 5-gallon gravity tank was carried under the starboard upper wing, dihedral was increased, and Rafwires replaced cables in the interplane bracing. A variable-incidence tailplane was fitted.

In this form the F.B.14 was tested at CFS, apparently as the subject of trials reports Nos.58 and 63, dated September 1916. It was still with the Test Flight at CFS on 7 September, 1916, when it was officially allotted to the Expeditionary Force in France. The allotment instruction concluded by requiring that the number A3505 'should be painted on machine before despatch overseas'. In fact, however, it appears that the F.B.14 never went to France, but its allotment to the E.F. was not formally cancelled until 23 January, 1917, at which date it was at the Vickers works.

A3505 was effectively the production prototype F.B.14 and is here seen at Orfordness.

All known photographs of A3505 depict it with completely new mainplanes that had straight raked tips and clear-view panels let into the roots of upper and lower wings; the upper mainplanes met centrally at a new cabane structure of two steel-tube N-struts. The cockpits were separated by a short top decking, on to which was built a large headrest for the pilot, who was also given a large opaque windshield in front. The top decking behind the observer's cockpit was retained, as was the straight-edged fin. At some time during the winter of 1916–17 A3505 was at Orfordness: it then had a Scarff ring mounting on the observer's cockpit, and came quite close to the final production form of the aircraft.

Initially the F.B.14 was considered for service in France, and in July 1916 RFC Headquarters expressed an intention to employ Armstrong Whitworth and Vickers Beardmore-powered types, (respectively the F.K.8 and F.B.14) if supplies of B.E.2cs were to run short. About November 1916 a further one hundred and fifty F.B.14s had been ordered as A8341—8490, being additional aircraft under Contract 87/A/453. A secret document dated 27 January, 1917, set down future policy for the Middle East Brigade, and included provision for the gradual replacement of all B.E.2c, 2d and 2e aircraft by Vickers F.B.14s. No doubt it was considered that the steel-tube framework of the F.B.14 fuselage

True production F.B.14 at Farnborough, a photograph dated 23 February, 1917.
(*RAE 2143; Crown copyright*)

made it specially suitable for service in hot countries. Thoughts of using the F.B.14 in France were still entertained in early 1917, for on 22 February, 1917, D.A.D.A.E. sent Trenchard two sets of drawings showing how bomb ribs were to be fitted under the lower wings of the F.B.14.

It was in February 1917 that the first production aircraft began to come forward. These were generally similar to A3505, but had no top decking behind the rear cockpit, and the fin was slightly enlarged and had a curved leading edge. Most significantly, a fixed Vickers gun was provided for the pilot, firing along the port side of the engine cylinders. The design of the engine-cowling panels differed in detail, and the headrest and decking behind the pilot's cockpit were abandoned.

The reasons for the abandonment of the F.B.14 have never been convincingly explained. The trouble again lies in the *Milestones* article in *Flight*, which stated:

F.B.14 A683 at Martlesham Heath with baffle plate over the fuselage louvres beside the cockpit. (*T. Heffernan*)

Vickers F.B.14A, with 150 hp Lorraine-Dietrich engine.

'Although 150 of these machines were contracted for, they were mostly delivered without engines, owing to trouble being experienced with the 160 hp Beardmore, and eventually 120 hp Beardmores were substituted.'

That is so largely nonsense that one wonders what trouble was experienced with the 160 hp Beardmore in the F.B.14. True, it was not so reliable as the 120 hp engine (of which only 400 were made), but by the end of 1916 only 317 of the 160 hp type had been delivered and were in great demand for the F.E.2bs and Martinsyde Elephants then fully operational in France. In October 1916 alone, four hundred F.E.2bs were ordered from G. & J. Weir, and the formation of No.100 Squadron with F.E.2bs for night bombing in February 1917 pointed the

Vickers F.B.14F with 140 hp R.A.F.4a engine, photographed at Martlesham Heath.
(*T. Heffernan*)

way ahead. It therefore seems possible that with such large demands in prospect the 160 hp Beardmore was simply withheld from the Vickers F.B.14. Few F.B.14s were ever engined; none went to the Middle East, but six went to Home Defence units in 1917; of the 94 that had been completed by 2 June, 1917, the great majority were delivered to store.

Alternative engines were fitted in three cases. The F.B.14A had a 150 hp Lorraine-Dietrich with flat frontal radiator, and had the headrest and windshield of A3505. It was tested at Martlesham in April 1917, but was reported to be 'tiring to fly, due to very bad lateral control'. In the following month Martlesham tested the F.B.14F, a conversion of A8391 to have the 140 hp R.A.F.4a engine in an unprepossessing installation. Its recorded performance was not good, yet Brancker told the Progress and Allocation Committee on 25 June, 1917, that it 'had given a remarkable performance'. He returned to the question on 17 July, when he said that the R.A.F.4a-powered F.B.14 would be very useful for service in the East. Col Whittington reminded the meeting that such was the shortage of R.A.F.4a engines that R.E.8s were having to be accepted without engines. Col Alexander visited Vickers' Crayford works that same day and reported to the Committee on 18 July that the firm had more work than they could cope with and were in no position to undertake the modification of the ninety-odd F.B.14s then in store. The Committee decided, notwithstanding Col Whittington's relevant reminder that R.A.F.4as were scarce, to send an F.B.14 to the Royal Aircraft Factory to be fitted with an R.A.F.4a, together with A8391. An F.B.14 was indeed sent to Farnborough late in July, and the F.B.14F went there on 21 July. Vickers set about producing the necessary drawings for the use of the Factory.

Vickers F.B.14D at Martlesham Heath.

Vickers apparently thought that the task of converting the F.B.14s was to be undertaken by the R.A.F., but the Factory protested that they had no instruction to undertake the task and could not do so without retarding work on the S.E.5 and F.E.9. Col Whittington repeated that there were no R.A.F.4as to spare for an F.B.14 conversion programme, and he intended to get a decision from Brancker. On 31 August Whittington stated flatly that R.A.F.4as were not available, and the question of fitting these engines to the F.B.14s should be dropped. The Committee was convinced and agreed to tell the Superintendent of the R.A.F. to do nothing further with the F.B.14. The F.B.14F, A8391, was last heard of at Farnborough on 13 October, 1917.

The F.B.14D, which had a 250 hp Rolls-Royce Mark IV engine, should properly have had a new and separate Vickers type number, for it owed very little to the basic F.B.14 design. It had two-bay wings of increased area and a full-length rounded top decking on the fuselage, which was lengthened. Evidently the F.B.14D was initially built as a private venture, for it had no official identity when tested at Martlesham Heath in April 1917, and it was not until 13 December, 1917, that it was allotted the serial number C4547 under the belated Contract No.A.S.22609/17. Martlesham had found the aircraft 'tiring to fly because of heavy lateral control, due to inadequate ailerons, and poor aileron control system.'

Although not acceptable as a production type the F.B.14D was sent to the experimental station at Orfordness, where it saw varied use. At one time its Vickers gun was mounted to fire upwards at 45 deg to the line of flight. While fitted with experimental gun sights, it rose to attack the formation of Gothas that bombed London on 7 July, 1917. The F.B.14D's pilot was Capt Vernon Brown, his observer Lt B. M. (later Sir Melville) Jones, and they pursued the Gotha formation across the North Sea to Zeebrugge. Unfortunately, their experimental gunsights proved to be useless in action and their gallant flight was not rewarded by the destruction of any of the Gothas. It was, however, the only known combat in which any F.B.14 variant participated.

In view of the statements that have been made about the origin of the F.B.14 design it is necessary to record that the unbuilt F.B.14C was a variant that would have had as its power unit the 200 hp B.H.P. engine. The fact that a new sub-type number had to be applied to the B.H.P.-powered version of the design makes it seem even less likely that the F.B.14 was originally supposed to have that engine.

The Vickers F.B.14 was not finally declared obsolete until Air Ministry Order No.896 of 1919 was published on 7 August, 1919.

F.B.14

F.B.14: 120 hp or 160 hp Beardmore
F.B.14A: 150 hp Lorraine-Dietrich
F.B.14D: 250 hp Rolls-Royce Mark IV
F.B.14F: 140 hp R.A.F.4a.

Two-seat fighter-reconnaissance.

Production F.B.14: Span 39 ft 6 in (upper), 33 ft (lower); length 28 ft 5 in; height 10 ft; wing area 427 sq ft.

F.B.14: Empty weight 1,662 lb; loaded weight 2,603 lb.

F.B.14: Maximum speed at sea level 99·5 mph, at 6,000 ft—90 mph, at 10,000 ft—84 mph; climb to 6,000 ft—16 min 35 sec, to 10,000 ft—40 min 50 sec; service ceiling 10,000 ft; endurance 3 hr.

Armament: One 0·303-in Vickers machine-gun, one 0·303-in Lewis machine-gun.

Manufacturer: Vickers.

Service use: *Home Defence*: Six F.B.14s were issued to Home Defence units in 1917; one of these may have been No.51 Sqn.

Serials: A676—727, A3505, A8341—8490 (of which A8392—8490 were probably not completed); C4547 was the solitary F.B.14D.

Vickers, F.B.19 Mk I with 100 hp Gnome Monosoupape engine but without clear-view panel in centre-section. (*E. B. Morgan*)

Vickers F.B.19

Despite the deficiencies of the Vickers E.S.1, the firm did not abandon the basic concept of a small, rotary-powered, single-bay biplane for fighting purposes. It reappeared in 1916 in a development of the E.S.1 designated F.B.19. The first prototype emerged as a stumpy little aircraft powered by a cleanly cowled 100 hp Gnome Monosoupape; it had no armament, and there was no clear-view panel in the centre-section. The fuselage sides were flat, suggesting that criticism of the E.S.1's lack of view had been heeded. This aircraft flew in the summer of 1916, and may have been the Monosoupape-powered F.B.19 that later had a large panel of Triplex glass let into the centre-section.

In July 1916 official serial numbers were allotted to three F.B.19s, followed by a fourth in August. The first two were A1968 and A1969, and it is possible that the Monosoupape aircraft was A1968. Official papers state that A1969 had a 110 hp Clerget engine and was fitted with a 'glass inlet to centre top plane and rising seat'. It is unlikely to have been armed, for, after having been confused with the E.S.1 7759 in official documents, it was allotted to the Expeditionary Force in France on 19 July, 1916, only for this allotment to be cancelled two days later. The allotment was evidently thus cancelled to enable A2122, the third F.B.19, to be substituted; and A2122, according to its allotment of 21 July, was 'of the flat-sided, short-fuselage type, and is fitted with glass roof, gun and interrupter gear'. It also had the 110 hp Clerget engine.

A2122 had been flown to the AID at Farnborough by Harold Barnwell on 18 July. Its supplanting of A1969 was conveyed to RFC Headquarters on 21 July, together with a brief description and a request for an assessment of the aircraft. This third F.B.19 did not go to France, however, for on 1 September a revised allotment consigned it to the Home Defence Wing. RFC Headquarters asked whether any other Vickers Scout was to be allotted as a replacement, but on 7 September A.D.A.E. replied that this was not intended. Presumably A2122 therefore joined the fourth F.B.19, A2992, which had been allotted to the 16th

Vickers F.B.19 Mk I with 110 hp Le Rhône engine and transparent centre-section. This aircraft is believed to have made its first flight on 16 November, 1916.

Home Defence Wing on 11 August. This last aircraft also had the 110 hp Clerget engine.

It is known that official acquisition of A1968, A1969 and A2122 was covered by Contract No.87/A/536, and this may have again been invoked in the case of A2992, for which special instructions of 8 August, 1916, provided for the allocation of its serial number. The RFC's possession of these four F.B.19s did not inspire a production order for that Service. Nevertheless, the design went into limited production in this form, the ensuing aircraft being delivered to the Russian Imperial Air Service; these apparently had the 110 hp Le Rhône 9J engine, and some survived or escaped the destructive intentions of a Royal Naval landing party to serve in Bolshevik markings after the revolution in Russia.

A second version of the design appeared in the autumn of 1916. On this variant the wings were given 1 ft 10 in of positive stagger in an attempt to improve the pilot's view. The fuselage structure had to be modified to permit the lower wing to be located farther aft than on the original F.B.19 design and to accommodate the forward position of the upper wing. A single Vickers gun was mounted in a deep trough on the port side of the fuselage; it was synchronised by the Vickers-Challenger interrupter gear, and its line of fire lay inside the circumference of the engine cowling, necessitating a small hole in the cowling itself. This variant was designated Vickers F.B.19 Mk II, its predecessor becoming the F.B.19 Mk I.

On 18 October, 1916, the Acting ADAE (Capt F. C. Jenkins) wrote to Brig-Gen Brooke-Popham to say that if the RFC were to order twelve Vickers Scouts with the 110 hp Clerget at once, all could be delivered by 20 November. Delaying the order until the end of October would result in delivery being considerably later. This letter evidently followed a suggestion made by Brooke-Popham while visiting England, that a Flight of the new Vickers single-seaters might be useful to the RFC, subject to approval of a sample aircraft. It was intended to send the first F.B.19 Mk II to France, and the aircraft was allotted the serial number A5174 during October 1916. Bad weather delayed its flight to France, however, and on 25 October Brooke-Popham asked DAE to place an order for twelve without waiting to see the prototype.

Not until 2 November did A5174 arrive at No.1 A.D., St-Omer. It flew on to No.2 A.D. at Candas next day, and on 10 November was flown to No.70 Squadron at Fienvillers by Capt W. J. C. K. Cochran Patrick. While flying it at Fienvillers Patrick looped the aircraft, and when he landed it was found that several of the ribs in the port lower wing had broken between the spars. Brooke-Popham wrote that day to D.A.E. pointing out that there seemed to be no proper compression members in the wing panels and asking for the AID's opinion and recommended remedy. He also intimated that solid ash ribs were to be fitted to A5174 on the chord of attachment of the interplane struts. The F.B.19 was returned to No.2 A.D. for the necessary repairs on 11 November.

Despite this alarming occurrence, Trenchard decided to take twelve aircraft and on 11 November Brooke-Popham advised D.A.E. accordingly. He went on to say:

> 'It is not considered that they fulfil our requirements as a fighting machine, but they might be useful as a stop gap owing to the shortage of fast aeroplanes.
>
> These machines must have some modifications introduced into the construction of the wings vide my letter on the subject of yesterday's date.
>
> It is presumed that these machines will be ready within the next four weeks. Unless they can be produced quickly, they will be of no value.'

On 14 November A/A.D.A.E. replied that an order for twelve F.B.19 Mk IIs had been placed and that the aircraft would have suitably modified wings. Vickers had promised delivery by 25 December, but that was before the need for the wing modifications had been assessed.

Six days later A.D.A.E. wrote to say that it had been found that A5174's wing panels had been structurally identical with those of the F.B.19 Mk I, no account having been taken of the additional stress on the compression ribs imposed by the marked stagger of the wings. The production aircraft were to have solid-web compression ribs with heavier flanges.

A5174, the prototype of the F.B.19 Mk II, photographed at No.2 A.D., Candas, 9 November, 1916.

A5174 went back to No.70 Squadron on 1 December, remaining ten days, presumably being assessed by squadron pilots. Their views have yet to be found, and must have made interesting reading, for only a few weeks previously the squadron had received the Sopwith Pup A626 for assessment. In any comparison the F.B.19 can hardly have matched the Pup. On 11 December, A5174 was returned to No.2 A.D., and eight days later it was sent back to England in a packing case.

As events turned out, A5174 was the only F.B.19 of any kind to go to France. The statistical table in Appendix VIII to Volume III of *The War in the Air* states that six F.B.19s went to France, but this can now be seen to have been a combination of A5174, the four E.S.1s (5127 and 7756—7758) and the F.B.12A A5210, for all were frequently named simply as Vickers Scouts in official documents, without differentiation. Indeed, the whole entry ostensibly relating to the F.B.19 in the table concerned is for this reason substantially inaccurate in several respects.

The contract for the twelve production F.B.19 Mk IIs was formally confirmed on 21 December, 1916, the aircraft being numbered A5225—5236. The standard engine was, as on A5174, the 110 hp Clerget 9Z, but A5230 had a 110 hp Le Rhône when it was tested at Martlesham in May 1917. With either engine the aircraft's performance was undistinguished.

A5234, one of the Vickers F.B.19 Mk IIs of No.111 Squadron, Palestine. (*Ray Vann*)

During the campaign in Palestine, the first Battle of Gaza demonstrated the advantage held by the German and Turkish forces because they employed modern aircraft whose performance outstripped that of the RFC's B.E.2cs and Martinsyde Elephants that opposed them. Aerial activity in the second Battle of Gaza was less critical, but the Palestine situation apparently inspired the War Office to send the available Bristol M.1Bs and all twelve F.B.19 Mk IIs to the Middle East Brigade in June 1917. The first RFC squadron to receive the F.B.19 was No.47 in Macedonia, its first arriving on 23 June, 1917. In Palestine, a few F.B.19s were allotted to, or initially maintained by, No.14 Squadron at Deir al Balah, which had A5231, A5234 and A5236 by 14 July. This was probably a temporary expedient while No.111 Squadron was forming, for the new unit subsequently operated both Bristol M.1Bs and Vickers F.B.19 Mk IIs, having five of the latter type on its operational strength on 27 October, 1917. They proved to be largely ineffective: in December 1917 Gen Brancker, then in command of the Middle East Brigade, wrote to Gen John Salmond: 'The Bristol Monoplanes and Vickers Bullets are not very much good except to frighten the Hun; they always seem to lose the enemy as soon as he starts manoeuvring.' It is unlikely that the F.B.19 remained operational much beyond the end of 1917, but a few were used for instructional purposes in 1918, notably at the School of Aerial Fighting at Heliopolis.

F.B.19

Mk I: 100 hp Gnome Monosoupape, 110 hp Clerget 9Z, 110 hp Le Rhône 9J
Mk II: 110 hp Clerget 9Z, 110 hp Le Rhône 9J

Single-seat fighting scout.
Span 24 ft; length 18 ft 2 in; height 8 ft 3 in; wing area 215 sq ft.
Weights and Performance:

	F.B.19 Mk I	F.B.19 Mk II	
	Monosoupape	Clerget	Le Rhône
Weight empty, lb	900	890	892
Weight loaded, lb	1,485	1,475	1,478
Maximum speed, mph			
at 10,000 ft	102	98	98
at 15,000 ft	—	—	90
Climb (min and sec)			
to 10,000 ft	14 00	14 30	14 50
to 15,000 ft	— —	— —	37 10
Absolute ceiling, ft	17,500	16,500	17,000
Endurance	2¾ hr	3 hr	3¼ hr

Armament: One 0·303-in Vickers machine-gun.

Manufacturer: Vickers.

Service use: **F.B.19 Mk I**: *Home Defence*: Two of the F.B.19s were allotted to the 16th Home Defence Wing.*
F.B.19 Mk II: *Western Front*: No.70 Sqn; *Palestine*: Sqns 14 and 111; *Macedonia*: No.47 Sqn. *Training duties*: School of Aerial Fighting, Heliopolis; No.58 Training Sqn.

Serials: **Mk I**: A1968, A1969, A2122, A2992.
Mk II: A5174, A5225—5236.

*The table in Vol III of *The War in the Air* states that six F.B.19s were sent to Home Defence squadrons, one in 1916 and five in 1917, but this is probably another instance of E.S.1s and F.B.19s being grouped together.

The first prototype Vickers F.B.27, B9952, as it first appeared, with Hispano-Suiza engines and horn-balanced elevators. (*Vickers Ltd*)

Vickers F.B.27 Vimy

In mid-1917 the RFC did not favour heavy bombers because they were considered to be too vulnerable, but preferred day bombers like the D.H.4, believing the moral effect of day raiding to be greater than that of night attacks. Despite the grim demonstration of the effectiveness of heavy bombers used by day that was provided by the Gotha raid on London of 13 June, 1917, the Air Board at its meeting of 23 July decided that all orders for experimental heavy bombers should be cancelled. At this the Controller of the Technical Department immediately protested in robust terms, the question was re-opened a week later, and orders were placed for one hundred Handley Page O/400s and prototypes of new heavy bombers from Handley Page and Vickers.

The basic outline of the Vickers design was sketched out by Rex Pierson in the course of a discussion with Maj J. S. Buchanan at the Air Board; it had something of the Vickers F.B.7 and F.B.8 in its general proportions. Contract No.A.S.22689/1/17 dated 16 August, 1917, provided for three prototypes to which the serial numbers B9952—B9954 were allotted on 2 November; to the Vickers company the new bomber was the F.B.27. Its engines were intended to be either two 200 hp Hispano-Suizas or two 200 hp R.A.F.4ds. The choice of the Hispano-Suiza was extraordinary: contrary to what has been said about the Vimy there was at that time no surplus of the engine; it was scarce to the point of crisis. Yet the first F.B.27 prototype, B9952, was completed with Hispano-Suiza engines; the R.A.F.4d, though then in production at the Daimler works, was never numerous and was abandoned in January 1918.

B9952 emerged as an inelegant equal-span biplane with 1 deg of dihedral on the outer wing panels, a biplane tail unit with twin rudders but no fins, horn-balanced elevators and ailerons, and a four-wheel undercarriage. The Hispano engines were in mid-gap with long exhaust stacks passing upwards through the upper wing. The crew was to consist of two gunners and the pilot, the gunners having cockpits in the nose and abaft the wings. With Gordon Bell at the controls the F.B.27 made its first flight at Joyce Green on 30 November, 1917.

On 10 January, 1918, B9952 went to Martlesham Heath for official trials, by which time its radiators had been enlarged, and horizontal exhaust pipes had replaced the upright stacks. The 200 hp Hispano-Suiza was always a troublesome engine and those in the F.B.27 were no exception, the aircraft's trials being delayed by engine difficulties. B9952's tests were not pronounced complete until 27 April, 1918, by which time the second prototype, B9953, was also at Martlesham, having arrived on 25 April. This aircraft was designated Vickers F.B.27A and had two 260 hp Sunbeam Maori engines and different outer wings that carried plain but inversely tapered ailerons. The name Vimy had been allotted to the F.B.27 early in March 1918, and the Maori-powered version was officially designated Vimy Mk II. Its delivery to Martlesham was delayed by engine trouble.

To get the F.B.27 out of its assembly shed a ramp had to be dug to enable the aircraft to be pulled out of the dug-away floor.

Meanwhile, the first and only RFC contract for production of the type had been placed on 26 March, 1918. It was for 150 aircraft for which the serial numbers F701—850 were allotted on 9 April; these were to have B.H.P., Fiat, or Liberty engines. Initially the Vimy design had been passed for production with Fiat A.12bis engines under the official designation Vimy III.

All subsequent development was, of course, for the Royal Air Force. Martlesham's report of 4 May, 1918, stated that B9952 was to return to Vickers to have Rolls-Royce Eagle engines fitted, and it flew from Martlesham on 11 May. Eventually it was given two 260 hp Salmson 9Zm water-cooled radial engines and reappeared with new tailplanes and plain elevators, while the dihedral on the mainplanes was increased to 3 deg. In this form B9952 survived the war. Not so B9953, which crashed at Martlesham during the week ending 25 May owing to engine failure.

The third prototype, B9954, had two 300 hp Fiat A.12bis engines and a modified nose; otherwise it was originally identical with B9953, having the same inversely tapered ailerons. It arrived at Martlesham on 15 August, 1918, by which time it had been modified to have wings and horn-balanced ailerons like those of B9952, and its dihedral was increased to 3 deg. Its performance trials were given the highest grading of priority, for by then more production contracts had been given to several contractors, and no fewer than 765 aircraft were on order, with further contracts imminent.

Propeller breakages held up the trials of B9954, and indeed it was never fully

The first F.B.27 at Martlesham Heath.

evaluated, for it was destroyed on 11 September, 1918, when, in a crash said to have been caused by a stall after take-off, the bombs with which it had been loaded exploded. Two further prototypes had been ordered in August 1918 under the original contract for prototypes, and were allotted the serial numbers F9569—9570 on 21 August. As far as is known, only F9569 was built.

This was the first Vimy to have Rolls-Royce Eagle VIII engines, and these were housed in nacelles generally similar to those of B9954. The rudders of F9569 were somewhat larger than those of its predecessors, and it had no fins. This prototype arrived at Martlesham Heath on 11 October, and its tests were executed so quickly that the Vimy was able to fly to France to join the Independent Force, RAF, on 26 October. It arrived at No.1 A.S.D. at 17.00 hours that day and went to No.3 A.D. two days later. By 10 December it was back at Joyce Green and was re-allotted to the Midland Area.

The first production Vimy, F701, was a Fiat-powered aircraft that was held up for weeks while minor modifications were made. It did not go to Martlesham until 27 February, 1919, and remained there for several weeks. Production continued for some time after the end of the war, the great majority of the aircraft having the Rolls-Royce Eagle VIII installation and being designated Vimy IV.

Had the war continued, 24 squadrons of the RAF would have been equipped with Vimys, which would have been used both as bombers and as anti-submarine

First production Vimy, F701 with Fiat engines, at Martlesham Heath.
(*E. F. Cheesman*)

aircraft, replacing such widely differing types as the Handley Page O/400 and the D.H.6. It had been hoped that the first Vimy squadron would have been an anti-submarine unit becoming operational in November 1918, followed by one further such squadron each month until April 1919, with two more in May. These squadrons would have had Fiat-powered Vimys, whereas initial thinking on the bomber squadrons provided for Vimys with B.H.P. engines in ordinary RAF squadrons and Liberty-powered aircraft in the squadrons of the Independent Force. It was hoped that by May 1919 there would be five of the former units (from December 1918 onwards) and eleven of the latter (starting in 1919). The coming of peace removed the necessity for these developments, and F9569 was the only Vimy to get near the war.

F.B.27 Vimy

Two 200 hp Hispano-Suiza, two 260 hp Salmson 9Zm, two 260 hp Sunbeam Maori, two 300 hp Fiat A.12bis, two 360 hp Rolls-Royce Eagle VIII

Short-distance night bomber (Air Force Type VIII).

Span 68 ft; length 43 ft 6½ in; height 15 ft; wing area 1,300 sq ft.

Weights and Performance:

Aircraft	B9952	B9953	B9954	F9569
Weight empty, lb	5,420	6,735	6,934	7,101
Weight loaded, lb	9,120	10,300	10,808	12,500
Maximum speed, mph				
at sea level	90	—	98	103
at 5,000 ft	87	89	—	98
at 6,500 ft	85	85	—	95
Climb to (min and sec)				
5,000 ft	23 35	19 20	— —	21 55
6,500 ft	36 10	28 30	30 00	33 00
Service ceiling, ft	6,500	8,000	6,500	7,000
Endurance	3½ hr	4¼ hr	3¼ hr	11 hr

Armament: Provision was made for carrying a maximum total of 4,804 lb of bombs, but the normal load appears to have been of about 2,000 lb of bombs, or two torpedoes. Four 0·303-in Lewis machine-guns Mk III.

Manufacturer: Vickers.

Service use: *Western Front*: none with RFC; only one Vimy, F9569, went to the Independent Force in France in October 1918.

Serials: *Prototypes*: B9952—9954, F9569—9570. *Production*: F701—850, F2915—2934, F2996—3095, F3146—3195, F8596—8645, F9146—9295, H651—670, H4046—4195, H4725—4824, H5065—5139, H9413—9512, H9963, J251—300, J1941—1990, J7238—7247, J7440—7454, J7701—7705.

Voisin LA No.1865 at Farnborough. This aircraft was delivered to the Aircraft Park from Paris on 16 February, 1915, and was sent to England on 21 February, possibly to serve as a model for British production of the type. This photograph is dated 25 February, 1915.
(RAE 715; Crown copyright)

Voisin LA and LA.S

The Voisin brothers were among the greatest of France's aviation pioneers, and their standard biplane of 1908–09 did as much as any aircraft to establish European aviation, being both sturdy and easy to fly. As time went on Voisins of improved design appeared, one that was exhibited at the 1910 Paris Aero Salon having a quick-firing gun of imposing dimensions on the nacelle. Thus did the Voisins anticipate the development of combat aircraft; of equal importance from the military standpoint was their use of steel as a structural material.

By 1914 Gabriel Voisin (Charles Voisin was killed in a motoring accident in 1912) had evolved a practical pusher biplane of steel construction, and in June 1914 he demonstrated a biplane with a 100 hp Canton-Unné engine, armed with a 37-mm shell-firing gun. At that time the French Aviation militaire had two escadrilles, V.14 and V.21, that were equipped with Voisin biplanes. These were notable for having four-wheel tail-high undercarriages in which the rear wheels had brakes.

By impressing the Canton-Unné Voisins that were in production for Russia in 1914, the Aviation militaire was able quickly to form and equip a third Voisin escadrille, V.24. It was a Voisin of this unit that on 5 October, 1914, was the first Allied aircraft to shoot down a German aeroplane. This type of Voisin was designated Type III, alternatively Type LA, and, with its steel airframe and robust undercarriage, stood up well to the rigours of the war and the rugosities of the early operational airfields.

When the RFC began to acquire French aircraft by direct purchase in France, the types supplied eventually came to include a number of Voisin LAs. The first of these was given the RFC serial number 1856 and was delivered direct to No.4 Squadron from Paris on 1 February, 1915; it had a 130 hp Canton-Unné engine. No.1858 was delivered to No.4 Squadron four days later, and No.1860 arrived at the unit on 15 February, No.1864 next day. On 27 and 28 February respectively, 1867 and 1868 were delivered to No.5 Squadron, but were promptly transferred to No.16 Squadron on 1 March. In all, the RFC in France received twenty-six Voisins, the last being 5097, delivered in October 1915. Of these, nineteen saw operational use with squadrons in the Field; 1865 was sent to Farnborough for examination and assessment on 21 February, 1915, and 1899 was reported at Farnborough on 13 May, 1915. The other five (1889, 5010, 5011, 5013 and 5017) did not go further than No.1 Aircraft Park and probably provided spares for the others.

Two other French-built Voisins were supplied to the RFC. These were numbered 7458 and A5169; the former had originally had the French identity V.562 and received its RFC number late in December 1915, some time after the RFC had ceased to use the Voisin operationally in France. A5169 was a very belated allocation made in October 1916 to give an identity to the aircraft that had

This photograph may also be of Voisin LA No.1865 (*RAF Museum*)

served as a production model for the batch of fifty (4787—4836) ordered from the Dudbridge Iron Works early in 1915 under Contract No.94/A/134. The Dudbridge company were the British licensees for the manufacture of the Salmson (Canton-Unné) engines that powered the Voisin, and doubtless they provided the engines required for the 50 biplanes. Construction of the airframes has always been attributed to Savages of King's Lynn, and it is possible that the Dudbridge Iron Works may have sub-let the contract.

The ordering of 50 Voisins from a British contractor suggests that the RFC must have been sufficiently satisfied with the type to want more. There was momentary official hesitation in April 1915, when it was possible to compare the severely critical report on 1865 provided by the Royal Aircraft Factory with CFS's very favourable report on the first F.E.2a that was tested there at the end of March. On 8 April D.D.M.A. wrote to the Officer Commanding the RFC in France, inviting attention to the two reports and asking whether the order for 50 Voisins should go ahead. In a thoughtful and realistic reply on 11 April, probably written by Lt-Col F. H. Sykes, RFC Headquarters compared the attributes of the Voisin and F.E.2a, recognising the greater potential of the latter but concluding that the strength, carrying capabilities and, perhaps above all, the availability of

The Voisins of No.4 Squadron in the Field. (*RAF Museum*)

the Voisin, made it desirable to go on using the sturdy French pusher. Lt-Col D. S. MacInnes, then A.D.A.E., readily accepted this sensible view, and the order for the 50 Voisins went ahead.

Nevertheless, no British-built Voisin ever saw operational use in France, and the type did not remain long in use. It was withdrawn from No.4 Squadron in July 1915 and from No.7 in August and September; the last in the Field was 5097, delivered to No.12 Squadron from the Aircraft Park as late as 27 October, on which date it replaced 5066, which was sent to England. The operational career of 5097 was brief, for it, too, was sent to England on 4 November.

By then the Voisin could not be expected to survive in combat, for its normal defensive armament consisted of a single Lewis gun, wielded with great difficulty

Maintenance or instructional work on RFC Voisins, possibly at Farnborough.

by the observer, who occupied the rear seat in the nacelle. The gun was on an elevated mounting that placed it more or less directly above the pilot's head. Occasionally the Lewis was supplemented by other weaponry, but it proved unavailing on 21 July, 1915, when Corporal V. Judge and his observer, 2nd Lt J. Parker, of No.4 Squadron set out on an unescorted reconnaissance of Bapaume on 1858. In addition to the Lewis gun they had two revolvers and a rifle, but were brought down by German aircraft; Judge was made a prisoner of war, but Parker died of his wounds. In the bombing role the Voisin could carry two 100-lb bombs provided the pilot flew solo, as Lt H. F. Glanville of No.16 Squadron did on 9 May, 1915, when he bombed Don, continuing to the attack although wounded.

No.4811, one of the Voisins built in England by Savages Ltd.

Voisin LA.S No.8506, one of three transferred from the RNAS to the RFC in Mesopotamia, on the aerodrome at Ma'gil. (*A. E. Shorland, via Col Brian P. Flanagan*)

The RNAS also used Voisins, notably in Africa, the Aegean and Mesopotamia. Two, Nos.8506 and 8523, arrived at Basra on 17 January, 1916, to reinforce the RNAS in Mesopotamia. At that time the RNAS was short of pilots, the RFC short of aircraft, so it was decided to form a composite Flight for duty with the Tigris Corps. The aircraft, two Shorts and the two Voisins, set out for Ora on 31 January. One Short and one Voisin crashed en route; the other Voisin subsequently made three reconnaissance flights before engine trouble brought it down on 8 February, whereupon it was bogged for ten days. On 5 March, one of the Voisins, evidently 8523, was shot down and its crew killed, while the other was virtually discarded because its rate of climb was too poor. This was

8506, which had flown no fewer than 80 missions over besieged Kut-al-Imara. When the RNAS detachment withdrew from Mesopotamia on 29 June, 1916, two RNAS Voisins were handed over to the RFC. RNAS records show that three Voisins (8506, 8518 and 8523) were transferred to the RFC for the use of Force D, consequently it would seem that a third aircraft must have been there, possibly as a source of spares. These Voisins were of the LA.S type, in which the engine was a 140 hp Salmson B.9 mounted at a higher level and inclined slightly to provide downthrust. The additional S in the designation signified surélevé, a reference to the slightly elevated engine position.

Although all published references describe the Type LA.S as the Voisin Type 5, all known references in French official documents describe the aircraft as Voisin III LA.S.

Just how long the RFC retained 8506 and 8518 is unknown. Nearer home, some of the LA biplanes were used for training purposes in England after their withdrawal from operational service.

LA and LA.S
LA: 130 hp Salmson (Canton-Unné) M.9
LA.S: 140 hp Salmson (Canton-Unné) B.9

Two-seat reconnaissance-bomber.

Span 14·74 m (48 ft 4 in); length 9·5 m (31 ft 2 in); height 2·95 m (9 ft 8 in); wing area 49·65 sq m (534 sq ft).

Empty weight 950 kg (2,094 lb); loaded weight 1,350 kg (2,976 lb).

LA: Speed at sea level 105 km/h (65·3 mph); climb to 1,000 m—12 min, to 2,000 m—30 min; endurance 4½ hr.

LA.S: Speed at sea level 110 km/h (68·4 mph); climb to 1,000 m—10 min, to 2,000 m—24 min 30 sec; endurance 4 hr.

Armament: One 0·303-in (7·7-mm) Lewis machine-gun, supplemented by revolvers and a rifle as occasion demanded; up to 200 lb (91 kg) of bombs.

Manufacturers: Voisin; Savages (either by sub-letting or diversion from Dudbridge Iron Works).

Service use: *Western Front*: Sqns 4, 7, 12 and 16; two Voisin LA with No.5 Sqn for two days. *Mesopotamia*: Composite RNAS/RFC Flight at Ora; No.30 Sqn, RFC. *Training duties*: No.1 Reserve Aeroplane Sqn; No.8 Reserve Aeroplane Sqn; 4th Wing, Netheravon.

Serials: *Acquired in France*: 1856, 1858, 1860, 1864, 1865, 1867, 1868, 1876, 1877, 1879, 1883, 1889, 1890, 1898, 1899, 5001, 5010, 5011, 5013, 5014, 5017, 5025, 5026, 5028, 5066, 5097, 7458, A5169. *Built by Savages Ltd*: 4787—4836. *Transferred from RNAS*: 8506, 8518, 8523.

The Hon Charles Rolls's Wright biplane at Farnborough in June 1910. The figure second from the left is Mervyn O'Gorman. (*RAE 0615, Crown copyright*)

Wright biplanes

Preliminary announcements relating to the St Louis Universal Exposition of 1904 suggested that it would include a substantial and varied aeronautical section. At the suggestion of Col J. L. B. Templer, Superintendent of HM Balloon Factory, Lt-Col J. E. Capper, Commander of the Balloon Section, R.E., was sent to the USA to attend the Exposition. Capper found nothing aeronautical at St Louis, but on his own initiative he went to visit the Wright brothers in October 1904. On his return to England he submitted a report to the War Office, and in it he described his meeting with the Wrights.

The Wright brothers had offered their invention to the United States government but had been rebuffed. On 10 January, 1905, they wrote to Lt-Col Capper seeking his advice on how best to approach the British Government. When Capper referred the letter to the War Office the official reaction and subsequent action, by the Director of Fortifications and Works (Col R. M. Ruck), was swift, sensible and constructive. After some months of delay occasioned solely by the absence from Washington of the British Military Attaché there followed correspondence between him and the Wrights. In this the Wrights sought to impose impossible conditions and virtually refused to allow any War Office representative actually to see their aircraft; eventually the War Office, with complete justification, gave up, and on 8 February, 1906, the Director of Artillery wrote to the Wrights that their terms and conditions were unacceptable. The brothers reopened negotiations in May but went on to demand preposterous sums of money, so much so that even Capper had to recommend emphatically against acceptance. The negotiations and correspondence of 1906 were concluded by a memorandum of 9 November, 1906, sent by Maj N. Malcolm to the Military Attaché in Washington, in which the Attaché was asked to advise the Wright brothers that the Army Council had decided against purchasing the Wright aircraft.

According to correspondence published in *The Times* on 19 and 24 February, 1916, an offer of the Wright patents was made to the British Government in February 1907 by Mr Charles R. Flint of Flint & Co, New York, then a well-known financial company, apparently acting as an agent for the Wrights. The letter that was published on 24 February, 1916, written by Mr Laurance Lyon,* contained remarkably detailed quotations from the correspondence between Flint and the Rt Hon R. B. Haldane, the Secretary of State for War. On 4 March, 1907, Haldane had written to Flint:

> 'The War Office has not the least intention of entering into any agreement as to flying machines with anyone, or of giving them the slightest guarantee.'

According to Laurance Lyon a written proposal, made on behalf of the Wright brothers, was sent to Haldane on 10 April, 1907. This contained an offer of 'a preliminary exhibition' of flight, to be made in America by the Wrights in the presence of the British Ambassador or his nominee. The Wrights would

> '. . . pay all expenses of a full and complete demonstration made under the conditions to be fixed by contract. They also undertake to train a certain number of British subjects in the management of their aeroplanes. The Wrights being of English birth are anxious that the British Government should have the first advantage in this new and unique machine, which is exciting so much attention on the Continent.'

In his letter to *The Times* Lyon stated that Haldane replied to this offer on 12 April, 1907:

> 'I have nothing to add to my last letter to you. The War Office is not disposed to enter into relations at present with any manufacturer of aeroplanes.'

Lyon, probably unaware of the correspondence and abortive negotiations that had preceded Flint's approach to the War Office, righteously denounced Haldane's action in declining the offer of the Wright patents; but that offer, viewed through official eyes in the light of very recent experience of dealing with the Wrights, must have been seen as little more than an attempt to revive the subject through a different channel.

Interestingly, the remarkably well-informed Mr Lyon concluded his letter to *The Times* thus:

> 'It is fair to add that at the same time the matter was submitted to the Admiralty. It appeared to receive more careful consideration in that quarter, but on March 6, 1907, the then First Lord (the late Lord Tweedsmouth) wrote:—
> "I have consulted my expert advisers with regard to your suggestion as to the employment of aeroplanes, and I regret to have to tell you that, after careful consideration of the matter, the Board of Admiralty, while thanking you for so kindly bringing the proposals to their notice, are not of opinion that they would be of any practical value for the purposes of the Naval Service."'

Latter-day historians have not, it seems, commented on this letter from Mr

*Laurence Lyon (1875–1932) was a Canadian barrister who, after practising in Canada until 1905, resided in Paris and London. He was the proprietor of *The Outlook*, 1916–19, MP for Hastings 1918–21, and the author of several books.

Lyon, either in relation to the War Office or the Admiralty. It is perhaps uncertain whether the Wright brothers were themselves aware of Flint's approaches to the British authorities: if they were, they did not refer to them when they wrote again to the War Office on 10 April, 1908, claiming that their aircraft were now '... suitable for military scouting, being capable of carrying two men (an operator and an observer), and sufficient fuel for long flights.' The correspondence thus reopened proved inconclusive, and was not pursued by either side after Orville Wright's crash of 17 September, 1908, in which his passenger, Lt T. E. Selfridge, sustained fatal injuries.

In August 1908, a few weeks before Orville's disaster, Wilbur Wright had demonstrated the new and improved Wright Model A biplane in France. His flights astonished the aviation world and brought immediate demands from would-be purchasers of the aircraft. One of these was the Hon Charles S. Rolls who, according to Sir Walter Raleigh's account in *The War in the Air*, Vol I (page 156), offered in the autumn of 1908

> '... to bring to Farnborough a biplane of the Farman-Delagrange (*sic*) type, and to experiment with it on behalf of the Government, in return for the necessary shed accommodation.'

The correctness of that assertion is questionable, not least because in 1908 there was no aeroplane that could conceivably have been described as a Farman-Delagrange biplane. Rolls had first stated his desire to buy a Wright biplane as early as 11 June, 1908, but at that stage the Wrights had not determined where or by whom their aircraft were to be built in Britain and Rolls's intentions could not be conveniently realised. He may therefore have tried to buy a French aircraft as an alternative, but at that time he is more likely to have considered a Voisin (and Delagrange's Voisin was, unnecessarily and misleadingly, often described as a Voisin-Delagrange biplane).

According to Raleigh, Rolls's offer might have been officially accepted had not what Raleigh described as 'an accident to Mr Cody, caused by want of space' led to deferment of agreement. There can be little doubt that the accident in question was the heavy landing that terminated Cody's first truly successful flight on British Army Aeroplane No.1 on 16 October, 1908.

By February 1909 the Short brothers had entered into an agreement with the Wrights to build six Wright biplanes; the first of these was allotted to Rolls in fulfilment of his request of the previous year to purchase a Wright. In May 1909 it was reported in *Flight* that Rolls had 'generously placed his [Wright] machine at the disposal of the War Office. A shed is being erected for it by the military authorities, and a suitable ground for testing purposes has been provided.' The locale was not identified at that time, but *Flight* of 12 February, 1910, reported:

> 'A large shed is being erected on Hounslow Heath, which is, we understand, to accommodate the Wright flyer to be used by the army officers for their aeroplane training. It is probable that the Hon. C. S. Rolls will be actively identified with the initial training of the first flying pilots of the British Army.'

Rolls flew his aircraft for the first time early in October 1909, but a mishap that necessitated repairs prevented him from flying regularly until 1 November. Rolls made somewhat more than 200 successful flights, some with passengers, during the winter of 1909–10 until late February 1910. He towed the suitably dismantled Wright to London behind his car in time for it to be exhibited on the Royal Aero Club stand in the Olympia Aero Show that opened on 11 March, 1910.

During the Aero Show, Rolls sold his Wright biplane to the War Office on 16 March for the sum of £1,000: it was the first aeroplane to be acquired in this way by the British Army (though it cannot be claimed to have been the first aircraft owned by the Army, for it must give precedence to the Cody biplane of 1908, which was known as British Army Aeroplane No.1). The Wright biplane went first to Eastchurch after the Aero Show and was subsequently delivered to Farnborough in April, a movement that must have been accomplished very discreetly, for it seems to have escaped the notice of all the British aviation journals at the time.

On the evening of Monday 20 June, 1910, Charles Rolls visited the Balloon Factory and explained the Wright biplane to a group of officers of the Balloon section of the Royal Engineers. He continued this instruction next morning, when the biplane's engine was started up and the controls demonstrated. No flight was attempted but, reported *Flight* of 25 June, 1910. 'A starting rail has been laid down on Jersey Brow, and it was anticipated that practical flying would be commenced at the end of this week.'

No record yet found indicates that the Wright was ever flown at Farnborough or elsewhere. Rolls met his death on a Wright biplane at Bournemouth on 12 July, 1910, a fact that might well have decided the War Office not to fly the Army's Wright. It is uncertain how long the Wright survived in official hands. The files of today's Royal Aircraft Establishment contain a note in Mervyn O'Gorman's hand that was written in February or March 1911 listing the aeroplanes then at Farnborough:

'We have at present or shall shortly have:—
(1) Wright biplane, old type, Wright engine 25 hp
(2) a Blériot monoplane, old type, E.N.V. engine 60 hp
(3) a Farman type biplane made by Henry Farman, Gnome engine 7 cylinder rotary, 50 hp, 2nd type
(4) a Farman type biplane made by de Havilland, de Havilland engine, 40/45 hp
(5) a Farman type biplane made by Paulhan, Gnome engine 7 cylinder, rotary, 50 hp, improved type
(6) a Farman type biplane made by the Bristol Co., Renault engine, 60 hp.'

Furthermore, when the Army Estimates were published on 25 February, 1911, the section devoted to Aeronautics stated:

'There are now five aeroplanes available for Army work, of the Wright, Farman, Paulhan, Blériot and Havilland types respectively.'

These two documents suggest that the Wright was still on official charge late in February 1911.

In *Shorts Aircraft since 1900* (Putnam, 1967) C. H. Barnes states that the Army's Wright biplane went to Hounslow Barracks, a singularly odd move for which a logical reason is not easy to discern, unless there was official determination to justify the shed proposed for Hounslow Heath. The Wright's ultimate fate seems not to have been recorded, though Brig P. W. L. Broke-Smith, who was an instructor and acting Adjutant at Farnborough and had been selected for instruction at the time, recorded that it became warped and obsolete and was eventually written off.

A Wright biplane of a very different type arrived at Farnborough in September 1915. On 7 September, 1915, a telegram was sent to the War Office:

'Wright biplane received North Camp station today addressed Officer Commanding Aircraft Park, South Farnborough. No intimation having been received can you explain.'

It is not known whether the War Office ever provided an explanation, but on the same day a telegram originating from Sheerness arrived for the O.C. Aircraft Park: '60 hp Wright in covered railway van left here this morning for North Camp Station. Ogilvie.' The signatory could only have been Alec Ogilvie, in prewar years himself an exponent of early Wright biplanes, but the aircraft to which his telegram referred was a Wright of later date.

It was, amazingly, a variant of the biplane sometimes described as Wright Model F or Model H, similar to the aircraft that had been supplied to the US Signal Corps on 29 June, 1914, in an attempted response to an official US Army requirement for an armed and armoured aircraft. Despite its 90 hp Austro-Daimler engine it proved to be an abject failure. Although the US Army reluctantly purchased the 'Tin Cow', as the Wright was known, in March 1915, it swiftly consigned the aircraft to the scrap heap on 14 June, 1915.

The Wright Type F biplane at Farnborough in the autumn of 1915.
(*RAE 624; Crown copyright*)

Why or by whom it was decided to import and test an aircraft as hopelessly outdated and as militarily useless as the Wright F/H is unknown. Much of its technology dated from 1903, for the mainplanes and propulsion installation did not differ essentially from those of the Wright Flyer of December 1903. The aircraft delivered to Farnborough had only a 60 hp Wright engine mounted under a voluminous cowling in the nose of the fuselage, transmitting its power to the two pusher propellers via a long shaft and chain drive. The conventional fuselage supported a tailplane and biplane rudder at the rear. This can only have been the 60 hp Wright delivered by W. Cater & Co. under Contract No. A.2962.

It is not known whether this Wright was taken over by the Aircraft Park, the Royal Flying Corps Administration Wing, the Royal Aircraft Factory, or the Aircraft Inspection Department at Farnborough, but it was assembled and presumably was flown. It appears that at some time its radiator was greatly enlarged and repositioned behind the cockpit, which contained side-by-side seats for two.

The activities and fate of the Wright Model F/H at Farnborough, or anywhere else in England, remain unknown. It is doubtful whether it received an official serial number, but it deserves to be mentioned lest future historians be misled by references to it.

Short-built Wright biplane
27 hp Wright (made by Léon Bollée)

Two-seat biplane.
Span 40 ft; length 27 ft 9 in; wing area 590 sq ft.
Empty weight 885 lb.
Armament: None.

Manufacturer: Short Brothers.

Service use: Brief ground instructional use at HM Balloon Factory, Farnborough, June 1910.

Serial: No official number allotted, but when exhibited at Olympia in March 1910 the rudder bore Charles Rolls's personal marking C.S.R. No.2.

APPENDIX

Aircraft Manufacturers

The following are the correct full names of the aircraft manufacturers shown in abbreviated form in the data for individual aircraft types in the body of this book.

The Aeronautical Syndicate Ltd, Hendon.
Aéroplanes Farman, Paris (works at Camp de Châlons (Marne)).
Aéroplanes Henry et Maurice Farman, Billancourt (Seine).
Aéroplanes Hanriot, Rheims.
Aéroplanes Morane-Saulnier, Paris.
Aéroplanes Voisin, Issy-les-Moulineaux.
The Air Navigation Co Ltd: see Blériot & Spad Ltd.
The Aircraft Manufacturing Co Ltd (Airco), Hendon.
The Alliance Aeroplane Co Ltd, London.
Sir W. G. Armstrong, Whitworth & Co Ltd, Gosforth, Newcastle upon Tyne.
HM Army Aircraft Factory (later Royal Aircraft Factory), Farnborough, Hants.
Ateliers d'Aviation R. Savary et H. de la Fresnaye, Levallois-Perret.
The Austin Motor Co (1914) Ltd, Birmingham.
Avionnerie Kellner et ses Fils, Billancourt (Seine).
Barclay, Curle & Co Ltd, Whiteinch, Glasgow (some assembly only).
William Beardmore & Co Ltd, Dalmuir, Dunbartonshire.
Beatty School of Flying, Hendon and Cricklewood.
F. W. Berwick & Co Ltd, London.
The Birmingham Carriage Co, Birmingham.
Blackburn Aeroplane & Motor Co Ltd, Leeds.
The Blair Atholl Aeroplane Syndicate Ltd, London.
Blériot Aeronautics, Hendon.
Blériot Aéronautique, Levallois-Perret.
L. Blériot (Aeronautics) Ltd, Brooklands, Surrey (later Blériot & Spad Ltd).
Blériot & Spad Ltd (later The Air Navigation Co Ltd), Addlestone, Surrey.
Boulton & Paul Ltd, Norwich.
Borel: *see* Société anonyme des Aéroplanes Borel.
Breguet: *see* Société anonyme des Ateliers d'Aviation Louis Breguet.
Breguet Aeroplanes Ltd, London.
The British Caudron Co Ltd, Cricklewood and Alloa.
The British and Colonial Aeroplane Co Ltd, Bristol.
The British Deperdussin Aeroplane Co Ltd (formerly The British Deperdussin Syndicate Ltd) London.
The British Deperdussin Aeroplane Syndicate Ltd (later The British Deperdussin Aeroplane Co Ltd), London.
The Brush Electrical Engineering Co Ltd, Loughborough.
Canadian Aeroplanes Ltd, Toronto.
Caproni: *see* Società per lo Sviluppo dell'Aviazione in Italia.
Caudron Frères (later Aéroplanes Caudron), Rue, Lyon and Issy-les Moulineaux.
Clayton & Shuttleworth Ltd, Lincoln.
S. F. Cody, Farnborough, Hants.

The Coventry Ordnance Works Ltd, Coventry and London.
Cubitt Ltd, Croydon.
The Curtiss Aeroplane Co, Buffalo, NY, and Toronto.
The Curtiss Aeroplanes & Motors Co, Toronto.
The Daimler Co Ltd, Coventry.
Darracq: see Société anonyme Darracq.
A. Darracq & Co (1905) Ltd, London.
The Darracq Motor Engineering Co Ltd, London.
William Denny & Bros, Dumbarton (some assembly only).
Armand Deperdussin, Paris.
Deutsche Flugzeug-Werke GmbH, Lindenthal bei Leipzig.
Dudbridge Iron Works Ltd, Stroud, Gloucestershire.
The Eastbourne Aviation Co Ltd, Eastbourne.
Établissements Nieuport, Suresnes (Seine). (later Société anonyme des Établissements Nieuport).
The W. H. Ewen Aviation Co Ltd, Hendon (later The British Caudron Co Ltd).
The Fairey Aviation Co Ltd, Hayes, Middlesex.
Farman: see Aéroplanes Henry et Maurice Farman.
Richard Garrett & Sons, Leiston, Suffolk.
The Glendower Aircraft Co Ltd, London.
The Gloucestershire Aircraft Co Ltd, Cheltenham.
The Gosport Aviation Co Ltd, Gosport.
The Grahame-White Aviation Co Ltd, Hendon.
Handley Page Ltd, Barking, Essex, and Cricklewood.
Hanriot: see Aéroplanes Hanriot.
Harland & Wolff Ltd, Belfast.
Harris Lebus Ltd.
The Henderson Scottish Aviation Factory, Aberdeen.
Hewlett & Blondeau Ltd, Luton and London.
Hooper & Co Ltd, London.
R. L. Howard-Flanders Ltd, Richmond and Brooklands, Surrey.
Howard T. Wright, London.
Humber Ltd, Coventry
The Joucques Aviation Works, Willesden, London NW10.
The Kingsbury Aviation Co Ltd, Kingsbury and London.
Jacob Lohner and Co, Vienna.
Kellner: see Avionnerie Kellner.
Mann, Egerton & Co Ltd, Norwich.
Marsh, Jones & Cribb Ltd, Leeds.
Marshall Sons & Co Ltd, Gainsborough.
Martin and Handasyde (later Martinsyde Ltd), Brooklands, Surrey.
The Metropolitan Carriage, Wagon and Finance Co, Birmingham.
Morane-Saulnier: see Aéroplanes Morane-Saulnier.
Morgan & Co, Leighton Buzzard.
Napier & Miller Ltd, Old Kilpatrick (some assembly only).
D. Napier & Son, London.
National Aircraft Factory No.1, Waddon, Surrey.
National Aircraft Factory No.2, Heaton Chapel near Stockport.
National Aircraft Factory No.3, Liverpool.
Nieuport: see Établissements Nieuport and Société anonyme des Établissements Nieuport.
Nieuport & General Aircraft Co Ltd, Cricklewood.
Northern Aircraft Repair Depot, Coal Aston (conversions.)
Palladium Autocars Ltd, London.
Parnall & Sons Ltd, Bristol.
Louis Paulhan, St-Cyr.
The Phoenix Dynamo Manufacturing Co Ltd, Bradford.

Piggott Brothers and Co Ltd, London.
Planes Ltd, Freshfield, Lancashire.
Portholme Aerodrome Ltd, Huntingdon.
Ransomes, Sims & Jefferies, Ipswich.
Robey & Co Ltd, Lincoln.
A. V. Roe & Co Ltd, Manchester and Hamble.
Royal Aircraft Factory (later the Royal Aircraft Establishment), Farnborough, Hants.
Ruston, Proctor & Co Ltd (later Ruston, Hornsby & Co Ltd), Lincoln.
Frederick Sage & Co Ltd, Peterborough.
Angus Sanderson & Co, Newcastle upon Tyne.
S. E. Saunders Ltd, East Cowes, Isle of Wight.
Savages Ltd, Kings Lynn, Norfolk.
Louis Schreck, Argenteuil.
Short Brothers, Isle of Sheppey and, later, Rochester.
The Siddeley-Deasy Motor Car Co Ltd, Coventry.
Società per lo Sviluppo dell'Aviazione in Italia, Milan; factories at Vizzola Ticino and Taliedo.
Société anonyme Darracq, Suresnes (Seine).
Société anonyme des Aéroplanes Borel, Paris.
Société anonyme des Ateliers d'Aviation Louis Breguet, Douai.
Société anonyme des Établissements Nieuport, Paris. (factories at Argenteuil, Bordeaux, Ile de la Jatte, Issy-les-Moulineaux and Levallois-Perret).
Société anonyme française de Constructions aéronautiques, Levallois-Perret.
Société anonyme pour l'Aviation et ses Dérivés (SPAD), Paris.
Société Henry et Maurice Farman, Billancourt (Seine). (previously Aéroplanes Henry et Maurice Farman, Billancourt).
Société pour la Construction et l'Entretien d'Avions, Paris.
The Sopwith Aviation Co Ltd, Kingston-on-Thames, Surrey.
SPAD: *see* Société anonyme pour l'Aviation et ses Dérivés.
Standard Motor Co Ltd, Coventry.
Alexander Stephen & Sons, Linthouse, Glasgow (assembly only).
The Sunbeam Motor Car Co Ltd, Wolverhampton.
Vickers Ltd (Aviation Department), London. (factories at Crayford, Bexley Heath and Weybridge).
Vickers, Son and Maxim Ltd, Erith, Kent.
Voisin: *see* Aéroplanes Voisin.
Vulcan Motor & Engineering Co (1906) Ltd, Southport.
Waring & Gillow Ltd, London.
G. & J. Weir Ltd, Cathcart, Glasgow.
Wells Aviation Co Ltd, London.
Westland Aircraft Works, Yeovil, Somerset.
Weston Hurlin & Co, London.
Whitehead Aircraft Ltd, Richmond, Surrey.
Wolseley Motors Ltd, Birmingham.

INDEXES
AIRCRAFT

A.D. Navyplane 589
Aerial Wheel 3, 4–5
Aeronautical Syndicate Valkyrie
   35–37, 136
Airco
  D.H.1 and 1A 38–41, 49, 449
  D.H.2 39, 41–46, 49, 61, 66,
    164, 360, 361, 423, 431, 432,
    434, 435, 544, 585
  D.H.3 46–49, 83, 84
  D.H.4 48, 49–61, 67, 73, 75,
    77, 78, 79, 89, 173, 262, 332,
    423, 424, 425, 531, 602
  D.H.5 61–66, 67, 436, 480,
    516, 550
  D.H.6 66–71, 605
  D.H.7 84
  D.H.9 59, 69, 72–78, 79, 80,
    535
  D.H.9A 69, 76, 78–83
  D.H.10, 10A, 10C Amiens 48,
    76, 80, 81, 82, 83–88
  D.H.15 Gazelle 82
Albatros
  D I and D II 490
  D III 173, 366
  Two-seater 497
Antoinette monoplane 271
A.R.1 88–90
ARL 1A.2 88
ARL 2A.2 88
Armstrong Whitworth
  Biplane (N513) 107
  F.K.2 90–93
  F.K.3 93–99, 100, 516
  F.K.7 and 8 51, 97, 100–105,
    106, 177, 367, 592
  F.K.8 experimental 103, 104

F.K.9 and 10 106–109
Triplanes 106
Austin A.F.T.3 Osprey 548
Aviatik 40, 291
Avro
  Types E and Es xxii, xxiii, 6,
    24, 110–112, 287
  Type G 3, 6, 7, 26
  500 and 502 110, 111, 112
  504 xiii, xvi, xxii, xxiii, 24, 98,
    112–120, 121, 122, 210, 233,
    267, 274, 350, 568
  521 115, 118, 120–122
  529 and 529A 84

Beatty
  Beatty pusher biplane 125–128
  Beatty-Wright biplane 123–128
Blériot
  XI xi, xvi, xx, xxii, xxiii, 3, 4,
    8–10, 36, 120, 128–132, 134,
    135, 140, 219, 253, 290
  XII xiii, 37, 133–135, 341,
    614
  XXI 3, 8–10, 37, 135–137
  Parasol xvi, 131, 132, 138–141
Borel monoplane 4, 10
Boulton & Paul Bobolink 548
Breguet
  biplanes 142–147
  G3 144, 145
  L1 142, 143, 315
  L2 144, 145, 146
  U2 3, 11–13
Bristol
  Boxkite xi, 148–151, 371, 375,
    483, 614
  B.R.7 156

Bristol—*cont.*
  Coanda monoplanes 3, 14,
    153–156
  F.2A   169–174, 178, 502, 544
  F.2B   55, 102, 103, 171, 173,
    174–180, 282, 391, 430, 502,
    516, 531
  G.B.75   159–160
  Gordon England biplanes 3, 13,
    14, 153, 155
  M.1A, 1B, 1C   180–184, 314,
    535, 600, 601
  P.B.8   151
  Prier monoplane 151–153
  S.2A   168–169
  Scouts B, C and D   161–167,
    272, 278, 296, 298, 360, 467,
    490, 578
  T.B.8   151, 156–158, 159
  Type T   149
  R.2A   169, 170
  R.2B   170
British Nieuport
  B.N.1   548
  Nighthawk   541
Burgess-Wright biplane   509

Caproni Ca.1   185–189
Caudron
  biplane   126
  45 hp Anzani biplane   189–191
  C, D, E   191
  CRB   197
  G.3   191–194, 195, 196, 275
  G.4   194–196
  R.11   196–198
Cody
  British Army Aeroplane No. 1
    614
  III   15, 198, 199, 200
  IV   4, 15, 16, 198
  V   xv, 4, 16, 27, 198–200, 345
Coventry Ordnance Works
  Military Trials biplane (Chenu)
    3, 16–18, 30
  Military Trials biplane (Gnome)
    3, 16–18

Curtiss
  C   204–208
  H-1   204, 205
  H-4   204, 205
  H-7   205
  H-12   262
  JN-1   201
  JN-3   71, 201–204, 205, 209,
    210, 212
  JN-4   209, 213
  JN-4A   209–211, 214
  JN-4 (Canadian)   211–215
  N   201
  R-2   201

De Havilland
  D.H.82A Tiger Moth   454
  D.H.98 Mosquito   59
Deperdussin
  monoplanes 17, 215–220, 315
  British
    60 hp Anzani   37
    100 hp Anzani 3, 18–21, 38,
      217
    70 hp Gnome   218
    100 Gnome   3, 217, 218
  French
    60 hp Anzani   215
    100 hp Gnome   3, 4, 155,
      217
D.F.W. Mars monoplane   3, 29
Dorand tractor biplane   88
Dunne
  D.5   221
  D.8   221–223

Etrich monoplane   15, 21, 198

Fabre floatplane   339
Fairey
  F.1   489
  IIIA   284
Farman, Henry
  Biplane, Nos. 208 and
    209   226–227
  F.20   xii, xvi, xvii, xx, xxii,
    xxiii, 226, 229–234, 252, 254,
    255, 352, 397, 398

F.20 hybrid at Ismailia 233
F.22 227
F.27 234–238
F.40 88
Type III xi, 36, 148, 223, 564
Type III militaire 223–226, 614
Wake up, England biplane
  227–228, 252
Farman, Maurice
  S.7 Longhorn xxii, xxiii, 3, 22,
    23, 160, 238–241, 260, 290,
    493
  S.11 Shorthorn xix, xxii, xxiii,
    241–246, 260, 275, 290, 351
  S.11 Shorthorn seaplane 242,
    247
F.B.A. Flying-boat 247–249, 487
Felixstowe
  F.2A 59, 80, 82
  F.3 59, 80, 82, 263
  F.5 80, 82
Flanders
  B.2 3, 23, 24, 574
  F.2 249
  F.3 249, 250
  F.4 23, 219, 249–251
Fokker
  Dr I 104
  E I 42, 361, 366, 381, 384, 385,
    432

Gotha 103, 171, 396, 508, 521, 539,
  596, 602
Grahame-White
  Pusher Biplane 252, 258
  School Biplane 252–253
  Type VII Popular Biplane 252,
    254–255, 256
  Type VIIc Popular Passenger
    Biplane 252, 254, 255–256,
    258, 259, 260
  Type VIII 252, 256–258
  Type XV 259–261
  Type 18 489

Halberstadt fighter 366, 490

Handley Page
  O/100 89, 208, 261, 262, 263,
    264, 266, 489
  O/400 59, 76, 80, 82, 261–268,
    414, 602, 605
  Type E 123
  Type F 4, 25, 26
  V/1500 44, 76, 82, 457, 602
Hanriot monoplane 3, 26–28
Harper monoplane 3, 28, 29
Hewlett & Blondeau Bomber 489
Howard Wright biplane 269–270

Lohner biplane 3, 29
L.V.G. C V 478

Martin-Handasyde
  Military Trials monoplane 3,
    30, 31
  monoplanes 3, 276
  RFC monoplane 30, 219,
    271–272
Martinsyde
  G.100/G.102 51, 56, 97,
    276–280, 305, 360, 389, 392,
    594, 600
  F.3 175, 280–285
  F.4 Buzzard 282–285, 541, 549
  R.G. 175, 280
  S.1 272–276, 298, 350
  Single-seat fighter, 190 hp
    Rolls-Royce 280
Mersey monoplane 3, 31, 32
Morane-Saulnier
  Type AC 182, 312–314
  Type AI 312
  Type BB 290, 294, 296, 301,
    303–306, 307, 308, 323, 360
  Type G 285–289
  Type G-19 138, 289
  Type H 285–289
  Type I 300–303, 313, 314
  Type L 138, 289–292, 294, 296
  Type LA 290, 292–296, 301,
    306, 307, 309, 310, 311, 360
  Type N 180, 278, 280, 290,
    294, 296–303, 307, 313, 314

Morane-Saulnier—*cont.*
   Type P   295, 304, 306–312
   Type U   313
   Type V   300–303, 313, 314

Nieuport
   11 and 21   319–320, 326
   12 and 20   320–326, 490, 514
   16   45, 326–328, 329
   17   319, 328–333, 436
   17B   319
   17bis   319, 334
   23 and 23bis   332–333, 337
   24 and 24bis   332, 334–336, 337
   27   327, 332, 337–338
   Monoplanes 37, 219, 252, 315–319
   Nighthawk 541

Paulhan biplane 223, 338–341, 371, 614
Piggott
   biplane 4, 32, 33
   monoplane 32
Planes Ltd biplane 31

Royal Aircraft Factory
   A.E.3   267
   B.E.1   226, 317, 341–344, 353, 370, 371
   B.E.2   xv, 156, 270, 344–354, 370, 371, 377, 441
   B.E.2a   xii, xvi, xvii, xxii, xxiii, 114, 131, 240, 251, 255, 287, 344, 346–354, 398
   B.E.2b   xxiii, 344, 350–354
   B.E.2c   xvii, xx, xxiii, 51, 90, 91, 92, 98, 160, 166, 169, 171, 235, 236, 239, 294, 354–370, 380, 382, 383, 385, 386, 387, 389, 394, 399, 404, 458, 472, 487, 592, 600
   B.E.2d   51, 361, 462, 592
   B.E.2e   51, 360, 361, 389, 394, 395, 460, 461, 592
   B.E.2f   51, 366, 367, 368
   B.E.2g   51, 366, 367, 368

B.E.3   137, 370–375, 377, 464
B.E.4   137, 370–375, 377
B.E.5   345, 347
B.E.6   270, 345
B.E.7   375–376, 377
B.E.8 and 8a   xvi, xvii, xxii, xxiii, 113, 377–381, 484
B.E.9   381–385
B.E.9a   384
B.E.12   89, 98, 171, 280, 385–397, 448, 455, 457, 458
B.E.12a   367, 392–394
B.E.12b   394–397
B.S.1 (S.E.2)   440, 464, 465
B.S.2 (R.E.1)   440
F.E.1   xiii, 398, 614
F.E.2   398, 399
F.E.2a   39, 95, 381, 397–403, 405, 410, 607
F.E.2b   xv, 39, 40, 48, 55, 68, 69, 95, 97, 360, 361, 403–419, 420, 436, 469, 594
F.E.2c   402, 414–417
F.E.2d   ii, 53, 262, 410, 414, 419–426, 428, 429, 544, 553
F.E.2h   417–419
F.E.4   47, 426–430
F.E.4a   430
F.E.4b   430
F.E.8   xiv, 431–436, 544, 585, 587
F.E.9   436–440, 595
F.E.10   475
H.R.E.2   374, 445
R.E.1   xix, xxii, xxiii, 354, 355, 440–444, 445, 447
R.E.3   445
R.E.4   445
R.E.5   xxii, xxiii, 95, 400, 444–451
R.E.7   51, 389, 451–458
R.E.8   51, 100, 101, 169, 177, 178, 367, 388, 458–464
R.E.8a   460
R.E.9   462
R.T.1   See Siddeley R.T.1
S.E.1   341, 371

S.E.2 440, 464–468, 493, 495, 497
S.E.3 466, 468
S.E.4 376, 445, 467, 468, 469
S.E.4a 416, 468–471
S.E.5 xvi, 332, 460, 471, 536, 544, 555, 562, 595
S.E.5a 66, 176, 177, 183, 311, 395, 460, 471–481, 517, 531, 536, 542, 558

Short
  Bomber 489–490
  No. 18 482
  S.26 482
  S.27 482
  S.28 482
  S.32 482, 498
  S.33 482
  S.36 484
  S.38 486
  S.41 484
  S.45 484
  S.62 486–487
  S.80 498
  S.81 485
  School biplane 482–483
  Seaplane, Admiralty Type 184 489
  Seaplane, Admiralty Type 827 487–488, 609
  Tractor biplane 378, 484–485
Siddeley R.T.1 462
Sommer monoplane 143
Sopwith
  Baby seaplane 89
  Bat Boat 491
  Camel (F.1) xvi, xvii, 78, 117, 266, 311, 497, 507, 508, 517, 518, 522–535, 544, 545, 546, 549, 550, 551
  Dolphin (5F.1) 176, 177, 197, 523, 535–544, 556, 562
  Dragon 541, 548
  Hawker's Runabout 509, 510, 511, 512
  LCT, the 1½ Strutter 88, 168, 267, 489, 490, 499–509, 514, 519, 525, 526, 528
  Pup 322, 325, 326, 509, 512–519, 522, 523, 544, 600
  Pup (Admiralty Type 9901/9901a) 513, 522
  Rhino 177
  Salamander (T.F.2) 550–552
  Schneider seaplane 1914 495
  Sigrist Bus 499
  SL.T.B.P. 510
  Snipe (7F.1) 284, 543, 544–549
  Sparrow 509–512
  S.S.1 Tabloid xvii, xxii, xxiii, 272, 273, 443, 493–498
  T.F.1 533, 551
  Tractor biplane (80 hp Gnome) xxii, 352, 491–493
  Triplane 106, 326, 519–521, 544, 555
  Sopwith-Wright tractor biplane 491
Spad
  Type A 381
  7.C 1 313, 460, 472, 521, 539, 553–558, 559, 561, 563
  12.Ca 1 559–560
  13.C 1 338, 477, 536, 541, 542, 556, 558, 560, 561–564
Staaken Giant 396
Stedman bomber projects 264

Tarrant Tabor 195
Taube 568

Vickers
  Boxkite 564–565
  Early gun-carrying biplanes 566–568
  E.F.B.1, 2 and 3 566
  E.S.1 576–581, 597, 600, 601
  F.B.5 400, 404, 568, 569–573, 582, 589
  F.B.7 and 7A 574–576, 602
  F.B.8 602
  F.B.9 404, 572, 573, 582–584, 588

Vickers—*cont.*
    F.B.12   585–590, 600
    F.B.14   367, 588, 590–596
    F.B.19   184, 597–601
    F.B.27 Vimy 76, 80, 268, 414, 602–605
    Nos. 2, 3, 4 and 5 monoplanes 564
    No. 6 monoplane 3, 33, 34
    No. 8 monoplane 34
    Pumpkin 565
Voisin
    biplane 340, 342, 613

    gun-carrying pusher 606
    LA   400, 606–610
    LA.S   606–610

Weston Hurlin biplane 28
Wight Bomber 489
Wright
    biplane xiii, 123, 125, 126, 611–616
    Type F/H   615, 616

AERO-ENGINES

A.B.C.
  60 hp 6
  100 hp 23, 24
  320 hp Dragonfly 548
  35 hp Gnat 511
Antoinette 65 hp 30, 31, 271, 272
Anzani
  25 hp 128, 129
  28 hp xi, 132
  35 hp 33, 254, 255
  45 hp 132, 190, 191
  60 hp 215, 216, 217, 219, 220, 256, 258
  80 hp 193, 194
  100 hp 18–21, 194, 196, 220, 588, 589, 590
  125 hp 589
A.R.1 (see also Bentley B.R.1) 184, 528, 545
Austro-Daimler
  90 hp 615
  120 hp 15, 16, 21, 29, 40, 46, 187, 198, 199, 200, 276, 400, 401, 403, 446

Beardmore
  120 hp 39, 40, 46, 48, 49, 95, 96, 97, 98, 99, 100, 169, 186, 187, 188, 189, 276, 278, 279, 280, 399, 403, 410, 411, 414, 418, 451, 453, 454, 455, 458, 591, 594, 596
  160 hp 48, 49, 100, 105, 278, 280, 404, 410, 411, 414, 418, 420, 456, 458, 590, 592, 594, 595, 596
Beatty 50 hp 125, 128
B.H.P.
  200 hp 49, 50, 51, 60, 72, 84, 500, 590, 596
  230 hp (see also Galloway Adriatic) 52, 53, 55, 56, 177, 179, 280, 603
Bentley
  B.R.1 524, 525, 527, 529, 530, 533, 546
  B.R.2 546, 549, 552
Brotherhood 200 hp 489

Canton–Unné—see Salmson (Canton-Unné)
Chenu
  75 hp 30, 31, 271
  110 hp 16–18, 30
Clerget
  80 hp 7Z 113, 119, 157, 158, 380, 381, 470, 471, 518
  110 hp 9Z 45, 46, 66, 106, 109, 122, 167, 169, 184, 320, 321, 325, 434, 436, 484, 485, 499, 503, 504, 508, 519, 522, 573, 577, 578, 581, 597, 598, 600, 601
  130 hp 9B 108, 109, 117, 118, 119, 184, 319, 323, 334, 502, 504, 508, 521, 523, 525, 526, 527, 528, 529, 530, 531, 533, 545, 579
  140 hp 9Bf 530, 533
  200 hp 11E 546, 547, 552
Curtiss
  90 hp OX-5 70, 71, 203, 205, 210, 211, 212, 215, 240
  170 hp VX 205, 206, 207, 208

Daimler-Mercedes
  70 hp 13, 14
  90 hp 156
Dansette-Gillet 143
de Dion 80 hp 241, 242
de Havilland 40 hp Iris 614

E.N.V.
  50 hp   133
  60 hp Type F   110, 133, 134, 135, 269, 270, 347

Fiat
  100 hp A.10   186, 189
  260 hp A.12   56, 58, 60, 75, 76, 77, 262
  300 hp A.12bis   264, 265, 603, 604, 605

Galloway
  230 hp Adriatic 55, 56, 57, 60, 72, 73, 84
  500 hp Atlantic 82, 83

Gnome
  50 hp Omega 35, 36, 37, 110, 112, 114, 124, 126, 127, 128, 129, 130, 132, 143, 148, 149, 151, 153, 226, 253, 254, 256, 290, 315, 316, 319, 339, 341, 372, 373, 375, 398, 487, 509, 510, 511, 512, 565
  70 hp Gamma 8, 10, 25, 26, 34, 129, 130, 132, 135, 137, 143, 218, 220, 227, 228, 255, 256, 259, 315, 317, 319, 371, 372, 375, 377, 398, 482, 483, 485, 491, 565
  75 hp Monosoupape A   159, 160
  80 hp Lambda 10, 41, 113, 119, 129, 130, 131, 132, 138, 141, 143, 153, 156, 157, 158, 161, 162, 167, 185, 194, 221, 222, 229, 234, 256, 260, 261, 289, 290, 373, 375, 378, 381, 467, 468, 469, 470, 471, 491, 493, 498, 499, 518
  100 hp Delta 185, 466
  100 hp Monosoupape B-2   41, 44, 45, 46, 66, 115, 118, 119, 169, 249, 322, 323, 325, 379, 380, 381, 431, 434, 436, 495, 498, 512, 515, 516, 518, 532, 533, 566, 568, 569, 572, 573,
574, 576, 577, 578, 580, 581, 582, 584, 586, 587, 589, 590, 597, 601
  100 hp Omega-Omega 13, 14, 16–21, 26, 28, 140, 218, 219, 220, 315, 317, 319, 464, 465, 468
  140 hp Gamma-Gamma 375, 376, 465
  150 hp Monosoupape N   532, 533, 545
  160 hp Lambda-Lambda 375, 376

Green
  30–35 hp   35, 36, 37
  40–50 hp   37
  60 hp   6, 7, 28, 29, 36, 37, 221, 249, 260
  100 hp   39, 200, 397, 400, 401, 403

Gyro
  50 hp   123, 128
  60 hp   123

Hart
  35 hp   585
  150 hp   585, 586, 587, 588, 590

Hispano-Suiza
  150 hp 8Aa   107, 170, 171, 172, 173, 361, 362, 366, 369, 471, 473, 480, 481, 553, 554, 558
  180 hp 8Ab   481, 554, 558
  200 hp 8Ba, 8Bb, 8Bd   51, 72, 107, 175, 176, 177, 178, 179, 196, 197, 262, 311, 395, 396, 436, 439, 440, 460, 462, 473, 475, 476, 477, 478, 479, 481, 535, 536, 542, 543, 544, 559, 561, 564, 602, 603, 605
  215 hp 8Bda   198
  220 hp 8Bc/Be   481, 544, 561, 564
  220 hp 8Cb   559, 560
  300 hp 8Fb   282, 284, 285, 542, 543

Isaacson 45 hp   32

Le Rhône
- 60 hp   260
- 80 hp 9C   113, 117, 119, 132, 158, 162, 166, 167, 188, 194, 195, 196, 234, 260, 290, 291, 292, 296, 298, 302, 309, 310, 312, 319, 326, 470, 471, 512, 516, 518, 585, 586, 590
- 110 hp 9J   44, 45, 46, 64, 66, 108, 109, 117, 118, 119, 167, 184, 298, 301, 302, 304, 306, 308, 310, 312, 313, 314, 319, 323, 325, 326, 327, 328, 332, 333, 434, 436, 508, 517, 518, 525, 528, 529, 530, 531, 532, 533, 545, 572, 573, 579, 581, 584, 588, 589, 590, 598, 600, 601
- 120 hp 9Jb   314, 332, 333, 335, 336
- 130 hp 9Jby   336, 338

Liberty
- 290 hp Liberty 8   80
- 350 hp Liberty 12N   268
- 400 hp Liberty 12   74, 80, 81, 82, 83, 85, 86, 87, 268, 603

Lorraine
- 150 hp   105, 594, 595, 596
- 240 hp 8A   88
- 240 hp 8Bb   282, 285

Mercedes 100 hp   21

N.E.C. 50 hp   5
Nieuport 28 hp   318, 319

Renault
- 60 hp   147, 148, 149, 151, 342, 343, 344, 347, 353
- 70 hp   22, 23, 39, 40, 90, 93, 99, 145, 146, 147, 156, 193, 194, 249, 251, 345, 347, 353, 354, 356, 369, 370, 382, 385, 398, 441, 444
- 80 hp   71, 241, 242, 246, 247, 353, 575, 576
- 160 hp   88

- 190 hp 8Gd and 8GDx   88, 90
- 220 hp   264, 430, 456, 458

R.E.P.   143

Rolls-Royce
- 190 hp Mk I (190 hp Falcon I)   52, 170, 171, 173, 174, 179, 280, 430, 455, 456, 458
- 190 hp Mk II (220 hp Falcon II)   174, 175, 179
- 190 hp Mk III (275 hp Falcon III)   174, 176, 177, 178, 179, 280, 285
- 250 hp Mk I (255 hp Eagle I)   49, 50, 52, 53, 60, 261, 410, 419, 421, 423, 424, 426, 428, 429, 430, 490
- 250 hp Mk II (266 hp Eagle II)   60, 264
- 250 hp Mk III (284 hp Eagle III)   53, 54, 60, 419, 423, 426, 456, 457, 458
- 250 hp Mk IV (284 hp Eagle IV)   53, 54, 60, 268, 423, 424, 426, 595, 596
- 275 hp Mk I (322 hp Eagle V)   53, 54, 57, 60, 74, 75, 423, 424, 426
- 275 hp Mk II (322 hp Eagle VI)   53, 54, 60, 264, 423, 426
- 275 hp Mk III (322 hp Eagle VII)   53, 60, 264
- 285 hp Falcon experimental 285
- 375 hp Eagle VIII   54, 58, 59, 60, 78, 79, 80, 81, 82, 84, 85, 86, 87, 267, 268, 603, 604, 605

Royal Aircraft Factory
- 90 hp R.A.F.1a   67, 71, 92, 93, 95, 96, 98, 99, 355, 357, 364, 369, 382, 384, 385
- 105 hp R.A.F.1b   98, 364, 365, 369
- 120 R.A.F.2   380, 381
- 200 hp R.A.F.3a   52, 55, 56, 60, 262, 417, 426, 427, 429, 430, 456, 458
- 140 hp R.A.F.4   383, 384, 385, 448

629

Royal Aircraft Factory—*cont.*
  140 hp R.A.F.4a 46, 51, 98, 105, 385, 387, 389, 390, 395, 396, 417, 448, 454, 455, 456, 458, 460, 461, 463, 595, 596
  200 hp R.A.F.4d 105, 176, 177, 462, 463, 602
  150 hp R.A.F.5 411, 417, 418, 427, 428, 429, 430
  170 hp R.A.F.5b 417

Salmson (Canton-Unné)
  85 hp 145, 146, 147
  100 hp 606
  110 hp 11, 13, 146, 147
  130 hp M.9 143, 607, 610
  135 hp 234
  140 hp B.9 237, 610
  150 hp 188
  160 hp 237
  260 hp 9Zm 603, 605
Siddeley
  230 hp Puma 55 56, 59, 60, 72, 73, 75, 76, 77, 83, 84, 85, 87, 177, 417, 418
  Puma (high compression) 57
Smith 150 hp Static 589
Sunbeam
  200 hp 107

200 hp Arab 72, 176, 177, 178, 179, 462, 477, 480, 481
320 hp Cossack 261, 262, 263, 268
150 hp Crusader 206, 448, 488
260/275 hp Maori 264, 265, 266, 267, 268, 603, 605
225 hp Mohawk 455, 456, 458, 489, 490
200 hp Saracen 72

Viale 70 hp 33
Vickers 50 hp Vickers-Boucier radial 565

Wolseley
  60 hp 341, 342, 343, 344
  150 hp W.4A Python I 480, 481, 558
  180 hp W.4A Python II 481, 558
  200 hp W.4B Adder 477, 481
  200 hp W.4A* Viper 479, 480, 481, 542
Wright
  27 hp 616
  35 hp 124
  40 hp 123
  60 hp 615

ARMAMENT

Armour 359, 360, 368, 441, 444, 470, 524, 533, 550, 551

Bombs
  10-lb 156
  16-lb H.E.R.L. 236
  20-lb 353, 368, 388, 426, 488, 504, 551
  20-lb H.E. Hales 451
  25-lb H.E. Cooper 66, 74, 105, 179, 412, 418, 426, 478, 481, 508, 519, 534, 544, 552, 558
  40-lb 418
  65-lb 488
  100-lb H.E.R.L. 379, 394, 504
  112-lb H.E.R.L. 50, 60, 73, 78, 278, 369, 393, 397, 412, 418, 463, 489, 490
  200-lb 50
  230-lb H.E.R.F.C. 58, 60, 77, 78, 82, 87, 278, 368, 369, 414, 418, 490
  336-lb H.E.R.A.F. 357, 392, 414, 451, 454, 458
  350-lb 357
  500-lb 448, 457
  599-lb R.A.F. 426
  10-kg Anilite 558
  120-mm (French) 90
  450-lb depth charge 89
  Hales grenade 276
  Petrol bomb 233
  Powder bomb 276
  R.A.F. jettisonable petrol tank 452
  Rifle grenade 161, 162

Bomb sights
  Equal Distance 505
  R.A.F. Gyro-stabilised 51, 206, 413, 417
  R.A.F. Periscopic 457

Guns
  (a) Heavy guns
    Davis 6-pdr 395, 418
    37-mm Hotchkiss 559, 560
    Vickers 1-pdr 398, 414, 415, 418, 574, 575, 576
    Vickers 1-pdr Mk III 207, 208
    Vickers pom-pom 413, 414
  (b) Machine-guns
    Hotchkiss 232, 234, 296, 301, 397
    Lewis 40, 42, 44, 46, 48, 49, 60, 71, 78, 83, 86, 87, 90, 93, 99, 100, 108, 109, 114, 115, 118, 119, 122, 162, 164, 167, 169, 170, 173, 179, 195, 206, 208, 215, 233, 234, 237, 242, 244, 246, 261, 263, 274, 276, 277, 278, 280, 285, 291, 296, 301, 303, 305, 306, 308, 309, 311, 312, 322, 325, 327, 328, 329, 330, 332, 333, 335, 336, 338, 357, 358, 359, 368, 369, 385, 388, 389, 391, 393, 394, 397, 400, 401, 402, 403, 409, 413, 418, 420, 421, 423, 424, 426, 427, 428, 430, 431, 432, 435, 436, 437, 440, 450, 451, 453, 457, 458, 459, 460, 461, 463, 470, 471, 472, 473, 475, 479, 480, 481, 488, 490, 499, 500, 503, 506, 507, 508, 516, 517, 519, 531, 533, 534, 537, 539, 541, 542, 544, 548, 549, 550, 551, 557, 570, 572, 573, 584, 587, 590, 596, 605, 608, 610

Guns—*cont.*
    Lewis, double-yoked 90, 97, 105, 179, 268, 280, 406, 413, 423, 480, 507
    Maxim, 0.303-in 398
        0.45-in 44, 413, 418
    Rexer 231, 232, 234, 397, 398
    Vickers, 0.303-in 57, 60, 63, 64, 66, 78, 83, 90, 100, 105, 108, 109, 165, 166, 167, 170, 173, 177, 179, 181, 182, 184, 214, 215, 230, 231, 232, 234, 280, 285, 301, 303, 308, 309, 310, 312, 313, 314, 321, 324, 325, 330, 332, 335, 358, 368, 369, 388, 389, 391, 397, 457, 458, 459, 460, 462, 463, 472, 475, 479, 481, 499, 502, 503, 507, 508, 512, 519, 521, 523, 524, 534, 535, 544, 545, 549, 551, 552, 553, 558, 559, 560, 561, 564, 566, 567, 568, 570, 579, 581, 593, 596, 598, 601
  (c) Other guns
    Carbine 132, 226, 353, 364
    Pistol 132, 161, 162, 194, 246, 353, 451, 498, 609, 610
    Rifle xiii, 132, 161, 162, 194, 233, 246, 353, 369, 451, 468, 572, 573, 609, 610
    Signal pistol 53
Gun Mountings
  Albemarle 358
  Anderson (F.E.2 No. 4 Mks I and II) 52, 405, 406, 408, 420, 421, 422
  Anderson arch (F.E.2 No. 10 Mk I) ii, 406, 420, 422, 424
  B.E.2c No. 10 Mk I 357
  Bowan & Williams 550
  Clark (F.E.2 No. 4 Mk IV) ii, 405, 421, 423, 424
  Dixon-Spain ii, 421, 425
  Eeman 280, 479, 480
  Elevating (D.H.5) 62, 63
  Elevating (Vickers F.B.12A) 587
  Etévé (Nieuport) 169, 170, 325, 499, 501, 503
  F.E.2 No. 2 Mk I 400, 401, 404
  F.E.2 No. 2 Mk II 404
  F.E.2 No. 4 Mks I, II, III, IV ii, 405, 406, 408, 421, 423, 424
  Foster 118, 119, 327, 329, 330, 333, 336, 338, 475, 507
  Garros-Hue 286, 290, 296
  Martinsyde No. 5 Mks I, II 277, 278
  Medlicott 358
  Nieuport—see Etévé
  R.E.7 No. 3 Mk I barbette 456
  Rising pillar 40, 43
  Scarff Ring 52, 55, 95, 99, 100, 171, 173, 175, 214, 215, 310, 457, 461, 503, 584, 592
  Strange 114, 358, 368
  Vickers No. 1 and No. 2 570
  Vickers No. 3 Mk I 582
  Vickers No. 3 Mk II 583
  Vickers ring mounting 503
Gun Sights
  Aldis 278, 471, 479
  Bellieni 389
  Hutton 409
  Parallel motion 567
  Periscopic 389, 550
  Ring and bead 391, 479
Prideaux ammunition-belt links 461, 516, 555
Rockets
  Le Prieur 45, 328, 330, 333, 364, 368, 369, 392, 395, 397
Synchronising mechanisms
  Airco 64

Alkan 301, 308, 327
Armstrong Whitworth 101
Arsiad 101, 165, 167, 324, 460
Birkigt 553
Constantinesco 64, 65, 101, 184, 284, 462, 475, 506, 531
Ross 505, 506
Scarff-Dibovsky 321, 502, 503
Sopwith-Kauper 64, 184, 506, 526, 529
Vickers-Challenger 165, 166, 167, 280, 389, 457, 460, 462, 502, 503, 579, 585, 598

Other weapons
    Darts/Fléchettes 497, 498
    Depth charge, 450-lb 89
    Fiery Grapnel 353, 369
    Hand grenade 497
    R.A.F. jettisonable fuel tank bomb 452
    R.L. Tube 53
    Torpedo 605

PEOPLE

Ainslie, Capt E. M. L. 65
Allabarton, Lt S. F. 554
Allen, Capt C. R. W. 153, 374
Allen, Lt Denis, RN 550
Anderson, Capt E. V. 493
Anderson, Lt W. 424
Andrews, Lt Cdr RN 78
Armstrong, Lt D. V. 181
Arthur, Lt C. E. F. 78, 535
Arthur, Lt Desmond 347, 373
Ashmore, Brig-Gen E. B. 383, 406
Asquith, Rt Hon Herbert xiii

Babington, Sqn Cdr J. T. 264
Baldwin, Maj J. E. A. 416
Balfour, Lt H. H. (Lord Balfour of Inchrye) 122
Ball, Capt Albert 165, 327, 329, 472, 475
Bannerman, Bt Maj Sir Alexander xiii, 339
Barber, Horatio 35–37
Barfield, Lt J. C. H. 571
Baring, The Hon Maurice 187, 384
Barnes, C. H. 614
Barnwell, Frank S. 158, 161, 163, 171, 180
Barnwell, Harold 495, 567, 576, 578, 597
Barratt, Lt A. S. 141
Barrington-Kennett, Lt B. H. xiii, 150, 317
Barrington-Kennett, Lt V. A. 379
Barry, Lt C. C. 497
Bayly, Lt C. G. G. 112
Beatty, George W. 123
Beatty, Lt Col W. D. 188, 511
Béchereau, Louis 559
Becke, Maj J. H. W. 92, 348, 446
Beevor, 2/Lt C. F. 565
Bell, C. Gordon 20, 28, 30, 271, 444, 484, 486, 602

Bell-Irving, Lt A. D. 298
Benbow, Lt E. L. 435, 587
Bentley, W. O. 546
Bettington, Lt C. A. 153, 155
Bielovucic, J. 26–28
Bier, Leutnant H. 15, 21, 198
Billing, Noel Pemberton xvi
Birdwood, Lt H. F. 305
Birkigt, Marc 471, 553
Bishop, Capt W. A. 480, 536
Bispham, 2/Lt D. C. 127
Blackburn, Col H. 560
Blaschke, Leutnant von 29
Blériot, Louis xi, 10, 35, 128, 133, 134, 342
Blondeau, Gustav 564
Blood, Capt Bindon 470
Borel, Gabriel 10
Borton, Lt A. E. 291
Bottieau, Col 186
Brancker, Maj-Gen W. Sefton xix, xxii, 50, 51, 175, 177, 181, 184, 186, 247, 308, 354, 355, 364, 365, 384, 429, 430, 466, 473, 521, 588, 589, 601
Brandon, 2/Lt A. de B. 359
Breeze, Lt 43, 382
Breguet, Louis 142
Briggs, 2/Lt S. P. 388
Broadsmith, H. E. 117
Brock, W. L. 286, 287
Broke-Smith, Capt P. W. L. xiii, 240, 371, 483, 614
Brooke-Popham, Brig.-Gen H. R. M. xiii, 51, 63, 74, 102, 187, 188, 208, 304, 305, 307, 308, 309, 371, 382, 383, 389, 404, 405, 410, 414, 417, 420, 421, 427, 439, 459, 471, 476, 503, 504, 527, 539, 542, 556, 560, 561, 567, 578, 587, 598, 599
Brown, Capt Vernon 197, 521, 574, 596

Brownell, Capt R. J. xvii
Buchanan, Lt W. J. 89, 90
Buchanan, Maj J. S. 602
Buck, Lt G. S. 562
Burke, Lt Col C. J. xiii, 211, 223, 224, 225, 343, 348
Burroughs, Lt J. E. G. 374
Busby, Capt Vernon 44, 457
Busby, 2/Lt H. E. 126
Busk, Edward T. 350, 354, 355, 442, 443, 444
Busteed, Harry 153, 154, 155
Buxton, 2/Lt G. B. 331

Caddell, Brig-Gen W. B. 530
Caillé, Albert 339
Cairns, 2/Lt D. S. 367
Caldwell, 2/Lt K. L. 523
Cambray, Lt W. C. ii, 425
Cammell, Lt R. A. xiii, 35, 37, 134, 135, 136, 137
Campbell-Heathcote, Maj R. 454
Cantacuzene, Prince 159
Capel, Capt A. J. 182, 580
Capper, Lt-Col J. E. 611
Caproni, Gianni 185, 186
Carbery, Lord 161
Carey, 2/Lt A. S. 502
Carmichael, Lt G. I. 378, 379
Carr, Maj R. H. 287, 318, 527, 528
Carter, Air Mech 493
Carter, Maj A. D. 539, 542
Caudron, René 197
Chadwick, Capt 541
Challenger, G. H. 34
Chambenois, Marcel 10
Chanteloup, Pierre 197
Chavannes, Fernand 559
Chelmsford, Lord 237
Chidson, 2/Lt M. R. 568
Cholmondeley, Capt R. 161, 567
Churchill, Winston S. 82
Clark, Flt Cdr Dalrymple, RN 121
Clark, Lt J. W. G. 439
Coanda, Henri 153, 156, 159
Cockburn, G. B. xi, 136, 144

Cody, S. F. 15, 16, 198–200, 221, 613
Cogan, Lt F. J. L. 486
Colmore, G. C. 482
Conner, Capt D. G. xiii, 136, 270, 317
Conran, Lt E. L. 140, 232
Cook, 2/Lt W. W. 368
Cooke, Col 371
Cooke, Lt I. 209
Cooper, Capt R. A. 187
Corballis, Capt E. R. L. 416, 560
Corbett, Flt Sub Lt The Hon A. C. 320
Cortinez, Lt Armando 184
Courtney, Sgt F. T. 289
Cowan, Lt S. E. 43
Creagh, R. P. 157
Cronyn, Lt R. H. 214
Cure, Sir Edward Capel 185
Curtiss, Glenn 204

Dallas, Maj R. S. 480
Davies, Lt H. R. 460
Davis, C. F. Lan 112
Dawes, Capt G. W. P. 348
Debussy 147
de Courcelles, Adjudant 560
de Crespigny, Capt H. V. Champion 582
de Havilland, Geoffrey xv, 38, 41, 46, 49, 51, 61, 67, 73, 74, 84, 143, 144, 224, 270, 340, 341, 342, 343, 344, 345, 371, 372, 373, 375, 398, 443, 447, 464, 465, 466
Delage, Gustave 323
Delagrange, Léon 613
de Marmier, Lionel 559
Deullin, Albert 559
Deville, Paul 196
Dickey, Lt Col Philip S. 80
Dickson, Capt Bertram xi, xiv, 152
Dismore, Air Mech F. 486
Ditchburn, Sgt 332
Dixon-Spain, Capt G. 421
Dodds, Lt A. 477
Dorand, Col 88

Douglas, Lt W. Sholto (MRAF Lord Douglas of Kirtleside) 384, 504
Dowding, Maj H. C. T. 383, 384
Duigan, J. R. 28, 110
Dukinfield-Jones, A. 320
Dunn, Capt F. G. 195, 206, 277, 336
Dunne, Lt J. W. xi, 221
Dunville, Sgt 52
Dutton, Murray 122
Duval, Col 477

Eckley, Lt A. 335
Edwards, Stoker 484
Elder, Capt. W. L., RN 201, 489
Ellis, Lt P. C. 324
England, E. C. Gordon 13, 14
Ericson, F. G. 207, 214
Esnault-Pelterie, Robert 186
Etches, F. E. 287

Fabre, Henri 338, 339
Farman, Henry 343
Farren, W. S. 382, 427, 473
Fawcett, H. 125
Felix, Commandant 221, 222
Fenwick, Robert C. 31, 32
Fielder, Lt 563
Flanders, R. L. Howard 23, 249, 251, 574
Fleury, Col 186
Flint, Charles R. 612, 613
Flynn, Flt Sgt J. M. 541
Folland, H. P. 426, 440, 464, 468, 473
Fonck, Capitaine René 559
Foot, Capt E. L. 581
Forbes, Capt E. W. 410
Foster, Sgt R. G. 327, 388
Fox, Capt A. G. 130, 131, 138, 140, 226, 228, 343, 371, 492
Franke, Vzfw 522
Fraser, Lt S. McK. 308
Freeston, Charles L. 95
Fry, Wg Cdr W. M. 543, 563
Fuller, Capt E. N. 202
Fulton, Capt E. J. 276
Fulton, Lt Col J. D. B. xi, xii, xix,
xx, 110, 128, 134, 143, 148, 151, 201, 202, 216, 270, 315, 339, 340, 443, 444, 483, 484, 486

Garland, Capt Ewart 87
Garnett, Lt W. H. Stuart 122
Garros, Roland 285, 286, 290, 296
Gaskell, Lt L. da C. Penn 233
Gates, Richard T. 229, 256
Geard, Cpl F. 378
George V, H.M. King 382, 427, 484
Gerrard, Maj E. L. 226, 375, 483, 484, 486
Gethin, Capt P. E. L. 388
Gibbs, Lt L. D. L. xi
Gilbert, Eugène 296
Gillett, Capt F. W. 542
Glanville, Lt H. F. 609
Glen, Lt D. A. 384
Godoy, Lt Dagoberto 184
Godwin, Lt C. C. 324
Goodden, Maj F. W. xvi, 382, 398, 400, 419, 427, 429, 431, 432, 435, 444, 447, 459, 466, 469, 473, 475, 569
Gordon, Lt J. R. 179
Gouin, Lt 138
Gould, Lt R. G. 388
Gower E. L. 140
Gower, Lt W. E. 423
Grace, 2/Lt F. A. D. 103
Grahame-White, Claude xi, 22, 68, 133, 134, 223, 227, 228, 229, 287, 318, 319, 338
Granet, Col 186
Grant, Lt D. Lyall 420
Gray, Maj 78
Grebby, Lt R. J. P. 335
Green, Frederick M. 341, 344, 371, 382
Grenfell, Lt E. O. 379
Grey, C. G. xv, xvi, 13, 96, 134, 242, 348, 498
Grey, Sqn Crd Spenser D. A., RN 491, 497
Gröschler, Uffizier 297
Grosset, Lt W. E. 522

Guillaux, Maurice 297, 307
Guynemer, Capitaine Georges 559

Hadrill, 2/Lt G. C. T. 515
Haig, Field Marshal Sir Douglas 66, 74, 266, 520
Haldane, Rt Hon R. B. xiii, 612
Halford, Maj F. B. 72, 410
Hall, J. Laurence 111, 112
Hamel, Gustav 8, 285, 287, 288
Hamilton, Capt Patrick W. 21, 216, 217
Hammond, Lt A. W. 104, 105
Hampton, Lt P. R. 179
Handasyde, George H. 30, 271
Hanriot, René 26
Hardwick, A. Arkell 26
Harper, A. Monnier 28
Harvey-Kelly, Lt H. D. 198, 352
Hawker, Harry G. 491, 492, 493, 494, 495, 497, 499, 509, 512
Hawker, Maj L. G. 43, 162, 163, 383
Hay, Lt Stephen xiv
Haynes, Ewart 398
Hazell, Maj T. F. 480
Hearle, F. T. 371
Henderson, Brig-Gen Sir David xix, xx, 187, 345
Herbert, Lt P. W. L. 349
Hewlett, Mrs Hilda 564
Higgins, Brig-Gen J. F. A. 161, 439, 466, 495, 497
Higgins, Lt Col J. W. 195
Hill, Capt G. T. R. 438, 439
Hill, Capt Roderic M. 181
Hippert, Feldwebel 424
Hirschauer, Général 186
Hoare, Brig-Gen C. G. 71, 213
Holder, Lt F. D. 171, 197
Holt, Capt F. V. 150, 151, 486
Honnett, Capt F. W. 507
Horsfall, Capt E. D. 100
Hotchkiss, Lt E. 155
Howard Wright 270
Howitt, R. (probably Lt J. R. Howett) 578

Hubbard, Lt T. O'Brien 221, 222, 258, 484, 567
Hucks, Capt B. C. 87, 130, 131, 285, 411, 431
Hue, Jules 286, 296
Hughes-Chamberlain, 2/Lt J. L. M. de C. 582
Humphreys, Lt G. N. 486
Hurlin, Mr 28
Huskinson, Capt P. 539
Hynes, Lt G. B. xiii, 143, 144

Immelmann, Leutnant Max 305
Innes-Ker, Capt Lord Robert 186, 187, 188, 193, 236, 293, 304, 306, 307, 324
Isaacson, R. J. 32

Jannus, A. 205
Jenkins, Lt Col F. Conway 201, 261, 461, 598
Johnston, John 81
Johnston, P. A. 125
Jones, Capt J. I. T. 480
Jones, Lt B. M. 382, 437, 596
Jones, Lt Trafford ii, 409, 410, 421
Joubert de la Ferté, Capt P. B. 131, 287
Joynson-Hicks MP, William 252
Judge, Cpl V. 609
Jullerot, Henri 149
Junor, Capt K. W. 479

Keary, Lt 563
Kemp, Ronald 398
Kenworthy, John, 370, 375, 431
Kenworthy, R. W. 125
King, L. L. 125
Knight, Archibald 565
Kny, Cecil E. 21, 29
Konnecke, Vzfw 525
Koolhoven, Frederick 90, 91, 100, 106, 107

Lamb, Col 186
Lambe, Wg Capt C. L. 107
Lanchester, F. W. 178

Langford-Sainsbury, Lt T. A. 543
Law, Bill 255
Lawrence, Maj G. A. K. 452, 504
Laycock, Col 134
Ledeboer, 2/Lt J. H. 528
Lee, Col 182
Lee, 2/Lt H. M. 65
Legh, Flt Lt Peter 106
Lepère, Capitaine G. 88
Lewis, Cecil x, 293, 472
Lewis, Maj D. S. 303
Lewis, Sapper 371
Liddell, Capt J. A. 450
Linford, 2/Lt R. D. 425
Littlewood, 2/Lt S. C. T. 420, 421
Long, 2/Lt G. R. 554
Longcroft, Brig-Gen C. A. H. 348, 349, 350, 359, 459, 466
Longmore, Lt A. M., RN 110, 483
Loraine, Capt E. B. xiii, 128
Loraine, Maj Robert xi
Lowery, Capt A. M. 181, 523, 587
Lucas, Keith 382
Ludlow-Hewitt, Lt Col E. R. 165, 166, 379
Lyon, Laurance 612, 613

McArthur, Capt L. W. 502
McClean, F. K. 482, 484
McClintock, 2/Lt J. L. 539
McCudden, Maj J. T. B. 45, 140, 141, 467, 477, 480, 497, 513
McDonald, Air Mechanic 375
McDonald, Leslie F. 33, 34
McElroy, Capt G. E. H. 480
MacInnes, Lt Col D. S. 469, 608
McKeever, Lt A. E. 175
McLeod, Lt A. A. 104
Macmillan, Wg Cdr Norman 504
Mai, Vzfw Josef 477
Maitland, Capt E. M. 269, 270
Malcolm, Maj C. J. 409
Malcolm, Maj N. 611
Mallinckrodt, Leutnant Friedrich 514
Malone, Lt C. J. L'Estrange, RN 491
Manning, W. O. 16

Mannock, 2/Lt Edward 330, 480
Mansfield, Maj W. H. C. 423
Manton, Marcus D. 260
Mapplebeck, Lt G. W. 131
Martin, Cpl 568
Martin, H. P. 30, 273
Mary, H.M. Queen 484
Massy, Capt H. D. 225
Maxwell-Pike, Capt R. 42
Mayo, Capt R. H. 49, 50, 459
Mercanti, Arturo 185, 186
Merrick, Maj G. C. 486, 487
Mills, Maj 62
Mitchell, A. E. 125
Mitchell, Maj W. 388
Moineau, René 11, 12
Money, R. R. 383, 578
Moore-Brabazon, Maj J. T. C. 437
Moorhouse, Lt W. B. Rhodes 11, 352
Morane, Léon 10, 285, 304
Morris, 2/Lt E. C. 425
Morris, 2/Lt Lionel 411
Mottershead, Sgt Thomas 423
Moullin, Lt O. M. 379
Moult, 2/AM A. 502
Mouser, Mr 512
Müller, Offstvtr Max 502
Murray, 2/Lt G. 103
Muspratt, Lt K. K. 473
Musson, Lt F. W. 171

Nash, Air Mech 221
Nethersole, Maj M. H. B. 66
Newall, Lt Col C. L. N. 56
Newcomb, 2/Lt M. 513
Nicholas, Capt C. H. 587
Niéport, Edouard de 315
Nixon, Lt L. J. 525
Noel, Louis 229
Noorduyn, R. B. C. 91
Norman, Capt G. H. 437
Nungesser, Charles 334

Ogilvie, Wg Cdr Alec 106, 107, 108, 550, 551, 615
O'Gorman, Mervyn xviii, xxi, 134,

O'Gorman, Mervyn—*cont.*
  135, 224, 339, 340, 341, 343, 345, 363, 441, 444, 466, 611, 614
Ohrt, Lt F. M. 529
Osipenko, Mavreky 260

Page, Frederick Handley 263
Pagny, M. 26
Paine, Capt Godfrey, RN 345, 382
Palmer, Lt C. W. 305
Parfitt, Air Mech H. E. 378, 379
Parke, Lt Wilfred, RN 6, 26
Parker, 2/Lt J. 609
Parkes, 2/Lt G. A. H. 331
Parr, S. C. 31
Pashley, Cecil 5
Patheiger, Leutnant 297
Patrick, Capt W. J. C. K. Cochran 327, 561, 599
Paulhan, Louis 338, 339
Pearce, Mr 427
Peck, 2/Lt R. H. 450
Pellegrino, Tenente Ernesto 187
Pendavis, 2/Lt W. 389
Pennell, 2/Lt E. R. 278
Percival, N. S. 221, 222
Perreyon, Edmond 8, 9, 136
Perry, Lt E. W. Copland 378, 379
Peters, Mr 343
Petre, Edward 25, 272
Petre, Henry A. 25, 275, 276
Pettigrew, Lt G. T. 330
Pickles, Flt Lt Sidney 201
Piercey, Flt Sgt 62, 560
Pierson, R. K. 577, 602
Pinsard, Armand 553
Pitcher, Brig-Gen D. le G. 73, 175, 511, 523
Pixton, C. Howard 13, 14, 154, 155, 495
Pizey, Collyns 148
Platts, Ralph 5
Playfair, Capt P. H. L. 578
Pollard, 2/Lt E. M. 305
Pope-Hennessy, Maj 195
Portal, Lt C. F. A. 274

Porte, Lt John C., RN 18, 19, 20, 204, 206
Poulet, Etienne 197
Powell, Lt F. J. 432
Powell, Lt L. F. 175
Power, Sgt-Maj W. B. 206, 208, 384
Pretyman, Capt G. 232
Prévost, Maurice 18
Prier, Pierre 151
Proctor, Capt W. A. Beauchamp 480
Prodger, Clifford B. 125

Raleigh, Maj G. H. 191
Raleigh, Sir Walter xii, xvi, xviii, 144, 344, 613
Raynham, F. P. 17, 23, 120, 250, 251
Read, Lt M. 9
Redrup, C. B. 585, 587, 588
Reece, Lt C. M. 502
Rees, Maj L. W. B. 45, 579, 580
Régnier, Col 553
Reilly, Maj H. L. 275
Reynolds, Lt H. R. P. xiii, 136, 150, 152, 191, 270
Richthofen, Rittmeister Manfred, Freiherr von 173, 424, 436
Ridley, 2/Lt C. A. 294
Ridley, Sgt 435
Rigby, Sgt 374
Roberts, Lt H. J. W. 127
Robinson, Capt W. Leefe 173, 359
Roche-Kelly, W. 125
Rodwell, Capt J. T. 388
Roe, A. V. 564
Rogers-Harrison, Lt L. C. 200
Rolls, The Hon Charles S. xi, 611, 613, 614, 616
Ross, Capt 505
Rowell, Capt H. R. 550
Ruck, Col R. M. 611
Ryan, Mr 82

St-Quentin, Capitaine 195
Salmet, Henri 4
Salmet, Mme 131
Salmond, Brig-Gen W. G. H. 182
Salmond, Maj-Gen J. M. 74, 75,

202, 232, 383, 402, 484, 551, 552, 601
Samson, Lt C. R., RN 486
Samuel, Lt J. R. 460
Sanday, Maj W. D. S. 538, 539, 541
Santos Dumont, Alberto 289, 343
Saulnier, Raymond 10, 285, 304
Schaefer, Leutnant K. 424
Schmidt, Leutnant 331
Scholte, Maj F. L. 561
Schubert 477
Scott, Sgt E. 74
Seely, Col J. E. B. 1, 37, 189, 228, 229, 252, 257, 319, 446, 447, 486
Selfridge, Lt T. E. 613
Sharp, Air Mech 375
Shelley, R. C. 510, 512
Sheppard, Lt RN 355
Short brothers 613
Short, Lt C. W. 308
Siddeley, J. D. 72
Sigrist, Frederick 499, 522
Sippe, S. V. 26, 28
Skerritt, Air Mech 352
Slack, Robert 129
Smith, Herbert 535, 542
Smith, S. Heckstall 343
Smith, 2/AM F. 179
Smith-Barry, Maj R. R. 117, 379, 484
Soames, Lt A. H. L. 191
Sopwith, T. O. M. 16, 484, 491, 509, 551
Soreau, Rodolphe 186
Sotham, Lt R. C. 337
Sowrey, Capt F. 359, 562
Spratt, Norman C. 91, 383, 445, 446, 497
Spuy, Lt K. R. van der 236
Stammler, Col 186
Staton, Capt W. E. 179
Stedman, Lt Cdr E. W. 264
Stellingwerf, Lt 260
Stevens, Capt F. D. ii, 425
Stewart, Capt Oliver 277, 451, 457, 587

Strange, Lt Louis A. 114, 233, 273, 274, 401, 402, 497
Stringer, Lt 563
Strohl, Adjudant 553
Sturgess, George 4, 5
Stutt, W. E. 444, 469
Sutcliffe, 1/AM 571
Swaby, Sydney T. 31, 32
Sykes, Herbert 125, 495
Sykes, Lt Col F. H. xviii, xx, 345, 443, 491, 492, 567, 607

Tabuteau, Maurice 149
Tedder, Capt A. W. 506
Tempest, Maj W. J. 416
Templer, Col J. L. B. 611
Tennant, Maj J. E. 579
Tholozan, Marquis de Lareinty 242
Thomas, George Holt 22, 38, 63, 339
Thompson, W. P. 31, 32
Thrupp, Lt T. C. 528
Thurston, Lt (later Dr) A. P. 473, 475
Tizard, Capt H. T. 527, 536
Townshend, Maj Gen C. V. F. 275
Trenchard, Maj Gen H. M. xxi, 49, 50, 51, 63, 74, 75, 79, 92, 93, 180, 181, 187, 193, 211, 240, 261, 266, 278, 282, 287, 291, 293, 304, 306, 314, 320, 324, 326, 364, 365, 385, 391, 402, 404, 405, 406, 410, 414, 419, 420, 426, 439, 453, 454, 459, 472, 489, 504, 506, 513, 517, 520, 523, 527, 529, 535, 536, 553, 554, 572, 582, 593, 599
Trudeau, Lt G. A. H. 563
Tulloch, 2/Lt K. E. 367
Turner, 2/Lt R. P. 297
Tutschek, Hauptmann Adolf, Ritter von 331
Tweedmouth, Lord 612

Valentine, Capt James 153, 187, 188, 287
Védrines, Jules 18, 20
Vere-Bettington, Maj A. 324
Verrier, Pierre 22

Vincent, Jesse G. 80
Virgilio, G. 125
Voisin, Charles 606
Voisin, Gabriel 606
Vyvyan, Capt V., RN 247, 248, 266

Wadham, Lt V. H. N. 291
Wagner, 2/Lt E. G. S. 44
Wainwright, 2/Lt B. M. 299
Waldron, Maj F. F. 348, 352
Walker, Air Mech 384
Wallace, Capt G. P. 235
Warburton, 2/Lt E. D. 173
Ware, Air Mech 221
Warneford, Flt Sub Lt R. A. J. 290
Waterfall, Lt V. 112
Waters, S. J. 426
Watkins, Lt H. E. 269
Watkins, Lt L. P. 395
Webb-Bowen, Brig-Gen T. I. 423, 486

Weir, Sir William 73, 75, 78, 79, 266
Weiss, José 25
Westminster, Duke of 134, 340, 342
Wheatley, Capt C. W. C. 588
White, AVM H. G. 457
White, Sir George 148
Whittington, Col C. W. 430, 511, 526
Wild, 2/Lt H. 74
Wilhelm, Kaiser xiii
Wilson, Lt C. W. 493
Wingate-Grey, Lt A. G. 334
Wood, Maj H. F. 565, 574, 582, 587
Woodhouse, Lt J. W. 164
Woolley, Lt D. B. 424
Wright brothers 35, 611, 612, 613
Wright, Howard 16, 269, 270
Wright, Orville 35, 611, 612, 613
Wright, Warwick 16
Wright, Wilbur 35, 611, 612, 613
Wyness-Stuart, Lt A. 21, 217